Bernard Gittelson, according t
Industry, was Director of the Committee on Discrimination in Employment and the Committee on Commerce and Industry for New York State. He was then President of one of the largest international public relations firms for twenty years, representing the West German government, the Common Market and large corporations all over the world. In 1968 he created Time Pattern Research Institute, and he now heads Biorhythm Computers, Inc., devoting most of his time to research, writing and lecturing on Biorhythm.

Also by Bernard Gittelson:

BIORHYTHM 1975–1977

Bernard Gittelson

Biorhythm

A Personal Science

Futura Publications Limited
A Futura Book

A Futura Book

First published in Great Britain by
Futura Publications Limited in 1978
Revised edition 1979

ISBN 0 7088 1612 6

Printed in Great Britain by
Hazell Watson & Viney Ltd
Aylesbury, Bucks

Futura Publications Limited
110 Warner Road
Camberwell, London SE5

This book is dedicated to George Thommen who inspired me, taught me, and handed me the torch to spread the truth about biorhythm, and to Rosalind, Louise, and Steven, who stood by me in my highs, lows, and critical periods, and without whom this book would not have been possible.

For a complete explanation of how to chart your own biorhythm curves, see pages 59-64 of the text.

Contents

Notes to the Second Edition 9

INTRODUCTION 12
Biorhythm: Theory and Practice 14
True or False? 17

THE RHYTHMS OF LIFE 21
A Day in the Life 23
Night and Day, Day and Night 26
Rhythms in Sickness and Health 30
That Time of Month 33
Beneath the Silvery Moon 35
Time and Life 38

THE DISCOVERY OF BIORHYTHM 39
Freud's Nose 42
The Third Biorhythm 46
What Causes Biorhythms? 47
Getting the Theory Right 52
The Numbers Game 59

CAN YOU PROVE IT? 66
Death in Zurich 69
Some Far-Sighted Americans 74
Rhythm on Wheels 80
Rhythm While You Work 87
The Sporting Life 93

BIORHYTHM IN ACTION 99
The Rhythms of Plays and Players 100
The Phases of the Mind 108
Biorhythm Abroad 111
The Biorhythm Underground 119

THE FUTURE OF BIORHYTHM 127
The Roots of Biorhythm 128
Working It Out 131

BIBLIOGRAPHY 140

BIORHYTHM CHARTS OF THE FAMOUS AND INFAMOUS 144

COMPATIBILITY CHARTS 157

APPENDIX A—BIRTH CHARTS 171

APPENDIX B—BIORHYTHM CHARTS 335

Notes to the Second Edition

Just as we went to press, I received some additional information which adds to my conviction that the theory of Biorhythm deserves further attention and serious study.

I have just seen the official report of the National Transportation Board on the Eastern Air Lines crash at Charlotte, North Carolina, in which 71 people died. The report states that ". . . the flight crew engaged in conversations not pertinent to the operation of the aircraft . . ." If you read their actual words, you would learn that the conversation was full of trivia, and that they certainly were not on the alert during the approach to land. The pilot and co-pilot were both suffering from low physical and emotional rhythms, and the pilot was intellectually low as well.

Another new report concerns the London subway crash of February 29, 1975, in which 41 people died. The train failed to stop at the end of the Moorgate station and crashed into the escape tunnel at the end of the line. The inquiry could find nothing wrong with the train, and there was no apparent reason why the brakes would not have operated. In charting the motorman's biorhythm, we found that February 29th was a critical day in his emotional cycle.

I recently received a letter from a Marquette University High School Psychologist who had heard me on a talk show. She tested 24 senior high school boys. Tests were given to each subject at the low, high, and critical points of their physical cycle. The results were: "The indication is thus that the variables of physical motor skill and intellectual alertness are very significantly influenced by the high as opposed to either the low or critical days."

Since the publication of BIORHYTHM: A PERSONAL SCIENCE, I have received letters and reports from all over the world: university and laboratory research covering biorhythm and heart attacks, strokes; tennis, golf, baseball, football, and other sports; suicides, deaths, accidents, use in school systems—to name just a few.

Therefore, George Thommen and I have renewed publication of the BIORHYTHM NEWSLETTER, bringing up-to-date research, both pro and con, to those who want to know what's new about biorhythm.

Naturally, many individuals and firms have tried to exploit the new interest in biorhythm with products that range from the serious and accurate to carnival merchandise and simply inaccurate and unproven new theories based on biorhythm theory as originated by Fleiss, Swoboda, and Teltscher.

Biorhythm is relatively young—more research has been done in the last ten years than in the previous 80. You can prove whether biorhythm works or does not work for you, but any new generalized findings must be backed by years of valid research and tens of thousands of supporting data.

Biorhythm is not a fad or a game any more than an electro-cardiogram, a stethoscope, or a blood analysis is a game. Data must be found, studied, and interpreted to be valuable. Biorhythm does not always work, but very few things do. It is worth the effort if biorhythm works for you—you will be better able to cope with life's problems and you will understand the problems of others a little better.

Introduction

On the evening of November 11, 1960, a retired Swiss importer named George Thommen was interviewed on the "Long John Nebel Show," a radio talkathon based in New York City. What Thommen had to say sounded surprising to most people and incredible to some. However, the strangest thing Thommen said was in the form of a warning. He cautioned that Clark Gable, who was then in the hospital recovering from a heart attack suffered six days before while filming *The Misfits* with Marilyn Monroe, would have to be very careful on November 16. On that date, explained Thommen, Gable's "physical rhythm" would be "critical." As a result, his condition would be unstable, putting him in danger of a fatal relapse.

Few listeners took Thommen's warning seriously; Gable and his doctors were probably unaware of it. On Wednesday, November 16, 1960, Clark Gable suffered an unexpected second heart attack and died. His doctor later admitted that the actor's life might have been saved if the needed medical equipment had been in place beside his bed when he was stricken a second time.

Pure coincidence? Maybe. But George Thommen's advice was neither a random prediction nor an occult speculation. He was not speaking off the top of his head, gazing into a crystal ball, plotting the influences of the stars, or claiming psychic powers. Instead, as a leading student of the emerging science of biorhythm, Thommen had made detailed calculations showing that for Gable, one of the three general biological rhythms that characterize human life would be changing from a high to a low phase on November 16. For healthy people, this kind of shift need not be dangerous; even for a man recuperating in the hospital, it was not necessarily a matter of life and death. However, the rhythm shift did indicate that Gable would be more than usually susceptible that day, and that the hospital should take appropriate precautions just in case something happened. If only Gable's doctors had known of biorhythm and its implications for their famous patient, that second heart attack might not have proved fatal.

"If only . . ." There are few more common or poignant phrases. We speak those words when we have accidents, miss opportunities, witness disasters, or whenever we are regretfully surprised by events in our own lives or in those of others. We would speak them much less often if we had a deeper understanding of the roots of human behavior and the causes of human frailty; some way to increase our knowledge of what people are likely to do and what may happen to them. It is exactly this kind of knowledge and understanding that biorhythm offers.

Biorhythm: Theory and Practice

The basics of biorhythm are easy to understand. In its simplest form, the theory states that from birth to death each of us is influenced by three internal cycles—the physical, the emotional, and the intellectual. The physical cycle takes 23 days to complete, and it affects a broad range of physical factors, including resistance to disease, strength, coordination, speed, physiology, other basic body functions, and the sensation of physical well-being. The emotional cycle governs creativity, sensitivity, mental health, mood, perceptions of the world and of ourselves, and, to some degree, the sex of children conceived during different phases of the cycle. It takes 28 days to come full circle. Finally, the intellectual cycle, which takes place over a 33-day period, regulates memory, alertness, receptivity to knowledge, and the logical or analytical functions of the mind.

On the day of birth, each of the cycles starts at a neutral baseline or zero point. From there, it begins to rise in a positive phase, during which the energies and abilities associated with each cycle are high. Gradually declining, the cycles cross the zero point midway through their complete periods—11½ days from the point of origin for the 23-day physical cycle, 14 days for the 28-day emotional cycle, and 16½ days for the 33-day intellectual cycle. For the balance of the period each rhythm is in a negative phase in which energies are recharged and our physical, emotional, and intellectual capabilities are low, or at least somewhat diminished. We pick up increasing amounts of energy as the negative phase continues until, at the end of each cycle, the zero point is recrossed into the positive phase, and the whole process begins again. (See Figure 1—Biorhythmic Cycles in the First Month of Life.)

Since the three cycles last for different numbers of days, they very rarely coincide and cross the baseline at exactly the same time (only at birth and every 58 years plus 67 or 68 days thereafter). Therefore, we are usually influenced by mixed rhythms. Some will be high while others are low; some will cross the neutral point while others have many days to go until they reach the same level; to make an even finer distinction, one rhythm may be in a stronger part of the positive phase (or a weaker part of the negative phase) than others that are going through the same phase. The result is that our behavior —from physical endurance to creativity to performance on academic examinations—is a composite of these differing rhythms. We seldom have absolutely wonderful or absolutely terrible days. We have up days, down days, and a good many in-between days, but every day can be understood in terms of a particular and almost unique combination of the three basic cycles.

Biorhythms seem to affect behavior in a peculiar way. Our weakest and most vulnerable moments are not those of the negative phases, as you might suspect. Rather, they occur when each cycle crosses the base line, switching from positive to negative or *vice versa*, and it is at these times that we can expect ourselves to be in the most danger. It appears that at these points the rhythms that guide our lives—and on whose regularity we depend as much as on the steady pulse of the heart—become unstable. They seem temporarily out of step, as though uncertain of their true direction and movement or unsettled by the ebb and flow of energy. These days of cross-over from one phase to another are called *critical days*. Students of biorhythm often compare them to the moments at which we switch a lightbulb on or off, since it is then that the bulb is most likely to burn out, not when the switch is in one of its two positions or phases.

As the Clark Gable case shows, critical days can be very important. On physically critical days, we are most likely to have accidents, catch colds, and suffer all types of bodily harm, including death. Quarrels, fights, depressions, and senseless frustration are typical of emotionally critical days. When the intellectual rhythm is at the critical point, we can expect bad judgment, difficulty in expressing things clearly, and a general resistance to learning anything new or remembering what we already know. The point is that by calculating and studying your biorhythms in advance, you can know what to expect and can do much to avoid the worst. For example, businessmen familiar with biorhythm make a point of refusing to sign crucial contracts on emotionally and (especially) intellectually critical days. Athletes in all sports are well-advised to play with extra care—or to avoid play altogether—on physically critical days, when they are very vulnerable to injury. All of us could avoid unnecessary arguments simply by exerting a little more self-control on emotionally critical days. Days when not one, but two or all three cycles are critical require special prudence—although probably not to the extent practiced by one follower of biorhythm, who swears that on triply critical days he stays in bed and tries not to move, feel, or think at all!

Planning for critical days may be the most significant application of biorhythm, since this can literally make the difference between life and death. But it is less than half the story, since critical days make up only 20% of the days of your life. The remaining 80% are mixed days, whose character is varied and ambiguous. You can do much to regulate your future by making plans that harmonize with your biorhythmic profile for non-critical days. It makes sense, for instance, to try to set records on days when all three cycles are near their peaks and you have as much energy and ability as possible. Conversely, knowing that all three cycles are near the bottom of their negative phases is a clear indication that you cannot count on turning in an outstanding performance. Other examples of how the study of biorhythm can improve your life are obvious, although more precise analysis of the implications of a particular biorhythmic profile is possible only with experience and a more complete knowledge of the theory.

True or False?

At this point, skepticism about biorhythm and eyebrows raised in vigorous disbelief are normal. Although the theory is not supernatural and requires no "leap of faith" to accept, it *is* startling. Both its substance and its ramifications may seem outrageous—or at least foolish—when first discussed. The object of this book is to give you enough information to decide for yourself whether or not biorhythm makes sense. Along the way, there are some fascinating stories to be told.

Actually, the theory of biorhythm is little more than an extension and generalization of the enormous amount of research that scientists have already done on the many biological rhythms and cycles of life. From the migrations of swallows and the feeding patterns of oysters to the levels of hormones in human blood and the patterns of sleep, life can be defined as regulated time. Countless rhythms, most of them fairly predictable, can be found in even the simplest of our bodily functions. Even the smallest component of our bodies, the cell, follows several clearly defined cycles as it creates and uses up energy. As Gay Luce put it in her book, *Body Time*, "We must be constructed out of time as certainly as we are constructed of bones and flesh."

At this stage in its development, research on biorhythm is not at all comparable to the rigorous and painstaking studies that have been made of smaller biological cycles. This is partly because the length of biorhythm cycles is so much greater than the length of most biological rhythms. Phenomena like changes in blood chemistry and sleep patterns can be measured in minutes or hours, and this makes them easier to study than biorhythms which take tens of days to complete. Also, the physical, emotional, and intellectual cycles relate to such complex behavior—made up of hundreds and maybe thousands of subtle physical and psychological changes over time—that studying them with any degree of scientific rigor is extraordinarily difficult. Also, unlike other long-term cycles such as seasonal migrations, the three great biorythms do not always produce a predictable result. They render humans *likely*, but not *sure*, to behave in particular ways, and scientists prefer to work with concrete, totally reliable phenomena.

However, there is nothing in the biorhythm theory that contradicts scientific knowledge. Biorhythm theory is totally consistent with the fundamental thesis of biology, which holds that all life consists of the discharge and creation of energy, or, in biorhythmic terms, an alternation of positive and negative phases. In addition, given that we are subject to a host of smaller but nonetheless finely regulated biological rhythms, it seems reasonable that larger, longer rhythms will also come into play. Those rhythms may depend on vast numbers of the most discrete cycles that science has already proven to exist; or they may depend in part on external cues, such as geomagnetism and light, many of which have been shown to influence the smaller cycles. But until we can perform strictly controlled studies of how and why biorhythm works, and until many other researchers have been able to replicate these studies, we will have to base the case for biorhythm on purely empirical research.

Fortunately, there is a large and growing body of such research. The European scientists who discovered biorhythm during the early 1900's based their conclusions on literally thousands of individual cases. This kind of clinical work gives the theory a firm foundation. Since then, as understanding and application of biorhythm have spread throughout the world, much more evidence has been amassed. Doctors, government agencies, and corporations in many countries, particularly Switzerland and Japan, have applied biorhythm with great effectiveness to reduce the risk of death in operations, in automobile accidents, and from industrial hazards. In the United States, members of organizations such as the National Safety Council, the Flight Safety Foundation, the U.S. Air Force, and N.A.S.A. have begun to consider the theory, and some of them have conducted research and issued reports that apparently confirm it. For obvious reasons, insurance companies have shown a special interest in biorhythm; available studies of their accident and death statistics in light of the victim's birth date and biorhythmic profile on the day of the mishap have done much to bolster the case for biorhythm. On a less convincing but still valuable level, thousands of individuals who use biorhythm to guide their lives have made reports that show its usefulness and have offered suggestions for fruitful types of new research.

Ultimately, however, the most convincing studies of biorhythm are those you can do yourself. By working out your own biorhythm chart and biorhythm profiles for particular days, and then comparing them with your experience of up and down days, of illness and health, of success and failure, you will be able to judge for yourself. Since the mathematics of biorhythmic computation can be difficult, we have provided in an appendix a simple method that you can use to plot your individual biorhythms accurately, from day to day and month to month. When you have learned more about the foundations, proofs, and applications of biorhythms, try out the theory in your own life. It could keep you from having to say, "If only . . ."

The Rhythms of Life

What is life? In basic biological terms, it is the cycle of energy production, storage, and discharge. This general cycle defines all living organisms, from one-celled algae to man. In very simple forms of life like the algae, the stages of the energy cycle are relatively clear. We understand fairly well how each stage functions, how it is linked with the next, and how all stages combine to produce the phenomenon of life. However, the more complex the organism, the less we understand. In man—the most complicated form of life—there are so many subtle, elaborate biological cycles that scientists have only begun to sketch in the nature of the larger cycles of human life that we call biorhythms.

What makes it so difficult to understand man's biological rhythms is the fact that he is a highly organized form of life. A bacterium consists of one cell. Man is made of many different kinds of cells which are structured into various kinds of tissues—nerve, muscle, fat, skin, and bone—which in turn compose the organs of the body. Each cell has a number of biological rhythms, and its "behavior" affects and is affected by the rhythms of all other cells. Literally millions of these rhythms contribute to even such an apparently simple rhythm as the heartbeat.

A great deal of reliable research has been done on the biological rhythms that characterize life. Since the days of Ancient Greece, scientists have sought to order human behavior in terms of time. They have observed and recorded regular rhythms in basic body functions such as respiration and pulse rate. More importantly, they have noted that such complicated behavior as sleep, illness, and sensory perception also seem to follow regular—and therefore predictable—rhythms. Doctors since Hippocrates, the founder of medicine, have known that health waxes and wanes cyclically, and that therapy will be more or less effective depending on when it is administered.

It was not until recently, however, that scientists began to measure man's biological cycles with great accuracy, to test the reliability of those cycles under laboratory conditions, and to hypothesize and prove theories that can account for our rhythms. Within the last twenty-five years, literally thousands of books and papers dealing with biological times have been published all over the world. In discussing this research, it is important to remember two things. First of all, although many biological cycles have been described in painstaking detail, there is severe disagreement about the forces which regulate these cycles. Do we have some sort of internal biological clock that ticks away independently of all outside forces and causes our bodies and minds to behave rhythmically? Or is rhythmic behavior simply a response to external cues, such as light or temperature? Even if we do have a biological clock, are its hands and its speed predetermined by forces within us, or are they set early in life by other environmental cues? We might think of researchers in biological rhythm as divided into three schools: those who believe in the pre-set clock, the cued clock, and environmental cues. At this time, there is no way to tell which school is right, although some of the evidence suggests that all three may be partially correct. The point is that although we know a good deal about the characteristics of biological cycles, the most exciting mystery lies in unravelling the roots of cyclical behavior.

The second point to keep in mind is that the three great biorhythms—physical, emotional, and intellectual—are the least-studied of all human cycles. Despite the potential importance of these rhythms, most scientists have preferred to work with shorter cycles which are more easily analyzed. This is understandable; it is hard to test and measure large general cycles, and even harder to trace their roots back through a succession of smaller and smaller cycles. So the bulk of convincing research about the nature and causes of biorhythm is still to come. When it is done, it will be based on the existing body of work about smaller, more discrete cycles, without which no true understanding of biorhythm is possible.

An intelligent view of the three central biorhythms depends on an understanding of the other, less crucial ways in which time runs through and structures our lives. Only by coming to see how and why time is of the essence in life can we comprehend the significance of biorhythm theory.

A Day in the Life

The most obvious statement that we can make about biological rhythms is that all nature follows the twenty-four-hour cycle of a day. All creatures, from the oyster to the elephant, follow a daily pattern of work and rest, sleep and activity. Although some may act like owls, sleeping during the day and working at night, and others like larks, following the opposite schedule, we all seem to live in twenty-four-hour segments.

Obvious though it is, the observation is also profound, for science has shown that many essential functions other than work and rest schedules are characterized by a twenty-four-hour day. Body temperature, blood pressure, brain activity, hormone levels, and a host of other factors obey this rhythm, which scientists call circadian, from the Latin for "around *(circa)* the day *(dies)*." Certain body functions like brain waves also show *infradian* rhythms, which take less than a day to complete. Others, especially those related to health and illness, demonstrate a *supradian* cycle, rising and falling over spans of time greater than a day. However, most measured cycles are circadian.

The best known example of how important circadian rhythms are is found in the phenomenon of jet lag. When a person arrives very quickly in a part of the world where the clock is significantly ahead of or behind his customary time, he suffers from more than a consciousness of disorientation. His body and mind are in fact out of step with his new schedule because they have not had time to adjust. A Californian in Paris, for example, finds that for the first few days he is not hungry when it is time to eat, not sleepy when it is time to sleep, and all too relaxed when it is time for work. His basic body functions will gradually shift from West Coast to French time, but until they do his mood and mental performance will both be below par. Today, when Presidents and other high officials travel to distant countries, their itineraries usually include a day or more of uncommitted time during which they can begin to make the needed circadian adjustments to bring their minds and bodies up to par. Similarly, many multinational corporations now make it a rule that executives travelling abroad adjust for a few days after arrival before making any important decisions.

Jet lag is very important to pilots, since they so frequently cross time zones and must be at peak performance during flight. Since it is essential for them to know at what hours of the day their performance will be highest and how long it will take them to adjust to certain time shifts, much research has been done on pilot's circadian rhythms. Since the mid-nineteen-sixties, Dr. K. E. Klein and his associates at the Institute for Flight Medicine in Bad Godesberg, Germany, have been collecting information about how a wide range of physiological factors (temperature, heart activity, blood composition, etc.) fluctuate during the course of a day and have tried to relate them to measured daily cycles in performance. Klein's group found that mental performance was highest between about two and four o'clock in the afternoon, when reaction times are quickest and physical coordination is best. The worst time for these types of performance was in the early morning, again between two and four o'clock. Although other studies designed to improve flight safety have indicated a performance peak between one and seven P.M., and a trough between two and six A.M., the symmetry is still striking: for one period every day we are at our best, and for one period at our worst, and both these periods fit neatly into a twenty-four-hour cycle.

Just as impressive were Klein's findings that the need for oxygen and the composition of blood—both prime determinants of a pilot's mental and physical performance at high altitudes—also varied greatly with a reliable daily rhythm and might be used to explain remarkable changes in performance.

Knowing when his performance is likely to be best, how can a pilot make sure that he is at work then, and not during a less favorable time? The simplest way, naturally, is to maintain a steady schedule that does not vary with local time or such normal cues as light and dark. On the basis of brain wave studies, Soviet scientists have suggested precisely such stability of scheduling for flight crews dealing with large time zone differences. But this kind of work/rest schedule can be awkward to maintain, and it doesn't take into account the schedules of the airlines.

How long must a pilot live on local time before he comes back into phase with the clock on the wall? According to researchers in The Netherlands, France, and the U.S.A., the answer is discouraging. Body temperature, heartbeat, and kidney functions can take anywhere from four to eleven days to synchronize with local time. Until they do, performance will not vary as it should, and the pilot's work schedule will have to reflect his gradual changeover from the original to the new time. Although a person's adaptibility may be greater than normal, it still seems that the safest—if least convenient—policy is to pretend that time zones do not exist, and maintain the same schedule regardless of geographic location.

Night and Day, Day and Night

Just how important circadian rhythms are can be seen in many studies of work/rest schedules. In factories that operate around the clock, in times of war, and in space flight, it may seem more efficient to ignore our normal daily rhythm. Many industries are obliged to schedule work at night, during normal human rest hours. Although this obviously disrupts normal life, most workers feel that they can adapt to night work relatively easily.

After all, what's the difference whether you sleep at night or during the day? There are two very important differences. First of all, studies of errors in subway systems, utilities, telephone exchanges, assembly plants, and other organizations that run around the clock show that most mistakes occur in the early hours of the morning. In other words, the latest night shift is the least efficient. Secondly, this inefficiency seems to be based on physical factors. Habitual night shift workers are more vulnerable to stress and stress-related diseases such as ulcers and high blood pressure. In fact, reversing the normal schedule may shorten life. Studies by Dr. Franz Halberg of the University of Minnesota have shown that mice on a reversed schedule have lives six percent shorter than mice on a normal schedule. Other studies by the military and by commercial airlines seem to support these findings by showing that "phase reversal," as a reversed work/rest schedule is called, inevitably results in reduced physical and mental efficiency, even if there is sufficient time for phase adjustment.

Clearly, we have some sort of internal work/rest rhythm with which it is dangerous to tamper. But is this rhythm necessarily a twenty-four-hour one that dictates a schedule of twelve hours of work and twelve hours of rest? Or is it just that we must alternate work and rest regularly, and in accordance with the cycle of light and dark? Studies of Russian and American astronauts and of military personnel throughout the world indicate that radical adjustments in the length of work and rest periods are just as difficult to adapt to as phase reversal. In the Navy, for example, it is traditional to take four-hour watches and four-hour sleep periods. Yet most sailors' performance is notably sub-normal on this schedule. Astronauts on four- or six-hour work schedules do not perform as well as those on more normal schedules.

Some of the most fascinating research on circadian

rhythms comes from modern cavemen—scientists and others who isolated themselves from external cues like light and social activity by living for long periods of time, often as much as two months, in caves or isolation chambers. Some cave-dwellers have gone underground with clocks, so that they could try to live a day shorter or longer than the normal one of twenty-four hours. Their success has been minimal. One group did report that they had managed to live a forty-eight-hour day, but they were uncomfortable doing it, and it is very likely that they would have failed if they had had any demanding work to do.

The point of most isolation studies is to find out just how long our normal day is when we are cut off from the things that we usually use to time ourselves: clocks, meal times, work schedules, light, temperature, and many other external cues. The results seem to support the view of those who believe in an internal clock. Despite individual differences of as much as one hour, virtually everybody in isolation soon settles down to a twenty-four- or twenty-five-hour rhythm. Without any external means of establishing a regular daily cycle, they do so anyway.

Most researchers support, if only partly, the internal clock theory. The evidence appears overwhelming, for in addition to studies of human beings in isolation, more dramatic work has been done with plants and animals. Studies of the circadian rhythm with which plants open and close their organs show that the rhythms are inherited and internal: it continues even when they are deprived of light, and different plant species have different, species-specific cycles. Animals that have been blinded experimentally also retain a circadian rhythm, and one scientist has even isolated a large cell in a very simple organism —the sea slug—that appears to control biological rhythm.

Still there is some evidence on the other side. The

major supporter of the "external cue" approach is Dr. Frank Brown of Northwestern University. He believes that humans are so extremely sensitive to a broad range of external cues that we need not hypothesize an internal clock in order to explain circadian rhythms. It may be that humans can detect changes in the earth's magnetism, in electrical fields, in gravitational fields, in various types of cosmic radiation, and even in the movement of other planets. Brown's viewpoint may be correct despite the isolation studies, since it is almost impossible to shield subjects from all of these influences. Moreover, studies at the Max Planck Institute in Germany have suggested that when we are protected from magnetic and electrical forces some of our physical functions lose regularity and establish non-circadian rhythms. Even empirical observations that we can all make lend support to the external cue theory. How else can we explain some people's ability to predict the weather by the way they feel, an animal's apparent ability to detect earthquakes long before they arrive, and the "sixth sense" which we all have to some degree and which enables us to detect the presence of other beings without any obvious signs that they are there?

Even when we think of cues less exotic than cosmic rays, such as alternation of night and day, there is some support for the view that circadian rhythms are a blend or interaction of an internal clock and external cues. In many of the isolation studies, people deprived of light cues experienced *desynchronization* of some body functions; that is, certain basic cycles—the rate at which potassium is excreted, the rate at which sodium is retained, and the rate at which such substances as hormones ebb and flow in the bloodstream—started to go out of synchronization with the roughly twenty-four-hour cycle of rest and activity that had been undisturbed by light deprivation. The rhythms of those cycles often became infradian, lasting three or four hours less than a full day, but they were resynchronized quickly when the subjects returned to the surface of the earth. Perhaps the truth of the matter is that there are certain forces within us, acquired through centuries of evolution and passed on by inheritance, that tend to impose a circadian rhythm on our behavior and physiology, while, at the same time, external cues do much to fine-tune these pre-set rhythms.

Rhythms in Sickness and Health

Since we are such creatures of time, and hundreds of our biochemical functions fluctuate regularly, a consciousness of natural rhythms is essential to medical treatment. Psychiatrists have realized this since Freud's day, when the classic manic-depressive cycle was first explored systematically. Unfortunately, doctors who deal with the body alone still tend to ignore rhythm in diagnosis and treatment, perhaps because we lack extensive evidence to show the causes of bodily cycles.

However, some research on the roots of biological rhythm is quite provocative. On the very basic level of the cell, distinct cycles are apparent. Cells on the surface of the eye of a kitten divide most quickly at 10:30 in the morning and at their lowest at 10:30 at night, in complete accord with the circadian cycle. Cells in the ear, mouth, skin, and eyes of rodents have been measured under strict laboratory control; all of them showed a basic circadian rhythm of reproduction and renewal through cell division. The famous Dr. Halberg and his associates have studied many different types of animal tissue: liver, kidney, glands, skin, intestines, and other organs. Each type of tissue showed a circadian rhythm, although the higher and lower parts of the cycle in one group of cells did not necessarily synchronize with the cycles of other groups. Very similar cycles of cell division have been observed in human skin and can be predicted for most other types of human tissue.

The potential significance of research on the rhythms of cell division is clear when we turn to cancer research. One British scientist working with the humble cockroach discovered that if the rhythm of the insect's brain was disturbed by grafting on parts of the brain of a cockroach with a diametrically opposed circadian rhythm, tumors formed. It appears that human cancer cells display no circadian rhythms at all; that is, they are completely out of step with the rest of the body. This seems reasonable when we recall that cancer is a disturbance of cell division in which cancerous cells multiply at an extraordinarily fast rate. When exposed to X-rays, however, cancer cells often acquire more and more regular rhythm. Therefore, if the formation of cancerous tumors has something to do with a disturbance of the body's circadian rhythms, we may look forward to the day when a whole range of methods will be applied to fight cancer through restoring cellular rhythm.

Other medical implications of cellular rhythm are enormous. In surgery, for example, we could ensure quicker recovery by operating at times just before the cycle of cellular reproduction picks up again. In prescribing drugs, we could be surer of their effectiveness if we were to measure the dose and time according to the body's rhythms, both the basic cell rhythms and the higher biochemical rhythms (such as the cycle of hormone release). In fact, diabetics and others who must take drugs that profoundly alter body chemistry respond much better if their dosages are based on a careful study of their individual biological rhythms. In the case of diabetics, who must take insulin to enable them to use the sugars in their blood, the insulin dose is usually administered once a day—and all at once. Research has shown that levels of insulin and blood sugar normally vary with a circadian rhythm so it seems to make more sense to give insulin according to that normal rhythm. Moreover, if insulin is administered all at once it can disrupt the normal cycles of certain crucial hormones, so carefully timed doses are very important.

Sometimes patients seem to have an inner sense of how their dosages of drugs should be spread out over the day. Addison's disease, for example, is a deficiency of an adrenal hormone, and it leaves sufferers feeling tired most of the time. The therapy is to give cortisol to make up for the hormone lack. However, if the cortisol is given all at once, patients still have periods of fatigue, since the dosage does not follow the normal hormone rhythms. Realizing this, two Parisian doctors, Alain Rheinberg and Jean Ghata, instructed their Addisonian patients to follow whatever schedule of treatment made them feel best. One of their patients decided that he felt best when he took one-third of his dose before sleeping and the rest in the morning. Strangely enough, that schedule created hormone levels almost exactly like those of a normal person with normal cortisol rhythms. Other patients, following their own intuitive schedules, often reproduced the circadian rhythm of adrenal hormones.

In extreme cases, where a patient needs a drug that is very dangerous in the effective dosage, timing can be crucial. The most dramatic illustration of this fact again comes from Dr. Halberg, who has found that mice do not die with uniform frequency. Rather, most of the mice injected at certain hours succumb, while those injected at other hours (when their circadian cycle is on the upswing) rarely die. Generally similar findings have been reported for non-bacterial drugs similar to those used in chemotherapy for humans. In short, our vulnerability to injection and to chemical imbalances varies greatly throughout the day and can mean the difference between life and death, between sickness and health. If you have to take a dangerous drug, you want to take as little of it as you can, as long as it has the required effect. If you know when your body is most responsive, you can take the dose then, and in a dose much smaller than you would need when you are most physically resistant.

That Time of Month

Moving up from the level of cells and of particular biochemical cycles, we find, not surprisingly, that many diseases show a rhythm too. In these cases, however, there are some important differences. Diseases usually show supradian rhythms, which take longer than a day to complete. Any victim of arthritis will verify this observation. Also, many diseases seem to recur, but their precise rhythm varies. Periodic peritonitis is an inflammation of the mucous membranes that line the stomach and intestinal tract. As the name implies, it is periodic, but its periods are not regular; they can vary from every two weeks to much longer periods, and we do not know why they should.

A clearer and much more interesting case can be made for a condition that is not a disease—the marked emotional swings that both men and women are subject to. In women, a great deal of research exists to show that emotional ups and downs are closely tied to the menstrual cycle. Many women are familiar with pre-menstrual tension. Some studies show that sixty percent of women report symptoms of premenstrual tension at some time, including anger, irritability, emotional instability, weeping, and the flare-up of chronic illnesses. These emotional states are not the result of a woman's knowing that her menstrual period is about to begin or of the shame that many cultures attach to a natural biological process.

Rather, emotional stress immediately prior to menstruation has a biological basis and occurs in a broad range of cultures.

Apparently, the levels of three important hormones—estrogen, progesterone, and aldosterone—peak right before menstruation (although the estrogen level also shows a peak in the middle of the monthly cycle). These hormones have profound effects on body chemistry, and on the nervous system in particular.

It is not as well-known that men also demonstrate a roughly twenty-eight-day mood cycle. Although little work has been done to explore this cycle, it has been noticeable for a long time. The seventeenth-century physician Sanctorius, noticing that his male patients seemed to have rhythmic mood changes remarkably like those of women, decided to find out if men's bodies also changed regularly. He weighed his patients daily for many months, hoping to find some obvious sign of rhythmic body change. According to his records, he was rewarded by finding monthly weight changes of between one and two pounds.

More systematic work on male mood changes was done by Dr. Rexford B. Hersey, an industrial psychologist, in the early 1930's. Convinced that management was making a mistake by overlooking the emotional behavior of workers in favor of their purely economic aspects, he began to observe workers in several factories. He interviewed workers and their families for many weeks and asked the workers to rate themselves emotionally. What he found really wasn't too surprising in light of women's known emotional cycles: men also demonstrated fairly regular mood swings, many of them geared to a roughly twenty-eight-day rhythm. During the high parts of the cycle, the men would be cheerful, joke with their co-workers, and be full of energy. As they approached the low point of the cycle, the men would gradually withdraw, become moody, and feel unable to do anything. In some cases, the low point of the cycle was marked by weight loss, loss of sleep, and other physical symptoms.

Contemporary scientists have extended Hersey's work in small ways. Dr. Christian Hamburger of Denmark suspected that the levels of some important male sex hormones that probably influence behavior fluctuated regularly. Since these hormones can be measured in the urine, Hamburger set out to prove his theory by collecting daily samples of his own urine for nearly two decades. When the samples were analyzed, they showed an almost exact monthly rhythm. Similar but more reliable work is now being done by Japanese doctors, who suspect that certain mental illnesses which seem to come and go in men over twenty-eight-day periods may be rooted in hormonal fluctuations.

Beneath the Silvery Moon

If monthly mood swings in both men and women are a product of varying levels of hormones in the blood, what determines the variation? Why do these hormones rise and fall on such an appparently regular schedule? One possible explanation that is attracting the attention of more and more scientists is the theory that the moon plays a crucial role in emotional and physical cycles.

Although the influence of the moon may sound like quackery, it really shouldn't. Since ancient times, a connection between human emotions and the moon has been thought likely. Thus we have the word "lunacy" to describe the sort of irrational behavior that the moon *(luna)* appears to provoke. As medicine became more sophisticated, the emotional power of the moon seemed no less likely than before. Even today, almost every psychiatric hospital reports that admissions swell remarkably at the time of the full moon. Bartenders know that when the moon is full, their patrons become more susceptible to liquor and more likely to start fights. Many police departments have stated that incidents stemming from mental imbalance—kleptomania, psychopathic murder, freak accidents, etc.—become more frequent around the time of a full moon.

Evidence for the power of the moon is even more convincing when we turn from psychiatry to biology. Many animal mating cycles are tied to the moon, as any resident of Southern California can tell you after observing the grunion fish slithering on the beaches to spawn only on the night after the full moon. Many other sea creatures, from lowly worms to more complicated fish, seem to catch a kind of mating fever when the moon is full.

In a renowned experiment, a marine biologist from Iowa transported some oysters from their home on the East Coast to his land-locked state. In the East, the oysters had opened and closed their shells to feed in strict accordance with the tides. In Iowa, there were no tides to guide their behavior, yet within a short time, the oysters were opening and closing their shells in perfect rhythm with what the tide should have been on the non-existent Iowa coast! The only way to explain this behavior is to assume that the oysters responded not just to tidal movement, but to the source of that movement: the moon.

How does the moon exercise its influence? The case of the exiled oysters suggests that gravitational pull is the principal means, but light may also be important. One recent set of experiments has shown that women's menstrual cycles can be made more regular if they sleep with a light in the bedroom during certain days of the cycle, and this method has been used successfully to insure conception in women whose normal cycles are too irregular for them to conceive. Given the well-known effects of light on many other kinds of cycles, it is only a little far-fetched to think that the light of the moon shining through the window may be a factor in human conception.

Although some people seem to become unbalanced during the full moon, there is research to suggest that what is important is not the phase of the moon itself, but the phase of the moon at the time of an individual's birth. In short, just as biorhythm theory would suggest, emotional ups and downs are keyed to the day of birth, and it is the moon that links the two. In many women, for example, the menstrual cycle is determined by when the moon comes into the same phase that it was in when the woman was born.

Even more provocative evidence has been reported by Sandia Laboratories, whose staff of scientists and engineers usually does work on developing nuclear devices for the Atomic Energy Commission. In 1971, the staff turned from its ordinary work to publish a report called, "Intriguing Accident Patterns Plotted Against a Background of Natural Environment Features." What these researchers had done was to collect data about accidents in their own laboratories over a twenty-year period and to analyze them with the help of a computer. When they compared disabling injuries with the phases of the moon in which the victims were born, the Sandia group asserted "the possibility of a heightened accident susceptibility for people during the phase similar to that in which they were born, and for the lunar phase 180° away from that in which they were born." This finding is very exciting when you realize that the periods of time during which accidents were found to be most likely to occur correspond very well with the critical days described by biorhythm. Those days are the day of birth, when the emotional cycle begins, and the day about fourteen days after birth, when the emotional cycle crosses the baseline to enter the negative phase.

Time and Life

Whether or not the moon determines our mood swings, it is obvious that time is as much a part of our lives as breathing and sleeping. Although most of us go through life unaware of the thousand rhythms by which we live, time in the form of regular cycles affects our minds, bodies, and feelings. Science has shown beyond question that from the simplest level—the cell—to the most complex—our lives as a whole—rhythm is the rule. It should go without saying that if only we knew more about the important rhythms of our life, we could exercise much greater control over what happens to us. Understanding the rhythms of life would radically transform all our lives.

It was exactly this kind of understanding that two European scientists began to achieve about the turn of the nineteenth century.

The Discovery of Biorhythm

Shortly before 1900, in the two cities that were the poles of European scientific life—Berlin and Vienna—the discovery of biorhythm occurred. The timing is significant: in that same period Freud started to bring together his clinical experience with the mentally ill to create the theory of psychoanalysis. The time was clearly ripe for revelations like biorhythm and psychoanalysis, both of which offer ways to explain the hidden roots of our actions and our lives.

Unlike psychoanalysis, biorhythm theory had two "fathers": Dr. Hermann Swoboda, professor of psychology at the University of Vienna; and Dr. Wilhelm Fliess, an eminent nose and throat specialist in Berlin who was later to become president of the German Academy of Sciences. Like many other scientists on the verge of a breakthrough, Fliess and Swoboda were working along very similar lines but were largely unaware of each other.

It is striking evidence of the force of the nineteenth-century intellectual climate—and perhaps of the truth of biorhythm theory—that these two men, through painstaking but independent research, came to virtually identical conclusions.

Swoboda was apparently drawn toward biorhythm by a number of highly suggestive reports which appeared around 1897, the year in which he began his research. He was familiar with the philosophical speculation of Johann Friedrich Herbart, whose *Textbook on Psychology*, published between 1850 and 1852, reported seemingly rhythmic changes in mental states but failed to find an explanation for those variations. Swoboda was also intrigued by one of Fliess' early papers on bisexuality in men, and by John Beard's 1897 examination of the regularity of timing in pregnancy and birth.

Out of this ferment of ideas, Swoboda's research in biorhythm began to grow. At first, he concentrated on his own field—psychology—and did not look farther afield for confirmation of biorhythms. He was concerned with how man feels and thinks about himself. As Swoboda himself put it in a paper presented to the University of Vienna in 1900:

> Life is subject to consistent changes. This understanding does not refer to changes in our destiny or to changes that take place in the course of life. Even if someone lived a life entirely free of outside forces, of anything that could alter his mental and physical state, still his life would *not* be identical from day to day. The best of physical health does not prevent us from feeling ill sometimes, or less happy than usual.

Analyzing the experience of his patients, Swoboda noticed that dreams, ideas, and creative impulses seemed to recur with a very regular rhythm. This phenomenon is familiar to many artists, who find themselves going through dry spells and frenzies of creations with a predictable variation. He also noticed that new mothers began to show anxiety about their infants whenever a critical day occurred or was about to occur.

But Swoboda went much farther than this. A meticulous researcher, he began to keep detailed records on pain and swelling in his patients' tissues, focusing mainly on seemingly inexplicable inflammations. He also plotted the course of fevers, the onset and development of other illnesses, and the frequency with which people suffered heart attacks and bouts of asthma. Slowly but surely he began to see that all of these physical phenomena seemed to recur rhythmically, and that cycles of 23 and 28 days had the power to predict their recurrence.

Swoboda's discovery of these two basic biorhythms quickly led him to write a succession of distinguished and widely-popular books explaining and developing the idea of cycles in human life. First, in 1904, came a book with the imposing title of *The Periods of Human Life (in their psychological and biological significance).* In the very next year Swoboda published *Studies on the Basis of Psychology*, which further elaborated his work on creativity and the recurrence of dreams. As biorhythm theory began to exert an ever stronger fascination for doctors and the general public, Swoboda realized that work in the area would be much easier if there were some means to calculate critical days accurately and easily, without resorting to a long series of manual calculations which carried a high risk of error. So, in 1909, he made available a slide rule of his own design, with which anyone could find his critical days quickly and surely. Along with the slide rule, he published an instruction booklet, *The Critical Days of Man.*

Swoboda's greatest book was one of his last, a volume of almost 600 pages called *The Year of Seven.* Much of this work is devoted to proving biorhythm theory through a mathematical analysis of how the timing of births tends to be rhythmical and predictable from generation to generation within the same family. Clearly, Swoboda was trying to establish—if only empirically—that biorhythms are innate and inherited. However, *The Year of Seven* also contains an analysis of other significant life-events —illnesses, heart attacks, and deaths—that Swoboda thought were related to the genetic pattern of the family. By studying thousands of family histories, Swoboda found that these events show a periodicity—a tendency to happen on critical days—that first begins with birth.

Swoboda led an enormously productive life, regularly lecturing at both the University of Vienna and over the Vienna radio, and turning out a succession of papers on biorhythm, among them a study of the basically rhythmical nature of cancerous tissue. During the Russian occupation of Vienna in 1945, much of his research and scores of his papers were lost, but after the war he returned to his life's work and achieved remarkable distinction. The city of Vienna awarded him a special medal for his work on biorhythm and, in 1951, in the course of extending his professorship yet again, the University recognized his outstanding contributions by awarding him an honorary degree. In 1954, he published his last work, an expansion of his work on families called *The Meaning of the Seven Year Rhythm in Human Heredity*. Swoboda died in 1963 at the age of ninety, an extravagantly mustachioed man who had lived to see the discoveries of his youth spread and accepted among an ever-growing number of people.

Freud's Nose

Much less gratifying was the life of Wilhelm Fliess, Swoboda's contemporary and the co-discoverer of biorhythm. In many ways, Fliess achieved far greater prominence and worldly success than Swoboda. Born in 1859, he attended medical school at the University of Berlin and graduated as an otolaryngologist—a nose and throat specialist—about 1885. He went on to become a member of the Berlin Board of Health and, in 1910, president of the German Academy of Sciences. Despite Fliess' thriving and lucrative practice in Berlin, Vienna was the home of his most famous friend and patient—Sigmund Freud. We know relatively little about the relationship between these two pioneers, despite publication of the 184 letters that Freud wrote to Fliess between 1887 and 1902. But it is evident that Freud found certain of Fliess' theories fascinating and trusted him enough to submit to two operations at Fliess' hands.

In other, more important ways, Fliess' life probably seemed a failure to him. A scientist of widely varied interests, he found evidence in many fields to support both the 23-day physical rhythm and the 28-day emotional rhythm. In the fairly short time of sixteen years, he published four books to support his findings and spread the practice of biorhythm. Working with others, he developed and published elaborate mathematical tables to enable readers to check his analyses of case histories and to calculate their own biorhythms. Unfortunately, at the time of his death in 1928, Fliess had yet to see biorhythm widely accepted in his native land. His critics felt that the mathematical analyses in his works were confusing and irrelevant; readers were intimidated by the seemingly endless pages of calculations and proofs. In short, he could not communicate his discoveries clearly and convincingly. It is little wonder that both the public and the medical profession failed to recognize his fundamental contribution to biorhythm theory.

A close examination of Fliess' work shows that it is

compelling. He first began to suspect the existence of biorhythms when, as a practicing doctor, he had to confront one of the persistent puzzles of medicine: the variation in resistance to disease. Fliess was particularly concerned with children's ailments. In an attempt to discover why children exposed to the same diseases remained immune for varying periods of time and succumbed on widely different days, he collected extensive data about the onset of illness, fever, and death, and related them to the day of birth. His studies, as reported in lectures and papers published between 1895 and 1905, persuaded him that the best explanation of the ebb and flow of resistance to disease lay in the 23-day and 28-day cycles that are the foundation of biorhythm.

Like any dedicated researcher, Fliess was not satisfied with having observed these two rhythms after the fact; he sought an explanation or rationale for such periodicity. Why, he wondered, are the emotional and physical cycles so precise? Why do they affect our bodies and behavior so profoundly, and how have they developed over the course of human history? To answer these questions, Fliess turned to the fundamental level of the cell. Noting that very primitive organisms are unisexual or hermaphroditic, he reasoned that as life evolved into differentiated sexual forms, there still remained some vestiges of the earlier unisexual stage. In other words, a member of one sex still retains cells and even tissues characteristic of the opposite sex. Because of this admixture, human beings are bisexual in profound ways. Fliess pointed to the classical evidence of bisexuality—the fact that men retain nipples, even though such organs serve no function in the male. Contemporary research offers much more convincing proof of fundamental bisexuality by revealing that all of us come under the influence of both male and female sex hormones.

Fliess felt that the male and female cells of the human

body had definite and different rhythms. To the female cells he ascribed the 28-day emotional rhythm; to the male cells, the 23-day physical rhythm. He explained the fact that the menstrual cycle varies widely among women (ranging from about 26 days to about 33 days) as being the result of the combination of male and female rhythms. Since they are not of the same duration—that is, they are not in phase—one rhythm often acts to shorten or extend the other. Fliess also relied on his theory of cellular bisexuality to explain the fact that men also undergo emotional cycles quite similar to those of menstruation.

Having observed periodicity in illness, birth, and death, and having developed a basic theory to account for his findings, Fliess set out to accumulate as much evidence as he could. He made extensive studies of such inherited characteristics as lefthandedness, which he felt was connected with children conceived during their parent's emotional high points. He did the same sort of research on family trees, birth cycles, and times of death which had fascinated Swoboda. And he continued to elaborate the formidable mathematical apparatus that he felt was needed to calculate and make sense of biorhythm. To check the reliability of his extensive statistical tables, Fliess had the help of a mathematician. And, from the beginning, he had the advice not only of Freud (whose own notions about bisexuality were inspired by Fliess' work), but also that of another doctor, Hans Schlieper, who reviewed Fliess' *The Course of Life* in a 1909 publication called *Rhythms in Life*, and went on to write of his own encouraging work on biorhythm in *The Year in Space*.

The Third Biorhythm

While Swoboda and Fliess can be seen as the two fathers of biorhythm theory, a third researcher played a critical role in developing the theory as we know it today. Alfred Teltscher was an Austrian, like Swoboda, and as a doctor of engineering, he taught in Innsbruck. No doubt influenced by the stir that biorhythm was beginning to cause, he turned to his own experience with the intellectual performance of students to see if any rhythms were apparent there, as they seemed to be in emotional and physical performance.

Like most teachers, Teltscher had noticed that even his best students seemed to have both good and bad days. Their performance was by no means stable, even though they might sit at the head of the class. He began to collect information about how well Innsbruck high school and college students did on examinations, the dates of the exams, and the birthdates of the students. Then he made a thorough statistical search through his data, looking for any remarkable rhythms.

The results were not surprising to any student of biorhythm: a regular cycle of 33 days was found and later verified. Evidently, Teltscher felt, the brain's ability to absorb, manipulate, and express new ideas—and the student's apparent alertness and mental ability—are as subject to internal clocks as other cycles.

Teltscher did not begin his research with the same kind of theory that Fliess did, so he was not setting out to prove a hypothesis. Rather, he pored over his figures in an attempt to pick out *any* apparent rhythm. The weakness of this approach was that when he succeeded in finding such a rhythm, he was unable to explain why it existed or what the causes of the intellectual cycle might be. He then turned to his colleagues, some of whom were medical men like Fliess and Swoboda. They suggested that the rhythm he had observed might be regulated by secretions of certain glands that affect the brain. In light of what we know today about the profound effect on the brain of such glands as the pineal and the thyroid, that early guess doesn't seem too far-fetched.

What Causes Biorhythms?

Clearly, the three pioneers of biorhythm theory had different and insufficiently developed ideas about the roots of the phenomena that so impressed them. Even today, the precise workings of the human mind, body, and perceptions which determine the three great rhythms of life are obscure, although some progress has been made. To understand the theory, however, we must try to understand what ideas and theories of causation led Fliess, Swoboda, and Teltscher to their conclusions.

Part of the background of biorhythm was nineteenth-century "scientism"—the belief that, after a century of breathtaking progress in scientific research, scientific method and the scientific approach could solve all the problems of life, even the most perplexing and elusive. Remember that the nineteenth century, which was just drawing to a close when Swoboda and Fliess began their research, had been witness to such discoveries as anasthesia, sterile operating procedures, the germ theory of disease, and the existence of radioactivity. On the industrial front, progress was even more impressive: mass production, the steam and internal combustion engines, and a host of technological breakthroughs had transformed the economies of all European nations. Progress through science and technology began to sound like a commonplace, and to look as though it would never stop.

To a degree, part of this attitude is still with us today, since technocrats argue that the only barrier to solving *all* human problems is a scientific scheme sufficiently subtle and ingenious.

It was only natural that the scientific approach—called "scientism" in its more extreme forms—should be extended beyond medicine and industry to the areas of human behavior and social organization. The systematic study of psychology and its separation from philosophy, with which it was allied in most European universities, first began in the late 1800's, as did the study of sociology and the wholesale application of statistics to any behavioral phenomenon about which researchers could collect information.

The results were sometimes absurd. Intrigued by the idea of system and the beauty of statistics, many so-called "scientists" of the nineteenth and early twentieth centuries came up with comprehensive and usually harebrained explanations of the universe and proposals to bring man into the golden age. Perhaps the most impressive was Auguste Comte's detailed, complete breakdown of all the fields of knowledge which should be explored in order to perfect humanity. Works like H. G. Wells' *History of the World*—in *one* volume—were more reputable, but they still partook of the same boundless faith in the ability of science to comprehend all aspects of life.

At first glance, the work of Fliess, Swoboda, and Teltscher is no different from that of so many other would-be scientists whose names are lost to us. They too began with entirely empirical observations, and not with any basic hypothesis which was subject to proof or disproof. Despite very persuasive observations made after the fact, the real test of the theories developed by these three men lies in two areas: first, the ability of biorhythm theory to accurately predict future behavior; and second, the presence of a workable rationale to explain what empirical studies suggest to be true.

As to the second requirement, neither Swoboda nor Teltscher developed an adequate rationale. They were content to accumulate evidence for an inexplicable phenomenon. However, Fliess did go a long way with his notions about bisexuality. No contemporary scientist would concur with Fliess about the bisexuality of individual cells or about our ability to identify a cell as either masculine or feminine in type. Neither would Fliess' concentration on the supposedly "generative" cells of the mucous membrane found in the nose win approval today. But even the greatest skeptic would have to agree that the general notion of bisexuality is sound; that the combination of male and female hormones found in every human influences behavior profoundly; and that those hormones, along with others, give a regular rhythm to our feelings, physical abilities, and mental capabilities. So, like many true pioneers, Fliess was wrong in his particulars, but very right in his general approach.

Regarding the first requirement—the ability of biorhythm to predict behavior—there is a real problem of interpretation. The three great rhythms are interdependent. None of them is so strong that it overwhelms the other two; they always act in concert to affect us. True, on critical days there is a good chance that the rhythm or rhythms showing temporary instability will dominate, but never completely. On an emotionally critical day, for example, it sometimes happens that the strength of the physical and intellectual rhythms neutralizes any threat. This is even more likely on non-critical, or mixed, days which are the ones which occur most often. If all three rhythms are in the low (or recharging) phase, you are not likely to perform at your peak. But exactly how far below your best you will in fact perform remains an area of controversy and uncertain interpretation. Only experience with your own biorhythm chart can teach you what is likely to happen on mixed days.

The consequences of the interpretative problem are obvious. In most physical sciences, precise mathematical formulas can be used to predict exactly what will happen—how one chemical will react with another to form an end product, or how a critical mass of plutonium will behave. The speed, heat, and other aspects of the process are known, and even the slightest variation from the prediction is cause to re-examine the theory. Even in less precise areas such as animal behavior, we can describe the likelihood of an event mathematically, and deviations from the behavior described will take us back to the drawing board in an attempt to refine the theory. However, since biorhythm depends heavily on interpretation, we cannot prove or disprove it so easily. We have to look at a greater number of cases. The most convincing ones will be those that involve critical days, since they are the least ambiguous area of the theory. Mixed days will be valuable when they also involve extremes—two or three rhythms at the peak of the high phase or the bottom of the low phase on the same day. It is at such extreme points that biorhythm theory can make the strongest predictions about how people are likely to behave.

Unfortunately, there is a further complication. Just because you are experiencing a critical day, or have two or three rhythms at a peak or in a trough, does not mean that what happens in your life will necessarily be good or bad. A lot depends on your immediate environment. Obviously, if you stay in bed during a critical day, disaster will have very little chance to strike. If you have no reason to perform outstandingly during a peak period, you will never know whether you had the potential to excel on that day. In this way, biorhythm theory is very like psychoanalytic theory. It gives us insight into our predispositions; into the ways in which we are likely to perform—all other factors being equal. But how we in fact behave and what actually happens to us depends on the setting, or on particular environmental features and events—on all other factors *not* being equal. To put it another way, both biorhythm and psychoanalytic theory describe internal states of thought, body, and perception which should be apparent to the individual, but which do not have practical consequences until they interact with external factors.

All this raises a very legitimate question: if biorhythm cannot make mathematically verifiable predictions, if it cannot say with 100 percent certainty that a person will behave in a precise way, how can it be proven? Since the predictions are so open to interpretation, particularly in mixed days, is biorhythm any different from astrology or numerology, where interpretation can be bent to make the predictions fit the case? There are a number of answers to this very basic question. First of all, if we take a great number of cases (people), work out their biorhythms, and follow them over time to see whether they behave *in general* as biorhythm would predict, then the theory will appear much more sound. Secondly, if we examine people in stressful or dangerous occupations—airline pilots and bus drivers, for example—and compare the events in their working lives with the predictions of biorhythm theory, we are much more likely to find a high correlation between theory and reality. And finally, there is a test that everybody can apply. Start keeping a diary of your own internal states—physical, emotional, and intellectual—over a month or more. *Don't* look at your biorhythm chart or table before you start, so that the diary will be blind to any predictions which might influence it. Then compare the results with the patterns the theory would have predicted. In a blind test like this, the general accuracy—or error—of biorhythm is immediately apparent.

Admittedly, even these methods of proof leave open the central question of rationale, or basic hypothesis. We would like to tie biorhythm directly to some measurable physical basis, as Fliess erroneously tried to relate it to the sexuality of individual cells. In Chapter I we learned that it may be possible to link our behavior to a vast number of physical factors, even when the behavior is as mysterious as emotional states or as subtle as ratiocination. However, no adequate explanation has been offered to date. Again, as in psychoanalytic theory, the acceptance of biorhythm must spring more from its ability to help people in practice and from its internal coherence than any clear and verifiable explanation of its roots within the body.

Getting the Theory Right

In order to make any valid test of biorhythm, two things are essential: a clear and detailed understanding of how the theory is used to indicate how people are likely to act, and some convenient means of making the somewhat elaborate calculations needed to plot biorhythms accurately. We will try to provide both of these essentials now, first with a review of the theory, and then with a brief history of the various means of calculation, along with an explanation of the condensed tables which appear in Appendix B.

The three biorhythms begin in the positive phase at the moment of birth and continue regularly thereafter until death. Why they begin punctually at birth is no mystery, although this fact confuses some people. An infant's emergence into the world is an instant of almost inconceivable trauma—for the first time, his senses are exposed to an unprotected, even harsh, environment; and for the first time, his basic life systems—the brain, the lungs, and the circulatory system—must function without support from the mother. The baby's body must often be shocked into functioning, as doctors do when they spank newborns to make them cry (and thereby inflate their lungs for the first time). Through a variety of means, the child quickly begins to function as an autonomous physical being. Part of this functioning is the beginning of the biorhythms, which are galvanized into action at the same moment as the other life systems, and by the same need for survival. No less than the nervous or muscular systems, biorhythms are basic to our first breath.

Of the three great rhythms, the physical is the shortest and the least ambiguous. As you will recall, it lasts for 23 days, and during the first half of the cycle (the first 11.5 days) physical aspects are ascending. We give off energy with a prodigality that peaks about the sixth day of each new physical cycle. It is during the first half of the cycle that we are strongest. We can work our hardest, and for longer periods of time. We are more resistant to disease, better coordinated, and generally in better physical condition.

During the second 11.5 days of the physical cycle, the body is recharging and energy is being accumulated. Naturally, we tire more easily during those days, need more rest, and are unlikely to behave vigorously. This is the most likely time for a physical slump, since we are weak and more vulnerable to disease. It is helpful to think of the recharging half of the physical cycle as a passive phase in contrast to the very active first half of the cycle. Generally depleted by all the energy thrown off in the preceding period, people in the second—or negative—phase are allowing their bodies to gather strength for a return to the physical "high." If they are in basically good physical condition, no mishaps or illnesses should be expected—simply a reduction in physical potential. It is only during the critical days of the cycle, when the rhythm is changing from positive to negative or *vice versa*, that the physical rhythm becomes so erratic and unstable that extra caution is an absolute necessity.

The emotional rhythm of 28 days is sometimes called the "sensitivity rhythm," since Fliess thought it was based in the "female" cells of nerve tissue, which he characterized as being more sensitive than the "male" cells of the muscles that he supposed regulated the physical rhythm. Again, it is useful to think of the emotional cycle as having a positive phase of 14 days and a negative phase of the same length, with a critical day starting the whole 28-day cycle, ending it, and marking its midpoint.

During the first, positive two-week phase, we are inclined to react in positive and constructive ways to most events. We get along better with other people and with ourselves, so the first half of the emotional cycle is the best time for undertaking projects requiring cooperation, a positive attitude, and creativity. Because the emotional cycle is a powerful and pervasive one, it can modify both the physical and the intellectual cycle more than you might expect. In a period of physical decline combined with an emotional peak, even athletic performance may be up to par, simply because of a very positive attitude. This is particularly true in "mind" sports such as running or golf, where concentration and will power are extraordinarily important. The strength of the emotional rhythm's influence is also apparent in intellectual and creative endeavors. It is reasonable to think that during the negative part of the intellectual cycle, it will be difficult to achieve any major insights or to produce ideas. However, if that same period is one of an emotional high, the ideas and insights may flow in a virtual tidal wave. This is a matter of attitude and of releasing the emotional sources of creativity. The emotional rhythm is the easiest of the three for an individual to keep track of in himself, since he is usually a better judge of his feelings than of his body and mind. Even skeptics about biorhythm often admit to an awareness of regular fluctuations in their emotions.

In its 14-day negative phase, the emotional rhythm can have very dangerous results. Since feelings affect judgment, the days when we are recharging our emotional powers can be poor ones on which to perform dangerous tasks that call for swift reactions and sound judgment. This is why industrial and transportation accidents are more likely to occur during a person's negative emotional phase. Since we tend to be testy, short-tempered, and negative on days of emotional recharging, these are not the best days for teamwork or for making job or family decisions. Critical days, of course, carry much more danger than the days of the negative phase. Due to the fact that all of the emotional minuses of the recovery phase are exaggerated, critical days of the emotional cycle leave us open to self-inflicted harm, violent arguments, and a disagreeableness which may seem entirely irrational to others. An emotionally critical day, particularly when it is combined with a critical day in either the physical or intellectual cycle, definitely calls for all available precautions.

Fortunately, it is extremely simple to forecast the day of the week on which such a day will fall. Since the emotional rhythm shifts from positive to negative (or vice versa) every two weeks and begins on the day of birth, the first emotionally critical day occurs two weeks after the day of birth. Every emotionally critical day from then until the day of death will occur at this two-week interval. If you were born on a Friday, every other Friday will be critical. (This is one of the easiest ways to check out biorhythm for yourself, since it needs no complex calculations. Simply find out the day on which you were born and ask yourself if that same day has accounted for many of your deepest mood swings.)

The relationship between the emotional rhythm and the menstrual cycle is intriguing but elusive. Since both the emotional and ideal menstrual cycles take 28 days to complete, and since the emotional effects of menstruation are well-known, many people feel that there is a natural relationship between the two. However, the onset of menstruation does not always occur on a critical day in the emotional rhythm, and this failure of the two cycles to correspond can be perplexing. Actually, several factors combine to desynchronize the two rhythms. First of all, as Fliess explained, the menstrual cycle varies in length from 23 to about 35 days (and sometimes even as long as 63 days) because it is affected by both the physical and the emotional biorhythms. If there were no physical rhythm at all, the menstrual cycle would be determined entirely by the emotional rhythm, and it would invariably last for 28 days, with a critical day marking the beginning of menstruation. As it is, the fact that the physical rhythm takes 23 days to complete means that it is almost always out of phase with the emotional rhythm, and this disjunction can affect the menstrual cycle, shortening and lengthening it at different stages of life.

Secondly, menstruation is subject to a number of external influences. It is not determined solely by an internal clock that begins to tick at the moment of birth. To some extent, the menstrual cycle is keyed to the phase of the moon under which a woman was born, as you read in Chapter I. To an apparently equal extent, menstruation can be regulated by light, as has been shown by experiments in which night lights timed to shine during the days of ovulation have been used to regulate erratic menstrual cycles. In sum, then, menstruation is regulated by a number of factors—emotional and physical biorhythms, the moon, and light patterns. It can also be controlled by the female hormones used in birth control pills. The menstrual cycle is therefore much more flexible and changeable than the emotional rhythm.

The intellectual cycle has received the least attention of the three, perhaps because it was not one of the first to be discovered, or perhaps because its impact on accidents and human survival is not as great as that of the other two. However, since life can be called a continual learning process, it is very valuable to know about the effects of the 33-day intellectual cycle.

Paradoxically, this least-studied of the biorhythms may be the first one for which science will provide a convincing rationale. As Teltscher's colleagues suggested, the brain is affected by many different hormones, particularly those secreted by the thyroid, pineal, and pituitary glands. Since so much is already known about hormonal effects, and since large research efforts are still underway, we may soon see the day when a basic, physiologically clear rhythmicity is discovered and directly related to the intellectual cycle.

Still, little is known about the intellectual rhythm today, aside from its most essential aspects—that the cycle takes 33 days to complete and is divided into a positive phase of 16½ days and a negative one of the same length. During the first phase, our minds are more open, our memories more retentive, and our ability to put together separate ideas to achieve an understanding is at its best. The positive phase is a good time for encountering new, unfamiliar situations that call for quick comprehension and adaptation such as a new job or a visit to a foreign country. Efforts at self-improvement through reading and education will also be more fruitful during this period.

During the 16½ day negative phase, people do not

become stupid or dull. They are simply less inclined to deal openly with new subjects or situations, mainly because they lack the high levels of intellectual energy needed to do so with ease. Most of us find it hard to resist closing our minds during this period so that we can "recharge the brain cells." Nor do we find it easy to concentrate or to take the time and trouble to think things through with maximum clarity. Again, these traits are not overwhelming during the negative phase of the intellectual cycle; they say more about our inclinations than about what we will actually do. The certainty of trouble on critical days of the intellectual cycle is much greater when the environment contains situations calling for intellectual acuity. If you have to analyze an important matter, make an important decision, or talk to others about matters of vital concern, it would be wise to avoid doing so on a critical day of your intellectual cycle.

The relative strength of the physical, emotional, and intellectual rhythms varies among individuals. It is foolish to generalize about which rhythms in which phases will dominate other rhythms and phases, and this problem naturally complicates the question of interpreting mixed days—a question already ambiguous enough. It seems that heredity and talent may do much to explain these differences. A "natural" athlete—i.e., someone who has inherited good coordination and physical strength—places much more emphasis on the physical cycle than most people do, and in the case of professional athletes this emphasis is even stronger. We would assume that the athlete's physical rhythm is more powerful in the first place, and will have more important effects on their life and work. Analogous examples can be drawn from people whose predilections and activities are mainly emotional (creative artists) or intellectual (mathematicians). What you are through inherited characteristics will strengthen one or another of your biorhythms; what you do because of your talents will give that rhythm more opportunities to affect your life.

The Numbers Game

Since biorhythm can be so important to people in so many ways, precise calculations of biorhythmic cycles have been a goal for many researchers since the days of Fliess and Swoboda. The need for precision is obvious: a difference of only 24 hours can mean the difference between an ordinary day and a critical one, and that difference, in turn, may affect matters of life and death. In most cases, knowing the day of birth is enough to make a precise biorhythm chart, but there are some cases where the hour of birth is important as well. This is not because biorhythm, like astrology, must calculate the complex interrelations of heavenly bodies at a particular time, but because people born very late at night have biorhythms more characteristic of the following day than of the day officially recorded as the day of birth. There have been several cases where biorhythm calculations always seemed to be a day behind life events, and the explanation has been that the person involved was born quite close to the beginning of the next day.

The matter of the hour of birth is simple to deal with, since all you have to do is push your chart forward a day. The problem seldom arises, anyway. The real difficulty, as Fliess found to his sorrow, is that the mathematical calculations of biorhythm can be forbidding. Few people have the patience to add up all the days they have lived since birth, to make the needed adjustments for leap years, to divide the resulting figure by the length of each of the cycles (23, 28, 33), and to use the remainder of each division as an indication of how far into the new cycle they have progressed on the day of calculation. Unless some record is kept every day thereafter, the same sort of calculation will have to be repeated later to avoid the possibility of error.

Recognizing this stumbling block, a number of researchers and engineers have tried to develop easy and widely available methods—and even machines—for making biorhythm calculations. Swoboda himself designed a paste-board slide rule for calculating critical days. Another pioneer was Alfred Judt, an engineers and mathematician from the North German port of Bremen. He was fascinated by the implications of biorhythm for team and individual sports, so he needed an easy way to calculate many different biorhythmic positions using only the athlete's birth date and the date of the contest. Since no such mechanism existed, he invented one: elaborate mathematical tables for calculating athletic performance almost instantly.

Of course, Judt's work was limited to the sports application for which it was designed. It was another engineer-mathematician, the Swiss Hans Frueh, who refined and generalized Judt's tables so that more people could use them. This was not enough for the ambitious Frueh. He designed and produced "biocards" on which individuals could chart their biorhythm calculations. He also designed a machine that could be set to calculate any individual's biorhythmic situation for a particular day. Called the Biocalculator and operated by hand cranking, it has been manufactured continuously in Switzerland since 1932. In time, other easy mechanical calculators were introduced. The first was the Biostar Electronic Calculator watch, made by Certina of Switzerland. The watch looks just like an ordinary one except for a window in the dial right above the center of the face, through which you can see three concentric bands, color-coded to the three biorhythms. Farthest from the center of the face is the physical band (red), then the emotional (blue). The intellectual band (green) is closest to the center. Each band passes through the window at a different speed and is marked off to correspond with the number of days in each cycle. Any jeweler can set the watch to an individual's birthdate. Since it is electronic, it needs no winding—just battery replacement every six months—and it ensures truly accurate readings. The watch will show precisely how far its owner has progressed through each new cycle, and with a few refinements: the numbers of the positive-phase days appear against the colored backgrounds of the bands; negative days show up against a silver background; critical days appear at the boundary between colored and silver; and there is even a small calendar window which shows not the date of the month, but the day of the week (Monday-1 through Sunday-7), since the irregularity of days in a month as well as the occurrence of leap years force people interested in biorhythm to ignore monthly units and concentrate on the invariable weekly unit instead.

One of the most sophisticated devices for calculating biorhythms was introduced recently by the Japanese, who are the most active practitioners of biorhythm today. It is a miniaturized desk-top computer called the Biocom 200. The operator punches in a birth date, then the month and year in which he is interested. In seconds the machine clicks out a small ticket which is a miniature biorhythm chart for the person and month in question. Japanese insurance and transportation companies, as well as the traffic police, have used this device to issue thousands of biorhythm charts to motorists.

A comparable machine was introduced in late 1973, when Olivetti—Italy's largest manufacturer of office machinery—unveiled a desk-top "Microcomputer" programmed to make almost instantaneous biorhythm calculations.

Rather than forcing biorhythm enthusiasts to invest in

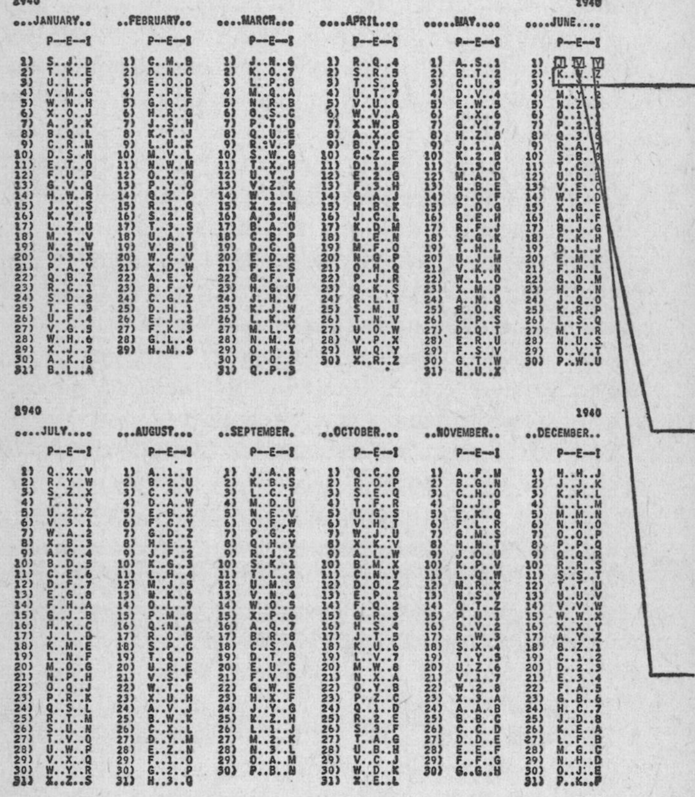

1940 | | | | | | 1940

Day	JANUARY P--E--I	FEBRUARY P--E--I	MARCH P--E--I	APRIL P--E--I	MAY P--E--I	JUNE P--E--I
1)	S..J..D	C..M..B	J..N..6	R..Q..4	A..S..1	J..W..Y
2)	T..K..E	D..N..C	K..O..7	S..R..5	B..T..2	K..X..Z
3)	U..L..F	E..O..D	L..P..8	T..S..6	C..U..3	L..[illegible]..1
4)	V..M..G	F..P..E	M..Q..A	U..T..7	D..V..4	M..Y..2
5)	W..N..H	G..Q..F	N..R..B	V..U..8	E..W..5	N..Z..3
6)	X..O..J	H..R..G	O..S..C	W..V..A	F..X..6	O..1..4
7)	A..P..K	J..S..H	P..T..D	X..W..B	G..Y..7	P..2..5
8)	B..Q..L	K..T..J	Q..U..E	A..X..C	H..Z..8	Q..3..6
9)	C..R..M	L..U..K	R..V..F	B..Y..D	J..1..A	R..A..7
10)	D..S..N	M..V..L	S..W..G	C..Z..E	K..2..B	S..B..8
11)	E..T..O	N..W..M	T..X..H	D..1..F	L..3..C	T..C..A
12)	F..U..P	O..X..N	U..Y..J	E..2..G	M..A..D	U..D..B
13)	G..V..Q	P..Y..O	V..Z..K	F..3..H	N..B..E	V..E..C
14)	H..W..R	Q..Z..P	W..1..L	G..A..J	O..C..F	W..F..D
15)	J..X..S	R..1..Q	X..2..M	H..B..K	P..D..G	X..G..E
16)	K..Y..T	S..2..R	A..3..N	J..C..L	Q..E..H	A..H..F
17)	L..Z..U	T..3..S	B..A..O	K..D..M	R..F..J	B..J..G
18)	M..1..V	U..A..T	C..B..P	L..E..N	S..G..K	C..K..H
19)	N..2..W	V..B..U	D..C..Q	M..F..O	T..H..L	D..L..J
20)	O..3..X	W..C..V	E..D..R	N..G..P	U..J..M	E..M..K
21)	P..A..Y	X..D..W	F..E..S	O..H..Q	V..K..N	F..N..L
22)	Q..B..Z	A..E..X	G..F..T	P..J..R	W..L..O	G..O..M
23)	R..C..1	B..F..Y	H..G..U	Q..K..S	X..M..P	H..P..N
24)	S..D..2	C..G..Z	J..H..V	R..L..T	A..N..Q	J..Q..O
25)	T..E..3	D..H..1	K..J..W	S..M..U	B..O..R	K..R..P
26)	U..F..4	E..J..2	L..K..X	T..N..V	C..P..S	L..S..Q
27)	V..G..5	F..K..3	M..L..Y	U..O..W	D..Q..T	M..T..R
28)	W..H..6	G..L..4	N..M..Z	V..P..X	E..R..U	N..U..S
29)	X..J..7	H..M..5	O..N..1	W..Q..Y	F..S..V	O..V..T
30)	A..K..8		P..O..2	X..R..Z	G..T..W	P..W..U
31)	B..L..A		Q..P..3		H..U..X	

1940 | | | | | | 1940

Day	JULY P--E--I	AUGUST P--E--I	SEPTEMBER P--E--I	OCTOBER P--E--I	NOVEMBER P--E--I	DECEMBER P--E--I
1)	Q..X..V	A..1..T	J..A..R	Q..C..O	A..F..M	H..H..J
2)	R..Y..W	B..2..U	K..B..S	R..D..P	B..G..N	J..J..K
3)	S..Z..X	C..3..V	L..C..T	S..E..Q	C..H..O	K..K..L
4)	T..1..Y	D..A..W	M..D..U	T..F..R	D..J..P	L..L..M
5)	U..2..Z	E..B..X	N..E..V	U..G..S	E..K..Q	M..M..N
6)	V..3..1	F..C..Y	O..F..W	V..H..T	F..L..R	N..N..O
7)	W..A..2	G..D..Z	P..G..X	W..J..U	G..M..S	O..O..P
8)	X..B..3	H..E..1	Q..H..Y	X..K..V	H..N..T	P..P..Q
9)	A..C..4	J..F..2	R..J..Z	A..L..W	J..O..U	Q..Q..R
10)	B..D..5	K..G..3	S..K..1	B..M..X	K..P..V	R..R..S
11)	C..E..6	L..H..4	T..L..2	C..N..Y	L..Q..W	S..S..T
12)	D..F..7	M..J..5	U..M..3	D..O..Z	M..R..X	T..T..U
13)	E..G..8	N..K..6	V..N..4	E..P..1	N..S..Y	U..U..V
14)	F..H..A	O..L..7	W..O..5	F..Q..2	O..T..Z	V..V..W
15)	G..J..B	P..M..8	X..P..6	G..R..3	P..U..1	W..W..X
16)	H..K..C	Q..N..A	A..Q..7	H..S..4	Q..V..2	X..X..Y
17)	J..L..D	R..O..B	B..R..8	J..T..5	R..W..3	A..Y..Z
18)	K..M..E	S..P..C	C..S..A	K..U..6	S..X..4	B..Z..1
19)	L..N..F	T..Q..D	D..T..B	L..V..7	T..Y..5	C..1..2
20)	M..O..G	U..R..E	E..U..C	M..W..8	U..Z..6	D..2..3
21)	N..P..H	V..S..F	F..V..D	N..X..A	V..1..7	E..3..4
22)	O..Q..J	W..T..G	G..W..E	O..Y..B	W..2..8	F..A..5
23)	P..R..K	X..U..H	H..X..F	P..Z..C	X..3..A	G..B..6
24)	Q..S..L	A..V..J	J..Y..G	Q..1..D	A..A..B	H..C..7
25)	R..T..M	B..W..K	K..Z..H	R..2..E	B..B..C	J..D..8
26)	S..U..N	C..X..L	L..1..J	S..3..F	C..C..D	K..E..A
27)	T..V..O	D..Y..M	M..2..K	T..A..G	D..D..E	L..F..B
28)	U..W..P	E..Z..N	N..3..L	U..B..H	E..E..F	M..G..C
29)	V..X..Q	F..1..O	O..A..M	V..C..J	F..F..G	N..H..D
30)	W..Y..R	G..2..P	P..B..N	W..D..K	G..G..H	O..J..E
31)	X..Z..S	H..3..Q		X..E..L		P..K..F

Birth chart for person born in 1940.

OCTOBER 1975 **PHYSICAL** OCTOBER 1975

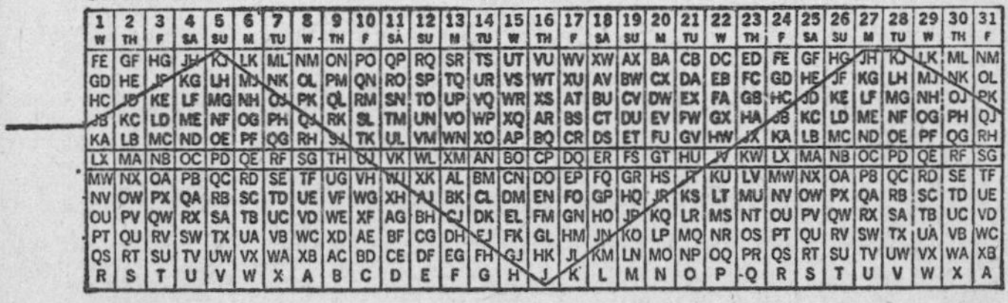

1 W	2 TH	3 F	4 SA	5 SU	6 M	7 TU	8 W	9 TH	10 F	11 SA	12 SU	13 M	14 TU	15 W	16 TH	17 F	18 SA	19 SU	20 M	21 TU	22 W	23 TH	24 F	25 SA	26 SU	27 M	28 TU	29 W	30 TH	31 F
FE	GF	HG	JH	KJ	LK	ML	NM	ON	PO	QP	RQ	SR	TS	UT	VU	WV	XW	AX	BA	CB	DC	ED	FE	GF	HG	JH	KJ	LK	ML	NM
GD	HE	JF	KG	LH	MJ	NK	OL	PM	QN	RO	SP	TQ	UR	VS	WT	XU	AV	BW	CX	DA	EB	FC	GD	HE	JF	KG	LH	MJ	NK	OL
HC	JD	KE	LF	MG	NH	OJ	PK	QL	RM	SN	TO	UP	VQ	WR	XS	AT	BU	CV	DW	EX	FA	GB	HC	JD	KE	LF	MG	NH	OJ	PK
JB	KC	LD	ME	NF	OG	PH	QJ	RK	SL	TM	UN	VO	WP	XQ	AR	BS	CT	DU	EV	FW	GX	HA	JB	KC	LD	ME	NF	OG	PH	QJ
KA	LB	MC	ND	OE	PF	QG	RH	SJ	TK	UL	VM	WN	XO	AP	BQ	CR	DS	ET	FU	GV	HW	JX	KA	LB	MC	ND	OE	PF	QG	RH
LX	MA	NB	OC	PD	QE	RF	SG	TH	UJ	VK	WL	XM	AN	BO	CP	DQ	ER	FS	GT	HU	JV	KW	LX	MA	NB	OC	PD	QE	RF	SG
MW	NX	OA	PB	QC	RD	SE	TF	UG	VH	WJ	XK	AL	BM	CN	DO	EP	FQ	GR	HS	JT	KU	LV	MW	NX	OA	PB	QC	RD	SE	TF
NV	OW	PX	QA	RB	SC	TD	UE	VF	WG	XH	AJ	BK	CL	DM	EN	FO	GP	HQ	JR	KS	LT	MU	NV	OW	PX	QA	RB	SC	TD	UE
OU	PV	QW	RX	SA	TB	UC	VD	WE	XF	AG	BH	CJ	DK	EL	FM	GN	HO	JP	KQ	LR	MS	NT	OU	PV	QW	RX	SA	TB	UC	VD
PT	QU	RV	SW	TX	UA	VB	WC	XD	AE	BF	CG	DH	EJ	FK	GL	HM	JN	KO	LP	MQ	NR	OS	PT	QU	RV	SW	TX	UA	VB	WC
QS	RT	SU	TV	UW	VX	WA	XB	AC	BD	CE	DF	EG	FH	GJ	HK	JL	KM	LN	MO	NP	OQ	PR	QS	RT	SU	TV	UW	VX	WA	XB
R	S	T	U	V	W	X	A	B	C	D	E	F	G	H	J	K	L	M	N	O	P	Q	R	S	T	U	V	W	X	A

OCTOBER 1975 **EMOTIONAL** OCTOBER 1975

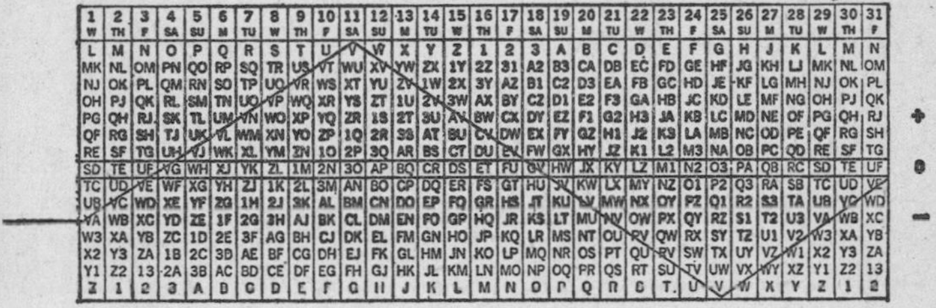

1 W	2 TH	3 F	4 SA	5 SU	6 M	7 TU	8 W	9 TH	10 F	11 SA	12 SU	13 M	14 TU	15 W	16 TH	17 F	18 SA	19 SU	20 M	21 TU	22 W	23 TH	24 F	25 SA	26 SU	27 M	28 TU	29 W	30 TH	31 F
L	M	N	O	P	Q	R	S	T	U	V	W	X	Y	Z	1	2	3	A	B	C	D	E	F	G	H	J	K	L	M	N
MK	NL	OM	PN	QO	RP	SQ	TR	US	VT	WU	XV	YW	ZX	1Y	2Z	31	A2	B3	CA	DB	EC	FD	GE	HF	JG	KH	LJ	MK	NL	OM
NJ	OK	PL	QM	RN	SO	TP	UQ	VR	WS	XT	YU	ZV	1W	2X	3Y	AZ	B1	C2	D3	EA	FB	GC	HD	JE	KF	LG	MH	NJ	OK	PL
OH	PJ	QK	RL	SM	TN	UO	VP	WQ	XR	YS	ZT	1U	2V	3W	AX	BY	CZ	D1	E2	F3	GA	HB	JC	KD	LE	MF	NG	OH	PJ	QK
PG	QH	RJ	SK	TL	UM	VN	WO	XP	YQ	ZR	1S	2T	3U	AV	BW	CX	DY	EZ	F1	G2	H3	JA	KB	LC	MD	NE	OF	PG	QH	RJ
QF	RG	SH	TJ	UK	VL	WM	XN	YO	ZP	1Q	2R	3S	AT	BU	CV	DW	EX	FY	GZ	H1	J2	K3	LA	MB	NC	OD	PE	QF	RG	SH
RE	SF	TG	UH	VJ	WK	XL	YM	ZN	1O	2P	3Q	AR	BS	CT	DU	EV	FW	GX	HY	JZ	K1	L2	M3	NA	OB	PC	QD	RE	SF	TG
SD	TE	UF	VG	WH	XJ	YK	ZL	1M	2N	3O	AP	BQ	CR	DS	ET	FU	GV	HW	JX	KY	LZ	M1	N2	O3	PA	QB	RC	SD	TE	UF
TC	UD	VE	WF	XG	YH	ZJ	1K	2L	3M	AN	BO	CP	DQ	ER	FS	GT	HU	JV	KW	LX	MY	NZ	O1	P2	Q3	RA	SB	TC	UD	VE
UB	VC	WD	XE	YF	ZG	1H	2J	3K	AL	BM	CN	DO	EP	FQ	GR	HS	JT	KU	LV	MW	NX	OY	PZ	Q1	R2	S3	TA	UB	VC	WD
VA	WB	XC	YD	ZE	1F	2G	3H	AJ	BK	CL	DM	EN	FO	GP	HQ	JR	KS	LT	MU	NV	OW	PX	QY	RZ	S1	T2	U3	VA	WB	XC
W3	XA	YB	ZC	1D	2E	3F	AG	BH	CJ	DK	EL	FM	GN	HO	JP	KQ	LR	MS	NT	OU	PV	QW	RX	SY	TZ	U1	V2	W3	XA	YB
X2	Y3	ZA	1B	2C	3D	AE	BF	CG	DH	EJ	FK	GL	HM	JN	KO	LP	MQ	NR	OS	PT	QU	RV	SW	TX	UY	VZ	W1	X2	Y3	ZA
Y1	Z2	13	2A	3B	AC	BD	CE	DF	EG	FH	GJ	HK	JL	KM	LN	MO	NP	OQ	PR	QS	RT	SU	TV	UW	VX	WY	XZ	Y1	Z2	13
Z	1	2	3	A	B	C	D	E	F	G	H	J	K	L	M	N	O	P	Q	R	S	T	U	V	W	X	Y	Z	1	2

\+ 0 −

OCTOBER 1975 **INTELLECTUAL** OCTOBER 1975

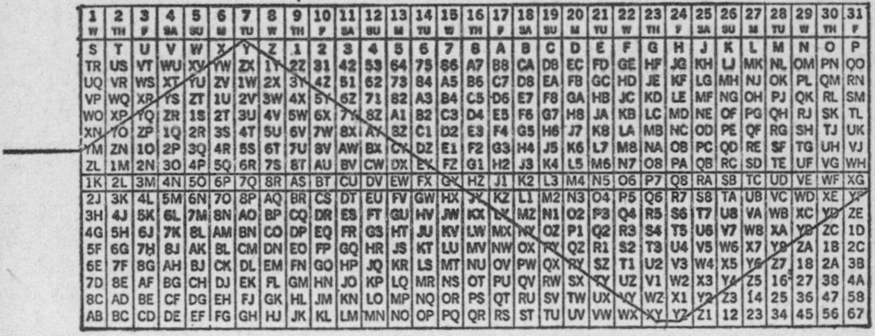

1 W	2 TH	3 F	4 SA	5 SU	6 M	7 TU	8 W	9 TH	10 F	11 SA	12 SU	13 M	14 TU	15 W	16 TH	17 F	18 SA	19 SU	20 M	21 TU	22 W	23 TH	24 F	25 SA	26 SU	27 M	28 TU	29 W	30 TH	31 F
S	T	U	V	W	X	Y	Z	1	2	3	4	5	6	7	8	A	B	C	D	E	F	G	H	J	K	L	M	N	O	P
TR	US	VT	WU	XV	YW	ZX	1Y	2Z	31	42	53	64	75	86	A7	B8	CA	DB	EC	FD	GE	HF	JG	KH	LJ	MK	NL	OM	PN	QO
UQ	VR	WS	XT	YU	ZV	1W	2X	3Y	4Z	51	62	73	84	A5	B6	C7	D8	EA	FB	GC	HD	JE	KF	LG	MH	NJ	OK	PL	QM	RN
VP	WQ	XR	YS	ZT	1U	2V	3W	4X	5Y	6Z	71	82	A3	B4	C5	D6	E7	F8	GA	HB	JC	KD	LE	MF	NG	OH	PJ	QK	RL	SM
WO	XP	YQ	ZR	1S	2T	3U	4V	5W	6X	7Y	8Z	A1	B2	C3	D4	E5	F6	G7	H8	JA	KB	LC	MD	NE	OF	PG	QH	RJ	SK	TL
XN	YO	ZP	1Q	2R	3S	4T	5U	6V	7W	8X	AY	BZ	C1	D2	E3	F4	G5	H6	J7	K8	LA	MB	NC	OD	PE	QF	RG	SH	TJ	UK
YM	ZN	1O	2P	3Q	4R	5S	6T	7U	8V	AW	BX	CY	DZ	E1	F2	G3	H4	J5	K6	L7	M8	NA	OB	PC	QD	RE	SF	TG	UH	VJ
ZL	1M	2N	3O	4P	5Q	6R	7S	8T	AU	BV	CW	DX	EY	FZ	G1	H2	J3	K4	L5	M6	N7	O8	PA	QB	RC	SD	TE	UF	VG	WH
1K	2L	3M	4N	5O	6P	7Q	8R	AS	BT	CU	DV	EW	FX	GY	HZ	J1	K2	L3	M4	N5	O6	P7	Q8	RA	SB	TC	UD	VE	WF	XG
2J	3K	4L	5M	6N	7O	8P	AQ	BR	CS	DT	EU	FV	GW	HX	JY	KZ	L1	M2	N3	O4	P5	Q6	R7	S8	TA	UB	VC	WD	XE	YF
3H	4J	5K	6L	7M	8N	AO	BP	CQ	DR	ES	FT	GU	HV	JW	KX	LY	MZ	N1	O2	P3	Q4	R5	S6	T7	U8	VA	WB	XC	YD	ZE
4G	5H	6J	7K	8L	AM	BN	CO	DP	EQ	FR	GS	HT	JU	KV	LW	MX	NY	OZ	P1	Q2	R3	S4	T5	U6	V7	W8	XA	YB	ZC	1D
5F	6G	7H	8J	AK	BL	CM	DN	EO	FP	GQ	HR	JS	KT	LU	MV	NW	OX	PY	QZ	R1	S2	T3	U4	V5	W6	X7	Y8	ZA	1B	2C
6E	7F	8G	AH	BJ	CK	DL	EM	FN	GO	HP	JQ	KR	LS	MT	NU	OV	PW	QX	RY	SZ	T1	U2	V3	W4	X5	Y6	Z7	18	2A	3B
7D	8E	AF	BG	CH	DJ	EK	FL	GM	HN	JO	KP	LQ	MR	NS	OT	PU	QV	RW	SX	TY	UZ	V1	W2	X3	Y4	Z5	16	27	38	4A
8C	AD	BE	CF	DG	EH	FJ	GK	HL	JM	KN	LO	MP	NQ	OR	PS	QT	RU	SV	TW	UX	VY	WZ	X1	Y2	Z3	14	25	36	47	58
AB	BC	CD	DE	EF	FG	GH	HJ	JK	KL	LM	MN	NO	OP	PQ	QR	RS	ST	TU	UV	VW	WX	XY	YZ	Z1	12	23	34	45	56	67

Biorhythm curves for October 1975 for person born June 1, 1940.

expensive hardware like the Biostar or the Biocom 200, contemporary proponents of the theory have developed much less expensive software. George Thommen, who pioneered biorhythm in the United States, has developed simplified calculation tables that require only a little mathematical facility to use; he has also introduced the "Dialgraf," a sort of circular slide rule for biorhythm which operates on the same principles as Swoboda's more orthodox straight slide rule. Thommen's most popular aid to biorhythmists is the "Cyclgraf"—a series of cards marked off into days so that anybody can use the accompanying biorhythm rulers (templates of each of the three cycles) to draw his own chart in the form of sine waves. B. J. Krause-Poray, Thommen's counterpart in Australia, has developed a Biorhythm Calendar. Computer-generated biorhythm charts, complete with day-to-day interpretations are available at Biorhythm Computers, Inc., 119 West 57 Street, New York 10019, which is a center for information on biorhythm as well as a source for do-it-yourself kits on biorhythm. Biolator—from Casio (a Japanese computer firm)—is a handheld pocket calculator with biorhythm programming ($35 with batteries and an AC adapter). With this calculator you put in today's year, month and date, then subtract your birth year, month and date, and push in a "bio" key. Three numbers will appear in the readout, and these numbers are compared to charts on the reverse side giving you your biorhythm reading for the day.

Although calculators are handy, they have the disadvantage of not having a print-out, nor can one easily see the biorhythm chart for planning and comparison. To overcome these problems, we have developed some very simple biorhythm tables, which appear in the Appendices. To use them, you don't have to make any computations. Simply look up the date, month, and year of your birth. Following it you will find three symbols, such as "J, V,

Y." The first symbol refers to the physical rhythm, and second to the emotional, and the third to the intellectual. Then turn to the second part of the tables, where you will find each month of the years 1976 to 1978 duplicated three times—once for each biorhythm—and numbered to correspond to the days of that month. You should not be confused by the swarms of symbols that appear within those columns: simply refer to the first symbol you found under your birthdate, locate the same symbol in the monthly table for the physical rhythm, and draw a line from column to column passing through your physical symbol. The result will be an accurate and highly readable sine wave chart of your physical rhythm for that month. Repeat the process for the monthly emotional and intellectual rhythm charts—drawing the line through your second and third symbols, respectively—and you will have a clear and complete biorhythm chart in three parts.

With this simplified method of calculating biorhythms, you can do your own research, finding out whether any accidents have happened to you on critical days and whether your high and low points in the last few months have corresponded to the peaks and valleys in your biorhythm charts. You can also plan future research, keeping a diary for a month or two and then checking with your charts to measure the degree of correspondence between your biorhythm state and the diary entries. However, before undertaking this kind of personal verification, you will probably want to know more about the evidence that has been accumulated over many years and in many different fields to support the theory of biorhythm. That is the subject of the next chapter.

Can You Prove It?

Like biorhythms themselves, research devoted to proving biorhythm theory has had its ups and downs. Shortly after the theory was first formulated, there was a flurry of interest among European researchers, especially in Germany and Switzerland. Although Swiss interest has persisted to this day, this first period of research was brief. It was followed by another flare-up of interest just before and during World War II, when industrial psychologists, deeply concerned with increasing efficiency and decreasing the accident rate in war-time plants, turned to biorhythm theory for answers. Much of this research persisted after the war. Biorhythm research in general received a big boost as commercial aviation came into its own: with the lives of so many people depending on the performance of airline pilots, researchers felt that biorhythmic analyses of the pilots might do much to reduce the incidence of fatal crashes. Similarly, as cities such as Tokyo developed elaborate road systems after World War II, became more congested, and relied increasingly on various forms of public transportation, there was an enormous increase in research on the effect of biorhythms on traffic accidents. Today, biorhythm research has spread well beyond the fields of industrial safety, airline crashes, and traffic accidents. Doctors have become more aware of how circadian rhythms affect our lives, and this has brought about greater medical interest in the larger cycles of biorhythm. As international television has contributed to world-wide sports mania, the application of biorhythm theory to athletic performance has become more common.

Our present knowledge of biorhythm can be roughly divided into two large categories. First of all, there are *retrospective* studies, which concentrate on analyzing past events to see how well they correspond to the theory. The events that are scrutinized range from large data series on insurance company accident claims to single cases, such as Clark Gable's heart attack. (See Figure 2.) The common characteristic of this work is that it deals with past events, and therefore includes few of the experimental controls or laboratory safeguards that scientists like to set up before doing a rigorous study of any theory. Still, researchers have compensated for this deficiency in various ways, and the results of retrospective analysis have been very impressive.

The second type of research might be called *experimental*. Whether the setting is that of a natural experiment (which takes place in the field and raises problems of experimental control) or that of a laboratory experiment (which has better controls, but raises the question of how well it replicates reality), all of these analyses involve applying biorhythm theory first, and then checking the results as events take place. This is what happens when a bus company issues biorhythm charts to its drivers and then compares their accident records for the periods before and after the drivers received the charts, or when a surgeon starts to use biorhythm to determine the best time for an operation and then compares his fatality record with his previous record or with the general mortality rate for the type of operation he is performing.

The big advantage of experimental studies, as opposed to retrospective ones, is that they allow before and after comparisons, and this is usually a good way of making sure that the results have been influenced by biorhythm and by nothing else. (If all factors in the experimental setting are the same for the pre-biorhythm and post-biorhythm periods, then any nonrandom changes must be attributed to the application of biorhythm.) Another advantage is that experimental studies leave much less latitude for manipulating the results—whether consciously or unconsciously. Since there is a significant element of interpretation in reading biorhythm charts, it can be all too easy for even the most honest researcher to bend his interpretation to fit the case. The researcher knows what has happened in the past, and he may—completely unconsciously—slant his interpretation. (A critical day, of course, is unambiguous, so no honest researcher will be trapped into an excessively favorable interpretation. However, mixed days allow a larger margin for error.)

The amount of retrospective research far outweighs the amount of experimental work for a very simple reason: it is much easier and cheaper to do. Any competent person can select past data and analyze them with the help of biorhythm calculations; only specialists with adequate funding and institutional cooperation can conduct good experimental work. There is a trade-off here between quantity and quality of work, and the results of retrospective analysis must therefore be viewed with more suspicion than the results of experiments. Still, the fact that almost all of the many studies of past events have come to similar conclusions is most provocative, and it is those studies we will examine first.

Death in Zurich

Hans Frueh, the Swiss engineer who invented the "Biocard" for displaying biorhythmic cycles, was very influential in making Switzerland the historical center of biorhythm research. Frueh himself cited many examples of the value of biorhythm in medicine, and he stimulated doctors and engineers in Germany and Switzerland to experiment with biorhythm theory in timing surgery and in preventing railroad accidents. Frueh's influence was so strong that a 1971 reissue of his most famous book, *The Triumph of Biorhythms*, carried an enthusiastic prefatory note by the renowned clinician Dr. Werner Zabel of Berchtesgaden, Germany. Zabel wrote:

> I have established the accuracy of this theory in every case I have investigated in my clinic, whether it was a case of death, of post-operative hemmorhaging, or of fatal results from anesthesia. And now I never advise my patients to undergo surgery when their biorhythm position is a counter-indication. In sum, the world of science cannot afford to bypass biorhythm any longer.

One of Frueh's most famous followers was Hans Schwing, who submitted a study of biorhythm to the Swiss Federal Institute of Technology in Zurich in 1939 in order to earn his doctorate in natural science. According to George Thommen, the pioneer of biorhythm in the United States, "[Schwing's] research report, in the form of a 78-page treatise, is probably one of the most precisely recorded analyses of the subject." Examination of Schwing's thesis bears out this assessment. Schwing actually performed two studies rolled into one. First he assembled data from insurance companies and the Swiss government which gave the date of birth of 700 people involved in serious accidents, along with the date of the accidents; and he related this information to critical days. Second, he obtained similar data about 300 people who had died in Zurich and related this to the occurrence of critical days in their physical and emotional rhythms.

Schwing's methods of correlating these data were complicated but reliable. First, he determined how often single and double critical days occur within one biorhythmic lifespan. (A biorhythmic lifespan is the period of 58 years and 66 or 67 days between birth, when the three biorhythms begin a new cycle at precisely the same time, and the day when all three biorhythms again cross the baseline at exactly the same time.) Schwing's calculations indicated that on slightly more than 20 percent of the days within the average biorhythmic lifespan—or on 4,327 days out of 21,253—at least one and often two of the three rhythms are crossing the baseline; the remaining 80 percent are characterized by mixed rhythms. Schwing also found that if he did the same sort of analysis for just the physical and emotional rhythms, critical days would account for slightly more than 15 percent of the days in a biorhythmic lifespan, while mixed rhythms take up 85 percent of the days. The logical conclusion to be drawn from these calculations is that if accidents and deaths occur *randomly*, and are not related to biorhythm, only 15 to 20 percent of all such events will fall on critical days (since critical days comprise only 15 to 20 percent of the total number of days). The vast majority of truly random accidents and deaths—or 80 to 85 percent—will occur on mixed days.

Of course, the accidents in question here do not include those that involve unavoidable mechanical failures or combinations of circumstances about which nobody could do anything. Biorhythm researchers in general have been very careful to limit themselves to accidents caused by human (and presumably avoidable) factors. Schwing, for example, excluded serious accidents such as two-car collisions in which the driver at fault could be hard to determine, and such cases as a mechanical failure in a sawmill that led to a worker's losing his hand despite his observation of all standard safety precautions. Instead, Schwing concentrated on cases such as single-car accidents, in which the judgment, reflexes, and mood of the driver almost always play a decisive role. In this way, he could be sure that whatever he found related to the effect of biorhythm on the accident victim, and not to freak events.

Schwing's next step was to calculate exactly how many of the events he was analyzing had occurred on critical days. The results were amazing. He found that of the 700 accidents, a total of 401 had occurred on singly, doubly, or (a once-a-year rarity) triply critical days; the balance of 299 accidents had occurred on mixed rhythm days. In other words, 57 percent of the people who had experienced serious accidents were undergoing a critical or changeover day, and only 43 percent were experiencing mixed rhythms. Chance would predict figures of 20 percent and 80 percent, respectively. Put another way, Schwing's results show that serious accidents are *five* times more likely to occur when you have a critical day than when you have a day of mixed rhythms. (You can figure this out easily. Since 57.3 percent of the accidents occurred on 20.4 percent of the days known to be critical, accidents were 2.8 times—57.3/20.4—more likely to happen on critical days than on all days. And since 42.7 percent of the accidents happened on 79.6 percent of the days known to be mixed-rhythm, accidents were only .54 times—42.7/79.6—as likely to happen on mixed-rhythm days as on all days. So, since 2.8 is more than five times greater than .54, we can expect accidents to happen to people five times more often on critical days than on mixed-rhythm days.)

Schwing's calculations for deaths in Zurich were even more impressive. Remember that in his work on deaths Schwing limited himself to just the two rhythms, the physical and the emotional, which he felt exerted a much more important influence on death than did the intellectual rhythm. This limitation meant that he called only 15.2 percent of the days in a biorhythmic lifespan critical, and 84.8 percent mixed-rhythm. But when he examined the 300 deaths, he found that 197—or 65.7 percent—occurred on a day that was either physically or emotionally critical or both. 103 of the deaths—or 34.3 percent—occurred on days of mixed rhythm. By doing the same sort of calculations shown above—(65.7/15.2)/(34.3/84.8)—you can see that death was almost *eleven* times more likely to occur on a critical day than on a mixed-rhythm day. No wonder the Swiss have taken biorhythm so seriously! And no wonder that one of Schwing's colleagues at the Swiss Federal Institute of Technology, Dr. Willy von Gonzenbach, said after reading Schwing's thesis that "Those ignorant of biorhythm will run full speed into danger."

Naturally, Schwing's astounding results led other European researchers to examine biorhythm closely. Some of them reported even more impressive evidence for the influence of biorhythm. One of these was Professor Reinhold Bochow of Humboldt University in Berlin. Bochow's most interesting study was a biorhythmic analysis of 497 accidents involving workers with agricultural machinery, mainly tractors. He used essentially the same approach as Schwing. The results: 24.8 percent of the accidents occurred on triply critical days, which make up only .3 percent of all days; 46.5 percent of the accidents occurred on doubly critical days, while chance would indicate a figure of about 3.3 percent; and 26.6

percent occurred on singly critical days, which account for only 14.2 percent of all days. In sum, only 2.2 percent of the agricultural accidents happened when the victims were experiencing days of mixed rhythm. Again working things out for yourself, you can see that Bochow's data show agricultural accidents as *171 times* more likely to happen on critical days than on non-critical days. That figure stretches belief to—and perhaps beyond—the breaking point; it is certainly much higher than generally accepted biorhythm probabilities. We must remember that peculiarities in the methods Bochow used to select his accident data and make his biorhythm calculations may have distorted his results. Alternatively, the nature of mechanized agriculture may make workers more susceptible to biorhythmic influence than other activities can. Both of these possibilities can be checked in Bochow's report, which appears in the 1954 issue of *The Scholarly Journal of Humboldt University*, number 6.

Bochow was only one of many European researchers who found what they felt was conclusive evidence of the importance of biorhythm. Others of note include Dr. Friedrich Pircher of Basle, Switzerland, who found that of 204 Swiss civil aviation plane crashes that he examined, almost 70 percent occurred on a day that was critical for the pilot; Colonel Wolfgang Karnbach of the Swiss Cadet Training Center, who analyzed 130 student pilot crashes and found that 91, or exactly 70 percent, happened when the pilot's rhythm was critical; and Otto Tope, Chief Engineer for the Department of Sanitation in Hanover, Germany, whose analysis of accidents among sanitation workers indicated that 83 percent of them occurred on critical days.

Some Far-Sighted Americans

During this early period—before and just after World War II—almost all biorhythm research was concentrated in Europe. One of the very few Americans to even touch on the general idea that internal clocks might regulate our moods and performance was Dr. Rexford B. Hersey of the University of Pennsylvania. An industrial psychologist, Hersey became interested in the fact that businessmen and industrial psychologists thought of the working man as an invariable, machine-like creature whose characteristics and moods were stable over time. Between 1928 and 1932, Hersey set out to determine just how stable the moods of factory workers actually were. He conducted intensive interviews with apparently average workers for three months, observed them many times a day, and asked them to fill out "mood questionnaires." He also interviewed their families and co-workers to make sure that his results were accurate. His findings were reported in *The Journal of Mental Science* in 1931 and also in a later book written with Dr. M. J. Bennet, *Worker's Emotions in Shop and Home*. Without any prejudice in favor of biorhythm, and seemingly without much knowledge of the theory, Hersey found that "emotional tone" varied regularly among his subjects, with a cycle of 33 to 36 days apparently the norm. This cycle length, of course, closely resembles the length of the intellectual cycle Teltscher discovered, so we must ask whether Hersey's research methods really led him to phenomena having only to do with the emotional, or whether his approach—which depended largely on the subject's ability to communicate with and articulate his thoughts to Hersey, his friends, and his family—inadvertently zeroed in on behavior we would place in the domain of the intellectual cycle. Whatever the answer, it is clear that Hersey had a deep appreciation of rhythmicity in human life, as he indicated in the preface to *Worker's Emotions in Shop and Home* by writing, "The realization that both blue and exultant moods often lie within us and are merely a passing phase of life, unjustified by external relationships, may remove the sting of our deepest depression and the arrogance of our most insolent elation."

Dr. Hobart A. Reimann, a professor at the Hahnemann Medical College and Hospital of Philadelphia, also turned up evidence of interest, although he was working in ignorance of biorhythm theory. Starting in 1948, Reimann did a great deal of pioneering work on diseases that last for periods of one to four weeks and that sometimes occur irregularly. Reimann identified the periodic nature of many conditions—fever, peritonitis, and synovitis among them. In a detailed account in the *Journal of the American Medical Association* for June 24, 1974, Reimann and his colleague, Dr. Richard V. McCloskey of the Albert Einstein Medical Center, reported on several cases of periodic fever which had baffled scores of doctors. In one case, they found that the patient periodically had a severe rise in temperature, complained of chills, aching, and sweating, and showed enlargement of the spleen. These incidents were repeated regularly, and one of the many proposed diagnoses was accurate. When the doctors noted that the fever episodes tended to occur every 23 days—the length of the physical cycle—and had no external causes, they wrote the condition down to periodic illness. In the same paper, Reimann and McCloskey report on another patient with very similar symptoms, except that in the second case the disease seemed to occur every 14 days, which might well correspond with the incidence of critical emotional days (since a critical day occurs at the end of each half of the 28-day emotional cycle). Many other similar cases (perplexing to physicians unaware of biorhythm) can be found in Reimann's 1963 monograph, *Periodic Diseases.*

Reimann is not the only American doctor who has shown interest in the medical applications of biorhythm. Since biorhythms seem to involve such profound changes in our internal states, the theory has been of significant interest to psychiatrists. Again, this interest should not be surprising, since the rhythmic nature of some mental illness has been acknowledged in names such as "manic-depressive cycle," and the uncanny effects of the seasons and the moon on patients have been noted through the centuries, although it has never been adequately explained. Dr. Lucien Sansouci, consultant to a state mental hospital in Rhode Island, has found that his patients often display behavior which synchronizes exactly with what their biorhythm charts would predict. Several of his schizophrenic patients shift into violent behavior patterns when their emotional cycles cross the base line, thus marking a critical day. One striking example was Duane, a schizophrenic boy whom Sansouci thought well enough to hold an outside job. When the doctor took his patient to a local store for a test by the personnel department, Duane's emotional and intellectual cycles were high, and he passed the test with flying colors. Soon afterwards, when his physical rhythm was critical and the other two cycles were low, Duane got into a serious fight with other employees and stalked off the job. By explaining the biorhythmic basis of Duane's behavior to the store manager, Sansouci won back the boy's job. Since then, both the doctor and the store manager have used Duane's charts to help him keep out of trouble and eventually to win a promotion.

Sansouci also has found that, "The 33-year-old ward 'baby' never fails to go into temper tantrums when her rhythms cross the critical line. I predict two days in advance when she will spell trouble." While Sansouci's experience is limited to a few cases, he is currently engaged in charting all the patients under his care and should soon develop some statistically interesting results.

Unfortunately, no American doctors have duplicated the extensive work done in Europe. In light of the possible benefits, this lack of curiosity and initiative is quite disappointing, and seems incomprehensible when we look at the results achieved overseas. In Locarno, Switzerland, for example, Dr. Fritz Wehrli has been using biorhythm in his private clinic for two decades. He plots patients' physical and emotional rhythms to determine when they will be best able to withstand the shock of surgery, and he has reported that in more than 10,000 operations performed with the help of biorhythm, there has not been one failure or complication. It is difficult not to speculate angrily about what would have happened to the more than 20 percent of American surgical patients who suffer complications if American hospitals were to study and apply biorhythm seriously.

Still the medical outlook in this country is not entirely bleak. One of the most active U.S. biorhythm researchers—Harold R. Willis, of the Department of Psychology at Missouri Southern State College—has recently done a fascinating study of hospital deaths in and around Joplin, Missouri. Willis selected a sample of 200 deaths occurring over a four-month period in 1973. To make sure that his sample was reliable, he excluded deaths of all persons over 70 years old, since he felt that birth records that far back might be unreliable. In contrast to the more traditional European researchers, Willis concentrated not just on the physically and emotionally critical days, but on intellectually unstable periods as well, for, as he reported, "people react to the type of cycle which is most peculiar to their personality or professional field. . . ." Willis' research included one final refinement: since medical technology is now sophisticated enough to keep people alive for one or two days after they would have died without the aids of machinery, Willis created two data categories, one for people who died exactly on a critical day, and one for people who died after "hospital effect" had kept them alive for up to two days after the critical day.

Willis' results seem unequivocal. Of the 200 cases studied, 112 people, or 56 percent, died on a critical day. An additional 23 cases of death, or 12 percent, occurring soon after a critical day were written down to "hospital effect," for a grand total of 68 percent. (Remember that chance would predict only 20 percent, since only 20 percent of all days are critical.) So, Willis concluded, 135 of his 200 cases of death could be linked to a critical day in each patient's cycle. Interestingly, another 21 cases (11 percent) expired one day before or after a critical day, when the rhythm in question was on the borderline between stability and instability. So a very liberal interpretation of Willis' results might be that 156 of his subjects, or 79 percent, died on a critical day or close to it.

Willis was sufficiently intrigued by his results to do a follow-up study in the summer of 1974. He used precisely the same method as before, although his sample was reduced to 120 cases in the second study. The results were remarkably consistent, however. 52 percent of his subjects (62 out of 130) died on critical days, 19 percent (23 out of 120) died during the days allotted to hospital effect, and 21 percent (25 out of 120) died one day before or after the critical day. If you pause to add up the results of both studies, you will see that of the 320 cases Willis studied, 54 percent of the subjects died on a critical day, 14 percent died on a hospital effect day, and another 14 percent died one day before or after a critical day, for a total of 82 percent. It is disturbing to speculate about how many of those lives might have been saved if special precautions had been taken on the critical day and the days immediately preceding and following it.

There is very little doctors can do about some diseases in their final stages. Cancer, for example, usually debilitates a patient so much that he may succumb to any number of complications that would be minor to other patients, and there is little that can be done to protect a terminal cancer patient from all of the potentially fatal maladies that may peak on a critical day. However, in the case of many other diseases, forewarned is forearmed. This is especially true of many types of heart attack and of suicide, and it is in these two areas that another portion of Willis' work is suggestive.

In both of his retrospective studies, Willis found a total of 78 victims of heart attacks. Of that number, 49 (63 percent) died on critical days, with deaths on physically, emotionally, and intellectually critical days about evenly distributed. In short, 63 percent of the people who died from heart attacks did so on a critical day when extra medical precautions would have had a decisive effect, had they been applied. The suicide sample was much too small (only eight cases) to be at all indicative, but Willis discovered that all but two of those cases died on critical days—a much higher percentage than we might expect for all causes of death. Perhaps the fact that five of these eight cases of death by suicide occurred on an emotionally critical day can explain the somewhat higher percentage here: suicide is a form of death largely determined by the victim's emotional state, and is less dependent than others on the general health of the body.

Individual cases of death, while much less significant than large scale studies, are equally thought-provoking. Aristotle Onassis for example, died of pneumonia on March 15, 1975. Was it simply coincidence that he was undergoing an emotionally critical day, or was his will to live—already weakened by the premature death of his son Alexander—fatally undermined by the crisis in his emotional biorhythm? Similar questions can be asked about other cases, including those of William Faulkner, Winston Churchill, Robert Frost, Douglas MacArthur, Gamal Nasser, J. Edgar Hoover, and Harry Truman, all of whom died exactly on critical days. Questions like these are especially poignant when they center on people who died by their own hands during emotionally critical days. The best-known cases, perhaps, are those of Marilyn Monroe and Judy Garland. Monroe was found dead of an overdose of sleeping pills on August 5, 1962, a physically critical day for her (see Figure 3). Garland died of similar causes on an emotionally critical night—June 21, 1969 (see Figure 4).

Rhythm on Wheels

To be sure, there is a significant problem in relying on correlations between death and critical days as a proof of biorhythm theory: it is almost impossible to determine when death occurred due to the inexorable process of disease, and when the subject's temporary vulnerability because of a critical day was the deciding factor. This problem is handled much more easily in retrospective studies of traffic accidents. Here again, Willis' work is the most substantial research done by an American. Examining data on 100 fatal single-car accidents in Missouri, and limiting themselves to cases in which the driver died, students under Willis' direction did a standard critical-day analysis. They found that 46 percent of those fatalities had occurred on a critical day, and another 11 percent had happened within 24 hours of a critical day, for a total of 57 percent. Because more than half of the single-car fatalities happened on emotionally critical days, or when the driver had a combination of an emotionally critical day with a critical day in one of the other two rhythms, Willis and his group concluded that the emotional cycle may have a greater impact on accidents than the other cycles, posibly because it affects judgment, as George Thommen had suggested.

A much larger, and therefore more reliable, sample was the subject of a study by the Japanese Military Police and the police of Fukuoka prefecture. According to Commander Shegeru Kawahara of Kasuga base, of 1166 self-caused traffic accidents, 59 percent happened on critical days. When Kawahara and his group turned their attention to three special accident categories which probably involve greater failures of judgment than most traffic accidents, they found even higher percentages. 65 percent of the accidents that happened while soldiers were on duty happened on critical days, 63 percent of those that involved a gross violation of traffic regulations took place on a critical day, and 66 percent of those that involved drunken driving happened on critical days. In comparison with what probability would predict, and with the general frequency of self-caused traffic accidents on critical days, these figures are quite impressive. In shorter reports, the Japan Biorhythm Association has detailed other retrospective studies of traffic accidents. A study completed by the Tokyo Metropolitan Police in 1971 indicated that 82 percent of all self-caused traffic accidents involved drivers experiencing a critical day. The Osaka Police approached the problem slightly differently. Reasoning that children had less highly developed defensive reactions than older and wiser pedestrians, they concentrated on children who had been injured by cars. Perhaps because the sample included incidents where the accident was in no way the child's fault, the percentage falling on critical days was somewhat lower than the Tokyo study would imply; but it was still a remarkable 70 percent.

Further work in the analysis of traffic accidents comes from Australia, where B. Krause-Poray and his associates asked the Queensland Police Department to supply them with the records of 100 randomly chosen single-car fatalities. Excluding those accidents in which only the passenger died, the Australian group found that 54 percent of the driver fatalities had taken place on a critical day of the driver's, and that an additional 25 percent had happened within one day of the critical day. True, the sample size is very small, but the findings are generally consistent with Willis' and slightly more conservative than the Japanese findings.

Airplane accidents offer an even more fruitful field for biorhythm research for two good reasons: because of flight recorders, the causes of most airplane accidents are known in minute detail, and we can be sure whether the pilot was at fault and eliminate "bogus" (mechanically caused) accidents from the sample; and secondly, airplane accidents are of such concern to governments everywhere that the data are more accurate and available than for most types of accidents.

We have already mentioned the two Swiss studies by Pircher and Karnbach, both of which revealed that about 70 percent of the accidents involving civil pilots and air cadets took place on the pilot's critical days. Some additional work has been done by Dr. Robert Woodham of the Guggenheim Aviation Safety Center. In a study of accidents with private planes, he reportedly found that almost 80 percent took place on a critical day for the pilot involved. So impressed was Dr. Woodham by this formal study that he wrote to George Thommen: "We regret that biorhythm theory has not been more generally accepted in this country, but like so many ideas that were discovered in Europe, it apparently has to make its way slowly before it is generally accepted. We feel that the principles are sound. . . . There appears to be a definite relation between the accident proneness of an individual on a particular date and his physiological cycles as determined by biorhythm theory."

While Woodham had neither the time nor money to make a more systematic study of the biorhythmic aspects of flight safety, some agencies of the U.S. Government have been more fortunate. Major Carl A. Weaver, Jr., of the U.S. Army Aviation School reported in 1974 that he had closely examined one-fourth of all Army aviation pilot-error accidents (those unrelated to mechanical failures) which had occurred over the previous two and a half years. Plotting the biorhythms of all the involved pilots, Weaver discovered that "Nearly half, or 49 percent, of the accidents occurred on a critical day." That figure excluded accidents that happened just before or just after critical days, so it can be taken as a very conservative one. Furthermore, Weaver found, pilot-error accidents were much more likely to happen on doubly-critical days, when two rhythms crossed the baseline at the same time, than on singly-critical days. (In 1972, however, a quick study by the Tactical Air Command of the U.S. Air Force found that of 59 aircraft accidents since 1969, only 13 had occurred on a critical day in the pilot's biorhythm. The results—23 percent of the accidents correlated with critical days—are only slightly higher than chance would predict and contrast strongly with most other known accidents. One explanation for this discrepancy may be that the TAC study included *both* pilot-error and unexplained accidents. Biorhythms, naturally, have no influence on accident factors other than pilot error, so the sample may have been inadequate not just in size, but in basic structure as well.)

Perhaps the most exciting evidence for the importance of biorhythm in flight safety was cited by Jerome F. Lederer, an internationally known safety expert and formerly the safety director for the National Aeronautics and Space Administration (NASA). Introduced to biorhythm theory by a friend, Lederer was highly skeptical, even after he learned that the Swiss government had used it successfully for many years. But then he saw a study of some 800 Army aviation accidents done by analysts at Fort Rucker. The degree of correlation between biorhythm theory and those 800 accidents so impressed Lederer that he took the matter before the National Transportation Safety Board. But there, as far as the public is concerned, the idea died—much as Woodham of the Guggenheim Center would have predicted.

In addition to these retrospective analyses of pilot error, there is a provocative study by Woods *et al.*, done under Willis' supervision at Missouri Southern State College. Attempting to find out exactly how the physical rhythm might affect pilot performance, Woods and his colleagues decided to measure the cyclical aspects of a very simple physical response: pushing a button. Each subject in this experiment was tested five times for his reaction speed. Each time, he was asked to push a button ten times in quick succession, and the speed of his reaction was precisely measured. The times were then related to the subjects' biorhythm charts. The results demonstrate one of the few laboratory-controlled proofs of biorhythm. On the average, the subjects took 5.13 seconds to push the button ten times. When their physical rhythm was in the negative phase, the time lengthened to 5.32 seconds; and when they were experiencing a physically critical day, the time was 5.41 seconds. As Willis himself concluded. "Considering this as the most simple movement that could be executed [by a pilot], it is reasonable to state that the reaction time of a pilot going through [a complex series of corrective reactions] would be considerably more, hence adding to the hazard." In short, if a pilot takes an average of .28 of a second longer than normal to push a button ten times on a critical day, he will take much longer than normal to perform *all* the actions needed to cope with flight emergencies.

One of the many striking examples of the role of biorhythm in airplane crashes occurred on December 1, 1974, when Trans World Airlines flight #514, approaching Dulles airport in Washington, D.C., descended too quickly and smashed into the Blue Ridge mountains, killing all 92 persons aboard. According to the testimony of Merle W. Dameron, a flight controller for 20 years, flight #514 had received clearance to land and seemed to be approaching the field normally. But then, Dameron said, his instruments showed that the pilot was already flying at 2,000 feet, much too low for the mountainous area that surrounds Dulles. Dameron queried the pilot for an altitude verification but never received a response. Moments later the plane crashed.

Obviously, the pilot and his crew were misreading the plane's altimeter or were unaware of the existence of the 1,764-foot mountain into which they crashed. But strangely enough, Dameron himself was unaware of the mountain—or had forgotten about it that day—and so the pilot never received a warning from the control tower.

Subsequent examination of the biorhythms of those involved in the accident revealed one of the worst biorhythmic situations imaginable. Dameron was undergoing a physically critical day (see Figure 5), which might well have affected his memory and almost certainly slowed his perceptions of the impending disaster and his ability to warn the pilot in time. Captain Richard I. Brock of flight #514 was in negative physical and emotional phases and was approaching an intellectually critical day (see Figure 6). Worst of all, Co-Pilot Leonard W. Kreschek, who might have warned Brock of the plane's dangerously low altitude, was negative in all three rhythms and had undergone a double-critical only two days before (see Figure 7). With a biorhythmic combination like that, an accident of some sort would have been difficult to avoid without extra precautions.

Other disturbing examples of airplane accidents related to biorhythms are legion. In January, 1975, pilots Robert D. Hatem and Richard N. White were killed when they slammed their private prop-jet into the radio tower of American University. Subsequent analysis showed that both men were in the grip of an emotionally critical day at the time (see Figures 8 and 9). On October 16, 1972, a private plane piloted by Don Jonz and carrying Representative Hale Boggs of Louisiana crashed in Alaska, leaving no survivors. On that day, Jonz's biorhythms were extremely depressed, with his mental and physical rhythms at their lowest points, and his emotional rhythm at the critical point.

When a Pan American cargo plane crashed at Boston's Logan Airport on November 3, 1973, killing all three members of the crew, it may well have been because both First Officer Gene Ritter and Flight Engineer Davis Melvin were in critical phases that undermined their judgment and reaction times (see Figure 10). Another example comes from a pilot with a major European airline. In September, 1974, his plane almost crashed when crew members misread the altimeter, thinking it said 5,000 feet when it actually said 7,500 feet. At the same time, the air traffic controller failed to correct this error. The result was an overshoot of the landing field that could have easily wrecked the aircraft on neighboring mountain peaks. Biorhythmic analysis revealed a combination of critical days even worse than those involved in flight #514. The captain was in the grip of an emotionally critical day, and his intellectual rhythm was depressed. The flight engineer had to contend with an intellectually critical day, when his capacity to absorb and interpret instrument data was impaired. Meanwhile, the controller was negative in all three rhythms. Only the co-pilot had all three rhythms in definite positive phases —but it was later learned that the co-pilot had food poisoning and was taking medication which was likely to affect his judgment and alertness. If airlines paid sufficient attention to biorhythm, it is conceivable that such incidents might become rarities—and that the 461 people who lost their lives in U.S. air crashes in 1974 might not have died.

Rhythm While You Work

While the influence of biorhythms in airplane accidents is impressive and dramatic, even more startling evidence has come from retrospective studies of how biorhythms affect industrial accidents. Industrial accidents may sound like a minor or unimportant field; but it is estimated that more than 6,000 people will *die* from injuries incurred at their place of work during 1976. On the average, 15 of every 100 manufacturing workers will lose time from their jobs due to one kind of injury or another. These are very serious numbers, and it is not surprising that so much biorhythm work in the U.S. —from that of Rexford Hersey on—has concentrated on damming the flood of occupational injury and death.

In studying the relationship between biorhythm and on-the-job accidents, researchers run into the same problems they face with all other kinds of accidents: i.e., determining when an accident was caused mainly by human factors, and when it was largely the result of unsafe working conditions or mechanical failures. Fortunately, almost all lost-time industrial accidents are reported on insurance and compensation forms, so it is relatively easy to pick out the human error accidents.

Perhaps the most experienced analyst of the effects of

biorhythm in industry is Russell K. Anderson, former president of Russell K. Anderson Associates, Safety Engineers. Over the years, Anderson has done a great deal of painstaking retrospective research that tends to confirm the importance of biorhythm in avoiding accidents. After his retirement, Anderson reported to George Thommen that he had first applied biorhythm analysis to only a small group of accidents, but that the results were so impressive that he extended the study for a two-year period, eventually investigating more than 300 accidents in four different settings: a metalworking plant, a chemical plant, a textile mill, and a knitting mill. Since all of the accidents studied were covered by the Workmen's Compensation Act and consequently had been described in great detail, Anderson was able to eliminate any accidents that might have been caused by other than human factors from his sample.

According to Anderson, almost 70 percent of the accidents had happened on a critical day of the victim's cycle. Some of the details Anderson reported underscored how the instability associated with critical days can upset the judgment and ruin the skills of even the most experienced worker. For instance, one power-press operator who had run his machine for over 20 years without injury was at his station on a day that was critical both emotionally and intellectually. Inexplicably, he reached behind the input guard of the press and then activated the machine. It came down with thousands of pounds of force on his hand, costing him three fingers. When he was later asked why he had done such an uncharacteristically dangerous thing, he was unable to reply. If he had known about his biorhythmic profile for that day, he might have stayed away from his machine and saved his hand.

Stimulated by finding a 70 percent incidence of accidents on critical days in the sample of over 300 workers, Anderson extended his analysis both in breadth and depth. By the end of 1974, he had analyzed more than 1200 human-error accidents and reported that the results were quite consistent with his earlier findings: the 70 percent figure remained unshaken. In an attempt to understand how biorhythmic fluctuations affect accidents more completely, Anderson tried to relate particular configurations and combinations of the three rhythms to the severity and frequency of occurrence of the mishap. Of all the accidents that happened on critical days, Anderson found, the greatest number occurred when *two* rhythms crossed each other at the baseline or zero point, or just below that point. In other words, when two rhythms are moving in opposite directions (when the physical rhythm is switching from positive to negative while the emotional rhythm is moving up from negative to positive, for example) and they cross when both are in critical or very close to critical, the individual appears to be violently pulled in opposite directions—creating a day of acute instability during which there is special danger. Not surprisingly, Anderson found that the most *severe* accidents also happened on days of contrary, critical rhythms. Our judgment and skills, it seems, are quite fallible when we are pulled in opposite directions by our biorhythms and are especially vulnerable to that conflict because of the instability associated with critical and double-critical days.

Based on his own extensive experience with biorhythm analysis, Anderson has drawn three central conclusions about the ways in which biorhythmic influence shows up in industrial accidents: first of all, an individual will show marked changes during critical days; secondly, when an accident occurs on a critical day and the individual does not know that it was a critical day, he is unable to explain why the accident happened; and thirdly, the individual's physical abilities, mental capacity, and emotional tone all reflect critical days in relatively obvious ways. Coming from such a thorough researcher, these findings are indeed significant.

Anderson is not alone in the field of connecting biorhythm with industrial accidents. In late 1974, Cyrus B. Newcomb reported to the Accident Prevention Committee of the Edison Electric Institute on his own work for a utility company. Rather than limiting himself to human error accidents (which would have yielded a purer measure of biorhythmic influence but might also have raised problems of deciding when a human-mechanical accident was mainly the result of human error), Newcomb pulled 100 accidents from the company files at random. All of these were lost-time accidents—sufficiently serious to have kept the victim from working for some time. Since Newcomb's sample undoubtedly included a number of cases where biorhythm had no effect, we would expect his analysis to have yielded disappointing results. However, this was not the case: Newcomb found that 53 percent of his sample represented accidents that happened on critical days. As he pointed out, this result indicates that a company should be able to halve its accident rate by heeding biorhythm. Newcomb's analysis of the utility company was later extended to 100 of its vehicle accidents and revealed that 68 percent had taken place on critical days.

In recent years, biorhythm has attracted more attention from industrial safety experts than from any other group. After all, there are large savings to be had if industrial accidents can be reduced. Newcomb calculated that for his sample, the preventive use of biorhythm would have saved about $320,000. Not all of the results of analyses stemming from this interest have been positive, however. In one of the most extensive biorhythm studies made, the Workmen's Compensation Board of British Columbia analyzed 13,285 lost-time accidents that had happened during the first four months of 1971. Their conclusion: "The results indicate that accidents are no more likely to occur during critical periods than at any other time." In addition, one Ohio utility company explored 250 of its lost-time accidents and found that only 25 percent happened on critical days, which is only slightly higher than the 20 percent of accidents that would happen on critical days if biorhythms had no influence at all. While it is impossible to give an instant explanation of the failure of these studies to confirm the findings of Anderson, Newcomb, and others, two potential sources of error are obvious. First, both studies included *all* accidents, not just human-error accidents, but also those involving unsafe working conditions and mechanical failures. Of course, this sampling error would reduce the percentage of accidents found to have occurred on critical days, and size of the reduction would depend on how many cases in the sample were not human-error accidents. Secondly, people with little experience in biorhythm analysis often make mistakes, such as failing to account for anomalies due to the hour of a victim's birth and simply miscalculating or misinterpreting the occurrence of critical days.

We cannot say for certain that the studies in British Columbia and the Ohio utility company were affected by such mistakes, but it is a real possibility, since the most statistically careful and sophisticated studies of biorhythm have yielded results in line with the Anderson and Newcomb work. This work was done by Dr. Douglas E. Neil and his associates at the Man-Machine Systems Design Laboratory at the Naval Postgraduate School in Monterey, California. The most distinguishing aspect of Neil's research work is his emphasis on upgrading the quality of biorhythm analysis by introducing more reliable ways of manipulating data. In his study of industrial accidents, for instance, Neil felt that, "examining only critical days results in a partial analysis of the total concept [of biorhythm]." To test the validity of the theory, he felt, it was necessary to look beyond critical days (the easiest method of verification) to see whether the influence of the biorhythm theory as a whole—with its critical days, positive and negative phases, and mixed rhythms—could be seen in the occurrence of accidents. To do this, Neil set up a two-stage analysis. First, he assembled data about 66 single-person human-error accidents in a Canadian forest products company and 127 randomly selected accidents in a U.S. aircraft maintenance facility and tested them to see if the accidents had a uniform distribution over all working days, i.e., if they had happened in a thoroughly random fashion. He found that they did not; accidents tended to occur more frequently on certain days and during certain periods. He then compared the distribution of these accidents over time to the distributions of critical, positive, and negative days in the biorhythms of the accident victims to see if there was any correspondence. There was, and Neil came up with some rather precise measures of how closely the accident distribution compared with the biorhythm distribution.

In analysing the 66 accident claims from the Canadian forest products firm, Neil found that he could confidently reject the hypothesis that accidents happened randomly, and with a uniform distribution. Looking at only the physical rhythm, he found that exactly twice as many accidents happened during the negative portion of the cycle as during the positive portion, and that this result was statistically significant.

With the feeling that his findings were "certainly suggestive," although far from conclusive, Neil decided to extend the analysis to the aircraft maintenance facility, and to look at all three biorhythms this time. The distribution of the 127 randomly selected accidents was found to be nonuniform, implying that some pattern was at work. A statistically significant relationship was again found between the accidents and the physical cycle. This time, however, Neil also discovered a significant relationship between accidents and the intellectual cycle, although the emotional cycle, for unexplained reasons, was not seen to influence accidents as much as previous research would suggest. Looking at critical days in all three rhythms, Neil also found a significant relationship, since the number of lost-time accidents happening on critical days was much higher than probability would predict.

The Sporting Life

Much non-academic interest in biorhythm has centered on athletic performance. Because sporting competition is full of challenge to the individual, it provides an excellent opportunity for biorhythms to come into play, since biorhythmic influence is often hidden in ordinary settings devoid of the challenging or dangerous. Also, because an athlete in competition is supposed to be performing at his peak, it is much safer to assume that factors other than biorhythms—physical condition, motivation, training, etc.—have all been brought to an optimum state, and that the difference between success and failure is the result of differences in the athlete's three fixed rhythms. And finally, the uncertainties of sport—the inability to predict how an athlete or team will perform—are endlessly fascinating.

Some of the most dramatic examples of the role of biorhythms in competition comes from analyses of boxing matches, where the fact that only two athletes are involved in the competition puts the influence of their biorhythms into high relief. Examples are numerous. After defeating Floyd Patterson for the Heavyweight Boxing Championship in Stockholm, Ingemar Johansson had a rematch with Patterson in New York on June 20, 1960. On that day, Johansson's physical and emotional rhythms were in the negative phase, while Patterson's were almost at the peak of the positive phase. Most sportswriters, unaware of the biorhythms of the fighters, strongly predicted a Johansson victory; but Patterson won. Muhammad Ali's March 31, 1973, fight with Ken Norton was an upset that shattered both Ali's standing and his jaw. To biorhythm analysts, it was no surprise, since Ali was suffering from a critical day in both his physical and emotional rhythms. Ali's performance a year and a half later was just as consistent with biorhythm theory. On September 25, he met George Foreman in the ring in Zaire and defeated him with a combination of tactics and endurance. Looking at the biorhythm charts of both fighters, we can see that Foreman was going through a physical low, and on the 26th, one day after the fight, he was to come out of an emotional low and into a critical day. His performance during the fight clearly mirrored physical debilitation and the lack of judgment associated with emotional weakness and criticality. On the other hand, Ali was high in both the physical and emotional rhythms, a situation that undoubtedly helped him to succeed in his punishing, defensive strategy.

A triple-critical day can be extraordinarily dangerous for boxers. It was for Benny (The Kid) Paret, who was knocked out by an enraged Emile Griffith on March 24, 1962, a triple-critical day for Paret. The Kid never recovered consciousness, partly because his physical, emotional, and intellectual rhythms were too weak to sustain Griffith's onslaught. Paret died ten days later, on the very next critical day of his biorhythm.

Retrospective analyses of baseball players also provide examples of biorhythm's importance to athletic performance. In 1927, for example, Babe Ruth set a new record by driving in 60 home runs in one season. Analysis shows that all but 13 of those home runs were hit on days when Ruth was at or near a physical or emotional peak. In short, relying on biorhythm charts alone, somebody who knew nothing about Ruth's condition during the different days of the 1927 season would have been 79 percent accurate in predicting his home-run batting performance.

Baseball is a team sport, and since the task of tracing the influence of all the players' biorhythms on team performance is forbiddingly complex, few researchers have had the fortitude to set up the analysis. There are some interesting individual cases, however. For example, on March 18, 1962, Mickey Mantle made a base hit and was running toward first when he suddenly and inexplicably collapsed. The fact that March 18 was a critical day in Mantle's emotional biorhythm may have had something to do with it.

Baseball pitching presents a more ambiguous case. When Bob Hambley, a computer scientist, calculated the charts of pitchers in 11 no-hit games, he found that nine of the 11 took place on the pitcher's critical day, and that most of these criticals were in the physical rhythm. George Thommen has suggested that the pitchers may have been so "charged up that they disregarded all caution"; but the fact that they were able to achieve excellent control of the ball on days when they were physically unstable is hard to rationalize.

Another computer scientist, John P. Higginson, has charted the rhythms of six members of the Chicago Cubs and compared them with the players' performances. He was so impressed by the high correlations he found that he feels confident in predicting whether an individual player should be benched or allowed to play in a particular game.

As the case of the 11 no-hit baseball games indicates, exactly how a biorhythmic profile will influence an athletic performance can be a matter of debate. Nowhere is this uncertainty clearer than in the work of Dr. Philip A. Costin, a former medical director for the Canadian Air Force and Regional Surgeon for Quebec. Aware of the implications of biorhythm, Costin has studied the theory for many years. In one of his studies, Costin examined the biorhythmic profiles of Canadian soldiers who had performed acts of heroism. He found that most of those acts—in fact, 88 percent—had taken place on critical days. He explained to one reporter that, "Essentially, heroism requires a person to throw caution to the winds, to take long chances, putting one's life in real jeopardy. In other words, the hero was not following a normal behavior pattern." Costin also found that soldiers were more likely to desert and to commit suicide on critical days than on normal days. How can heroism and desertion or suicide (another form of desertion) all be influenced by the instability of critical days? Costin feels that the link among these three unusual actions is the rashness or instability of people undergoing critical days. Heroism calls for ignoring the likely consequences of one's actions, and so do desertion and suicide.

The value of applying biorhythm to athletic performance is clear in three other cases: those of Evel Knievel, Mark Spitz, and Franco Harris. Knievel's attempt to rocket across the Snake River Canyon might not be called an athletic endeavor, of course. Nonetheless, it did call for finely tuned perception and coordination in order to avoid disaster. On September 8, 1974, the day scheduled for his flight, Evel Knievel's physical rhythm was critical. A crowd of 20,000 people had gathered on the heights of the Snake River Canyon to watch him being strapped into his Sky-Cycle, a primitive steam-driven rocket. A jet blast of steam sent Keievel soaring up his 108-foot-long takeoff ramp. But even before he reached the end, a small parachute opened. The rocket's thrust was blunted, and Knievel drifted to a crash landing on the near side of the canyon. After being picked up, Knievel said, "I don't know what happened." It is possible that he inadvertently activated the only control in his cabin—a lever designed to release his parachute. If so, that mistake would have been so uncharacteristic that it could be explained only by Knievel's adverse biorhythmic position.

Mark Spitz's case is even more obvious. In ten days during late August and early September, 1972, Spitz became the first person to win seven gold medals during one session of the Olympic games. Spitz's biorhythm chart shows that he was in a physically and emotionally high period during all ten days (see Figure 11). Would he have been able to achieve his swimming victories, where success hinges on seconds, if he had not been biorhythmically strong? His earlier, disappointing performances in Mexico City imply that he would not have.

The case of Franco Harris is even clearer. A running back for the Pittsburgh Steelers, Harris made the "Catch of the Century" when he received a pass against the Oakland Raiders during the final seconds of the December 23, 1972, playoffs. On that day, Harris' physical, emotional, and intellectual rhythms were all at their peak (see Figure 12). That Harris' superplay was not simply the result of his superb training and ability, but also owed something to biorhythm, can be seen by looking at his performance on December 10, less than two weeks earlier. On that day, Harris was undergoing a rare triply critical day. For the first time in a seven-week period, Harris failed to gain 100 yards or more, even though he was playing against Houston, the worst team in the league.

In light of many retrospective analyses of events in medicine, psychiatry, industrial accidents, airplane and traffic accidents, and sports, biorhythm appears to work.

However, you should remember that these are studies of past events and leave lots of room for researchers to bias their results, wittingly or unwittingly. Certainly, this research is highly suggestive; but it is far from conclusive. For more reliable results, we must turn to cases where biorhythm theory has been tested in practice or in the laboratory. Fortunately, there is a substantial and growing body of such work.

Biorhythm in Action

With all of the persuasive evidence in favor of biorhythm which has come from individual cases and from a host of larger-scale retrospective analyses, it is quite surprising that the theory has not taken the fields of industrial safety, traffic control, medicine, and athletics by storm. After all, applied biorhythm seems capable of making great savings in lives and in money, and of giving competitors in any field inestimable advantages. Perhaps the reason why biorhythm has yet to be applied as extensively in the United States as it is in Switzerland and Japan is a two-fold one: first, controlled laboratory studies that would clinch the case for the theory are few and far between. As we mentioned before, such studies are complicated and costly, and naturally do not appeal to skeptical scientists who prefer to limit their work to smaller, more discrete phenomena. Except for two researchers, no one has done laboratory experiments to test the validity of biorhythm theory in the United States.

Secondly, industry tends to be wary of any technique, such as biorhythm, which is only partly explained and smacks of the occult. Impressed by the results achieved in other countries, several U.S. firms have quietly—even secretly—begun to experiment with biorhythm. For the most part, though, they refuse to talk officially about their work, and the information that *is* available comes from inside sources who refuse to be identified. This attitude seems unreasonable, since experiments are nothing more than experiments—they do not expose a company's prestige and standing to criticism. As more and more people discover that applied biorhythm works, industry's attitude may change. Until then, however, research in biorhythm will continue to suffer from inadequate sharing of knowledge and a dearth of information about the extent of industrial experience and its results. Bear in mind when reading this chapter the fact that much of the U.S. industrial work we describe has been kept under wraps and may even be denied by the companies in question; but we have checked the sources of these data and believe them to be generally reliable.

The Rhythms of Plays and Players

The only known laboratory research on biorhythm in the United States comes from two researchers we have mentioned previously: Harold R. Willis of Missouri Southern State College, and Dr. Douglas E. Neil of the Naval Postgraduate School in Monterey, California. Willis' work is not as rigorous as Neil's, but it is more extensive. In addition to his work with hospital deaths, accident admissions, medical incidents, and single-car fatalities, Willis has joined with two of his students to do a predictive study of the Missouri Southern State football team. The first part of the study called for verifying biorhythmic predictions of how the individual players would perform. During spring football practice, one of Willis' students rated the performances of 24 backfield and end candidates on a ten-point scale in each of six categories—effort, execution of task, mistakes, attitude, endurance, and self-control—and then derived an overall rating. Each candidate was rated on every day he performed over a 27-day period. Curves were then drawn of each player's performance and superimposed over his biorhythm chart for the period. In the majority of cases, Willis found a very close correspondence between the two curves. In the final spring practice game, Willis reported, biorhythmic predictions of player performance were about 60 percent accurate. Given a ten-point overall performance scale, that figure is 50 percent more accurate than chance predictions would have been.

Encouraged by these results, Willis and Jan Case, one of his students, extended this work. They drew charts for 38 football players during the 1972 season, completing a total of 335 player/game charts. On the Thursday preceding each weekend game, Case would make biorhythmic predictions of individual player performance and give them to the football coach. Since team strategy depends not just on individual performance, but on interactions among team members and on the strengths and weakness of the opposing team, Case did not tell the coach how to use the information she gave him. On the Tuesday following each game, the Coach would watch the game film and give Case his own evaluation of how well each player had performed. Using these evaluations as a check on her predictions, Case found that the biorhythm method was accurate in 77 percent of the cases. She also found that of 13 injuries during the season, 69 percent occurred on critical days. Although exactly how the football coach used the data he received every

Thursday is unknown, the record of the MSSC team speaks for the effectiveness of biorhythm in sports: in 1972, for the first time, the football team remained undefeated and won its divisional championship. This was not just the result of superior talent: during the next year, with much the same roster of players, the MSSC team lost more than half of its games.

While Willis' football experiments may have seemed bizarre to people unfamiliar with biorhythm, he was simply continuing work done in Switzerland, where biorhythm has been accepted and applied for many years. The Swiss have brought special sophistication to the application of biorhythm to athletics. For example, Jack Gunthard, Swiss National Coach for Gymnastics, reported to Willis that he regularly used biorhythm to predict the performance of his athletes. At the World Championships in Gymnastics, held in Ljublana, Yugloslavia, Gunthard's biorhythmic predictions proved 82 percent accurate. At the Swiss National Championships, he used the same method of prediction and achieved an accuracy of 79 percent. Gunthard's most interesting work comes from his two years' experience with biorhythm, which has enabled him "to find out with which gymnasts my forecasts almost always, mostly, or rarely are true. In this way I can classify them as 'rhythmists' or 'non-rhythmists.' And this in turn gives me valuable advice about training, and especially competition. When choosing members of a team and in doubt about a decision, I always resort to biorhythm as a decisive factor." This suggestion that biorhythm influences some people more than others in a particular type of endeavor (such as gymnastics) is novel, but not inconsistent with the general theory; for, as we saw before, heredity helps to determine one's sensitivity to the three different rhythms, and the individual ways we develop to deal with the influence of biorhythms also modify their effects.

Another Swiss well-versed in applying biorhythm to sports is Helmut Benthaus, coach of the Basel Soccer Team, which is the Swiss championship team. Like Gunthard, Benthaus began his work by making biorhythmic performance predictions for his players and comparing their actual performance to their forecast. Again like Gunthard, he was sufficiently impressed by the results to begin using biorhythmic calculations regularly as an aid in developing lineups for particular games and in handling his players. Benthaus has not used biorhythm crudely, since he believes that a player's class will always show —that a first class player at a biorhythmic low point will always prevail over a second class player at a biorhythmic peak. He also feels that the teamwork which develops over many games between two or more players on the team should not be broken up because of poor biorhythmic profiles. So, instead of structuring his lineup strictly along biorhythmic lines, he uses each player's chart as a guide to training. When a player is in a physical high, Benthaus prescribes hard training to condition him; he forbids such exercise when the player's chart shows him to be physically negative and therefore vulnerable to injury or overexertion. In games, Benthaus uses the player information derived from his biorhythm calculations to reinforce his weaker sports and to leave the stronger parts of his team more exposed than usual. Because he fears his players may acquire superiority or inferiority feelings if they know about their rhythms on a particular day, Benthaus avoids enlightening them. So far, the record of the Basel team implies that Benthaus' approach is a productive one.

Although Gunthard, Willis, Benthaus and many others have used biorhythm calculations to make predictions about individual and team performance, that approach can be a dangerous one. The potentials that biorhythms describe may never come into play unless the individual's situation activates them. In competitive sports, where athletes are supposed to go "all out," the situation usually does activate biorhythmic potentials; yet a host of unforeseen factors can intervene. A good example is the running of the 1975 Indianapolis 500 automobile racc. In an article written for *Argosy* magazine before the race, John Brooks examined the biorhythms of some of the leading contenders and came up with some interesting predictions: of all the drivers he charted, two seemed to be predestined to turn in outstanding performances. One was Bobby Unser, who was at peaks in his physical and intellectual cycles. The other was Gordon Johncock, who was, if anything, in an even more favorable biorhythmic position, with all three rhythms in high. The outlook was much less favorable for the rest, among them A.J. Foyt, whose physical rhythm was in low, even though he was emotionally high and presumably capable of making the fine split-second judgments that successful automobile racing requires.

On the basis of biorhythms alone, one would have predicted a first for Johncock, a second for Unser and a possible third for the many drivers whose mixed biorhythmic positions were more or less equivalent. However, this is not the way things turned out. Rain shortened the race to only 174 laps instead of 200, and engine problems took their toll along with accidents. For 47 laps, despite his biorhythmic troubles, Foyt led the pack. Johncock led for the first seven laps, then ran second to Foyt, but was forced to retire because of mechanical troubles. Unser, driving a careful and shrewd race, found himself toward the end in a battle for first with Johnny Rutherford, whose biorhythmic profile showed him to be physically and intellectually negative, with his emotional rhythm alone in positive. The final results: Unser first, Rutherford second, and Foyt third. Brooks' predictions in *Argosy* may have been entirely accurate for the individual drivers, but they were modified by the uncontrollable factors which play such a large role in motor sports.

Perhaps the most sophisticated and accurate use of biorhythm to predict the results of athletic competitions was the product of Nancy Lee Roberts and Michael R. Wallerstein, a designer-writer team from Los Angeles. Writing in *Human Behavior*, they told how they had developed a video-graphic technique called "bio-curve display," which enabled people to watch the individual curves of football players. More significantly, they stated their belief that "a team is neither a sum nor an average [of individual bio-curves] but a creature entirely different from thc players who comprise it." So Wallerstein and Roberts developed some team bio-curves and applied them to the November 18, 1972, match between the University of Southern California (USC) and the University of California at Los Angeles (UCLA). That experiment was very successful. They also computed team bio-curves for the Los Angeles Rams and the Washington Redskins, as well as for their opponents. Their work culminated in an accurate prediction for the January 17, 1973, Superbowl, which pitted the Redskins against the Miami Dolphins.

In order to protect the originality of their work until it can be verified absolutely, Wallerstein and Roberts have not revealed precisely how they compute team biorhythms, except that they use a "complicated, weighted juxtaposition of player birthdates." However, they have described some of the considerations that they use to construct the composite curve. One of these is leadership. If a team leader or star player has a favorable bio-curve, and manages to fulfill it in action, he may inspire other team members to fulfill their own potential or even to exceed it. Therefore, heavier weight probably should be placed on the team leader's curve than on the other players' rhythms. Secondly, Wallerstein and Roberts found that if many individual players had favorable biorhythmic profiles, overall team performance would be much *lower* than expected. On the other hand, if most players had negative or fairly neutral biorhythms on a game day, the curve for the team as a whole tended to be *more favorable* than expected. The explanation, according to the authors, is that individuals who are not at the peak play better as a team—they are more willing to help each other and to cooperate—while players undergoing biorhythmic highs ignore teamwork at the expense of the overall effort.

Thirdly, Wallerstein and Roberts discovered that the direction of team curves could be as important as their position in the negative or positive phase of biorhythms. If a team's curve was headed downwards from a triple high, for example, actual performance would most likely be weaker than expected. A fourth element was that the biorhythmic situation of a team's coach might be decisive, since he makes the vital strategic decisions about the plays and players to be used. Wallerstein and Roberts suspected that there was a "distinct 'productivity' correlation between a coach's bio-curve and the curve of the team or unit under his control."

Finally, the California writers raised a basic methodological problem. They felt that the height and depth of the three biorhythms—the peaks and valleys of the oscillation, which are known as the *amplitude*—might be irregular. In other words, exactly how positive a positive phase is, or how negative a negative phase, can vary for the same individual at different times. Other researchers had suggested that amplitude might vary with age, since a young person's bio-curves could be expected to be more vigorous than those of an older subject. Wallerstein and Roberts took this idea one step further by stating that a player who over-extended himself during a physical negative might thereby diminish the strength of his next physical high, and by implying that a player who did not exhaust all the strength available to him during a high period might have a less severely negative physical rhythm in the period immediately following. While this idea could profoundly alter the ways in which we think about biorhythms, since it implies that a regular sine-wave is an inaccurate way of visualizing biorhythmic fluctuations, Wallerstein and Roberts finessed the issue in their own work. Since clearly superior performances tended to occur during the positive phase, the traditional method of displaying biorhythmic changes could still be used to predict how well players and teams will do, although not with the greatest precision.

Taking all of the considerations into account, Wallerstein and Roberts computed team curves and made predictions before games. The record was stunning. Before the Los Angeles Rams' 1972 season, they computed the team curve for the offensive unit on each scheduled game day, and then submitted it to a local sportswriter to keep until the end of the season. Of ten games played during that season, only one—a 10–16 loss to Denver—turned out to be different from the researchers' predictions. Wallerstein and Roberts then charted both the offensive and defensive units of USC and UCLA for the November 18, 1972, playoff. Both of USC's units were at a high point for the season, while UCLA's team curve showed an intellectually critical day for the offense and a physical decline from a week-old peak for the defense. Precisely as predicted, UCLA lost by a wide margin.

The culmination of the Wallerstein-Roberts work was the preparation of team bio-curves for the Superbowl on the "Today Show." The Redskin defense for the Superbowl showed only a slight improvement over a triple low of a few days before. The offense showed a decline from a triple peak two weeks earlier, with a rhythmic position very mixed between the positive and negative phases. Some individual offensive leaders were charted to have triple peaks on Superbowl day, which led Wallerstein and Roberts to predict a possible "come-from-behind" victory for the Redskins. (As they later admitted, that prediction may have been influenced by the fact that both of them were ardent Redskin fans.) The Miami Dolphins, on the other hand, showed much stronger team rhythms for both offense and defense. Against their own prejudices and "better judgment," Wallerstein and Roberts were forced by their own calculations to predict a Miami victory, and that is exactly how things turned out.

The Phases of the Mind

Although the Wallerstein-Roberts work is clearly one of the most elaborate examples of applied research in biorhythm, the authors themselves admitted that it was far from a perfect experiment. There were too many factors beyond their control—the coach's decisions, the condition of the players, even the weather—to enable them to make very detailed predictions or to say with certainty that it was mainly biorythm, rather than uncontrolled factors, that influenced the outcome of the many games they charted. It was toward this problem of experimental control that Dr. Douglas E. Neil addressed himself in his work at the Naval Postgraduate School.

As in his work on industrial accidents, in his laboratory experiment Neil was most concerned with testing the *general* validity of biorhythm theory rather than the predictive powers of any particular configuration of biorhythms, such as critical days or triple lows In his own words, "Generally, the approach [others have] used has involved the examination of isolated events in relation to specific segments of a particular cycle or cycles. The results have been interpreted as suggesting a degree of predictive association between the event(s) of interest and the theoretical biorhythm. However, as a result of the subjectivity involved in the interpretation of events, and the tendency to limit investigations to a specific segment of the hypothesized cycles, the results have not provided sufficient information upon which to evaluate the theory or the concept it purportedly represents."

To solve this problem, Neil repeated his former approach, which had three steps: 1) chart the distribution of real-life performance; 2) match that chart with the individual's biorhythms; and 3) measure how closely the observed distribution (i.e., actual performance) corresponds with the distribution predicted by biorhythm theory. To measure the similarity of the two distributions, Neil used three different, generally accepted statistical tests: the Chi-Square-Contingency Test, the Chi-Square-Coefficient of Contingency Test, and the Chi-Square-Goodness-of-Fit Test. Because he was working with a very small sample in this particular experiment, Neil supplemented these major tests of statistical validity with two other methods often used to make sure that the results of an experiment have not been distorted by the small number of observations: the Fischer Test for Exact Possibility, and the Kolmogorov-Smirnov Test for Distributional Form.*

Neil was very careful in setting up his experiment. He took four graduate students at the Naval Postgraduate School and examined their performance in 15 different courses taken over a 14-month period. All of the courses taken were highly quantitative, so grades received on examinations were not as subjective as they would have

*For a more complete explanation of these tests, see S. Siegel, *Nonparametric Statistics*, McGraw-Hill, New York, 1956.

been in non-quantitative courses. (A right answer and a wrong answer to a problem in mathematics could be clearly distinguished, and different graders will agree on the grades awarded; this is not at all true for essay examinations in subjects such as English and History.) Neil developed a six-point rating scale, ranging from "well above average" all the way down to "critical." Average referred both to the class average and to the individual's average in this case. When an individual turned in a performance rated "well above average" that term meant that his exam score exceeded his own record and the record of the class as a whole. Finally, to take differences in intellectual ability into account, Neil had his subjects rate the challenge of the examination in question. This was an important refinement because biorhythmic predictions most often come true in situations that challenge the individual. By the time he finished refining his sample, Neil had 89 observations of academic performance that he could match with predictions based on the intellectual biorhythm.

Armed with his battery of statistical tests and with his well-controlled experimental data, Neil proceeded to analyze the relationship of observed academic performance to performance as predicted by biorhythm. He made three important findings. First of all, average academic performance tended to occur at many points along an individual's intellectual rhythm. In short, the distribution of average performance was uniform, and therefore was unaffected by intellectual fluctuations. Neil interpreted this result as a confirmation of his theory that biorhythmic influence predominates only in situations that are highly challenging to the individual.

Secondly, Neil found that above average academic performance was *not* uniform or random. It clearly tended to occur in challenging situations and when a student's intellectual biorhythm was in the positive phase of the cycle. In fact, of the 46 performances that Neil recorded as above average, 36—or more than 75 percent—occurred during the positive phase of the student's intellectual cycle.

The distribution of below average performance was more ambiguous, since it seemed to be more uniform than above average performance, i.e., it tended to occur at more points of an individual's cycle. However, further analysis showed that the distribution of poor academic performance was quite complicated and probably reflected the influence of all three biorhythms, not just the intellectual one. Moreover, Neil found, of the 43 below average performances turned in by his subjects, above 70 percent happened during the negative phase of the intellectual cycle.

While Neil's work should not be seen as conclusive proof of the validity of biorhythm theory, it is very suggestive. Neil has also made two invaluable contributions to biorhythm research. He has developed a rigorous statistical model for analyzing the theory; and he has shown that it can be used successfully to support the biorhythm theory, whether it is applied to closely controlled experimental data or to data from the field, such as industrial accidents.

Biorhythm Abroad

While work on biorhythm in the laboratory has been slow and infrequent, corporations in the U.S. and elsewhere appear to have turned to the theory in ever-increasing, if unadvertised, numbers. Some have been using biorhythm for decades, while others have introduced biorhythm into their operations only within the last few years. In almost every case, it seems that the results have been profitable enough to ensure continued application of the theory.

In Switzerland, where Hans Frueh did so much to advance the theory of biorhythm by inventing the Bio-Card and Bio-Calculator, municipal and national authorities appear to have been applying biorhythm for many years. Perhaps the best-known of these groups is Swissair, which reportedly has been studying the critical days of its pilots for almost a decade. When either a pilot or co-pilot is experiencing a critical day, he is not allowed to fly with another experiencing the same kind of instability. According to informed sources, the result has been that Swissair has had *no* accidents on the flights to which biorhythm has been applied. Meanwhile, the Zurich Municipal Transit Company, which operates trolleys and buses within the city, has also been using biorhythm to warn its drivers and conductors of critical days. According to most reports, the accident rate per 10,000 kilometers had been slashed by about 50 percent within one year of the application of biorhythm. Similar results have been reported by the municipal transit system of Hanover, Germany.

While the Swiss may justly claim to have been the first to see and realize the benefits of biorhythm in reducing accidents, the Japanese have surely been the first nation to apply biorhythm on a large scale. In fact, the overwhelming majority of evidence about applied biorhythm comes from Japan. The interest in Japan originated in 1965, when Thommen sent a copy of his book *Is This Your Day?* to Kichinosuke Tatai, M.D., Professor of Physiology at the Institute of Public Health in Tokyo. Tatai used the calculation tables and charting equipment Thommen had introduced, found them practical, and formed the Japan Biorhythm Center. Within a few years and with Tatai's promotional flair, the biorhythm theory grew into popular public interest. His groundwork was enlarged in 1970 when Yujiro Shirai and his associates arranged for the translation of the new edition of *Is This Your Day?* and formed the Japan Biorhythm Association. Shirai became an active and successful lecturer and consultant. He was instrumental in bringing the application of the biorhythm theory to the attention of thousands of industries and originated the teaching of the biorhythm theory by traffic police and defensive driving courses.

Another partial cause may be that through the Biocom 200 printing computer, manufactured by Takachiho Koheki Co., Ltd. of Tokyo, the Japanese have inexpensive and almost instantaneous access to highly accurate biorhythm charts. Whatever the cause, more than 5,000 Japanese companies now use biorhythm to control their accident rates, and many of those companies are among the leaders in their fields. The roster of companies that use the theory in some aspect of their operations reads like an honor roll of Japanese business: Fuji Heavy Industries, Mitsubishi Heavy Industries, Hitachi, Bridgestone Tire, and Asahi Glass and Procelain (the largest Japanese crockery manufacturer). Utility companies in Tokyo, Chubu, and Tohaku have also found biorhythm valuable enough to use on a regular basis, as have a number of other municipal service companies. Several of Japan's largest insurance companies have discovered that they can reduce the accident rates of their customers—and thereby the amount they must pay in claims—by using appropriate biorhythm charts to covered clients. Among this group are Japan Fire Insurance, Tokyo Marine and Fire Insurance Taisho Fire Insurance, Yasuda Fire Insurance. Life insurance companies have also found biorhythm useful in reducing their claims. Led by such companies as Mitsui Life, Meiji Life, Asahi Life, and Fukoka Life, they regularly issue biorhythm charts to the people they insure in the belief that this strategy will reduce payments made for injuries, accidents, and sudden death. However, the heaviest users are probably the Japanese transportation companies, a group that includes Japan Express, Seino Transport, Yamato Transport, Japan Traffic Taxi, International Automobile Taxi, and Odakyu Taxi.

It is one of these companies—the Ohmi Railway Company—which has provided the most widely-cited example of biorhythm in action. Contrary to the implications of its name, Ohmi Railway operates a fleet of buses and taxis in Kyoto and Osaka, two of Japan's most densely populated cities. Having learned that the cities of Zurich and Hanover had been able to reduce the accident rates for their municipal transportation systems by using biorhythm, Ohmi decided to investigate its own accident record. The company's findings were startling. Of 331 accidents that had occurred over the five years between 1963 and 1968, 59 percent had occurred on a driver's critical day, or on the day immediately preceding or following the critical day. Over the same period, for the 212 most accident-prone drivers, about 61 percent of the recorded accidents had happened on a critical day or on the day immediately before or after (called the "half-periodic" or "half-critical" day because differences in time of birth can bring such days into the critical period). Interestingly, Ohmi also found that the physical rhythm was much more important than the emotional or intellectual rhythms. Accidents did not tend to happen on days that were critical both emotionally and intellectually, but happened mostly on days that combined physical and emotional criticals. This observation suggests that reaction time—as Willis had hypothesised—is a central factor in transportation accidents, since reaction time will be most influenced by changes in the physical rhythm. It also implies that when slowed reaction time is combined with impaired judgment—characteristic of emotionally critical periods—the likelihood of accidents is especially high.

In 1969, Ohmi decided that the seemingly inevitable rise of its accident rate could be stopped only by applying biorhythm. The company's straightforward strategy was to issue a special warning card to each of its more than 700 drivers whenever the driver's bio-curves indicated a critical day. This simple device has had impressive results. In the first year, the accident rate for Ohmi vehicles dropped almost 50 percent. The rate continued to decline in each subsequent year, despite increased traffic and a sharply rising accident rate for Japan as a whole. By 1973, Ohmi's bus division had achieved an unprecedented record of 4 million kilometers traveled without one reported accident.

Similar experiences from Japan are numerous. According to Yujiro Shirai, a chief representative of the Japan Biorhythm Association, the Meiji Bread Company of Saitama Prefecture decided to apply biorhythm because it had been suffering from the same sharply rising accident rate that has plagued all of Japan. In the first full year of warning its drivers to be extra careful on critical days, Meiji Bread managed to reduce the size of its losses from traffic accidents to 45 percent of the amount incurred in the preceding year—a savings of more than $10,000. Seibu Transport Company took a different tack, again according to Shirai. In order to keep accurate, easily retrieved records of all the goods it contracts to carry, Seibu had entirely computerized its filing system. This approach worked well except for one thing—the young girls whom Seibu employed because of their manual dexterity to punch the computer entries were making hundreds of costly mistakes. Occasionally, the company found itself unable to tell a customer the location of goods consigned for shipment because the computer file was defective. To solve this serious problem, Seibu turned to making biorhythm charts for each of its computer operators. Whenever one of them was subject to a critical day, the company would transfer her to some post other than keypunching. Within six months, the incidence of errors had been reduced by 35 percent from its level over the preceding months.

Equally gratifying results have been obtained by Dr. Kichinosuke Tatai, director of the Japan Biorhythm Laboratories in Tokyo, and one of Shirai's fiercest competitors in spreading the use of biorhythm. Tatai has written that the taxi section of Kokusai Automobile Co., Ltd. of Tokyo discovered that its drivers persistently had accidents on their emotionally critical days. To arrest this trend, Kokusai followed Ohmi's lead. In 1968, the year before biorhythm was applied, Kokusai registered 153 taxi accidents. In 1969, the first year it used biorhythm to warn its drivers of potentially dangerous days, the company registered 124 accidents—85 percent of the previous year—even though the number of cabs and the kilometers traveled had grown. Persisting in its application of biorhythm, Kokusai reported only 107 accidents in 1970 (70 percent of the 1968 record), and 83 taxi accidents in 1971 (54 percent of the 1968 record). This declining rate has continued, even though the number of kilometers travelled has increased.

According to Tatai, much the same success accompanied the application of biorhythm by Hiragishi Taxi Co., Ltd. of Sapporo, which had 165 drivers manning 65 vehicles. The company began applying biorhythm in the last three months of 1972, and the results were immediate. From October through December, 1972, Hiragishi reported 38 accidents, about 49 percent of the 78 accidents registered for the last three months of the preceding year, when biorhythm was not being used to warn drivers of their critical days.

Much of the success the Japanese have had with biorhythm—and much of the reason for the theory's broad adoption in Japan—can be tied to the ingenious ways in which the Japanese have applied the theory. Biorhythm charts are frequently used as a promotional tool. Large insurance companies prepare charts not just for those they insure, but also for those to whom they hope to sell insurance. The biocurve graphs are used as a way to get past the front door of the potential customer. Nissan Motors, which has been called the General Motors of Japan, has used the same technique with great effect. The five Nissan showrooms in Tokyo that offered free personal biorhythm charts for a limited time to anyone who walked in the door recorded their greatest customer traffic—and the greatest customer traffic for all automobile showrooms in Japan—during that period.

Most inventive, however, are the ways in which the Japanese transportation companies have inveigled their drivers to take extra precautions on biorhythmically dangerous days. When a driver has a critical day, some companies will hang a folded paper crane (familiar to any student of *origami*, the Japanese art of paper folding) from the dashboard of his vehicle, color-coded to the biorhythm that is critical on that day. A driver in a double or triple critical day may spend it with a miniature flock of paper cranes in his cab. The Japanese mail and telegraph services fly small triangular flags from the front fenders of motorcycle delivery men operating on critical days. Color-coded like the cranes to the rhythm that is critical, the flags act as a warning to the driver and to other drivers and pedestrians that extra precautions are wise. A few taxi companies hang a color-coded but empty candy box from the rear-view mirrors of drivers on their critical days. If the driver finishes his shift without an accident, he can trade in the empty candy box for a full one as a gift for his family.

The greatest force in the wave of biorhythm applications in Japan is the work done by the traffic police and other traffic safety organizations. Stimulated by Thommen's book, *Is This Your Day?*, the Tokyo Metropoliton Police published a study in 1971 which indicated that more than 80 percent of the traffic accidents reported during the previous year had taken place on the driver's critical day. At about the same time, the Osaka Police published a biorhythmic study of more than 100 traffic accidents involving child pedestrians. 70 percent of those young pedestrians had been injured on their critical days, when faulty judgment and slow reactions left them at the mercy of Japan's vicious traffic. Shirai continued this work, using the Biocom 200 computer to analyze accidents in the prefectures of Chiba and Kanegawa. In both cases, his results confirmed those of the Tokyo police—more than 80 percent of the accidents happened on the driver's critical day. Shirai also analyzed traffic accidents involving children under ten years of age for the same prefectures. He explained his finding that 85 to 90 percent of those accidents happened on the victims' critical days by writing, "As people grow older, they develop preventive abilities against danger. Children, on the contrary, act only on instinct, that is, under their own biorhythms." In short, children are more influenced by their biorhythms and are thus less able to protect themselves.

As a result of such work by the Japanese police and other organizations, biorhythm has become a prominent feature of the motoring scene in Japan. Traffic Safety organizations and courses in defensive driving have been set up by the traffic police of many prefectures. In 12 of these, each member or student receives regular, personalized biorhythm charts printed out by the Biocom 200. Gumma prefecture alone operates a dozen minicomputers for this program. Before the application of biorhythm, Gumma prefecture had an unusually high accident rate; but now over 250,000 drivers in the prefecture receive biorhythm reports, the accident rate has plummeted, and Gumma is considered to have the best traffic safety program in Japan. Kanegawa and Chiba prefectures, where Shirai did his studies, have adopted similar systems. In several other areas, everyone receiving a driver's license for the first time or renewing a license receives a personalized biorhythm chart, as does anyone involved in an accident. The result has generally been a significant drop in the accident rate. This development has not escaped the ever-vigilant Japanese insurance companies, who have helped to sponsor biorhythmically based driver safety courses and have even urged that every worker in Japan receive a bio-curve graph in his pay envelope.

The Biorhythm Underground

In light of the Japanese and Swiss experience, it may appear inexplicable that American government and industry have not been quicker to experiment with biorhythm. The U.S. has certainly experienced nothing like the biorhythm craze of Japan. However, there may well be more interest in biorhythm than meets the eye. Until a theory like biorhythm is proven beyond the slightest shadow of a doubt, government and industry will be reluctant to reveal whether or not they are experimenting with the method. This does not mean that large organizations are not in fact trying out the theory; they are probably wary of attaching their names to such an unorthodox concept. Given this attitude, the fact that members of such distinguished groups as the Flight Safety Foundation, the National Safety Council, NASA, the Tactical Air Command, the National Institutes of Mental Health, and Bell Telephone Laboratories have admitted even to an awareness of and interest in biorhythm is significant. And those admissions may disguise an involvement with biorhythm more active than we know.

A few of the biorhythm experiments conducted by American industry are well-known, if only because they began several years ago. Perhaps the most famous and significant was that reported by National Lead Company (now known as NL Industries). Partly as a result of George Thommen's lectures, the company's Titanium Division decided in 1965 to run a small experiment on biorhythm. The Division took three groups—the rigging department, the millwright shop, and the pipeshop—and used the pipeshop as a control group. All three groups received intensive instruction and supervision on safety. But biorhythm curves were plotted only for workers in the rigging and millwright areas. This information was given to the foremen in those departments. For workers with critical days, the foremen would stress safety, give extra supervision, and assign less dangerous tasks. On the other hand, foremen in the pipeshop would shift workers from dangerous jobs and give special safety warnings when they felt it was necessary—an essentially random procedure modified by intent and experience. In short, the only difference between the safety programs in the rigging department and millwright shop and the one in the pipeshop was the guidance of biorhythm analysis.

As so often happens, the results were striking. The biorhythmically based safety program was introduced to the rigging department in July, 1965, directly after that group had shown a six-month increase in the injury rate. By the end of 1965, the riggers had reduced their injury rate by 18 percent over the rate in the period immediately preceding. Continued throughout 1966, the program helped the riggers cut the injury rate to 58 percent of what it had been in the previous year. For the 18 months as a whole (a period that included more than two million man-hours in the rigging department), accidents were reduced by 40 percent over the preceding 18-month period. The millwright shop turned in a good though not impressive performance. The biorhythmically based safety program was introduced there at the beginning of 1966, after several months of showing a rising accident rate. By the end of the year, the millwrights had reduced their accident rate by 4 percent over the previous year.

But what of the pipeshop? This is an important question since critics of biorhythm rightfully contend that *any* new, systematic safety program, whether or not it is based on biorhythm, will reduce the injury and accident rate. It is also true that any experiment—no matter what its characteristics—will change the study group's behavior simply due to their knowledge that an experiment is being conducted. As one Pennsylvania psychologist summarized it: "If you tell a group of factory workers that you have some way of figuring out dangerous days, then remind them to be super-careful on those days, you will probably cut your accident rate, even if your way of figuring out those days is throwing darts at a calendar." If biorhythm has no basic validity, we would then expect the riggers, the millwrights, and the pipeshop to have achieved roughly similar accident rates.

This is not what happened. True, during the first six months of the experiment, the accident rate in the pipeshop declined. However, after the initial effort wore off, the uselessness of predicting "dangerous days" at random became apparent. For the 18-month period as a whole—from July, 1965, through December, 1966—accidents among the pipeshop workers *increased* by 28 percent over the preceding 18 months, following the rising trend that had been typical of the rigging and millwright departments before the introduction of the biorhythmically based safety program. It should come as no surprise that when the Executive Safety Committee of the Titanium Division reviewed the pilot project, it reported a "significant success . . . and a practical and effective means of preventing injuries." As one Committee member later observed, "I don't know why or how the program worked. All I can tell you is: it works."

Possibly impressed by the NL Industries experience, a Proctor & Gamble paper products factory in Green Bay, Wisconsin, decided to do its own experiment with biorhythm in the early 1970's. Like the Titanium Division at NL Industries, the P&G plant had been suffering from a high accident rate; retrospective analysis convinced some supervisors that biorhythm might help to reduce it. As one of the plant's safety engineers told a reporter:

> "We did preliminary research and discovered that in the past year 42 percent of all the self-induced (human error) accidents in our plant fell on critical days of the employees involved. Since critical days occur only about six times a month, or 20 percent of the calendar days, we could say that 42 percent of the self-induced accidents falling on only 20 percent of the days (our plant operates seven days a week) is a terribly impressive correlation, better than 1000 to 1 odds. . . . As a result, we selected a group of 200 employees for whom we've calculated critical days for the next two months. We've mailed this information to their homes, along with explanatory literature, in the hope that when each employee's critical day comes around, he'll take the proper precautions. The employees were invited to participate in the program on a voluntary basis, and it appears that 90 percent will be cooperating fully. After two months, we'll check the accident rate, and if the reduction turns out to be substantial, we'll continue using biorhythm analyses indefinitely."

Although the same safety engineer later reported a reduction in the accident rate as the result of this experiment, Proctor & Gamble denied both the experiment and its results.

United Airlines has been much more open about its experiences with biorhythm. In November, 1973, United began to use its IBM 360 computer in San Francisco to track the biorhythms of its more than 28,000 ground crew and maintenance workers. The computer was programmed to produce two-month bio-curve charts for each worker, plus a summary chart for each foreman. The foreman's chart showed the name of each employee undergoing a critical day, and which rhythm or rhythms would be critical on that date. Both printouts were delivered to United's ground crews at National Airport in Washington, D.C., from November, 1973, through November, 1974. According to Vern Parker, industrial engineering manager for United, maintenance workers on critical days would take extra precautions and discuss safety procedures with their supervisors. With this very simple plan—and no shifting of job assignments—the National Airport staff was able to cut its accident and injury rates for the 12-month period by more than half. Impressed by that record, United has expanded its computer program to include flight crews, although pilots have not yet received charts.

Not to be outdone by United, several other U.S. airlines have begun to investigate biorhythms, although none of them seem to have been frank about their interest. A computer team at Allegheny Airlines' Arlington, Virginia, complex has explored the theory and has run a pilot study of a group of aircraft maintenance workers in Pittsburgh. However, Allegheny disavows its activity in this field, as do Continental Airlines, Pan American, and Trans World Airlines, all of which have considered using the theory to reduce accidents and errors on the ground—and in the air. If we take the many retrospective analyses that show strong relationships between biorhythm and aircraft crashes as a guide, the airlines might be wise to apply biorhythmic safety programs immediately. Yet most of them appear to be exercising excessive caution and secrecy. As a spokesman for the Airline Pilots Association explained to one reporter:

> "Very few people—particularly in the airlines—are going to admit that they're studying biorhythms. It's kind of controversial. People think, 'Nobody but a kook would study that.' But let me tell you—we know most airlines are looking at it, and some of them are very excited about it. But the official airline attitude is always going to be, 'I don't buy a curve unless there's a set of equations to prove it.' "

Other companies seem to have noticed the flurry that biorhythm is causing among the airlines. Bell Telephone Laboratories has confirmed that it is studying biorhythm as one possibility for improving productivity, not just of workers, but of scientists and engineers as well. Although Robert Bailey of Bell's Human Technology Division, which was commissioned to do the study, had said that, "If there's something to it, I haven't found it yet," the Bell work is still in its early stages. It has covered fewer than 300 employees, although charts for almost 2,000 Bell workers were distributed as part of a safety promotion campagn. Unfortunately, no efforts were made to find out whether those charts had any impact on the accident and error rates of the people who received them. Biorhythm Computers, Inc., has done employee charts for more than 40 firms, including leaders in the insurance, lumber, and machinery businesses. Although Biorhythm Computer's clients have insisted on confidentiality, some of them have reported that simply telling employees when they are undergoing a critical day can reduce human errors and their costly consequences by up to 55 percent. Information leaked from the six computer firms that sell biorhythm charts indicates that their customers have included two leading insurance groups—The Truck Underwriters Association and Allstate Insurance. Both organizations deny their involvement with biorhythm, needless to say.

Official denials of the validity of biorhythm, in fact, seem to grow in violence and number as the theory gains more adherents. When consulted about biorhythm, three of the most distinguished students of circadian and other biological rhythms have been quick to dismiss the theory and to distinguish it from their own work. The most eminent of these scientists, Dr. Franz Halberg of the University of Minnesota Medical School, said of George Thommen's work with biorhythm: "He is talking about immutable, fixed rhythms, yet the rhythms we know of are actually wobbly. They vary quite a bit, though they vary predictably. As to any similarity with my own work, it's like Smith and Schmidt. We have only the name in common." Dr. John Hastings of Harvard simply stated: "This is not a serious subject being studied by serious scientists." However, Colin Pittendrigh, professor of biological sciences at Stanford University, seems to have been more offended by the notion of biorhythm than either Halberg or Hastings. When asked to comment about the theory, he said, "I consider this stuff an utter, total, unadulterated fraud. I consider anyone who offers to explain my life in terms of 23-day rhythms a numerological nut, just like somebody who wants to explore the rhythms of pig-iron prices to 14 decimal places." The attitude of the scientific establishment in general is best summarized by a National Institutes of Mental Health paper that classifies biorhythm as a mythology. However, as Douglas Kelley, a statistician for the National Safety Council, has pointed out: "When chemistry was at the state where biorhythm is today, it was called alchemy. But alchemy became chemistry, and within 50 years research may do the same for biorhythm."

To dismiss biorhythm theory out of hand, as many

scientists have done, is a remarkably unscientific reaction. Too little is really known about the theory and too little research has been done to refine it for anyone to be able to make a definitive judgment; those who have gone so far as to make judgments should at least understand the theory before condemning it. A more balanced attitude is shown by Dr. Bertram Brown, director of the National Institutes of Mental Health, who holds that, "These biorhythms seem to have a lot of validity. They help to explain in part everything from having a bad week to exciting scientific things like the varied effects medications have when administered at different times." Whether Brown's view will be justified, of course, depends on future research in biorhythm.

The Future of Biorhythm

Biorhythm theory is still in the earliest stages of its development. We have come a long way from the work of Fliess, Swoboda, and Teltscher in developing the theory and accumulating evidence to support it. We have begun to do rigorously controlled experimental studies to test the theory's reliability, but many important questions remain unanswered: Why do biorhythms exist and what is their physical basis? Through what channels do they influence us? How do they interact with the thousands of other biological cycles that have been shown to regulate life? How do individuals differ in their susceptibility to biorhythmic influence? What factors intervene to change the regularity and strength of the three great rhythms at different stages of life and under different environmental circumstances? What strategies will be the most productive in helping us adapt our lives in biorhythmic cycles? What are the most sensible ways to use biorhythm in industry, medicine, flight, sports, and the many fields where it can be helpful?

The purpose of this chapter is to suggest some answers to these questions and to point out future directions for biorhythm research and application. The field is wide open today and full of exciting possibilities; we hope that the future will see many of these possibilities realized. Until they are, though, the ultimate question of whether biorhythm theory is true or false must remain unanswered, at least by any overwhelming body of scientifically acceptable evidence.

However, as we suggested earlier, there is one way in which everyone can answer that question for themselves: by keeping a "blind diary" of emotional, physical, and intellectual fluctuations in their own life, and later comparing it with fluctuations in their three cycles. (See Figure 13 for an example of this.) If biorhythm does *not* seem to work for you, you may be one of those rare individuals who are "arhythmic" and do not respond fully to internal cycles. If it *does* work out (which seems more likely, given our knowledge of the theory), you will have acquired a new tool to guide your life away from catastrophe and toward success. Remember that only you can decide the merits of biorhythm; with the charting method provided in the Appendix, the mechanics of that decision should be quite easy.

The Roots of Biorhythm

The greatest weakness in biorhythm theory—and the field that calls most immediately for further research—is the question of exactly why it appears to work. No adequate explanation has even been mentioned, much less proven. This is not, of course, an unusual situation in an emerging science. Despite decades of painstaking investigation, for example, we still do not know why we are subject to circadian rhythms or to many of the smaller physiological cycles that combine to create circadian rhythms. Since we are so ignorant about phenomena that are much easier to understand than biorhythms—and which may form the basis of biorhythms—we can hardly expect to know much about the roots of the three great rhythms themselves.

A search through the scientific literature on the subject produces only a few tantalizing fragments. The Sandia Laboratories report mentioned in the second chapter raised the possibility that the phase of the moon under which a person is born may increase his accident-proneness each time it reappears. Since the lunar cycle corresponds to the 28-day emotional cycle, perhaps the moon is an important regulator of the emotional rhythm. Evidence of lunar influence on menstrual cycles and on emotional cycles in both men and women makes this possibility appear reasonable. Yet proof of the moon's role in the emotional biorhythm would still leave us in the dark about precisely *how* that role is played. Perhaps we have a subtle sensitivity to the same gravitational force with which the moon "pulls in" the tides. If so, our physical and intellectual biorhythms may respond to other, now unsuspected gravitational forces. Along the same lines, scientists who study biological rhythms have suggested that geomagnetism may be an important regulator of many circadian cycles. It may be that the earth's magnetism plays a role in biorhythms as well.

On the other hand, biorhythms may be rooted largely in internal forces rather than in anything outside our bodies, although, as with circadian rhythms, it is most likely that biorhythms result from a combination of external cues and internal clocks. One such internal clock was mentioned more than a decade ago in the *New York State Journal of Medicine*. In an article entitled "The

Direct Current Control System: A Link Between Environment and Organism," Drs. Robert O. Becker, Charles H. Bachman, and Howard Friedman described the results of an experiment in which they measured the charging capacity of the brain to transport electrical current. They wrote:

> Since the cranial direct current potential appeared to be particularly important . . . in the state of consciousness and level of irritability in the human being, the possibility that it was the controlling mechanism for biologically cyclic behavior was considered. In a very preliminary study, the transcranial d.c. potential of two normal subjects and two schizophrenic subjects was determined daily for a period of two months. *A definite cyclic pattern was evident in all four subjects, with a periodicity of approximately twenty-eight days and with all four following similar cycles.* (emphasis added)

If electrical potential is so important to people's moods, and if it does vary regularly over a 28-day period, it might be the central mechanism of the emotional biorhythm.

Alternatively, biorhythm may turn out to depend on one of the most basic biological processes known to science—the syntheses of DNA (dioxyribonucleic acid) and RNA (ribonucleic acid) by which all cells reproduce themselves and regulate their metabolisms. A recent paper by Dr. A. Ehret reported the discovery of small bodies called *chronons* situated in the complex DNA molecule. Ehret speculated that these bodies regulate the speed with which RNA is synthesized and therefore might also regulate a number of physiological and circadian rhythms. If chronons do have these effects, it appears possible that different chronon configurations could interact to influence the periodicity of the basic, 23-day physical biorhythm, and perhaps of the other two rhythms as well.

Working It Out

The mystery of precisely what causes biorhythms may take decades to solve. In the meantime, there are some easier qustions about the theory that are almost as important. Given that the three biorhythms exist and are the products of certain highly regular phenomena, how do the rhythms themselves affect different individuals and their behavior? As Jack Gunthard of the Swiss National Gymnastic Team saw it, some people are "rhythmists" and some are "non-rhythmists"; or, to put it another way, some people appear to be more sensitive to biorhythms than others. Why is this so, and what tools might we use to distinguish reliably between these two different types of people? Is it simply that people develop different ways of dealing with biorhythms, and that some of these methods effectively mask biorhythmic effects? Or is it that the strength of biorhythms—the amplitude of the sine-waves used to represent the curves—varies in different individuals, and also for the same individual at different times? These questions relate to three areas of research: the regularity of biorhythms; the interaction of biorhythms with heredity; and the characteristics of situations (environments) in which biorhythmic effects are most likely to appear.

The traditional position is that biorhythms are absolutely regular and do not vary—even by as little as one second—from the moment of birth to the moment of death. This axiom was the basis for the early work in biorhythm, and many contemporary adherents of the theory still cling to it. However, even rock-ribbed traditionalists admit that the effects of biorhythms—the ways in which they are manifested—may show slight irregularities, simply because of intervention by such factors as general physical condition and environmental stress. Less conservative researchers suspect that biorhythms themselves—like the overwhelming majority of natural rhythms—are less precise than they were originally thought to be, although not so erratic as to invalidate biorhythm calculations and charts. As Douglas Neil summed it up, "Our [biorhythm] studies are showing that there's definitely something there. . . . Biorhythms may be similar to circadian rhythms. But there are probably individual differences, just as there are in menstrual cycles." Neil's position sounds reasonable, especially when we remember that some biorhythm research has shown heightened vulnerability to accidents and disease not just on the crtical days, but also on the half-periodic days that immediately precede and follow each critical day.

The problem of biorhythm and heredity is a very tangled one because so much early research was devoted to finding out whether the rhythms themselves were inherited and conformed to the timing of one's ancestor's rhythms. Some of this same research also focused on whether the biorhythmic positions of one's parents could determine which of their traits could be inherited, including sex. Both of these questions are moot today. As we presently understand it, biorhythm theory holds that the rhythms begin at the moment of birth, and therefore their timing cannot be inherited, unless the time of birth is itself inherited (a possibility Fliess and Swoboda explored with inconclusive results). Although the question of how biorhythms affect inherited traits is still open, it is less basic—and therefore less interesting—than many other questions about the theory. Of more immediate concern is the problem of how the traits we inherit from our parents combine with our changing biorhythmic profiles to affect our performance. No solid work has been done in this area at all, perhaps because scientists are so uncertain as to how general traits such as athletic ability are related to particular combinations of genes. When we have succeeded in mapping the intricate pathways through which intelligence, strength, and other general traits are inherited, we will be in a position to clear up the matter of biorhythm and heredity.

The question of biorhythm and environment is more obvious. As we have emphasized in previous chapters, the effects of biorhythm are not usually apparent unless the individual encounters a situation likely to bring them out —for example, an automobile emergency on a critical day, or an athletic contest on a day of triple highs or triple lows. Neil's work with academic performance clearly indicated that unless a situation is challenging in some way, biorhythmic effects will be masked beneath an apparently average performance. The interesting problem is to find out *how* challenging the environment must be to bring biorhythms into full play. In other words, what is the *challenge threshold*, the point below which biorhythms will be masked and above which they will appear? This threshold is something that we can measure with acceptable precision, and it would be an invaluable aid in indicating when biorhythms demand our attention and when we may safely ignore them. Until a challenge threshold is established for many kinds of situations, we will have difficulty specifying what can be done to adapt ourselves to biorhythms. It is simple to make general prescriptions, such as avoiding hard exercise during physically negative periods or taking extra precautions against accidents on critical days. No such general descriptions, though, can help us to fulfill to the limit the potential that biorhythm holds for improving our lives. To do that, we must understand—almost viscerally—how the thousands of different combinations of the three rhythms affect us. For example, on a day when the physical and intellectual rhythms are ascending in the positive phase, while the emotional rhythm is descending through the negative phase, how should we approach work? Is it enough to guard against moodiness and irritability, while giving free play to our minds and bodies and pushing ourselves to the limit in those areas? Or would it be wiser to concentrate on maintaining emotional equilibrium while restraining ourselves physically and intellectually, so that our surging physical and mental energies will not be misdirected by the emotional low?

The ramifications of the problem are numerous. Is it true, as some athletic trainers feel, that keeping a man from playing or doing hard training during his physically negative phase will add to his performance in the positive phase? Even if that is true, might it be wiser to restrain a player from spending all of his energies during the positive phase so that he will have a reserve to draw on if he must play during his negative phase? The problem of how to use the "energy bank" that bio-curves describe is one that bothers all types of administrators —airline executives, military leaders, foremen, and doctors, as well as coaches. It is hard to solve, and raises real moral issues, when the administrator has confidential knowledge that one of his employees, working on a dangerous job, is undergoing a critical day. Is it better to keep the knowledge private and simply caution the worker? Or is there a moral obligation to tell the worker that he is critical? Speaking more pragmatically, will a warning and extra caution be enough to prevent accidents, as the United Airlines experience indicates? Or is it better, no matter what the inconvenience, to shift the worker to a less dangerous task? The answers to these and similar questions will depend on the development of an applied discipline: industrial biorhythm. If doctors are to use biorhythm effectively, applied medical biorhythmics must be developed.

Behind these questions of application lies a problem we have mentioned many times before: interpretation.

Certain interpretations are unambiguous. We know what a critical day is, when it occurs, and how it may affect us under stressful circumstances. Much the same can be said for triple highs or lows. Mixed rhythm days are much less clear, and to interpret them correctly we need to develop some guideposts. Two potentially useful guideposts have already been mentioned by biorhythm researchers. Wallerstein and Roberts, who did the successful composite bio-curves for football teams, found that the *direction* in which a rhythm was moving could be as important—and perhaps more important—than whether the rhythm was above or below the zero line. Might the same be true for individual bio-curves? If so, it would greatly simplify the task of extending bio-curve analysis beyond the elementary "high-critical-low" trichotomy. Several researchers who have dealt with accidents and biorhythm have suspected that days when two rhythms cross each other while going in opposite directions—regardless of whether they cross in the positive or negative phases—are potentially dangerous days. The reason is that days of "opposed crossing," much like critical days, may involve rhythmic instability or confusion; a declining emotional rhythm, for instance, can make a person *feel* listless, even though a physical rhythm ascending at the same time actually is giving him increasing energy. When these two curves reach the same level, they cross, and it may be at this point that the conflict and confusion in direction is most apparent. Confirmation of this hypothesis would also add needed depth to the interpretation of mixed days.

To summarize, a more *systematic approach* to the interpretation of bio-curves is needed. Biorhythm analysis will never be an exact science, one in which certain combinations of the three rhythms will always produce the same results. Because individual differences and the influence of the environment are so important, the analysis of bio-curves is somewhat like medical diagnosis or psychoanalysis, where particular symptoms or behavior appears to indicate the patient's condition and calls for certain therapies. Unfortunately, at its present stage of development biorhythm interpretation depends too much on the experience and wisdom of the analyst and too little on a general, widely accepted classification or typology of rhythmic positions and combinations. Unless experienced analysts can come together and agree at least on the terms of biorhythm interpretation, the theory will be open to the same charges levelled against astrology: that the leeway for interpretation is so wide as to eliminate the theory's claim to accurate predictions (you can always change an interpretation to fit the facts).

A similar problem involves the methods biorhythm researchers have used to test the theory. As Neil rightly pointed out, it is not enough to do critical-day analyses, correlating accidents, deaths, and other mishaps with the critical days of the victims. True, this approach does produce impressive results, and it is very relevant to many of the most useful applications of biorhythm. However, it cannot provide proof of the theory as a whole, since critical-day analysis is limited to a very small portion of the theory—the 20 percent of the days in an individual's life which are critical. The other 80 percent of the days have not been correlated with biorhythms for the most part, yet they must be if the theory is to gain broader acceptance. Exactly how we can measure the degree to which up days, down days, and in-between days correspond to bio-curves is somewhat perplexing. Blind diaries are a good method, but a limited one, since they can only be used case-by-case and with the full cooperation of the subject. Neil's approach, which calls for developing an objective performance scale, rating individual behavior on that scale over long periods of time, and then comparing the results with bio-curves, may be the most fruitful since it tests biorhythm theory at all performance points—not just the worst ones. However, to get all researchers to use a similar approach, we will have to establish a *standard procedure for biorhythm analysis.* It need not involve the tortuous statistical tests that Neil used (mainly because his sample was so small); it must involve techniques which are clear, widely accepted, and capable of testing the theory as a whole. Otherwise, biorhythm research will remain primarily a haphazard collection of empirical studies, few dealing with the whole theory, and none comparable with another.

The research needs we have discussed—exploring the roots of biorhythm, the regularity of the three rhythms, the role of heredity and environment, developing disciplines of applied biorhythm, establishing a systematic approach to interpretation, and agreeing on a standard procedure for analysis—spring from the fact that biorhythm, as a science, is still in its infancy. These needs will be met as biorhythm develops into a full-blown science; however, maturation takes time. Meanwhile, the potential benefits of applying biorhythm are so great that we can expect to see a host of new applications. Many of these have been sketched out by Willis and other students of biorhythm, while others have already been realized on a small scale.

In industry, the opportunities are limitless. In addition to reducing accident and injury rates, biorhythm can help people to achieve their full potential on the job—if employers assign challenging tasks during positive phases. Shared knowledge of biorhythm can improve management-worker relations and increase the productivity of work-groups at all levels. Wallerstein and Roberts feel that group productivity can be tracked as accurately as football performance, and this kind of knowledge could be very helpful in structuring work-groups and assigning tasks.

Whenever a strategic decision must be made in government, biorhythm analysis could be of value. Especially in foreign relations and negotiations, it would be very helpful to know the biorhythmic profiles of one's own diplomats and their opposite numbers on days scheduled for discussions, or even to schedule negotiations to exploit biorhythmic advantages! When top-level officials must make crucial decisions they would be wise to wait for positive days in their emotional and intellectual rhythms. This is also true of military commanders, although the exigencies of battle would probably limit them to avoiding important decisions on critical days, if possible. Biorhythm could be of more help in military operations as a guide to the times at which personnel and units will be at their performance peaks for high-risk operations. Tasks could be assigned in the military, as in business, according to the biorhythmic profiles of the group.

We have already discussed several medical applications: using biorhythms to schedule operations, to guide intensive care for heart attack victims, and to anticipate the behavior of mental patients. It might also be wise for surgeons and other doctors engaged in high-risk procedures to schedule their work around their biorhythmic peaks, or at least to avoid operating on their critical days.

The applications of Wallerstein and Roberts, Willis, Gunthard, Benthaus, and many others have already indicated what biorhythm can do for athletics. Wallerstein and Roberts envision even broader applications: "We hope to apply bio-curve [analysis] towards the training of athletics, and also to introduce it as a new form of sports statistics. Possibly there will come a day when line-up engineering becomes commonplace; where athletes can look to fewer injuries! where little children can learn that losing comes from having fewer capabilities on certain days than on others." For the millions of people who bet on sports, the day of applied biorhythm is actually here already—Jimmy the Greek is using bio-curves to help him pick winners!

Finally, the most meaningful applications of biorhythm are the personal ones. With a knowledge of your own biorhythm profile and some experience in interpreting it, you can reap huge benefits—from business success to improved interpersonal relationships to better health—and even to a longer life.

Bibliography

Anderson, Russell K. "Biorhythm—Man's Timing Mechanism." Park Ridge, Ill.: American Society of Safety Engineers, 1972.

Ault, Michael, and Kinkade, Kenna. "Biorhythm Analysis of Single Car Fatalities." Joplin, Mo.: Missouri Southern State College, 1973.

Bochow, Reinhold. "Der Unfall im landwirtschaftlichen Betrieb." *Wissenschaftliche Zeitschrift der Humbolt University*, no. 6, 1954/1955.

Brady, Timothy. "Biorhythm What?" Langley Air Force Base, W. Va.: *TAC Attack*, March, 1972.

Case, Jan. "Predictive Powers in Bio-Rhythm Analysis in the Performance of Football Players." Joplin, Mo.: Missouri Southern State College, 1972.

Coates, Lloyd D. *et al.* "A Study of Bio-Rhythmic Cycles,

Astrological Forecasts, and Personal Evaluations." Joplin, Mo.: Missouri Southern State College, 1972.

Fliess, Wilhelm. *Der Ablauf des Lebens*. Leipzig-Vienna: Deuticke, 1906.

———. *Von Leben und vom Tod*. Jena: Diederichs, 1909.

———. *Das Jahr im Lebendigen*. Jena: Diederichs, 1918.

———. *Zur Periodenlehre*. Leipzig: Ebenda, 1925.

Frueh, Hans. *Von der Periodenlehre zur Biorhythmenlehre*. Zurich-Leipzig: Wegweiser, 1939–1942.

———. *Rhythmenpraxis*. Zurich: Frueh, 1943.

———. *Kraft, Gesundheit und Leistung*. Basserdorf: Frueh, 1946.

———. *Deine Leistungskurve*. Basserdorf-Zurich: Frueh, 1953.

———. *Triumph der Lebensrhythmen*. Buedingen-Gettenbach: Lebensweiser, 1953.

Klein, Marcia. "Biorhythm in the Prediction of Heart Attacks Suffered by American Business Men." Joplin, Mo.: Missouri State Southern College, 1973.

Luce, Gay Gaer. *Body Time—Biological Rhythms in Human and Animal Physiology*. New York: Dover, 1971.

Neil, Douglas E. "Biorhythms and Industrial Safety."

Monterey, California: Naval Postgraduate School (unpublished).

Neil, Douglas E.; Giannotti, Louis J.; and Wyatt, Thomas A. "Statistical Analysis of the Theory of Biorhythms." Monterey, California: Naval Postgraduate School (unpublished).

Schlieper, Hans. *Der Rhythmus des Lebendigen.* Jena: Diedrichs, 1909.

———. *Das Jahr im Raum.* Jena: Diederichs (undated).

Schwing, Hans. *Über Biorhythmen und deren technische Anwendung.* Zurich: Leemann, 1939.

Senzaburo, Oka. "The Purpose of Driver's Self Control." Hikone, Japan: Ohmi Railway Co., 1969.

Siegel S. *Nonparametric Statistics.* New York: McGraw-Hill, 1956.

Swain, A.D.; Altman, J.W.; and Rook, L.W., Jr. "Human Error Qualification." Albuquerque, N.M.: Sandia Corp., 1963.

Swoboda, Hermann. *Die Perioden des menschlichen Lebens in ihrer psychologischen und biologischen Bedeutung.* Leipzig-Vienna: Deuticke, 1904.

———. *Studien zur Grundlegung der Psychologie.* Leipzig-Vienna: Deuticke, 1905.

———. *Die Kritischen Tage des Menschen.* Leipzig-Vienna: Deuticke, 1909.

———. *Das Siebenjahn.* Leipzig-Vienna: Orion, 1917.

———. *Die Bedeutung des Siebenjahn Rhythmus für die menschliche Vererbung.* Florence, Italy: Industria Tipografica Fiorentina, 1954.

Thommen, George S. *Is This Your Day?* New York: Crown, 1973 (revised edition).

Tope, Otto. *Biorhythmische Einflüsse und ihre Auswirkung in Fuhrparkbetrieben.* Hanover, Germany: Staedtehygiene, 1956.

Wallerstein, Michael R., and Roberts, Nancy Lee. "All Together on the Bio-Curve." *Human Behavior,* April, 1973.

Ward, Ritchie R. *The Living Clocks.* New York: Knopf, 1971.

Willis, Harold R. "Biorhythm—Phantom of Human Error." Dallas, Texas: Ling-Temco-Vought Corp., 1969.

———. "Biorhythm and Its Relationship to Human Error." Santa Monica, Cal.: *Proceedings of the Sixteenth Annual Meeting of the Human Factors Society,* 1972.

———. "The Effect of Biorhythm Cycles—Implications for Industry." Miami Beach, Fla.: *Proceedings of the American Industrial Hygiene Conference,* 1974.

Biorhythm Charts of the Famous and Infamous

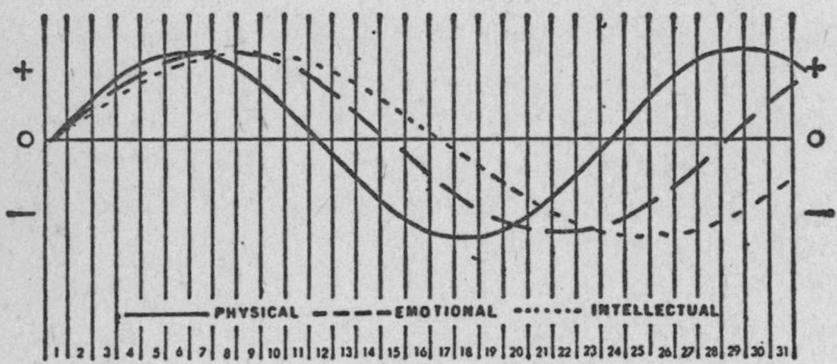

Figure 1. Biorhythmic cycles in the first month of life.

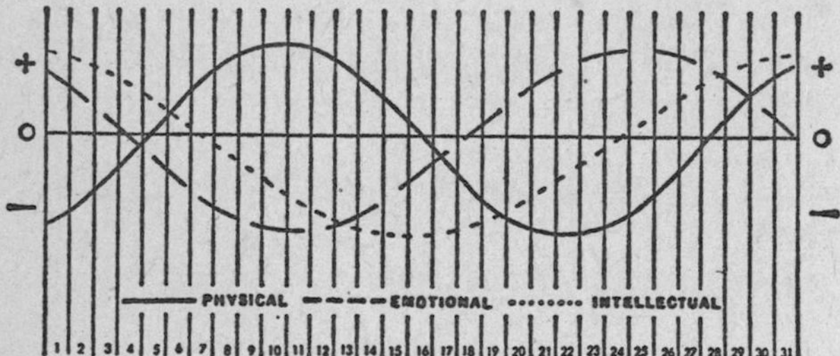

Figure 2. Clark Gable suffered a heart attack on November 5, 1960, when his physical rhythm was critical. He didn't die until November 16, when he suffered a second heart attack on his next physically critical day.

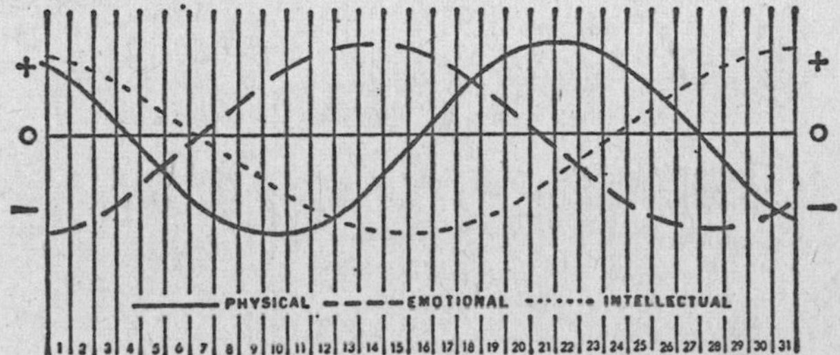

Figure 3. Marilyn Monroe was found dead of an overdose of sleeping pills on August 5, 1962, when she was physically critical and emotionally low.

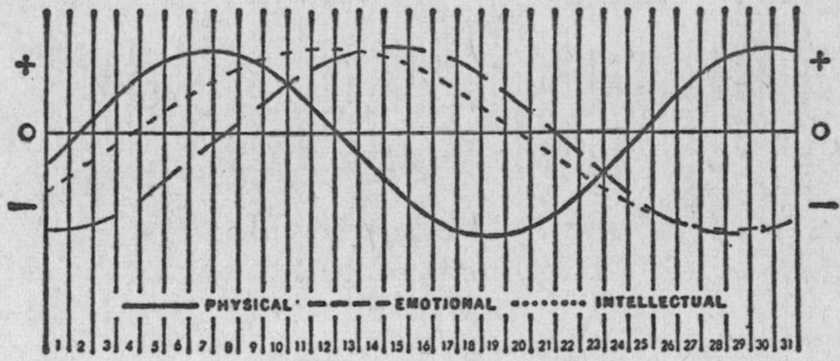

Figure 4. Judy Garland died of a drug overdose on the night of June 21, 1969, when she was emotionally critical, with both physical and intellectual rhythms in low.

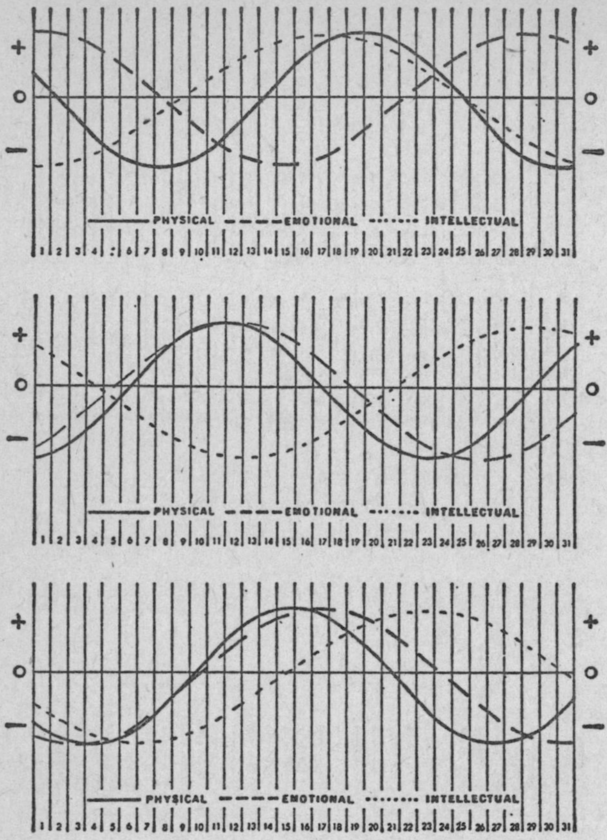

Figures 5, 6, and 7. On December 1, 1974, TWA Flight #514 crashed into a low mountain outside Dulles Airport in Washington, D.C. Flight Controller Merle W. Dameron was physically critical on that day, which may have made him slow to warn the flight crew of its dangerous descent. Captain Richard Brock was physically and emotionally negative, while Co-Pilot Leonard Kreschek had just recovered from a double-critical and was low in all three rhythms.

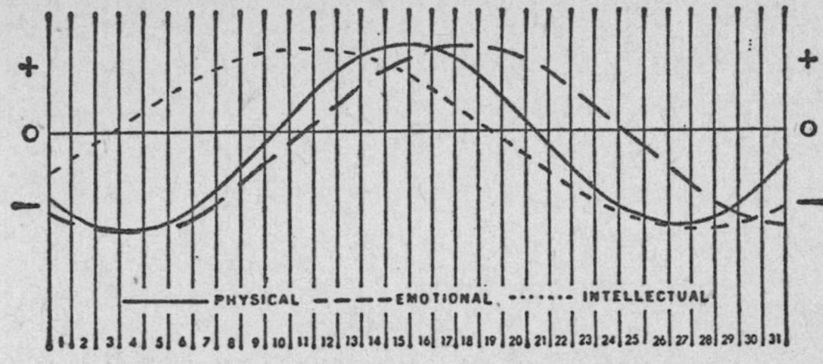

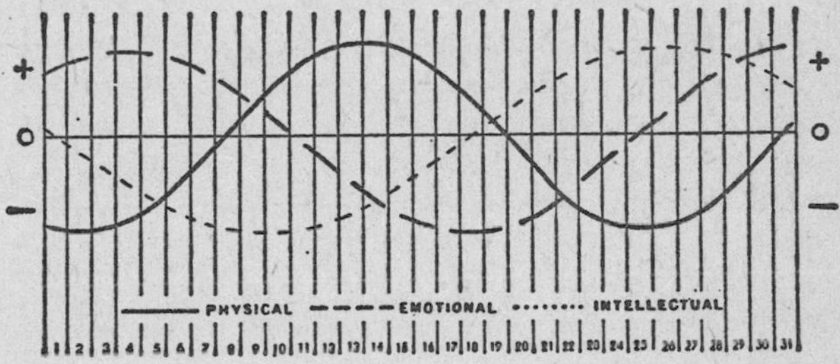

Figures 8 and 9. On January 25, 1975, a private plane crashed into the radio mast of American University in Washington, D.C. Both pilot Richard N. White and Co-Pilot Robert D. Hatem were emotionally critical on that day.

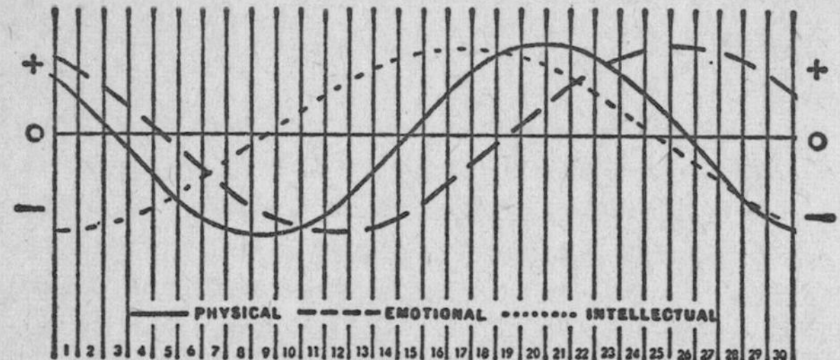

Figure 10. On November 3, 1973, a Pan American cargo plane crashed at Logan Airport in Boston, killing all three crew members. Davis Melvin, the Flight Engineer, was physically critical, as was Gene Ritter, the Co-Pilot.

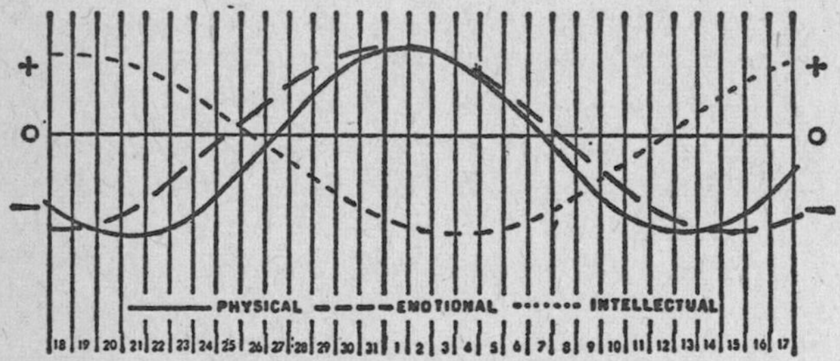

Figure 11. Mark Spitz won his record-breaking seven Olympic Gold Medals on days that were highly favorable in his physical and emotional rhythms.

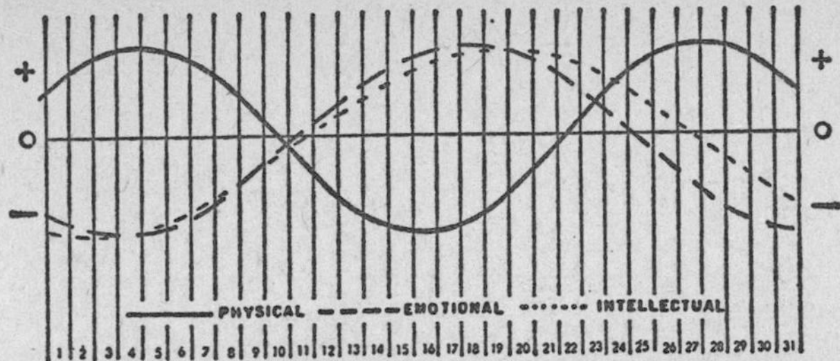

Figure 12. On December 10, 1972, Franco Harris of the Pittsburgh Steelers failed to gain 100 yards for the first time in seven weeks. He was suffering from a triple critical on that day. On December 23, less than two weeks later, he made the "Catch of the Century." On that day, he was in a triple positive phase.

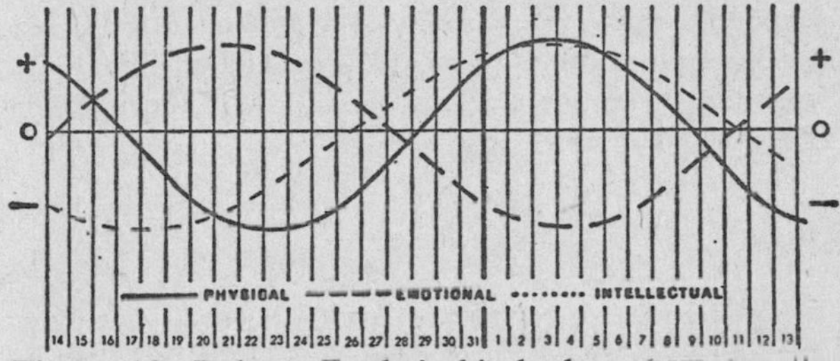

Figure 13. Robert Goulet's biorhythm chart for the month during which he kept a "blind diary," recording his feelings and condition without prior knowledge of his biorhythms for the month. Most of the month was spent on vacation, and Goulet was usually elated and relaxed. But on July 17, a physically critical day, during which he attended a birthday party for Diahann Carroll, he wrote, "Pleasant, but for some unknown reason, I'm sad. Ten years from now, most of us will not be here." On August 9, another physical critical, he wrote, "Didn't sleep much last night. Tired today." On August 11, a double critical in his intellectual and emotional rhythms, Goulet recorded no depression. However, the next day he wrote, "Have several fever blisters. Soon we shall all be dead."

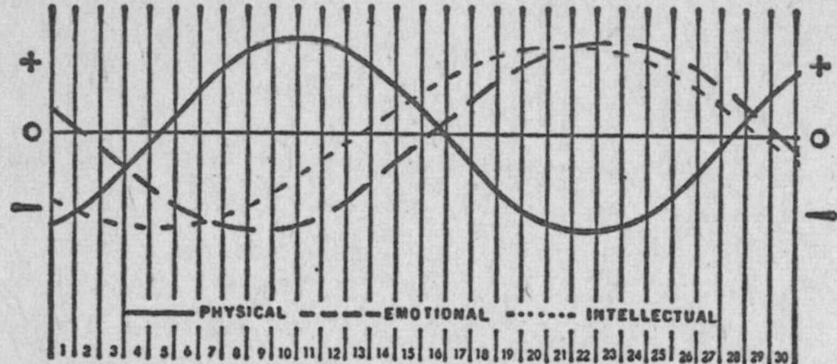

Figure 14. Gerald Ford's decision to grant Richard M. Nixon an unconditional pardon damaged Ford's presidency in the eyes of many. Ford made the decision on September 8, 1974, when he was intellectually and emotionally negative and may have been suffering from impaired judgment.

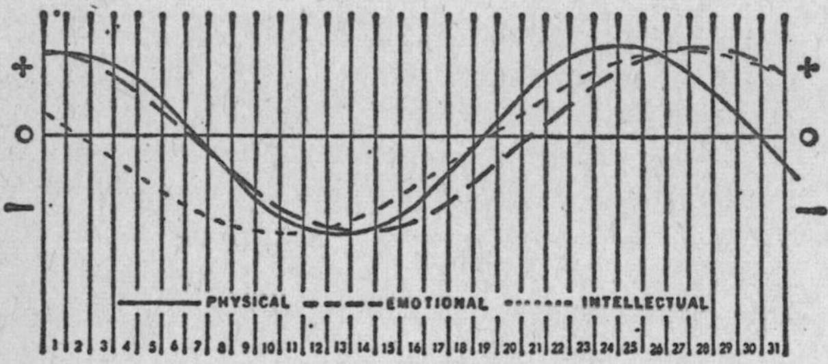

Figure 15. On the night of July 18, 1968, Senator Edward M. Kennedy became involved in the scandal knows as Chappaquiddick. His bizarre behavior can be explained by looking at his biorhythm chart, which shows him to have been physically and intellectually critical on that night.

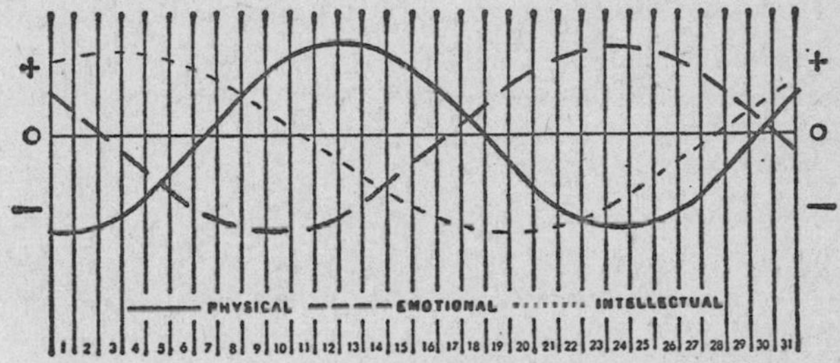

Figure 16. On October 6, 1973, President Anwar Sadat of Egypt made the decision to declare war on Israel. He was physically critical at the time.

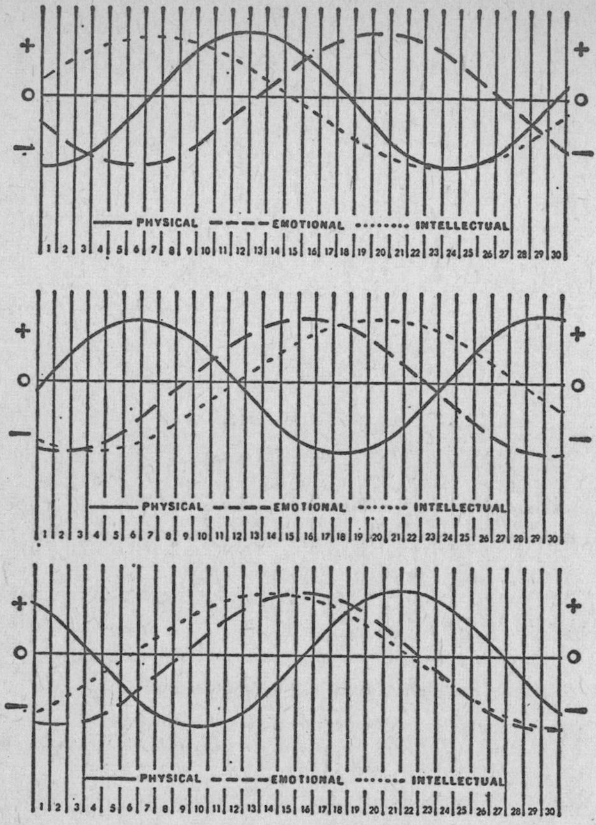

Figures 17, 18, and 19. Assassination by the mentally unstable has become a factor in American political life, and biorhythm may have something to do with it. On November 22, 1963, Lee Harvey Oswald was at the bottom of his physical and intellectual rhythms when he shot President John F. Kennedy. The next day, Oswald was killed by Jack Ruby, who was undergoing an emotionally critical day. When Sirhan Sirhan assassinated Robert F. Kennedy on June 5, 1968, he was intellectually critical and low in both his physical and emotional rhythms.

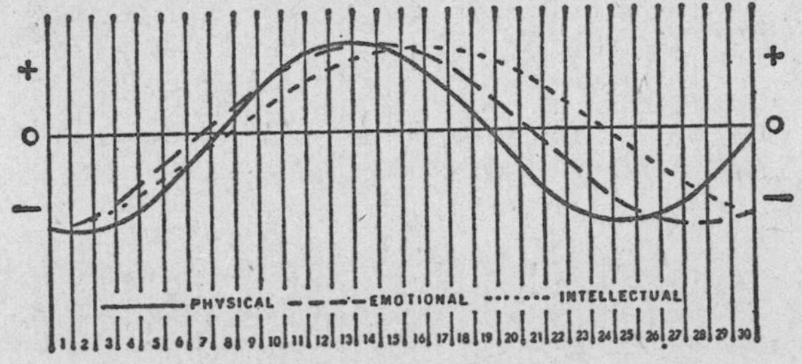

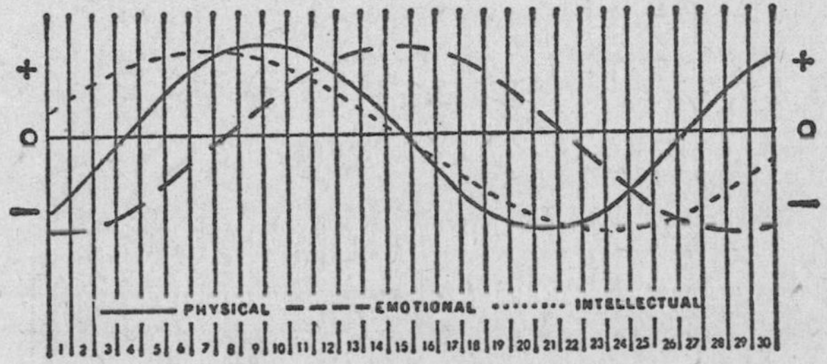

Figures 20 and 21. On June 7, 1971, an Allegheny Airlines flight crashed, killing 28 people. Co-Pilot James A. Walker later reported that Pilot George Eastridge "was not his usual self. I tried to take over the controls." As it turned out, Eastridge was intellectually critical and only a few hours away from a double critical in his emotional and physical rhythms. Walker was in little better condition since his emotional rhythm was critical.

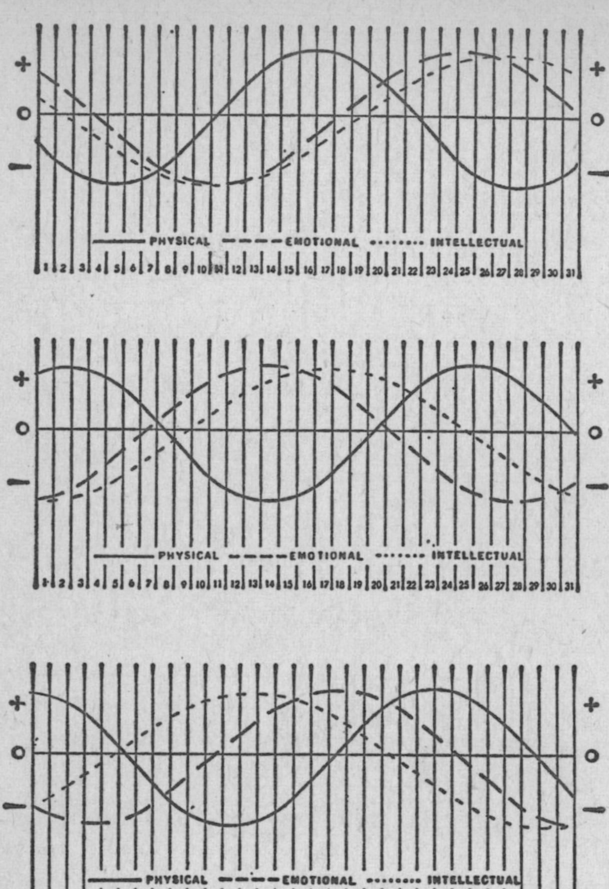

Figures 22, 23, and 24. Due to pilot error, a United Airlines flight crashed in Chicago on December 8, 1972, killing 43 passengers. Captain V. L. Whitehouse was low in all three rhythms; Co-Pilot W. O. Coble was undergoing a double critical in the physical and intellectual rhythms; and Flight Engineer B. J. Elder was low physically and emotionally.

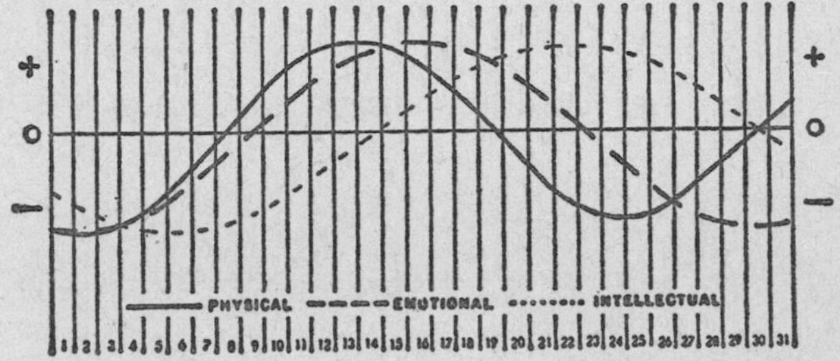

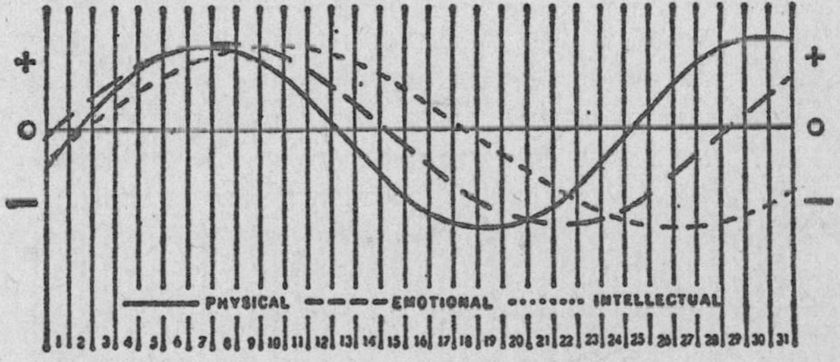

Figures 25 and 26. On the night of December 28, 1972, an Eastern Airlines flight crashed into the Florida Everglades because the crew had misjudged the plane's altitude. At the time, Captain Robert A. Loft was emotionally and physically low and only one day away from a double critical in the physical and intellectual rhythms. Co-Pilot A. J. Stockstill was emotionally critical and intellectually negative.

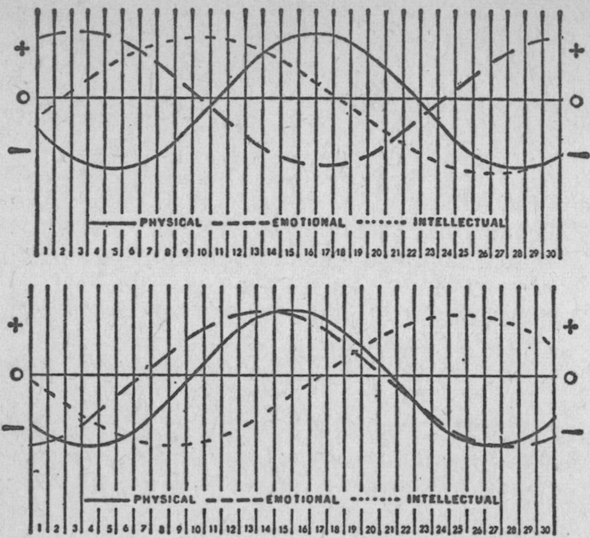

Figures 27 and 28. United Airlines Flight #475 landed at the wrong airport on June 10, 1973. Captain James Bosse was physically and emotionally critical and had abandoned his post to try to calm an excited passenger. Co-Pilot Charles Melbourne, who made the landing, was physically critical and intellectually negative.

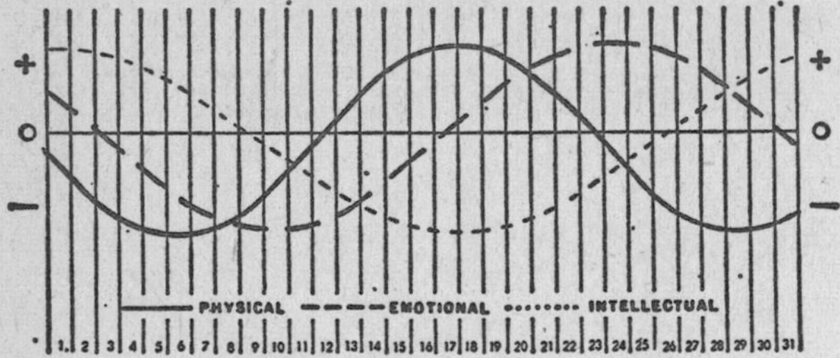

Figure 29. Captain Arvid Linke, who crashed an Ozark Airlines flight on the night of July 23, 1973, was physically critical and only one day away from an intellectually critical day.

INSTRUCTIONS FOR UNDERSTANDING COMPATIBILITY CHARTS

The charts on pages 159-169 will enable you to calculate your physical, emotional and intellectual compatibilities with anyone you choose—friends, lovers or mere acquaintances. Simply by knowing your birthdate and that of a given individual, you can see what percent of the time you are physically in tune, emotionally on the same level and intellectually similar. You may finally come to understand why she's not in the mood when you are. Why your brother or sister constantly gets on your nerves. Why you never see eye to eye with your best friend on any topic of discussion. The following instructions will show you how to compare compatibilities and, perhaps, open new social avenues for you.

1. Look up your biorhythm code in the birth charts provided in Appendix A.
2. Do likewise for the person in question.
3. On the physical compatibility chart, find your physical code letter descending vertically and match that with the person in question's code letter moving horizontally. (Example: If you were born on January 1st, 1950, your code is O-W-2. If your friend, lover, etc. was born on January 1st, 1951, his/her code is L-X-4. Looking at the physical compatibility chart, find O along the vertical axis. Move your finger horizontally from O until you reach the column with letter L at the top. The point where O and L intersect reads 74% or 74% physical compatibility. This means that 74% of the time you and the person in question are physically high together or physically low together. 26% of the time you are high while the other is low and vice versa.)

4. Follow the same procedure to attain compatibility percentages for your emotional and intellectual cycles. If you are emotionally compatible with someone, you and that person will experience good moods and bad moods during the same times. If you are not emotionally compatible, you may be crabby while the other is happy and vice versa. If you are intellectually compatible, you both will feel like attending a lecture or engaging in a serious discussion at the same times. If, however, you are not intellectually compatible, you will find that you will be seriously trying to concentrate or get a point across while the other is flighty and uncommunicative—and, of course, vice versa.

Preliminary research on biorhythmic compatibilities has found that:

1. People generally get along best when they have high compatibility percentages. That is, they are high at the same times and low at the same times.

2. People with divergent compatibilities can get along provided that each is willing and able to acquiesce to the will of the other as confrontations develop.

3. In a minority of cases, individuals actually prefer opposites and will get along quite well with someone having drastically different biorhythm cycles.

TRANSLATING BIORHYTHM PERCENTAGES

0 to 19%—No or extremely little compatibility.
20 to 39%—Sometimes compatible.
40 to 59%—Half the time compatible.
60 to 79%—Most of the time compatible.
80 to 100%—Always or almost always compatible.

PHYSICAL

	A	B	C	D	E	F	G	H	J
A	100	91	83	74	65	57	48	39	30
B	91	100	91	83	74	65	57	48	39
C	83	91	100	91	83	74	65	57	48
D	74	83	91	100	91	83	74	65	57
E	65	74	83	91	100	91	83	74	65
F	57	65	74	83	91	100	91	83	74
G	48	57	65	74	83	91	100	91	83
H	39	48	57	65	74	83	91	100	91
J	30	39	48	57	65	74	83	91	100
K	22	30	39	48	57	65	74	83	91
L	13	22	30	39	48	57	65	74	83
M	4	13	22	30	39	48	57	65	74
N	4	4	13	22	30	39	48	57	65
O	13	4	4	13	22	30	39	48	57
P	22	13	4	4	13	22	30	39	48
Q	30	22	13	4	4	13	22	30	39
R	39	30	22	13	4	4	13	22	30
S	48	39	30	22	13	4	4	13	22
T	57	48	39	30	22	13	4	4	13
U	65	57	48	39	30	22	13	4	4
V	74	65	57	48	39	30	22	13	4
W	83	74	65	57	48	39	30	22	13
X	91	83	74	65	57	48	39	30	22

PHYSICAL

	K	L	M	N	O	P	Q	R	S
A	22	13	4	4	13	22	30	39	48
B	30	22	13	4	4	13	22	30	39
C	39	30	22	13	4	4	13	22	30
D	48	39	30	22	13	4	4	13	22
E	57	48	39	30	22	13	4	4	13
F	65	57	48	39	30	22	13	4	4
G	74	65	57	48	39	30	22	13	4
H	83	74	65	57	48	39	30	22	13
J	91	83	74	65	57	48	39	30	22
K	100	91	83	74	65	57	48	39	30
L	91	100	91	83	74	65	57	48	39
M	83	91	100	91	83	74	65	57	48
N	74	83	91	100	91	83	74	65	57
O	65	74	83	91	100	91	83	74	65
P	57	65	74	83	91	100	91	83	74
Q	48	57	65	74	83	91	100	91	83
R	39	48	57	65	74	83	91	100	91
S	30	39	48	57	65	74	83	91	100
T	22	30	39	48	57	65	74	83	91
U	13	22	30	39	48	57	65	74	83
V	4	13	22	30	39	48	57	65	74
W	4	4	13	22	30	39	48	57	65
X	13	4	4	13	22	30	39	48	57

PHYSICAL

	T	U	V	W	X
A	57	65	74	83	91
B	48	57	65	74	83
C	39	48	57	65	74
D	30	39	48	57	65
E	22	30	39	48	57
F	13	22	30	39	48
G	4	13	22	30	39
H	4	4	13	22	30
J	13	4	4	13	22
K	22	13	4	4	13
L	30	22	13	4	4
M	39	30	22	13	4
N	48	39	30	22	13
O	57	48	39	30	22
P	65	57	48	39	30
Q	74	65	57	48	39
R	83	74	65	57	48
S	91	83	74	65	57
T	100	91	83	74	65
U	91	100	91	83	74
V	83	91	100	91	83
W	74	83	91	100	91
X	65	74	83	91	100

EMOTIONAL

	A	B	C	D	E	F	G
A	100	93	86	78	71	64	57
B	93	100	93	86	79	71	64
C	86	93	100	93	86	79	71
D	79	86	93	100	93	86	79
E	71	79	86	93	100	93	86
F	64	71	79	86	93	100	93
G	57	64	71	79	86	93	100
H	50	57	64	71	79	86	93
J	43	50	57	64	71	79	86
K	36	43	50	57	64	71	79
L	29	36	43	50	57	64	71
M	21	29	36	43	50	57	64
N	14	21	29	36	43	50	57
O	7	14	21	29	36	43	50
P	0	7	14	21	29	36	43
Q	7	0	7	14	21	29	36
R	14	7	0	7	14	21	29
S	21	14	7	0	7	14	21
T	29	21	14	7	0	7	14
U	36	29	21	14	7	0	7
V	43	36	29	21	14	7	0
W	50	43	36	29	21	14	7
X	57	50	43	36	29	21	14
Y	64	57	50	43	36	29	21
Z	71	64	57	50	43	36	29
1	79	71	64	57	50	43	36
2	86	79	71	64	57	50	43
3	93	86	79	71	64	57	50

EMOTIONAL

	H	J	K	L	M	N	O
A	50	43	36	29	21	14	7
B	57	50	43	36	29	21	14
C	64	57	50	43	36	29	21
D	71	64	57	50	43	36	29
E	79	71	64	57	50	43	36
F	86	79	71	64	57	50	43
G	93	86	79	71	64	57	50
H	100	93	86	79	71	64	57
J	93	100	93	86	79	71	64
K	86	93	100	93	86	79	71
L	79	86	93	100	93	86	79
M	71	79	86	93	100	93	86
N	64	71	79	86	93	100	93
O	57	64	71	79	86	93	100
P	50	57	64	71	79	86	93
Q	43	50	57	64	71	79	86
R	36	43	50	57	64	71	79
S	29	36	43	50	57	64	71
T	21	29	36	43	50	57	64
U	14	21	29	36	43	50	57
V	7	14	21	29	36	43	50
W	0	7	14	21	29	36	43
X	7	0	7	14	21	29	36
Y	14	7	0	7	14	21	29
Z	21	14	7	0	7	14	21
1	29	21	14	7	0	7	14
2	36	29	21	14	7	0	7
3	43	36	29	21	14	7	0

EMOTIONAL

	P	Q	R	S	T	U	V
A	0	7	14	21	29	36	43
B	7	0	7	14	21	29	36
C	14	7	0	7	14	21	29
D	21	14	7	0	7	14	21
E	29	21	14	7	0	7	14
F	36	29	21	14	7	0	7
G	43	36	29	21	14	7	0
H	50	43	36	29	21	14	7
J	57	50	43	36	29	21	14
K	64	57	50	43	36	29	21
L	71	64	57	50	43	36	29
M	79	71	64	57	50	43	36
N	86	79	71	64	57	50	43
O	93	86	79	71	64	57	50
P	100	93	86	79	71	64	57
Q	93	100	93	86	79	71	64
R	86	93	100	93	86	79	71
S	79	86	93	100	93	86	79
T	71	79	86	93	100	93	86
U	64	71	79	86	93	100	93
V	57	64	71	79	86	93	100
W	50	57	64	71	79	86	93
X	43	50	57	64	71	79	86
Y	36	43	50	57	64	71	79
Z	29	36	43	50	57	64	71
1	21	29	36	43	50	57	64
2	14	21	29	36	43	50	57
3	7	14	21	29	36	43	50

EMOTIONAL

	W	X	Y	Z	1	2	3
A	50	57	64	71	79	86	93
B	43	50	57	64	71	79	86
C	36	43	50	57	64	71	79
D	29	36	43	50	57	64	71
E	21	29	36	43	50	57	64
F	14	21	29	36	43	50	57
G	7	14	21	29	36	43	50
H	0	7	14	21	29	36	43
J	7	0	7	14	21	29	36
K	14	7	0	7	14	21	29
L	21	14	7	0	7	14	21
M	29	21	14	7	0	7	14
N	36	29	21	14	7	0	7
O	43	36	29	21	14	7	0
P	50	43	36	29	21	14	7
Q	57	50	43	36	29	21	14
R	64	57	50	43	36	29	21
S	71	64	57	50	43	36	29
T	79	71	64	57	50	43	36
U	86	79	71	64	57	50	43
V	93	86	79	71	64	57	50
W	100	93	86	79	71	64	57
X	93	100	93	86	79	71	64
Y	86	93	100	93	86	79	71
Z	79	86	93	100	93	86	79
1	71	79	86	93	100	93	86
2	64	71	79	86	93	100	93
3	57	64	71	79	86	93	100

INTELLECTUAL

	A	B	C	D	E	F	G	H	J
A	100	94	88	82	76	70	64	58	52
B	94	100	94	88	82	76	70	64	58
C	88	94	100	94	88	82	76	70	64
D	82	88	94	100	94	88	82	76	70
E	76	82	88	94	100	94	88	82	76
F	70	76	82	88	94	100	94	88	82
G	64	70	76	82	88	94	100	94	88
H	58	64	70	76	82	88	94	100	94
J	52	58	64	70	76	82	88	94	100
K	46	52	58	64	70	76	82	88	94
L	39	46	52	58	64	70	76	82	88
M	33	39	46	52	58	64	70	76	82
N	27	33	39	46	52	58	64	70	76
O	21	27	33	39	46	52	58	64	70
P	15	21	27	33	39	46	52	58	64
Q	9	15	21	27	33	39	46	52	58
R	3	9	15	21	27	33	39	46	52
S	3	3	9	15	21	27	33	39	46
T	9	3	3	9	15	21	27	33	39
U	15	9	3	3	9	15	21	27	33
V	21	15	9	3	3	9	15	21	27
W	27	21	15	9	3	3	9	15	21
X	33	27	21	15	9	3	3	9	15
Y	39	33	27	21	15	9	3	3	9
Z	46	39	33	27	21	15	9	3	3
1	52	46	39	33	27	21	15	9	3
2	58	52	46	39	33	27	21	15	9
3	64	58	52	46	39	33	27	21	15
4	70	64	58	52	46	39	33	27	21
5	76	70	64	58	52	46	39	33	27
6	82	76	70	64	58	52	46	39	33
7	88	82	76	70	64	58	52	46	39
8	94	88	82	76	70	64	58	52	46

INTELLECTUAL

	K	L	M	N	O	P	Q	R	S
A	46	39	33	27	21	15	9	3	3
B	52	46	39	33	27	21	15	9	3
C	58	52	46	39	33	27	21	15	9
D	64	58	52	46	39	33	27	21	15
E	70	64	58	52	46	39	33	27	21
F	76	70	64	58	52	46	39	33	27
G	82	76	70	64	58	52	46	39	33
H	88	82	76	70	64	58	52	46	39
J	94	88	82	76	70	64	58	52	46
K	100	94	88	82	76	70	64	58	52
L	94	100	94	88	82	76	70	64	58
M	88	94	100	94	88	82	76	70	64
N	82	88	94	100	94	88	82	76	70
O	76	82	88	94	100	94	88	82	76
P	70	76	82	88	94	100	94	88	82
Q	64	70	76	82	88	94	100	94	88
R	58	64	70	76	82	88	94	100	94
S	52	58	64	70	76	82	88	94	100
T	46	52	58	64	70	76	82	88	94
U	39	46	52	58	64	70	76	82	88
V	33	39	46	52	58	64	70	76	82
W	27	33	39	46	52	58	64	70	76
X	21	27	33	39	46	52	58	64	70
Y	15	21	27	33	39	46	52	58	64
Z	9	15	21	27	33	39	46	52	58
1	3	9	15	21	27	33	39	46	52
2	3	3	9	15	21	27	33	39	46
3	9	3	3	9	15	21	27	33	39
4	15	9	3	3	9	15	21	27	33
5	21	15	9	3	3	9	15	21	27
6	27	21	15	9	3	3	9	15	21
7	83	27	21	15	9	3	3	9	15
8	39	33	27	21	15	9	3	3	9

INTELLECTUAL

	T	U	V	W	X	Y	Z	1	2
A	9	15	21	27	33	39	46	52	58
B	3	9	15	21	27	33	39	46	52
C	3	3	9	15	21	27	33	39	46
D	9	3	3	9	15	21	27	33	39
E	15	9	3	3	9	15	21	27	33
F	21	15	9	3	3	9	15	21	27
G	27	21	15	9	3	3	9	15	21
H	33	27	21	15	9	3	3	9	15
J	39	33	27	21	15	9	3	3	9
K	46	39	33	27	21	15	9	3	3
L	52	46	39	33	27	21	15	9	3
M	58	52	46	39	33	27	21	15	9
N	64	58	52	46	39	33	27	21	15
O	70	64	58	52	46	39	33	27	21
P	76	70	64	58	52	46	39	33	27
Q	82	76	70	64	58	52	46	39	33
R	88	82	76	70	64	58	52	46	39
S	94	88	82	76	70	64	58	52	46
T	100	94	88	82	76	70	64	58	52
U	94	100	94	88	82	76	70	64	58
V	88	94	100	94	88	82	76	70	64
W	82	88	94	100	94	88	82	76	70
X	76	82	88	94	100	94	88	82	76
Y	70	76	82	88	94	100	94	88	82
Z	64	70	76	82	88	94	100	94	88
1	58	64	70	76	82	88	94	100	94
2	52	58	64	70	76	82	88	94	100
3	46	52	58	64	70	76	82	88	94
4	39	46	52	58	64	70	76	82	88
5	33	39	46	52	58	64	70	76	82
6	27	33	39	46	52	58	64	70	76
7	21	27	33	39	46	52	58	64	70
8	15	21	27	33	39	46	52	58	64

INTELLECTUAL

	3	4	5	6	7	8
A	64	70	76	82	88	94
B	58	64	70	76	82	88
C	52	58	64	70	76	82
D	46	52	58	64	70	76
E	39	46	52	58	64	70
F	33	39	46	52	58	64
G	27	33	39	46	52	58
H	21	27	33	39	46	52
J	15	21	27	33	39	46
K	9	15	21	27	33	39
L	3	9	15	21	27	33
M	3	3	9	15	21	27
N	9	3	3	9	15	21
O	15	9	3	3	9	15
P	21	15	9	3	3	9
O	27	21	15	9	3	3
R	33	27	21	15	9	3
S	39	33	27	21	15	9
T	46	39	33	27	21	15
U	52	46	39	33	27	21
V	58	52	46	39	33	27
W	64	58	52	46	39	33
X	70	64	58	52	46	39
Y	76	70	64	58	52	46
Z	82	76	70	64	58	52
1	88	82	76	70	64	58
2	94	88	82	76	70	64
3	100	94	88	82	76	70
4	94	100	94	88	82	76
5	88	94	100	94	88	82
6	82	88	94	100	94	88
7	76	82	88	94	100	94
8	70	76	82	88	94	100

Appendix A
Birth Charts

HOW TO CHART YOUR OWN BIORHYTHM CURVES

Look up the date of your birth in the Birth Charts (pages 172–333). Following it you will find three symbols, such as "J, V, Y." The first symbol refers to the physical rhythm, the second to the emotional, and the third to the intellectual. Then turn to the Biorhythm Charts (pages 336–407), where you will find each month of the years 1976 to 1978 duplicated three times—once for each biorhythm—and numbered to correspond to the days of that month. Do not be confused by the swarms of symbols that appear within the columns; simply refer to the first symbol you found under your birth date, locate the same symbol in the monthly table for the physical rhythm, and draw a line passing through your physical symbol. The result will be an accurate sine wave chart of your physical rhythm. Repeat the process for the monthly emotional and intellectual rhythm charts—drawing a line through your second and third symbols respectively—and you will have a complete biorhythm chart in three parts.

1895 1895

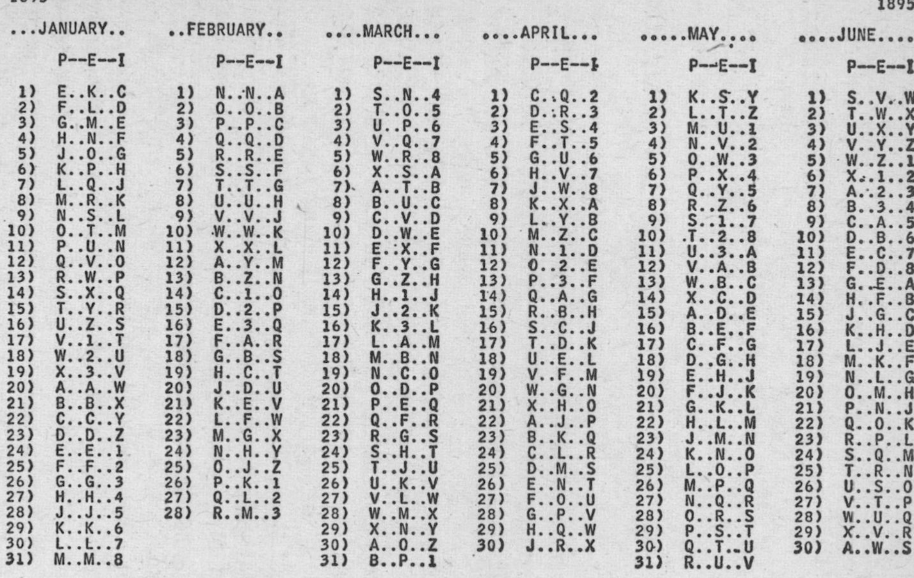

	...JANUARY..	..FEBRUARY..	....MARCH...	....APRIL...	.....MAY....	....JUNE....
	P--E--I	P--E--I	P--E--I	P--E--I	P--E--I	P--E--I
1)	E..K..C	N..N..A	S..N..4	C..Q..2	K..S..Y	S..V..W
2)	F..L..D	O..O..B	T..O..5	D..R..3	L..T..Z	T..W..X
3)	G..M..E	P..P..C	U..P..6	E..S..4	M..U..1	U..X..Y
4)	H..N..F	Q..Q..D	V..Q..7	F..T..5	N..V..2	V..Y..Z
5)	J..O..G	R..R..E	W..R..8	G..U..6	O..W..3	W..Z..1
6)	K..P..H	S..S..F	X..S..A	H..V..7	P..X..4	X..1..2
7)	L..Q..J	T..T..G	A..T..B	J..W..8	Q..Y..5	A..2..3
8)	M..R..K	U..U..H	B..U..C	K..X..A	R..Z..6	B..3..4
9)	N..S..L	V..V..J	C..V..D	L..Y..B	S..1..7	C..A..5
10)	O..T..M	W..W..K	D..W..E	M..Z..C	T..2..8	D..B..6
11)	P..U..N	X..X..L	E..X..F	N..1..D	U..3..A	E..C..7
12)	Q..V..O	A..Y..M	F..Y..G	O..2..E	V..A..B	F..D..8
13)	R..W..P	B..Z..N	G..Z..H	P..3..F	W..B..C	G..E..A
14)	S..X..Q	C..1..O	H..1..J	Q..A..G	X..C..D	H..F..B
15)	T..Y..R	D..2..P	J..2..K	R..B..H	A..D..E	J..G..C
16)	U..Z..S	E..3..Q	K..3..L	S..C..J	B..E..F	K..H..D
17)	V..1..T	F..A..R	L..A..M	T..D..K	C..F..G	L..J..E
18)	W..2..U	G..B..S	M..B..N	U..E..L	D..G..H	M..K..F
19)	X..3..V	H..C..T	N..C..O	V..F..M	E..H..J	N..L..G
20)	A..A..W	J..D..U	O..D..P	W..G..N	F..J..K	O..M..H
21)	B..B..X	K..E..V	P..E..Q	X..H..O	G..K..L	P..N..J
22)	C..C..Y	L..F..W	Q..F..R	A..J..P	H..L..M	Q..O..K
23)	D..D..Z	M..G..X	R..G..S	B..K..Q	J..M..N	R..P..L
24)	E..E..1	N..H..Y	S..H..T	C..L..R	K..N..O	S..Q..M
25)	F..F..2	O..J..Z	T..J..U	D..M..S	L..O..P	T..R..N
26)	G..G..3	P..K..1	U..K..V	E..N..T	M..P..Q	U..S..O
27)	H..H..4	Q..L..2	V..L..W	F..O..U	N..Q..R	V..T..P
28)	J..J..5	R..M..3	W..M..X	G..P..V	O..R..S	W..U..Q
29)	K..K..6		X..N..Y	H..Q..W	P..S..T	X..V..R
30)	L..L..7		A..O..Z	J..R..X	Q..T..U	A..W..S
31)	M..M..8		B..P..1		R..U..V	

1895						1895
	....JULY....	...AUGUST...	..SEPTEMBER.	..OCTOBER...	..NOVEMBER..	..DECEMBER..
	P--E--I	P--E--I	P--E--I	P--E--I	P--E--I	P--E--I
1)	B..X..T	K..1..R	S..A..P	B..C..M	K..F..K	R..H..G
2)	C..Y..U	L..2..S	T..B..Q	C..D..N	L..G..L	S..J..H
3)	D..Z..V	M..3..T	U..C..R	D..E..O	M..H..M	T..K..J
4)	E..1..W	N..A..U	V..D..S	E..F..P	N..J..N	U..L..K
5)	F..2..X	O..B..V	W..E..T	F..G..Q	O..K..O	V..M..L
6)	G..3..Y	P..C..W	X..F..U	G..H..R	P..L..P	W..N..M
7)	H..A..Z	Q..D..X	A..G..V	H..J..S	Q..M..Q	X..O..N
8)	J..B..1	R..E..Y	B..H..W	J..K..T	R..N..R	A..P..O
9)	K..C..2	S..F..Z	C..J..X	K..L..U	S..O..S	B..Q..P
10)	L..D..3	T..G..1	D..K..Y	L..M..V	T..P..T	C..R..Q
11)	M..E..4	U..H..2	E..L..Z	M..N..W	U..Q..U	D..S..R
12)	N..F..5	V..J..3	F..M..1	N..O..X	V..R..V	E..T..S
13)	O..G..6	W..K..4	G..N..2	O..P..Y	W..S..W	F..U..T
14)	P..H..7	X..L..5	H..O..3	P..Q..Z	X..T..X	G..V..U
15)	Q..J..8	A..M..6	J..P..4	Q..R..1	A..U..Y	H..W..V
16)	R..K..A	B..N..7	K..Q..5	R..S..2	B..V..Z	J..X..W
17)	S..L..B	C..O..8	L..R..6	S..T..3	C..W..1	K..Y..X
18)	T..M..C	D..P..A	M..S..7	T..U..4	D..X..2	L..Z..Y
19)	U..N..D	E..Q..B	N..T..8	U..V..5	E..Y..3	M..1..Z
20)	V..O..E	F..R..C	O..U..A	V..W..6	F..Z..4	N..2..1
21)	W..P..F	G..S..D	P..V..B	W..X..7	G..1..5	O..3..2
22)	X..Q..G	H..T..E	Q..W..C	X..Y..8	H..2..6	P..A..3
23)	A..R..H	J..U..F	R..X..D	A..Z..A	J..3..7	Q..B..4
24)	B..S..J	K..V..G	S..Y..E	B..1..B	K..A..8	R..C..5
25)	C..T..K	L..W..H	T..Z..F	C..2..C	L..B..A	S..D..6
26)	D..U..L	M..X..J	U..1..G	D..3..D	M..C..B	T..E..7
27)	E..V..M	N..Y..K	V..2..H	E..A..E	N..D..C	U..F..8
28)	F..W..N	O..Z..L	W..3..J	F..B..F	O..E..D	V..G..A
29)	G..X..O	P..1..M	X..A..K	G..C..G	P..F..E	W..H..B
30)	H..Y..P	Q..2..N	A..B..L	H..D..H	Q..G..F	X..J..C
31)	J..Z..Q	R..3..O		J..E..J		A..K..D

CODES: P-PHYSICAL BIORHYTHM CURVE,E-EMOTIONAL EIORHYTHM CURVE,I-INTELLECTUAL BIORHYTHM CURVE

1896 1896

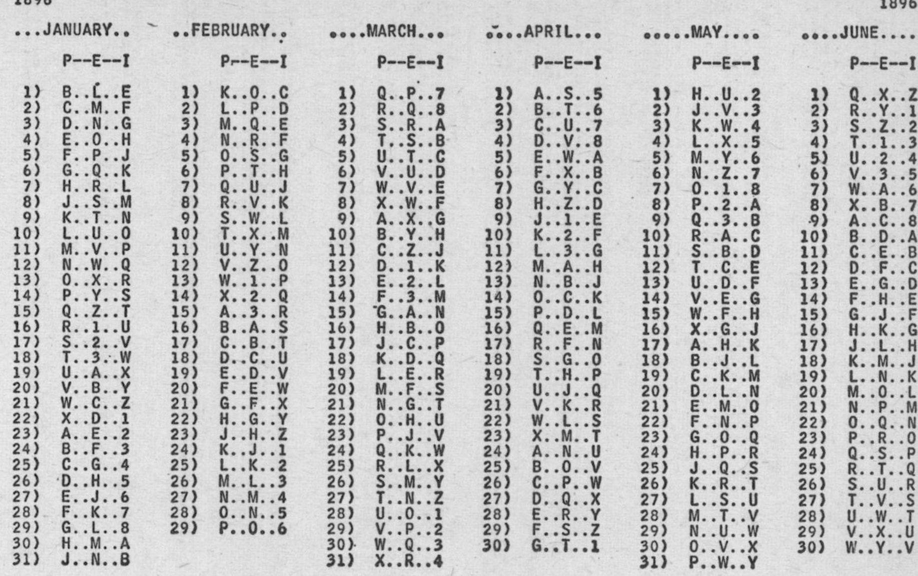

	...JANUARY..	..FEBRUARY..	....MARCH...	....APRIL...	.....MAY....	....JUNE....
	P--E--I	P--E--I	P--E--I	P--E--I	P--E--I	P--E--I
1)	B..L..E	K..O..C	Q..P..7	A..S..5	H..U..2	Q..X..Z
2)	C..M..F	L..P..D	R..Q..8	B..T..6	J..V..3	R..Y..1
3)	D..N..G	M..Q..E	S..R..A	C..U..7	K..W..4	S..Z..2
4)	E..O..H	N..R..F	T..S..B	D..V..8	L..X..5	T..1..3
5)	F..P..J	O..S..G	U..T..C	E..W..A	M..Y..6	U..2..4
6)	G..Q..K	P..T..H	V..U..D	F..X..B	N..Z..7	V..3..5
7)	H..R..L	Q..U..J	W..V..E	G..Y..C	O..1..8	W..A..6
8)	J..S..M	R..V..K	X..W..F	H..Z..D	P..2..A	X..B..7
9)	K..T..N	S..W..L	A..X..G	J..1..E	Q..3..B	A..C..8
10)	L..U..O	T..X..M	B..Y..H	K..2..F	R..A..C	B..D..A
11)	M..V..P	U..Y..N	C..Z..J	L..3..G	S..B..D	C..E..B
12)	N..W..Q	V..Z..O	D..1..K	M..A..H	T..C..E	D..F..C
13)	O..X..R	W..1..P	E..2..L	N..B..J	U..D..F	E..G..D
14)	P..Y..S	X..2..Q	F..3..M	O..C..K	V..E..G	F..H..E
15)	Q..Z..T	A..3..R	G..A..N	P..D..L	W..F..H	G..J..F
16)	R..1..U	B..A..S	H..B..O	Q..E..M	X..G..J	H..K..G
17)	S..2..V	C..B..T	J..C..P	R..F..N	A..H..K	J..L..H
18)	T..3..W	D..C..U	K..D..Q	S..G..O	B..J..L	K..M..J
19)	U..A..X	E..D..V	L..E..R	T..H..P	C..K..M	L..N..K
20)	V..B..Y	F..E..W	M..F..S	U..J..Q	D..L..N	M..O..L
21)	W..C..Z	G..F..X	N..G..T	V..K..R	E..M..O	N..P..M
22)	X..D..1	H..G..Y	O..H..U	W..L..S	F..N..P	O..Q..N
23)	A..E..2	J..H..Z	P..J..V	X..M..T	G..O..Q	P..R..O
24)	B..F..3	K..J..1	Q..K..W	A..N..U	H..P..R	Q..S..P
25)	C..G..4	L..K..2	R..L..X	B..O..V	J..Q..S	R..T..Q
26)	D..H..5	M..L..3	S..M..Y	C..P..W	K..R..T	S..U..R
27)	E..J..6	N..M..4	T..N..Z	D..Q..X	L..S..U	T..V..S
28)	F..K..7	O..N..5	U..O..1	E..R..Y	M..T..V	U..W..T
29)	G..L..8	P..O..6	V..P..2	F..S..Z	N..U..W	V..X..U
30)	H..M..A		W..Q..3	G..T..1	O..V..X	W..Y..V
31)	J..N..B		X..R..4		P..W..Y	

1896 1896

	....JULY....	...AUGUST...	..SEPTEMBER.	..OCTOBER...	..NOVEMBER..	..DECEMBER..
	P--E--I	P--E--I	P--E--I	P--E--I	P--E--I	P--E--I
1)	X..Z..W	H..3. U	Q..C..S	X..E..P	H..H..N	P..K..K
2)	A..1..X	J..A. V	R..D..T	A..F..Q	J..J..O	Q..L..L
3)	B..2..Y	K..B. W	S..E..U	B..G..R	K..K..P	R..M..M
4)	C..3..Z	L..C. X	T..F..V	C..H..S	L..L..Q	S..N..N
5)	D..A..1	M..D. Y	U..G..W	D..J..T	M..M..R	T..O..O
6)	E..B..2	N..E..Z	V..H..X	E..K..U	N..N..S	U..P..P
7)	F..C..3	O..F..1	W..J..Y	F..L..V	O..O..T	V..Q..Q
8)	G..D..4	P..G..2	X..K..Z	G..M..W	P..P..U	W..R..R
9)	H..E..5	Q..H..3	A..L..1	H..N..X	Q..Q..V	X..S..S
10)	J..F..6	R..J..4	B..M..2	J..O..Y	R..R..W	A..T..T
11)	K..G..7	S..K..5	C..N..3	K..P..Z	S..S..X	B..U..U
12)	L..H..8	T..L..6	D..O..4	L..Q..1	T..T..Y	C..V..V
13)	M..J..A	U..M..7	E..P..5	M..R..2	U..U..Z	D..W..W
14)	N..K..B	V..N..8	F..Q..6	N..S..3	V..V..1	E..X..X
15)	O..L..C	W..O..A	G..R..7	O..T..4	W..W..2	F..Y..Y
16)	P..M..D	X..P..B	H..S..8	P..U..5	X..X..3	G..Z..Z
17)	Q..N..E	A..Q..C	J..T..A	Q..V..6	A..Y..4	H..1..1
18)	R..O..F	B..R..D	K..U..B	R..W..7	B..Z..5	J..2..2
19)	S..P..G	C..S..E	L..V..C	S..X..8	C..1..6	K..3..3
20)	T..Q..H	D..T..F	M..W..D	T..Y..A	D..2..7	L..A..4
21)	U..R..J	E..U..G	N..X..E	U..Z..B	E..3..8	M..B..5
22)	V..S..K	F..V..H	O..Y..F	V..1..C	F..A..A	N..C..6
23)	W..T..L	G..W..J	P..Z..G	W..2..D	G..B..B	O..D..7
24)	X..U..M	H..X..K	Q..1..H	X..3..E	H..C..C	P..E..8
25)	A..V..N	J..Y..L	R..2..J	A..A..F	J..D..D	Q..F..A
26)	B..W..O	K..Z..M	S..3..K	B..B..G	K..E..E	R..G..B
27)	C..X..P	L..1..N	T..A..L	C..C..H	L..F..F	S..H..C
28)	D..Y..Q	M..2..O	U..B..M	D..D..J	M..G..G	T..J..D
29)	E..Z..R	N..3..P	V..C..N	E..E..K	N..H..H	U..K..E
30)	F..1..S	O..A..Q	W..D..O	F..F..L	O..J..J	V..L..F
31)	G..2..T	P..B..R		G..G..M		W..M..G

CODES: P-PHYSICAL BIORHYTHM CURVE, E-EMOTIONAL BIORHYTHM CURVE, I-INTELLECTUAL BIORHYTHM CURVE

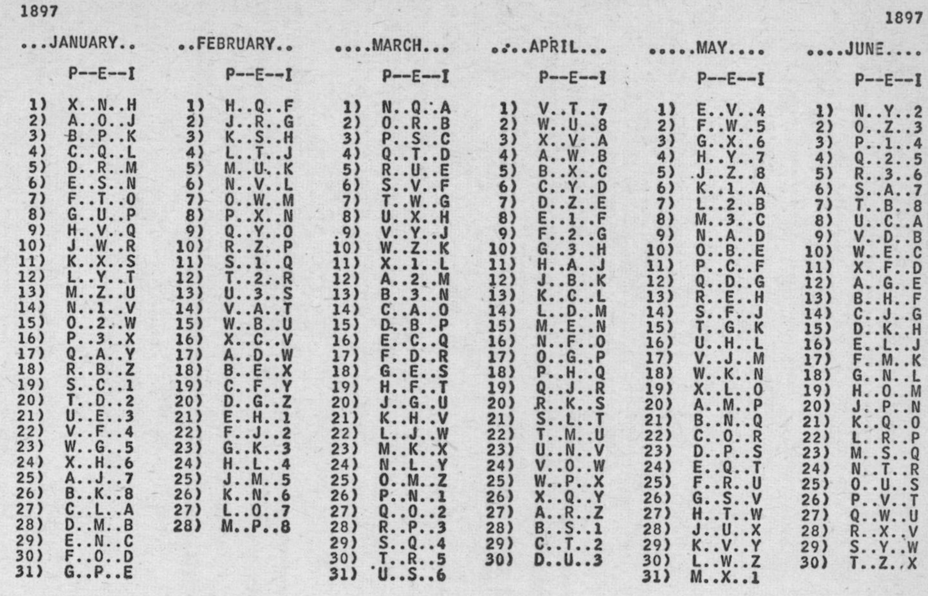

1897 | | | | | | 1897

	...JANUARY..	..FEBRUARY..	....MARCH...	....APRIL...	.....MAY....	....JUNE....
	P--E--I	P--E--I	P--E--I	P--E--I	P--E--I	P--E--I
1)	X..N..H	H..Q..F	N..Q..A	V..T..7	E..V..4	N..Y..2
2)	A..O..J	J..R..G	O..R..B	W..U..8	F..W..5	O..Z..3
3)	B..P..K	K..S..H	P..S..C	X..V..A	G..X..6	P..1..4
4)	C..Q..L	L..T..J	Q..T..D	A..W..B	H..Y..7	Q..2..5
5)	D..R..M	M..U..K	R..U..E	B..X..C	J..Z..8	R..3..6
6)	E..S..N	N..V..L	S..V..F	C..Y..D	K..1..A	S..A..7
7)	F..T..O	O..W..M	T..W..G	D..Z..E	L..2..B	T..B..8
8)	G..U..P	P..X..N	U..X..H	E..1..F	M..3..C	U..C..A
9)	H..V..Q	Q..Y..O	V..Y..J	F..2..G	N..A..D	V..D..B
10)	J..W..R	R..Z..P	W..Z..K	G..3..H	O..B..E	W..E..C
11)	K..X..S	S..1..Q	X..1..L	H..A..J	P..C..F	X..F..D
12)	L..Y..T	T..2..R	A..2..M	J..B..K	Q..D..G	A..G..E
13)	M..Z..U	U..3..S	B..3..N	K..C..L	R..E..H	B..H..F
14)	N..1..V	V..A..T	C..A..O	L..D..M	S..F..J	C..J..G
15)	O..2..W	W..B..U	D..B..P	M..E..N	T..G..K	D..K..H
16)	P..3..X	X..C..V	E..C..Q	N..F..O	U..H..L	E..L..J
17)	Q..A..Y	A..D..W	F..D..R	O..G..P	V..J..M	F..M..K
18)	R..B..Z	B..E..X	G..E..S	P..H..Q	W..K..N	G..N..L
19)	S..C..1	C..F..Y	H..F..T	Q..J..R	X..L..O	H..O..M
20)	T..D..2	D..G..Z	J..G..U	R..K..S	A..M..P	J..P..N
21)	U..E..3	E..H..1	K..H..V	S..L..T	B..N..Q	K..Q..O
22)	V..F..4	F..J..2	L..J..W	T..M..U	C..O..R	L..R..P
23)	W..G..5	G..K..3	M..K..X	U..N..V	D..P..S	M..S..Q
24)	X..H..6	H..L..4	N..L..Y	V..O..W	E..Q..T	N..T..R
25)	A..J..7	J..M..5	O..M..Z	W..P..X	F..R..U	O..U..S
26)	B..K..8	K..N..6	P..N..1	X..Q..Y	G..S..V	P..V..T
27)	C..L..A	L..O..7	Q..O..2	A..R..Z	H..T..W	Q..W..U
28)	D..M..B	M..P..8	R..P..3	B..S..1	J..U..X	R..X..V
29)	E..N..C		S..Q..4	C..T..2	K..V..Y	S..Y..W
30)	F..O..D		T..R..5	D..U..3	L..W..Z	T..Z..X
31)	G..P..E		U..S..6		M..X..1	

1897 | 1897

	....JULY....	...AUGUST...	..SEPTEMBER.	..OCTOBER...	..NOVEMBER..	..DECEMBER..
	P--E--I	P--E--I	P--E--I	P--E--I	P--E--I	P--E--I
1)	U..1..Y	E..A..W	N..D..U	U..F..R	E..J..P	M..L..M
2)	V..2..Z	F..B..X	O..E..V	V..G..S	F..K..Q	N..M..N
3)	W..3..1	G..C..Y	P..F..W	W..H..T	G..L..R	O..N..O
4)	X..A..2	H..D..Z	Q..G..X	X..J..U	H..M..S	P..O..P
5)	A..B..3	J..E..1	R..H..Y	A..K..V	J..N..T	Q..P..Q
6)	B..C..4	K..F..2	S..J..Z	B..L..W	K..O..U	R..Q..R
7)	C..D..5	L..G..3	T..K..1	C..M..X	L..P..V	S..R..S
8)	D..E..6	M..H..4	U..L..2	D..N..Y	M..Q..W	T..S..T
9)	E..F..7	N..J..5	V..M..3	E..O..Z	N..R..X	U..T..U
10)	F..G..8	O..K..6	W..N..4	F..P..1	O..S..Y	V..U..V
11)	G..H..A	P..L..7	X..O..5	G..Q..2	P..T..Z	W..V..W
12)	H..J..B	Q..M..8	A..P..6	H..R..3	Q..U..1	X..W..X
13)	J..K..C	R..N..A	B..Q..7	J..S..4	R..V..2	A..X..Y
14)	K..L..D	S..O..B	C..R..8	K..T..5	S..W..3	B..Y..Z
15)	L..M..E	T..P..C	D..S..A	L..U..6	T..X..4	C..Z..1
16)	M..N..F	U..Q..D	E..T..B	M..V..7	U..Y..5	D..1..2
17)	N..O..G	V..R..E	F..U..C	N..W..8	V..Z..6	E..2..3
18)	O..P..H	W..S..F	G..V..D	O..X..A	W..1..7	F..3..4
19)	P..Q..J	X..T..G	H..W..E	P..Y..B	X..2..8	G..A..5
20)	Q..R..K	A..U..H	J..X..F	Q..Z..C	A..3..A	H..B..6
21)	R..S..L	B..V..J	K..Y..G	R..1..D	B..A..B	J..C..7
22)	S..T..M	C..W..K	L..Z..H	S..2..E	C..B..C	K..D..8
23)	T..U..N	D..X..L	M..1..J	T..3..F	D..C..D	L..E..A
24)	U..V..O	E..Y..M	N..2..K	U..A..G	E..D..E	M..F..B
25)	V..W..P	F..Z..N	O..3..L	V..B..H	F..E..F	N..G..C
26)	W..X..Q	G..1..O	P..A..M	W..C..J	G..F..G	O..H..D
27)	X..Y..R	H..2..P	Q..B..N	X..D..K	H..G..H	P..J..E
28)	A..Z..S	J..3..Q	R..C..O	A..E..L	J..H..J	Q..K..F
29)	B..1..T	K..A..R	S..D..P	B..F..M	K..J..K	R..L..G
30)	C..2..U	L..B..S	T..E..Q	C..G..N	L..K..L	S..M..H
31)	D..3..V	M..C..T		D..H..O		T..N..J

CODES: P-PHYSICAL BIORHYTHM CURVE,E-EMOTIONAL BIORHYTHM CURVE,I-INTELLECTUAL BIORHYTHM CURVE

1898 1898

	...JANUARY..		..FEBRUARY..		....MARCH...		....APRIL...		.....MAY....		....JUNE....
	P--E--I		P--E--I		P--E--I		P--E--I		P--E--I		P--E--I
1)	U..O..K	1)	E..R..H	1)	K..R..C	1)	S..U..A	1)	B..W..6	1)	K..Z..4
2)	V..P..L	2)	F..S..J	2)	L..S..D	2)	T..V..B	2)	C..X..7	2)	L..1..5
3)	W..Q..M	3)	G..T..K	3)	M..T..E	3)	U..W..C	3)	D..Y..8	3)	M..2..6
4)	X..R..N	4)	H..U..L	4)	N..U..F	4)	V..X..D	4)	E..Z..A	4)	N..3..7
5)	A..S..O	5)	J..V..M	5)	O..V..G	5)	W..Y..E	5)	F..1..B	5)	O..A..8
6)	B..T..P	6)	K..W..N	6)	P..W..H	6)	X..Z..F	6)	G..2..C	6)	P..B..A
7)	C..U..Q	7)	L..X..O	7)	Q..X..J	7)	A..1..G	7)	H..3..D	7)	Q..C..B
8)	D..V..R	8)	M..Y..P	8)	R..Y..K	8)	B..2..H	8)	J..A..E	8)	R..D..C
9)	E..W..S	9)	N..Z..Q	9)	S..Z..L	9)	C..3..J	9)	K..B..F	9)	S..E..D
10)	F..X..T	10)	O..1..R	10)	T..1..M	10)	D..A..K	10)	L..C..G	10)	T..F..E
11)	G..Y..U	11)	P..2..S	11)	U..2..N	11)	E..B..L	11)	M..D..H	11)	U..G..F
12)	H..Z..V	12)	Q..3..T	12)	V..3..O	12)	F..C..M	12)	N..E..J	12)	V..H..G
13)	J..1..W	13)	R..A..U	13)	W..A..P	13)	G..D..N	13)	O..F..K	13)	W..J..H
14)	K..2..X	14)	S..B..V	14)	X..B..Q	14)	H..E..O	14)	P..G..L	14)	X..K..J
15)	L..3..Y	15)	T..C..W	15)	A..C..R	15)	J..F..P	15)	Q..H..M	15)	A..L..K
16)	M..A..Z	16)	U..D..X	16)	B..D..S	16)	K..G..Q	16)	R..J..N	16)	B..M..L
17)	N..B..1	17)	V..E..Y	17)	C..E..T	17)	L..H..R	17)	S..K..O	17)	C..N..M
18)	O..C..2	18)	W..F..Z	18)	D..F..U	18)	M..J..S	18)	T..L..P	18)	D..O..N
19)	P..D..3	19)	X..G..1	19)	E..G..V	19)	N..K..T	19)	U..M..Q	19)	E..P..O
20)	Q..E..4	20)	A..H..2	20)	F..H..W	20)	O..L..U	20)	V..N..R	20)	F..Q..P
21)	R..F..5	21)	B..J..3	21)	G..J..X	21)	P..M..V	21)	W..O..S	21)	G..R..Q
22)	S..G..6	22)	C..K..4	22)	H..K..Y	22)	Q..N..W	22)	X..P..T	22)	H..S..R
23)	T..H..7	23)	D..L..5	23)	J..L..Z	23)	R..O..X	23)	A..Q..U	23)	J..T..S
24)	U..J..8	24)	E..M..6	24)	K..M..1	24)	S..P..Y	24)	B..R..V	24)	K..U..T
25)	V..K..A	25)	F..N..7	25)	L..N..2	25)	T..Q..Z	25)	C..S..W	25)	L..V..U
26)	W..L..B	26)	G..O..8	26)	M..O..3	26)	U..R..1	26)	D..T..X	26)	M..W..V
27)	X..M..C	27)	H..P..A	27)	N..P..4	27)	V..S..2	27)	E..U..Y	27)	N..X..W
28)	A..N..D	28)	J..Q..B	28)	O..Q..5	28)	W..T..3	28)	F..V..Z	28)	O..Y..X
29)	B..O..E			29)	P..R..6	29)	X..U..4	29)	G..W..1	29)	P..Z..Y
30)	C..P..F			30)	Q..S..7	30)	A..V..5	30)	H..X..2	30)	Q..1..Z
31)	D..Q..G			31)	R..T..8			31)	J..Y..3		

1898 | 1898

	....JULY....	...AUGUST...	..SEPTEMBER.	..OCTOBER...	..NOVEMBER..	..DECEMBER..
	P--E--I	P--E--I	P--E--I	P--E--I	P--E--I	P--E--I
1)	R..2..1	B..B..Y	K..E..W	R..G..T	B..K..R	J..M..O
2)	S..3..2	C..C..Z	L..F..X	S..H..U	C..L..S	K..N..P
3)	T..A..3	D..D..1	M..G..Y	T..J..V	D..M..T	L..O..Q
4)	U..B..4	E..E..2	N..H..Z	U..K..W	E..N..U	M..P..R
5)	V..C..5	F..F..3	O..J..1	V..L..X	F..O..V	N..Q..S
6)	W..D..6	G..G..4	P..K..2	W..M..Y	G..P..W	O..R..T
7)	X..E..7	H..H..5	Q..L..3	X..N..Z	H..Q..X	P..S..U
8)	A..F..8	J..J..6	R..M..4	A..O..1	J..R..Y	Q..T..V
9)	B..G..A	K..K..7	S..N..5	B..P..2	K..S..Z	R..U..W
10)	C..H..B	L..L..8	T..O..6	C..Q..3	L..T..1	S..V..X
11)	D..J..C	M..M..A	U..P..7	D..R..4	M..U..2	T..W..Y
12)	E..K..D	N..N..B	V..Q..8	E..S..5	N..V..3	U..X..Z
13)	F..L..E	O..O..C	W..R..A	F..T..6	O..W..4	V..Y..1
14)	G..M..F	P..P..D	X..S..B	G..U..7	P..X..5	W..Z..2
15)	H..N..G	Q..Q..E	A..T..C	H..V..8	Q..Y..6	X..1..3
16)	J..O..H	R..R..F	B..U..D	J..W..A	R..Z..7	A..2..4
17)	K..P..J	S..S..G	C..V..E	K..X..B	S..1..8	B..3..5
18)	L..Q..K	T..T..H	D..W..F	L..Y..C	T..2..A	C..A..6
19)	M..R..L	U..U..J	E..X..G	M..Z..D	U..3..B	D..B..7
20)	N..S..M	V..V..K	F..Y..H	N..1..E	V..A..C	E..C..8
21)	O..T..N	W..W..L	G..Z..J	O..2..F	W..B..D	F..D..A
22)	P..U..O	X..X..M	H..1..K	P..3..G	X..C..E	G..E..B
23)	Q..V..P	A..Y..N	J..2..L	Q..A..H	A..D..F	H..F..C
24)	R..W..Q	B..Z..O	K..3..M	R..B..J	B..E..G	J..G..D
25)	S..X..R	C..1..P	L..A..N	S..C..K	C..F..H	K..H..E
26)	T..Y..S	D..2..Q	M..B..O	T..D..L	D..G..J	L..J..F
27)	U..Z..T	E..3..R	N..C..P	U..E..M	E..H..K	M..K..G
28)	V..1..U	F..A..S	O..D..Q	V..F..N	F..J..L	N..L..H
29)	W..2..V	G..B..T	P..E..R	W..G..O	G..K..M	O..M..J
30)	X..3..W	H..C..U	Q..F..S	X..H..P	H..L..N	P..N..K
31)	A..A..X	J..D..V		A..J..Q		Q..O..L

CODES: P-PHYSICAL BIORHYTHM CURVE, E-EMOTIONAL BIORHYTHM CURVE, I-INTELLECTUAL BIORHYTHM CURVE

1899 | | | | | | 1899

	...JANUARY..	..FEBRUARY..	....MARCH...	....APRIL...	....MAY....	....JUNE....
	P--E--I	P--E--I	P--E--I	P--E--I	P--E--I	P--E--I
1)	R..P..M	B..S..K	G..S..E	P..V..C	W..X..8	G..1..6
2)	S..Q..N	C..T..L	H..T..F	Q..W..D	X..Y..A	H..2..7
3)	T..R..O	D..U..M	J..U..G	R..X..E	A..Z..B	J..3..8
4)	U..S..P	E..V..N	K..V..H	S..Y..F	B..1..C	K..A..A
5)	V..T..Q	F..W..O	L..W..J	T..Z..G	C..2..D	L..B..B
6)	W..U..R	G..X..P	M..X..K	U..1..H	D..3..E	M..C..C
7)	X..V..S	H..Y..Q	N..Y..L	V..2..J	E..A..F	N..D..D
8)	A..W..T	J..Z..R	O..Z..M	W..3..K	F..B..G	O..E..E
9)	B..X..U	K..1..S	P..1..N	X..A..L	G..C..H	P..F..F
10)	C..Y..V	L..2..T	Q..2..O	A..B..M	H..D..J	Q..G..G
11)	D..Z..W	M..3..U	R..3..P	B..C..N	J..E..K	R..H..H
12)	E..1..X	N..A..V	S..A..Q	C..D..O	K..F..L	S..J..J
13)	F..2..Y	O..B..W	T..B..R	D..E..P	L..G..M	T..K..K
14)	G..3..Z	P..C..X	U..C..S	E..F..Q	M..H..N	U..L..L
15)	H..A..1	Q..D..Y	V..D..T	F..G..R	N..J..O	V..M..M
16)	J..B..2	R..E..Z	W..E..U	G..H..S	O..K..P	W..N..N
17)	K..C..3	S..F..1	X..F..V	H..J..T	P..L..Q	X..O..O
18)	L..D..4	T..G..2	A..G..W	J..K..U	Q..M..R	A..P..P
19)	M..E..5	U..H..3	B..H..X	K..L..V	R..N..S	B..Q..Q
20)	N..F..6	V..J..4	C..J..Y	L..M..W	S..O..T	C..R..R
21)	O..G..7	W..K..5	D..K..Z	M..N..X	T..P..U	D..S..S
22)	P..H..8	X..L..6	E..L..1	N..O..Y	U..Q..V	E..T..T
23)	Q..J..A	A..M..7	F..M..2	O..P..Z	V..R..W	F..U..U
24)	R..K..B	B..N..8	G..N..3	P..Q..1	W..S..X	G..V..V
25)	S..L..C	C..O..A	H..O..4	Q..R..2	X..T..Y	H..W..W
26)	T..M..D	D..P..B	J..P..5	R..S..3	A..U..Z	J..X..X
27)	U..N..E	E..Q..C	K..Q..6	S..T..4	B..V..1	K..Y..Y
28)	V..O..F	F..R..D	L..R..7	T..U..5	C..W..2	L..Z..Z
29)	W..P..G		M..S..8	U..V..6	D..X..3	M..1..1
30)	X..Q..H		N..T..A	V..W..7	E..Y..4	N..2..2
31)	A..R..J		O..U..B		F..Z..5	

1899 1899

	....JULY....	...AUGUST...	..SEPTEMBER.	..OCTOBER...	..NOVEMBER..	..DECEMBER..
	P--E--I	P--E--I	P--E--I	P--E--I	P--E--I	P--E--I
1)	O..3..3	W..C..1	G..F..Y	O..H..V	W..L..T	F..N..Q
2)	P..A..4	X..D..2	H..G..Z	P..J..W	X..M..U	G..O..R
3)	Q..B..5	A..E..3	J..H..1	Q..K..X	A..N..V	H..P..S
4)	R..C..6	B..F..4	K..J..2	R..L..Y	B..O..W	J..Q..T
5)	S..D..7	C..G..5	L..K..3	S..M..Z	C..P..X	K..R..U
6)	T..E..8	D..H..6	M..L..4	T..N..1	D..Q..Y	L..S..V
7)	U..F..A	E..J..7	N..M..5	U..O..2	E..R..Z	M..T..W
8)	V..G..B	F..K..8	O..N..6	V..P..3	F..S..1	N..U..X
9)	W..H..C	G..L..A	P..O..7	W..Q..4	G..T..2	O..V..Y
10)	X..J..D	H..M..B	Q..P..8	X..R..5	H..U..3	P..W..Z
11)	A..K..E	J..N..C	R..Q..A	A..S..6	J..V..4	Q..X..1
12)	B..L..F	K..O..D	S..R..B	B..T..7	K..W..5	R..Y..2
13)	C..M..G	L..P..E	T..S..C	C..U..8	L..X..6	S..Z..3
14)	D..N..H	M..Q..F	U..T..D	D..V..A	M..Y..7	T..1..4
15)	E..O..J	N..R..G	V..U..E	E..W..B	N..Z..8	U..2..5
16)	F..P..K	O..S..H	W..V..F	F..X..C	O..1..A	V..3..6
17)	G..Q..L	P..T..J	X..W..G	G..Y..D	P..2..B	W..A..7
18)	H..R..M	Q..U..K	A..X..H	H..Z..E	Q..3..C	X..B..8
19)	J..S..N	R..V..L	B..Y..J	J..1..F	R..A..D	A..C..A
20)	K..T..O	S..W..M	C..Z..K	K..2..G	S..B..E	B..D..B
21)	L..U..P	T..X..N	D..1..L	L..3..H	T..C..F	C..E..C
22)	M..V..Q	U..Y..O	E..2..M	M..A..J	U..D..G	D..F..D
23)	N..W..R	V..Z..P	F..3..N	N..B..K	V..E..H	E..G..E
24)	O..X..S	W..1..Q	G..A..O	O..C..L	W..F..J	F..H..F
25)	P..Y..T	X..2..R	H..B..P	P..D..M	X..G..K	G..J..G
26)	Q..Z..U	A..3..S	J..C..Q	Q..E..N	A..H..L	H..K..H
27)	R..1..V	B..A..T	K..D..R	R..F..O	B..J..M	J..L..J
28)	S..2..W	C..B..U	L..E..S	S..G..P	C..K..N	K..M..K
29)	T..3..X	D..C..V	M..F..T	T..H..Q	D..L..O	L..N..L
30)	U..A..Y	E..D..W	N..G..U	U..J..R	E..M..P	M..O..M
31)	V..B..Z	F..E..X		V..K..S		N..P..N

CODES: P-PHYSICAL BIORHYTHM CURVE,E-EMOTIONAL BIORHYTHM CURVE,I-INTELLECTUAL BIORHYTHM CURVE

1900 1900

	...JANUARY..	..FEBRUARY..	....MARCH...	....APRIL...	.....MAY....	....JUNE....
	P--E--I	P--E--I	P--E--I	P--E--I	P--E--I	P--E--I
1)	O..Q..O	W..T..M	D..T..G	M..W..E	T..Y..B	D..2..8
2)	P..R..P	X..U..N	E..U..H	N..X..F	U..Z..C	E..3..A
3)	Q..S..Q	A..V..O	F..V..J	O..Y..G	V..1..D	F..A..B
4)	R..T..R	B..W..P	G..W..K	P..Z..H	W..2..E	G..B..C
5)	S..U..S	C..X..Q	H..X..L	Q..1..J	X..3..F	H..C..D
6)	T..V..T	D..Y..R	J..Y..M	R..2..K	A..A..G	J..D..E
7)	U..W..U	E..Z..S	K..Z..N	S..3..L	B..B..H	K..E..F
8)	V..X..V	F..1..T	L..1..O	T..A..M	C..C..J	L..F..G
9)	W..Y..W	G..2..U	M..2..P	U..B..N	D..D..K	M..G..H
10)	X..Z..X	H..3..V	N..3..Q	V..C..O	E..E..L	N..H..J
11)	A..1..Y	J..A..W	O..A..R	W..D..P	F..F..M	O..J..K
12)	B..2..Z	K..B..X	P..B..S	X..E..Q	G..G..N	P..K..L
13)	C..3..1	L..C..Y	Q..C..T	A..F..R	H..H..O	Q..L..M
14)	D..A..2	M..D..Z	R..D..U	B..G..S	J..J..P	R..M..N
15)	E..B..3	N..E..1	S..E..V	C..H..T	K..K..Q	S..N..O
16)	F..C..4	O..F..2	T..F..W	D..J..U	L..L..R	T..O..P
17)	G..D..5	P..G..3	U..G..X	E..K..V	M..M..S	U..P..Q
18)	H..E..6	Q..H..4	V..H..Y	F..L..W	N..N..T	V..Q..R
19)	J..F..7	R..J..5	W..J..Z	G..M..X	O..O..U	W..R..S
20)	K..G..8	S..K..6	X..K..1	H..N..Y	P..P..V	X..S..T
21)	L..H..A	T..L..7	A..L..2	J..O..Z	Q..Q..W	A..T..U
22)	M..J..B	U..M..8	B..M..3	K..P..1	R..R..X	B..U..V
23)	N..K..C	V..N..A	C..N..4	L..Q..2	S..S..Y	C..V..W
24)	O..L..D	W..O..B	D..O..5	M..R..3	T..T..Z	D..W..X
25)	P..M..E	X..P..C	E..P..6	N..S..4	U..U..1	E..X..Y
26)	Q..N..F	A..Q..D	F..Q..7	O..T..5	V..V..2	F..Y..Z
27)	R..O..G	B..R..E	G..R..8	P..U..6	W..W..3	G..Z..1
28)	S..P..H	C..S..F	H..S..A	Q..V..7	X..X..4	H..1..2
29)	T..Q..J		J..T..B	R..W..8	A..Y..5	J..2..3
30)	U..R..K		K..U..C	S..X..A	B..Z..6	K..3..4
31)	V..S..L		L..V..D		C..1..7	

1900						1900
Day	....JULY.... P--E--I	...AUGUST... P--E--I	..SEPTEMBER. P--E--I	..OCTOBER... P--E--I	..NOVEMBER.. P--E--I	..DECEMBER.. P--E--I
1)	L..A..5	T..D..3	D..G..1	L..J..X	T..M..V	C..O..S
2)	M..B..6	U..E..4	E..H..2	M..K..Y	U..N..W	D..P..T
3)	N..C..7	V..F..5	F..J..3	N..L..Z	V..O..X	E..Q..U
4)	O..D..8	W..G..6	G..K..4	O..M..1	W..P..Y	F..R..V
5)	P..E..A	X..H..7	H..L..5	P..N..2	X..Q..Z	G..S..W
6)	Q..F..B	A..J..8	J..M..6	Q..O..3	A..R..1	H..T..X
7)	R..G..C	B..K..A	K..N..7	R..P..4	B..S..2	J..U..Y
8)	S..H..D	C..L..B	L..O..8	S..Q..5	C..T..3	K..V..Z
9)	T..J..E	D..M..C	M..P..A	T..R..6	D..U..4	L..W..1
10)	U..K..F	E..N..D	N..Q..B	U..S..7	E..V..5	M..X..2
11)	V..L..G	F..O..E	O..R..C	V..T..8	F..W..6	N..Y..3
12)	W..M..H	G..P..F	P..S..D	W..U..A	G..X..7	O..Z..4
13)	X..N..J	H..Q..G	Q..T..E	X..V..B	H..Y..8	P..1..5
14)	A..O..K	J..R..H	R..U..F	A..W..C	J..Z..A	Q..2..6
15)	B..P..L	K..S..J	S..V..G	B..X..D	K..1..B	R..3..7
16)	C..Q..M	L..T..K	T..W..H	C..Y..E	L..2..C	S..A..8
17)	D..R..N	M..U..L	U..X..J	D..Z..F	M..3..D	T..B..A
18)	E..S..O	N..V..M	V..Y..K	E..1..G	N..A..E	U..C..B
19)	F..T..P	O..W..N	W..Z..L	F..2..H	O..B..F	V..D..C
20)	G..U..Q	P..X..O	X..1..M	G..3..J	P..C..G	W..E..D
21)	H..V..R	Q..Y..P	A..2..N	H..A..K	Q..D..H	X..F..E
22)	J..W..S	R..Z..Q	B..3..O	J..B..L	R..E..J	A..G..F
23)	K..X..T	S..1..R	C..A..P	K..C..M	S..F..K	B..H..G
24)	L..Y..U	T..2..S	D..B..Q	L..D..N	T..G..L	C..J..H
25)	M..Z..V	U..3..T	E..C..R	M..E..O	U..H..M	D..K..J
26)	N..1..W	V..A..U	F..D..S	N..F..P	V..J..N	E..L..K
27)	O..2..X	W..B..V	G..E..T	O..G..Q	W..K..O	F..M..L
28)	P..3..Y	X..C..W	H..F..U	P..H..R	X..L..P	G..N..M
29)	Q..A..Z	A..D..X	J..G..V	Q..J..S	A..M..Q	H..O..N
30)	R..B..1	B..E..Y	K..H..W	R..K..T	B..N..R	J..P..O
31)	S..C..2	C..F..Z		S..L..U		K..Q..P

CODES: P-PHYSICAL BIORHYTHM CURVE,E-EMOTIONAL BIORHYTHM CURVE,I-INTELLECTUAL BIORHYTHM CURVE

1901 1901

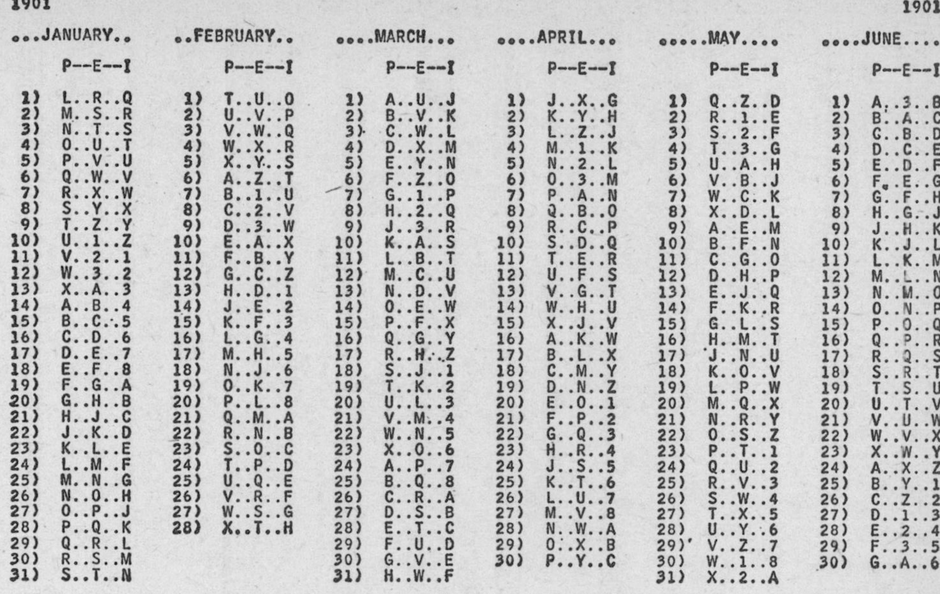

...JANUARY..		..FEBRUARY..		....MARCH...		....APRIL...		.....MAY....		....JUNE....	
	P--E--I		P--E--I		P--E--I		P--E--I		P--E--I		P--E--I
1)	L..R..Q	1)	T..U..O	1)	A..U..J	1)	J..X..G	1)	Q..Z..D	1)	A..3..B
2)	M..S..R	2)	U..V..P	2)	B..V..K	2)	K..Y..H	2)	R..1..E	2)	B..A..C
3)	N..T..S	3)	V..W..Q	3)	C..W..L	3)	L..Z..J	3)	S..2..F	3)	C..B..D
4)	O..U..T	4)	W..X..R	4)	D..X..M	4)	M..1..K	4)	T..3..G	4)	D..C..E
5)	P..V..U	5)	X..Y..S	5)	E..Y..N	5)	N..2..L	5)	U..A..H	5)	E..D..F
6)	Q..W..V	6)	A..Z..T	6)	F..Z..O	6)	O..3..M	6)	V..B..J	6)	F..E..G
7)	R..X..W	7)	B..1..U	7)	G..1..P	7)	P..A..N	7)	W..C..K	7)	G..F..H
8)	S..Y..X	8)	C..2..V	8)	H..2..Q	8)	Q..B..O	8)	X..D..L	8)	H..G..J
9)	T..Z..Y	9)	D..3..W	9)	J..3..R	9)	R..C..P	9)	A..E..M	9)	J..H..K
10)	U..1..Z	10)	E..A..X	10)	K..A..S	10)	S..D..Q	10)	B..F..N	10)	K..J..L
11)	V..2..1	11)	F..B..Y	11)	L..B..T	11)	T..E..R	11)	C..G..O	11)	L..K..M
12)	W..3..2	12)	G..C..Z	12)	M..C..U	12)	U..F..S	12)	D..H..P	12)	M..L..N
13)	X..A..3	13)	H..D..1	13)	N..D..V	13)	V..G..T	13)	E..J..Q	13)	N..M..O
14)	A..B..4	14)	J..E..2	14)	O..E..W	14)	W..H..U	14)	F..K..R	14)	O..N..P
15)	B..C..5	15)	K..F..3	15)	P..F..X	15)	X..J..V	15)	G..L..S	15)	P..O..Q
16)	C..D..6	16)	L..G..4	16)	Q..G..Y	16)	A..K..W	16)	H..M..T	16)	Q..P..R
17)	D..E..7	17)	M..H..5	17)	R..H..Z	17)	B..L..X	17)	J..N..U	17)	R..Q..S
18)	E..F..8	18)	N..J..6	18)	S..J..1	18)	C..M..Y	18)	K..O..V	18)	S..R..T
19)	F..G..A	19)	O..K..7	19)	T..K..2	19)	D..N..Z	19)	L..P..W	19)	T..S..U
20)	G..H..B	20)	P..L..8	20)	U..L..3	20)	E..O..1	20)	M..Q..X	20)	U..T..V
21)	H..J..C	21)	Q..M..A	21)	V..M..4	21)	F..P..2	21)	N..R..Y	21)	V..U..W
22)	J..K..D	22)	R..N..B	22)	W..N..5	22)	G..Q..3	22)	O..S..Z	22)	W..V..X
23)	K..L..E	23)	S..O..C	23)	X..O..6	23)	H..R..4	23)	P..T..1	23)	X..W..Y
24)	L..M..F	24)	T..P..D	24)	A..P..7	24)	J..S..5	24)	Q..U..2	24)	A..X..Z
25)	M..N..G	25)	U..Q..E	25)	B..Q..8	25)	K..T..6	25)	R..V..3	25)	B..Y..1
26)	N..O..H	26)	V..R..F	26)	C..R..A	26)	L..U..7	26)	S..W..4	26)	C..Z..2
27)	O..P..J	27)	W..S..G	27)	D..S..B	27)	M..V..8	27)	T..X..5	27)	D..1..3
28)	P..Q..K	28)	X..T..H	28)	E..T..C	28)	N..W..A	28)	U..Y..6	28)	E..2..4
29)	Q..R..L			29)	F..U..D	29)	O..X..B	29)	V..Z..7	29)	F..3..5
30)	R..S..M			30)	G..V..E	30)	P..Y..C	30)	W..1..8	30)	G..A..6
31)	S..T..N			31)	H..W..F			31)	X..2..A		

	1901					1901
	....JULY....	...AUGUST...	..SEPTEMBER.	..OCTOBER...	..NOVEMBER..	..DECEMBER..
	P—E—I	P—E—I	P—E—I	P—E—I	P—E—I	P—E—I
1)	H..B..7	Q..E..5	A..H..3	H..K..Z	Q..N..X	X..P..U
2)	J..C..8	R..F..6	B..J..4	J..L..1	R..O..Y	A..Q..V
3)	K..D..A	S..G..7	C..K..5	K..M..2	S..P..Z	B..R..W
4)	L..E..B	T..H..8	D..L..6	L..N..3	T..Q..1	C..S..X
5)	M..F..C	U..J..A	E..M..7	M..O..4	U..R..2	D..T..Y
6)	N..G..D	V..K..B	F..N..8	N..P..5	V..S..3	E..U..Z
7)	O..H..E	W..L..C	G..O..A	O..Q..6	W..T..4	F..V..1
8)	P..J..F	X..M..D	H..P..B	P..R..7	X..U..5	G..W..2
9)	Q..K..G	A..N..E	J..Q..C	Q..S..8	A..V..6	H..X..3
10)	R..L..H	B..O..F	K..R..D	R..T..A	B..W..7	J..Y..4
11)	S..M..J	C..P..G	L..S..E	S..U..B	C..X..8	K..Z..5
12)	T..N..K	D..Q..H	M..T..F	T..V..C	D..Y..A	L..1..6
13)	U..O..L	E..R..J	N..U..G	U..W..D	E..Z..B	M..2..7
14)	V..P..M	F..S..K	O..V..H	V..X..E	F..1..C	N..3..8
15)	W..Q..N	G..T..L	P..W..J	W..Y..F	G..2..D	O..A..A
16)	X..R..O	H..U..M	Q..X..K	X..Z..G	H..3..E	P..B..B
17)	A..S..P	J..V..N	R..Y..L	A..1..H	J..A..F	Q..C..C
18)	B..T..Q	K..W..O	S..Z..M	B..2..J	K..B..G	R..D..D
19)	C..U..R	L..X..P	T..1..N	C..3..K	L..C..H	S..E..E
20)	D..V..S	M..Y..Q	U..2..O	D..A..L	M..D..J	T..F..F
21)	E..W..T	N..Z..R	V..3..P	E..B..M	N..E..K	U..G..G
22)	F..X..U	O..1..S	W..A..Q	F..C..N	O..F..L	V..H..H
23)	G..Y..V	P..2..T	X..B..R	G..D..O	P..G..M	W..J..J
24)	H..Z..W	Q..3..U	A..C..S	H..E..P	Q..H..N	X..K..K
25)	J..1..X	R..A..V	B..D..T	J..F..Q	R..J..O	A..L..L
26)	K..2..Y	S..B..W	C..E..U	K..G..R	S..K..P	B..M..M
27)	L..3..Z	T..C..X	D..F..V	L..H..S	T..L..Q	C..N..N
28)	M..A..1	U..D..Y	E..G..W	M..J..T	U..M..R	D..O..O
29)	N..B..2	V..E..Z	F..H..X	N..K..U	V..N..S	E..P..P
30)	O..C..3	W..F..1	G..J..Y	O..L..V	W..O..T	F..Q..Q
31)	P..D..4	X..G..2		P..M..W		G..R..R

CODES: P-PHYSICAL BIORHYTHM CURVE,E-EMOTIONAL BIORHYTHM CURVE,I-INTELLECTUAL BIORHYTHM CURVE

1902 | 1902

Day	...JANUARY.. P--E--I	..FEBRUARY.. P--E--I	....MARCH... P--E--I	....APRIL... P--E--I	.....MAY.... P--E--I	....JUNE.... P--E--I
1)	H..S..S	Q..V..Q	V..V..L	F..Y..J	N..1..F	V..A..D
2)	J..T..T	R..W..R	W..W..M	G..Z..K	O..2..G	W..B..E
3)	K..U..U	S..X..S	X..X..N	H..1..L	P..3..H	X..C..F
4)	L..V..V	T..Y..T	A..Y..O	J..2..M	Q..A..J	A..D..G
5)	M..W..W	U..Z..U	B..Z..P	K..3..N	R..B..K	B..E..H
6)	N..X..X	V..1..V	C..1..Q	L..A..O	S..C..L	C..F..J
7)	O..Y..Y	W..2..W	D..2..R	M..B..P	T..D..M	D..G..K
8)	P..Z..Z	X..3..X	E..3..S	N..C..Q	U..E..N	E..H..L
9)	Q..1..1	A..A..Y	F..A..T	O..D..R	V..F..O	F..J..M
10)	R..2..2	B..B..Z	G..B..U	P..E..S	W..G..P	G..K..N
11)	S..3..3	C..C..1	H..C..V	Q..F..T	X..H..Q	H..L..O
12)	T..A..4	D..D..2	J..D..W	R..G..U	A..J..R	J..M..P
13)	U..B..5	E..E..3	K..E..X	S..H..V	B..K..S	K..N..Q
14)	V..C..6	F..F..4	L..F..Y	T..J..W	C..L..T	L..O..R
15)	W..D..7	G..G..5	M..G..Z	U..K..X	D..M..U	M..P..S
16)	X..E..8	H..H..6	N..H..1	V..L..Y	E..N..V	N..Q..T
17)	A..F..A	J..J..7	O..J..2	W..M..Z	F..O..W	O..R..U
18)	B..G..B	K..K..8	P..K..3	X..N..1	G..P..X	P..S..V
19)	C..H..C	L..L..A	Q..L..4	A..O..2	H..Q..Y	Q..T..W
20)	D..J..D	M..M..B	R..M..5	B..P..3	J..R..Z	R..U..X
21)	E..K..E	N..N..C	S..N..6	C..Q..4	K..S..1	S..V..Y
22)	F..L..F	O..O..D	T..O..7	D..R..5	L..T..2	T..W..Z
23)	G..M..G	P..P..E	U..P..8	E..S..6	M..U..3	U..X..1
24)	H..N..H	Q..Q..F	V..Q..A	F..T..7	N..V..4	V..Y..2
25)	J..O..J	R..R..G	W..R..B	G..U..8	O..W..5	W..Z..3
26)	K..P..K	S..S..H	X..S..C	H..V..A	P..X..6	X..1..4
27)	L..Q..L	T..T..J	A..T..D	J..W..B	Q..Y..7	A..2..5
28)	M..R..M	U..U..K	B..U..E	K..X..C	R..Z..8	B..3..6
29)	N..S..N		C..V..F	L..Y..D	S..1..A	C..A..7
30)	O..T..O		D..W..G	M..Z..E	T..2..B	D..B..8
31)	P..U..P		E..X..H		U..3..C	

1902 1902

Day	....JULY....	...AUGUST...	..SEPTEMBER.	..OCTOBER...	..NOVEMBER..	..DECEMBER..
	P--E--I	P--E--I	P--E--I	P--E--I	P--E--I	P--E--I
1)	E..C..A	N..F..7	V..J..5	E..L..2	N..O..Z	U..Q..W
2)	F..D..B	O..G..8	W..K..6	F..M..3	O..P..1	V..R..X
3)	G..E..C	P..H..A	X..L..7	G..N..4	P..Q..2	W..S..Y
4)	H..F..D	Q..J..B	A..M..8	H..O..5	Q..R..3	X..T..Z
5)	J..G..E	R..K..C	B..N..A	J..P..6	R..S..4	A..U..1
6)	K..H..F	S..L..D	C..O..B	K..Q..7	S..T..5	B..V..2
7)	L..J..G	T..M..E	D..P..C	L..R..8	T..U..6	C..W..3
8)	M..K..H	U..N..F	E..Q..D	M..S..A	U..V..7	D..X..4
9)	N..L..J	V..O..G	F..R..E	N..T..B	V..W..8	E..Y..5
10)	O..M..K	W..P..H	G..S..F	O..U..C	W..X..A	F..Z..6
11)	P..N..L	X..Q..J	H..T..G	P..V..D	X..Y..B	G..1..7
12)	Q..O..M	A..R..K	J..U..H	Q..W..E	A..Z..C	H..2..8
13)	R..P..N	B..S..L	K..V..J	R..X..F	B..1..D	J..3..A
14)	S..Q..O	C..T..M	L..W..K	S..Y..G	C..2..E	K..A..B
15)	T..R..P	D..U..N	M..X..L	T..Z..H	D..3..F	L..B..C
16)	U..S..Q	E..V..O	N..Y..M	U..1..J	E..A..G	M..C..D
17)	V..T..R	F..W..P	O..Z..N	V..2..K	F..B..H	N..D..E
18)	W..U..S	G..X..Q	P..1..O	W..3..L	G..C..J	O..E..F
19)	X..V..T	H..Y..R	Q..2..P	X..A..M	H..D..K	P..F..G
20)	A..W..U	J..Z..S	R..3..Q	A..B..N	J..E..L	Q..G..H
21)	B..X..V	K..1..T	S..A..R	B..C..O	K..F..M	R..H..J
22)	C..Y..W	L..2..U	T..B..S	C..D..P	L..G..N	S..J..K
23)	D..Z..X	M..3..V	U..C..T	D..E..Q	M..H..O	T..K..L
24)	E..1..Y	N..A..W	V..D..U	E..F..R	N..J..P	U..L..M
25)	F..2..Z	O..B..X	W..E..V	F..G..S	O..K..Q	V..M..N
26)	G..3..1	P..C..Y	X..F..W	G..H..T	P..L..R	W..N..O
27)	H..A..2	Q..D..Z	A..G..X	H..J..U	Q..M..S	X..O..P
28)	J..B..3	R..E..1	B..H..Y	J..K..V	R..N..T	A..P..Q
29)	K..C..4	S..F..2	C..J..Z	K..L..W	S..O..U	B..Q..R
30)	L..D..5	T..G..3	D..K..1	L..M..X	T..P..V	C..R..S
31)	M..E..6	U..H..4		M..N..Y		D..S..T

CODES: P-PHYSICAL BIORHYTHM CURVE,E-EMOTIONAL BIORHYTHM CURVE,I-INTELLECTUAL BIORHYTHM CURVE

	...JANUARY..	..FEBRUARY..	....MARCH...	....APRIL...	.....MAY....	....JUNE....
	P--E--I	P--E--I	P--E--I	P--E--I	P--E--I	P--E--I
1)	E..T..U	N..W..S	S..W..N	C..Z..L	K..2..H	S..B..F
2)	F..U..V	O..X..T	T..X..O	D..1..M	L..3..J	T..C..G
3)	G..V..W	P..Y..U	U..Y..P	E..2..N	M..A..K	U..D..H
4)	H..W..X	Q..Z..V	V..Z..Q	F..3..O	N..B..L	V..E..J
5)	J..X..Y	R..1..W	W..1..R	G..A..P	O..C..M	W..F..K
6)	K..Y..Z	S..2..X	X..2..S	H..B..Q	P..D..N	X..G..L
7)	L..Z..1	T..3..Y	A..3..T	J..C..R	Q..E..O	A..H..M
8)	M..1..2	U..A..Z	B..A..U	K..D..S	R..F..P	B..J..N
9)	N..2..3	V..B..1	C..B..V	L..E..T	S..G..Q	C..K..O
10)	O..3..4	W..C..2	D..C..W	M..F..U	T..H..R	D..L..P
11)	P..A..5	X..D..3	E..D..X	N..G..V	U..J..S	E..M..Q
12)	Q..B..6	A..E..4	F..E..Y	O..H..W	V..K..T	F..N..R
13)	R..C..7	B..F..5	G..F..Z	P..J..X	W..L..U	G..O..S
14)	S..D..8	C..G..6	H..G..1	Q..K..Y	X..M..V	H..P..T
15)	T..E..A	D..H..7	J..H..2	R..L..Z	A..N..W	J..Q..U
16)	U..F..B	E..J..8	K..J..3	S..M..1	B..O..X	K..R..V
17)	V..G..C	F..K..A	L..K..4	T..N..2	C..P..Y	L..S..W
18)	W..H..D	G..L..B	M..L..5	U..O..3	D..Q..Z	M..T..X
19)	X..J..E	H..M..C	N..M..6	V..P..4	E..R..1	N..U..Y
20)	A..K..F	J..N..D	O..N..7	W..Q..5	F..S..2	O..V..Z
21)	B..L..G	K..O..E	P..O..8	X..R..6	G..T..3	P..W..1
22)	C..M..H	L..P..F	Q..P..A	A..S..7	H..U..4	Q..X..2
23)	D..N..J	M..Q..G	R..Q..B	B..T..8	J..V..5	R..Y..3
24)	E..O..K	N..R..H	S..R..C	C..U..A	K..W..6	S..Z..4
25)	F..P..L	O..S..J	T..S..D	D..V..B	L..X..7	T..1..5
26)	G..Q..M	P..T..K	U..T..E	E..W..C	M..Y..8	U..2..6
27)	H..R..N	Q..U..L	V..U..F	F..X..D	N..Z..A	V..3..7
28)	J..S..O	R..V..M	W..V..G	G..Y..E	O..1..B	W..A..8
29)	K..T..P		X..W..H	H..Z..F	P..2..C	X..B..A
30)	L..U..Q		A..X..J	J..1..G	Q..3..D	A..C..B
31)	M..V..R		B..Y..K		R..A..E	

1903 | 1903

	....JULY....	...AUGUST...	..SEPTEMBER.	..OCTOBER...	..NOVEMBER..	..DECEMBER..
	P--E--I	P--E--I	P--E--I	P--E--I	P--E--I	P--E--I
1)	B..D..C	K..G..A	S..K..7	B..M..4	K..P..2	R..R..Y
2)	C..E..D	L..H..B	T..L..8	C..N..5	L..Q..3	S..S..Z
3)	D..F..E	M..J..C	U..M..A	D..O..6	M..R..4	T..T..1
4)	E..G..F	N..K..D	V..N..B	E..P..7	N..S..5	U..U..2
5)	F..H..G	O..L..E	W..O..C	F..Q..8	O..T..6	V..V..3
6)	G..J..H	P..M..F	X..P..D	G..R..A	P..U..7	W..W..4
7)	H..K..J	Q..N..G	A..Q..E	H..S..B	Q..V..8	X..X..5
8)	J..L..K	R..O..H	B..R..F	J..T..C	R..W..A	A..Y..6
9)	K..M..L	S..P..J	C..S..G	K..U..D	S..X..B	B..Z..7
10)	L..N..M	T..Q..K	D..T..H	L..V..E	T..Y..C	C..1..8
11)	M..O..N	U..R..L	E..U..J	M..W..F	U..Z..D	D..2..A
12)	N..P..O	V..S..M	F..V..K	N..X..G	V..1..E	E..3..B
13)	O..Q..P	W..T..N	G..W..L	O..Y..H	W..2..F	F..A..C
14)	P..R..Q	X..U..O	H..X..M	P..Z..J	X..3..G	G..B..D
15)	Q..S..R	A..V..P	J..Y..N	Q..1..K	A..A..H	H..C..E
16)	R..T..S	B..W..Q	K..Z..O	R..2..L	B..B..J	J..D..F
17)	S..U..T	C..X..R	L..1..P	S..3..M	C..C..K	K..E..G
18)	T..V..U	D..Y..S	M..2..Q	T..A..N	D..D..L	L..F..H
19)	U..W..V	E..Z..T	N..3..R	U..B..O	E..E..M	M..G..J
20)	V..X..W	F..1..U	O..A..S	V..C..P	F..F..N	N..H..K
21)	W..Y..X	G..2..V	P..B..T	W..D..Q	G..G..O	O..J..L
22)	X..Z..Y	H..3..W	Q..C..U	X..E..R	H..H..P	P..K..M
23)	A..1..Z	J..A..X	R..D..V	A..F..S	J..J..Q	Q..L..N
24)	B..2..1	K..B..Y	S..E..W	B..G..T	K..K..R	R..M..O
25)	C..3..2	L..C..Z	T..F..X	C..H..U	L..L..S	S..N..P
26)	D..A..3	M..D..1	U..G..Y	D..J..V	M..M..T	T..O..Q
27)	E..B..4	N..E..2	V..H..Z	E..K..W	N..N..U	U..P..R
28)	F..C..5	O..F..3	W..J..1	F..L..X	O..O..V	V..Q..S
29)	G..D..6	P..G..4	X..K..2	G..M..Y	P..P..W	W..R..T
30)	H..E..7	Q..H..5	A..L..3	H..N..Z	Q..Q..X	X..S..U
31)	J..F..8	R..J..6		J..O..1		A..T..V

CODES: P-PHYSICAL BIORHYTHM CURVE, E-EMOTIONAL BIORHYTHM CURVE, I-INTELLECTUAL BIORHYTHM CURVE

1904 1904

	...JANUARY..		..FEBRUARY..		....MARCH...		....APRIL...		.....MAY....		.....JUNE....
	P--E--I		P--E--I		P--E--I		P--E--I		P--E--I		P--E--I
1)	B..U..W	1)	K..X..U	1)	Q..Y..Q	1)	A..2..O	1)	H..A..L	1)	Q..D..J
2)	C..V..X	2)	L..Y..V	2)	R..Z..R	2)	B..3..P	2)	J..B..M	2)	R..E..K
3)	D..W..Y	3)	M..Z..W	3)	S..1..S	3)	C..A..Q	3)	K..C..N	3)	S..F..L
4)	E..X..Z	4)	N..1..X	4)	T..2..T	4)	D..B..R	4)	L..D..O	4)	T..G..M
5)	F..Y..1	5)	O..2..Y	5)	U..3..U	5)	E..C..S	5)	M..E..P	5)	U..H..N
6)	G..Z..2	6)	P..3..Z	6)	V..A..V	6)	F..D..T	6)	N..F..Q	6)	V..J..O
7)	H..1..3	7)	Q..A..1	7)	W..B..W	7)	G..E..U	7)	O..G..R	7)	W..K..P
8)	J..2..4	8)	R..B..2	8)	X..C..X	8)	H..F..V	8)	P..H..S	8)	X..L..Q
9)	K..3..5	9)	S..C..3	9)	A..D..Y	9)	J..G..W	9)	Q..J..T	9)	A..M..R
10)	L..A..6	10)	T..D..4	10)	B..E..Z	10)	K..H..X	10)	R..K..U	10)	B..N..S
11)	M..B..7	11)	U..E..5	11)	C..F..1	11)	L..J..Y	11)	S..L..V	11)	C..O..T
12)	N..C..8	12)	V..F..6	12)	D..G..2	12)	M..K..Z	12)	T..M..W	12)	D..P..U
13)	O..D..A	13)	W..G..7	13)	E..H..3	13)	N..L..1	13)	U..N..X	13)	E..Q..V
14)	P..E..B	14)	X..H..8	14)	F..J..4	14)	O..M..2	14)	V..O..Y	14)	F..R..W
15)	Q..F..C	15)	A..J..A	15)	G..K..5	15)	P..N..3	15)	W..P..Z	15)	G..S..X
16)	R..G..D	16)	B..K..B	16)	H..L..6	16)	Q..O..4	16)	X..Q..1	16)	H..T..Y
17)	S..H..E	17)	C..L..C	17)	J..M..7	17)	R..P..5	17)	A..R..2	17)	J..U..Z
18)	T..J..F	18)	D..M..D	18)	K..N..8	18)	S..Q..6	18)	B..S..3	18)	K..V..1
19)	U..K..G	19)	E..N..E	19)	L..O..A	19)	T..R..7	19)	C..T..4	19)	L..W..2
20)	V..L..H	20)	F..O..F	20)	M..P..B	20)	U..S..8	20)	D..U..5	20)	M..X..3
21)	W..M..J	21)	G..P..G	21)	N..Q..C	21)	V..T..A	21)	E..V..6	21)	N..Y..4
22)	X..N..K	22)	H..Q..H	22)	O..R..D	22)	W..U..B	22)	F..W..7	22)	O..Z..5
23)	A..O..L	23)	J..R..J	23)	P..S..E	23)	X..V..C	23)	G..X..8	23)	P..1..6
24)	B..P..M	24)	K..S..K	24)	Q..T..F	24)	A..W..D	24)	H..Y..A	24)	Q..2..7
25)	C..Q..N	25)	L..T..L	25)	R..U..G	25)	B..X..E	25)	J..Z..B	25)	R..3..8
26)	D..R..O	26)	M..U..M	26)	S..V..H	26)	C..Y..F	26)	K..1..C	26)	S..A..A
27)	E..S..P	27)	N..V..N	27)	T..W..J	27)	D..Z..G	27)	L..2..D	27)	T..B..B
28)	F..T..Q	28)	O..W..O	28)	U..X..K	28)	E..1..H	28)	M..3..E	28)	U..C..C
29)	G..U..R	29)	P..X..P	29)	V..Y..L	29)	F..2..J	29)	N..A..F	29)	V..D..D
30)	H..V..S			30)	W..Z..M	30)	G..3..K	30)	O..B..G	30)	W..E..E
31)	J..W..T			31)	X..1..N			31)	P..C..H		

1904 | 1904

	....JULY....	...AUGUST...	..SEPTEMBER.	..OCTOBER...	..NOVEMBER..	..DECEMBER..
	P--E--I	P--E--I	P--E--I	P--E--I	P--E--I	P--E--I
1)	X..F..F	H..J..D	Q..M..B	X..O..7	H..R..5	P..T..2
2)	A..G..G	J..K..E	R..N..C	A..P..8	J..S..6	Q..U..3
3)	B..H..H	K..L..F	S..O..D	B..Q..A	K..T..7	R..V..4
4)	C..J..J	L..M..G	T..P..E	C..R..B	L..U..8	S..W..5
5)	D..K..K	M..N..H	U..Q..F	D..S..C	M..V..A	T..X..6
6)	E..L..L	N..O..J	V..R..G	E..T..D	N..W..B	U..Y..7
7)	F..M..M	O..P..K	W..S..H	F..U..E	O..X..C	V..Z..8
8)	G..N..N	P..Q..L	X..T..J	G..V..F	P..Y..D	W..1..A
9)	H..O..O	Q..R..M	A..U..K	H..W..G	Q..Z..E	X..2..B
10)	J..P..P	R..S..N	B..V..L	J..X..H	R..1..F	A..3..C
11)	K..Q..Q	S..T..O	C..W..M	K..Y..J	S..2..G	B..A..D
12)	L..R..R	T..U..P	D..X..N	L..Z..K	T..3..H	C..B..E
13)	M..S..S	U..V..Q	E..Y..O	M..1..L	U..A..J	D..C..F
14)	N..T..T	V..W..R	F..Z..P	N..2..M	V..B..K	E..D..G
15)	O..U..U	W..X..S	G..1..Q	O..3..N	W..C..L	F..E..H
16)	P..V..V	X..Y..T	H..2..R	P..A..O	X..D..M	G..F..J
17)	Q..W..W	A..Z..U	J..3..S	Q..B..P	A..E..N	H..G..K
18)	R..X..X	B..1..V	K..A..T	R..C..Q	B..F..O	J..H..L
19)	S..Y..Y	C..2..W	L..B..U	S..D..R	C..G..P	K..J..M
20)	T..Z..Z	D..3..X	M..C..V	T..E..S	D..H..Q	L..K..N
21)	U..1..1	E..A..Y	N..D..W	U..F..T	E..J..R	M..L..O
22)	V..2..2	F..B..Z	O..E..X	V..G..U	F..K..S	N..M..P
23)	W..3..3	G..C..1	P..F..Y	W..H..V	G..L..T	O..N..Q
24)	X..A..4	H..D..2	Q..G..Z	X..J..W	H..M..U	P..O..R
25)	A..B..5	J..E..3	R..H..1	A..K..X	J..N..V	Q..P..S
26)	B..C..6	K..F..4	S..J..2	B..L..Y	K..O..W	R..Q..T
27)	C..D..7	L..G..5	T..K..3	C..M..Z	L..P..X	S..R..U
28)	D..E..8	M..H..6	U..L..4	D..N..1	M..Q..Y	T..S..V
29)	E..F..A	N..J..7	V..M..5	E..O..2	N..R..Z	U..T..W
30)	F..G..B	O..K..8	W..N..6	F..P..3	O..S..1	V..U..X
31)	G..H..C	P..L..A		G..Q..4		W..V..Y

CODES: P-PHYSICAL BIORHYTHM CURVE, E-EMOTIONAL BIORHYTHM CURVE, I-INTELLECTUAL BIORHYTHM CURVE

1905 1905

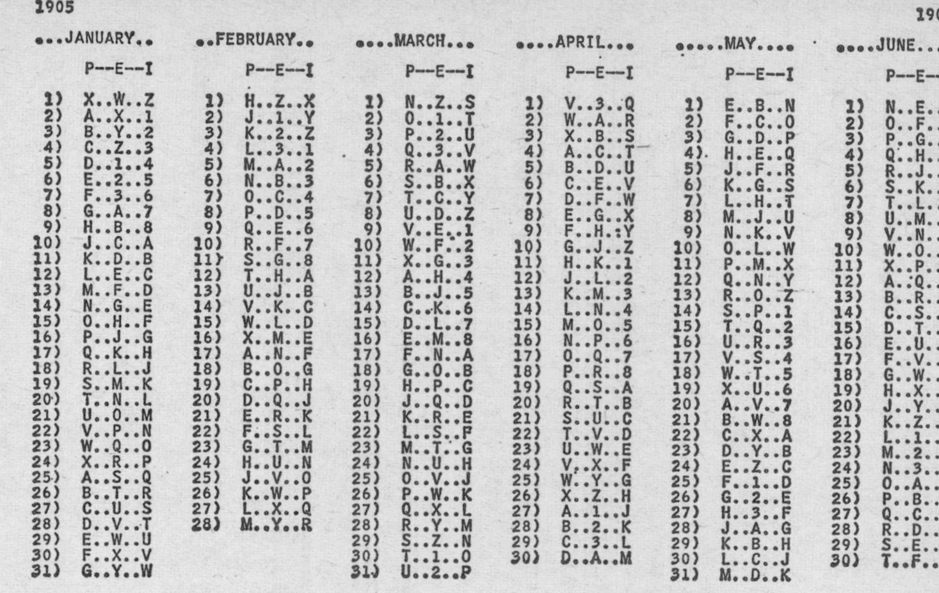

Day	...JANUARY..	..FEBRUARY..	....MARCH...	....APRIL...	.....MAY....	....JUNE....
	P--E--I	P--E--I	P--E--I	P--E--I	P--E--I	P--E--I
1)	X..W..Z	H..Z..X	N..Z..S	V..3..Q	E..B..N	N..E..L
2)	A..X..1	J..1..Y	O..1..T	W..A..R	F..C..O	O..F..M
3)	B..Y..2	K..2..Z	P..2..U	X..B..S	G..D..P	P..G..N
4)	C..Z..3	L..3..1	Q..3..V	A..C..T	H..E..Q	Q..H..O
5)	D..1..4	M..A..2	R..A..W	B..D..U	J..F..R	R..J..P
6)	E..2..5	N..B..3	S..B..X	C..E..V	K..G..S	S..K..Q
7)	F..3..6	O..C..4	T..C..Y	D..F..W	L..H..T	T..L..R
8)	G..A..7	P..D..5	U..D..Z	E..G..X	M..J..U	U..M..S
9)	H..B..8	Q..E..6	V..E..1	F..H..Y	N..K..V	V..N..T
10)	J..C..A	R..F..7	W..F..2	G..J..Z	O..L..W	W..O..U
11)	K..D..B	S..G..8	X..G..3	H..K..1	P..M..X	X..P..V
12)	L..E..C	T..H..A	A..H..4	J..L..2	Q..N..Y	A..Q..W
13)	M..F..D	U..J..B	B..J..5	K..M..3	R..O..Z	B..R..X
14)	N..G..E	V..K..C	C..K..6	L..N..4	S..P..1	C..S..Y
15)	O..H..F	W..L..D	D..L..7	M..O..5	T..Q..2	D..T..Z
16)	P..J..G	X..M..E	E..M..8	N..P..6	U..R..3	E..U..1
17)	Q..K..H	A..N..F	F..N..A	O..Q..7	V..S..4	F..V..2
18)	R..L..J	B..O..G	G..O..B	P..R..8	W..T..5	G..W..3
19)	S..M..K	C..P..H	H..P..C	Q..S..A	X..U..6	H..X..4
20)	T..N..L	D..Q..J	J..Q..D	R..T..B	A..V..7	J..Y..5
21)	U..O..M	E..R..K	K..R..E	S..U..C	B..W..8	K..Z..6
22)	V..P..N	F..S..L	L..S..F	T..V..D	C..X..A	L..1..7
23)	W..Q..O	G..T..M	M..T..G	U..W..E	D..Y..B	M..2..8
24)	X..R..P	H..U..N	N..U..H	V..X..F	E..Z..C	N..3..A
25)	A..S..Q	J..V..O	O..V..J	W..Y..G	F..1..D	O..A..B
26)	B..T..R	K..W..P	P..W..K	X..Z..H	G..2..E	P..B..C
27)	C..U..S	L..X..Q	Q..X..L	A..1..J	H..3..F	Q..C..D
28)	D..V..T	M..Y..R	R..Y..M	B..2..K	J..A..G	R..D..E
29)	E..W..U		S..Z..N	C..3..L	K..B..H	S..E..F
30)	F..X..V		T..1..O	D..A..M	L..C..J	T..F..G
31)	G..Y..W		U..2..P		M..D..K	

1905						1905
	....JULY....	...AUGUST...	..SEPTEMBER.	..OCTOBER...	..NOVEMBER..	..DECEMBER..
	P--E--I	P--E--I	P--E--I	P--E--I	P--E--I	P--E--I
1)	U..G..H	E..K..F	N..N..D	U..P..A	E..S..7	M..U..4
2)	V..H..J	F..L..G	O..O..E	V..Q..B	F..T..8	N..V..5
3)	W..J..K	G..M..H	P..P..F	W..R..C	G..U..A	O..W..6
4)	X..K..L	H..N..J	Q..Q..G	X..S..D	H..V..B	P..X..7
5)	A..L..M	J..O..K	R..R..H	A..T..E	J..W..C	Q..Y..8
6)	B..M..N	K..P..L	S..S..J	B..U..F	K..X..D	R..Z..A
7)	C..N..O	L..Q..M	T..T..K	C..V..G	L..Y..E	S..1..B
8)	D..O..P	M..R..N	U..U..L	D..W..H	M..Z..F	T..2..C
9)	E..P..Q	N..S..O	V..V..M	E..X..J	N..1..G	U..3..D
10)	F..Q..R	O..T..P	W..W..N	F..Y..K	O..2..H	V..A..E
11)	G..R..S	P..U..Q	X..X..O	G..Z..L	P..3..J	W..B..F
12)	H..S..T	Q..V..R	A..Y..P	H..1..M	Q..A..K	X..C..G
13)	J..T..U	R..W..S	B..Z..Q	J..2..N	R..B..L	A..D..H
14)	K..U..V	S..X..T	C..1..R	K..3..O	S..C..M	B..E..J
15)	L..V..W	T..Y..U	D..2..S	L..A..P	T..D..N	C..F..K
16)	M..W..X	U..Z..V	E..3..T	M..B..Q	U..E..O	D..G..L
17)	N..X..Y	V..1..W	F..A..U	N..C..R	V..F..P	E..H..M
18)	O..Y..Z	W..2..X	G..B..V	O..D..S	W..G..Q	F..J..N
19)	P..Z..1	X..3..Y	H..C..W	P..E..T	X..H..R	G..K..O
20)	Q..1..2	A..A..Z	J..D..X	Q..F..U	A..J..S	H..L..P
21)	R..2..3	B..B..1	K..E..Y	R..G..V	B..K..T	J..M..Q
22)	S..3..4	C..C..2	L..F..Z	S..H..W	C..L..U	K..N..R
23)	T..A..5	D..D..3	M..G..1	T..J..X	D..M..V	L..O..S
24)	U..B..6	E..E..4	N..H..2	U..K..Y	E..N..W	M..P..T
25)	V..C..7	F..F..5	O..J..3	V..L..Z	F..O..X	N..Q..U
26)	W..D..8	G..G..6	P..K..4	W..M..1	G..P..Y	O..R..V
27)	X..E..A	H..H..7	Q..L..5	X..N..2	H..Q..Z	P..S..W
28)	A..F..B	J..J..8	R..M..6	A..O..3	J..R..1	Q..T..X
29)	B..G..C	K..K..A	S..N..7	B..P..4	K..S..2	R..U..Y
30)	C..H..D	L..L..B	T..O..8	C..Q..5	L..T..3	S..V..Z
31)	D..J..E	M..M..C		D..R..6		T..W..1

CODES: P-PHYSICAL BIORHYTHM CURVE, E-EMOTIONAL BIORHYTHM CURVE, I-INTELLECTUAL BIORHYTHM CURVE

1906						1906
	...JANUARY..	..FEBRUARY..	....MARCH...	....APRIL...	.....MAY....	....JUNE....
	P--E--I	P--E--I	P--E--I	P--E--I	P--E--I	P--E--I
1)	U..X..2	E..1..Z	K..1..U	S..A..S	B..C..P	K..F..N
2)	V..Y..3	F..2..1	L..2..V	T..B..T	C..D..Q	L..G..O
3)	W..Z..4	G..3..2	M..3..W	U..C..U	D..E..R	M..H..P
4)	X..1..5	H..A..3	N..A..X	V..D..V	E..F..S	N..J..Q
5)	A..2..6	J..B..4	O..B..Y	W..E..W	F..G..T	O..K..R
6)	B..3..7	K..C..5	P..C..Z	X..F..X	G..H..U	P..L..S
7)	C..A..8	L..D..6	Q..D..1	A..G..Y	H..J..V	Q..M..T
8)	D..B..A	M..E..7	R..E..2	B..H..Z	J..K..W	R..N..U
9)	E..C..B	N..F..8	S..F..3	C..J..1	K..L..X	S..O..V
10)	F..D..C	O..G..A	T..G..4	D..K..2	L..M..Y	T..P..W
11)	G..E..D	P..H..B	U..H..5	E..L..3	M..N..Z	U..Q..X
12)	H..F..E	Q..J..C	V..J..6	F..M..4	N..O..1	V..R..Y
13)	J..G..F	R..K..D	W..K..7	G..N..5	O..P..2	W..S..Z
14)	K..H..G	S..L..E	X..L..8	H..O..6	P..Q..3	X..T..1
15)	L..J..H	T..M..F	A..M..A	J..P..7	Q..R..4	A..U..2
16)	M..K..J	U..N..G	B..N..B	K..Q..8	R..S..5	B..V..3
17)	N..L..K	V..O..H	C..O..C	L..R..A	S..T..6	C..W..4
18)	O..M..L	W..P..J	D..P..D	M..S..B	T..U..7	D..X..5
19)	P..N..M	X..Q..K	E..Q..E	N..T..C	U..V..8	E..Y..6
20)	Q..O..N	A..R..L	F..R..F	O..U..D	V..W..A	F..Z..7
21)	R..P..O	B..S..M	G..S..G	P..V..E	W..X..B	G..1..8
22)	S..Q..P	C..T..N	H..T..H	Q..W..F	X..Y..C	H..2..A
23)	T..R..Q	D..U..O	J..U..J	R..X..G	A..Z..D	J..3..B
24)	U..S..R	E..V..P	K..V..K	S..Y..H	B..1..E	K..A..C
25)	V..T..S	F..W..Q	L..W..L	T..Z..J	C..2..F	L..B..D
26)	W..U..T	G..X..R	M..X..M	U..1..K	D..3..G	M..C..E
27)	X..V..U	H..Y..S	N..Y..N	V..2..L	E..A..H	N..D..F
28)	A..W..V	J..Z..T	O..Z..O	W..3..M	F..B..J	O..E..G
29)	B..X..W		P..1..P	X..A..N	G..C..K	P..F..H
30)	C..Y..X		Q..2..Q	A..B..O	H..D..L	Q..G..J
31)	D..Z..Y		R..3..R		J..E..M	

1906					1906
....JULY....	...AUGUST...	..SEPTEMBER.	..OCTOBER...	..NOVEMBER..	..DECEMBER..
P--E--I	P--E--I	P--E--I	P--E--I	P--E--I	P--E--I
1) R..H..K	1) B..L..H	1) K..O..F	1) R..Q..C	1) B..T..A	1) J..V..6
2) S..J..L	2) C..M..J	2) L..P..G	2) S..R..D	2) C..U..B	2) K..W..7
3) T..K..M	3) D..N..K	3) M..Q..H	3) T..S..E	3) D..V..C	3) L..X..8
4) U..L..N	4) E..O..L	4) N..R..J	4) U..T..F	4) E..W..D	4) M..Y..A
5) V..M..O	5) F..P..M	5) O..S..K	5) V..U..G	5) F..X..E	5) N..Z..B
6) W..N..P	6) G..Q..N	6) P..T..L	6) W..V..H	6) G..Y..F	6) O..1..C
7) X..O..Q	7) H..R..O	7) Q..U..M	7) X..W..J	7) H..Z..G	7) P..2..D
8) A..P..R	8) J..S..P	8) R..V..N	8) A..X..K	8) J..1..H	8) Q..3..E
9) B..Q..S	9) K..T..Q	9) S..W..O	9) B..Y..L	9) K..2..J	9) R..A..F
10) C..R..T	10) L..U..R	10) T..X..P	10) C..Z..M	10) L..3..K	10) S..B..G
11) D..S..U	11) M..V..S	11) U..Y..Q	11) D..1..N	11) M..A..L	11) T..C..H
12) E..T..V	12) N..W..T	12) V..Z..R	12) E..2..O	12) N..B..M	12) U..D..J
13) F..U..W	13) O..X..U	13) W..1..S	13) F..3..P	13) O..C..N	13) V..E..K
14) G..V..X	14) P..Y..V	14) X..2..T	14) G..A..Q	14) P..D..O	14) W..F..L
15) H..W..Y	15) Q..Z..W	15) A..3..U	15) H..B..R	15) Q..E..P	15) X..G..M
16) J..X..Z	16) R..1..X	16) B..A..V	16) J..C..S	16) R..F..Q	16) A..H..N
17) K..Y..1	17) S..2..Y	17) C..B..W	17) K..D..T	17) S..G..R	17) B..J..O
18) L..Z..2	18) T..3..Z	18) D..C..X	18) L..E..U	18) T..H..S	18) C..K..P
19) M..1..3	19) U..A..1	19) E..D..Y	19) M..F..V	19) U..J..T	19) D..L..Q
20) N..2..4	20) V..B..2	20) F..E..Z	20) N..G..W	20) V..K..U	20) E..M..R
21) O..3..5	21) W..C..3	21) G..F..1	21) O..H..X	21) W..L..V	21) F..N..S
22) P..A..6	22) X..D..4	22) H..G..2	22) P..J..Y	22) X..M..W	22) G..O..T
23) Q..B..7	23) A..E..5	23) J..H..3	23) Q..K..Z	23) A..N..X	23) H..P..U
24) R..C..8	24) B..F..6	24) K..J..4	24) R..L..1	24) B..O..Y	24) J..Q..V
25) S..D..A	25) C..G..7	25) L..K..5	25) S..M..2	25) C..P..Z	25) K..R..W
26) T..E..B	26) D..H..8	26) M..L..6	26) T..N..3	26) D..Q..1	26) L..S..X
27) U..F..C	27) E..J..A	27) N..M..7	27) U..O..4	27) E..R..2	27) M..T..Y
28) V..G..D	28) F..K..B	28) O..N..8	28) V..P..5	28) F..S..3	28) N..U..Z
29) W..H..E	29) G..L..C	29) P..C..A	29) W..Q..6	29) G..T..4	29) O..V..1
30) X..J..F	30) H..M..D	30) Q..F..B	30) X..R..7	30) H..U..5	30) P..W..2
31) A..K..G	31) J..N..E		31) A..S..8		31) Q..X..3

CODES: P-PHYSICAL BIORHYTHM CURVE, E-EMOTIONAL BIORHYTHM CURVE, I-INTELLECTUAL BIORHYTHM CURVE

1907 | 1907

	..JANUARY..		..FEBRUARY..		....MARCH...		....APRIL...		.....MAY....		....JUNE....
	P--E--I		P--E--I		P--E--I		P--E--I		P--E--I		P--E--I
1)	R..Y..4	1)	B..2..2	1)	G..2..W	1)	P..B..U	1)	W..D..R	1)	G..G..P
2)	S..Z..5	2)	C..3..3	2)	H..3..X	2)	Q..C..V	2)	X..E..S	2)	H..H..Q
3)	T..1..6	3)	D..A..4	3)	J..A..Y	3)	R..D..W	3)	A..F..T	3)	J..J..R
4)	U..2..7	4)	E..B..5	4)	K..B..Z	4)	S..E..X	4)	B..G..U	4)	K..K..S
5)	V..3..8	5)	F..C..6	5)	L..C..1	5)	T..F..Y	5)	C..H..V	5)	L..L..T
6)	W..A..A	6)	G..D..7	6)	M..D..2	6)	U..G..Z	6)	D..J..W	6)	M..M..U
7)	X..B..B	7)	H..E..8	7)	N..E..3	7)	V..H..1	7)	E..K..X	7)	N..N..V
8)	A..C..C	8)	J..F..A	8)	O..F..4	8)	W..J..2	8)	F..L..Y	8)	O..O..W
9)	B..D..D	9)	K..G..B	9)	P..G..5	9)	X..K..3	9)	G..M..Z	9)	P..P..X
10)	C..E..E	10)	L..H..C	10)	Q..H..6	10)	A..L..4	10)	H..N..1	10)	Q..Q..Y
11)	D..F..F	11)	M..J..D	11)	R..J..7	11)	B..M..5	11)	J..O..2	11)	R..R..Z
12)	E..G..G	12)	N..K..E	12)	S..K..8	12)	C..N..6	12)	K..P..3	12)	S..S..1
13)	F..H..H	13)	O..L..F	13)	T..L..A	13)	D..O..7	13)	L..Q..4	13)	T..T..2
14)	G..J..J	14)	P..M..G	14)	U..M..B	14)	E..P..8	14)	M..R..5	14)	U..U..3
15)	H..K..K	15)	Q..N..H	15)	V..N..C	15)	F..Q..A	15)	N..S..6	15)	V..V..4
16)	J..L..L	16)	R..O..J	16)	W..O..D	16)	G..R..B	16)	O..T..7	16)	W..W..5
17)	K..M..M	17)	S..P..K	17)	X..P..E	17)	H..S..C	17)	P..U..8	17)	X..X..6
18)	L..N..N	18)	T..Q..L	18)	A..Q..F	18)	J..T..D	18)	Q..V..A	18)	A..Y..7
19)	M..O..O	19)	U..R..M	19)	B..R..G	19)	K..U..E	19)	R..W..B	19)	B..Z..8
20)	N..P..P	20)	V..S..N	20)	C..S..H	20)	L..V..F	20)	S..X..C	20)	C..1..A
21)	O..Q..Q	21)	W..T..O	21)	D..T..J	21)	M..W..G	21)	T..Y..D	21)	D..2..B
22)	P..R..R	22)	X..U..P	22)	E..U..K	22)	N..X..H	22)	U..Z..E	22)	E..3..C
23)	Q..S..S	23)	A..V..Q	23)	F..V..L	23)	O..Y..J	23)	V..1..F	23)	F..A..D
24)	R..T..T	24)	B..W..R	24)	G..W..M	24)	P..Z..K	24)	W..2..G	24)	G..B..E
25)	S..U..U	25)	C..X..S	25)	H..X..N	25)	Q..1..L	25)	X..3..H	25)	H..C..F
26)	T..V..V	26)	D..Y..T	26)	J..Y..O	26)	R..2..M	26)	A..A..J	26)	J..D..G
27)	U..W..W	27)	E..Z..U	27)	K..Z..P	27)	S..3..N	27)	B..B..K	27)	K..E..H
28)	V..X..X	28)	F..1..V	28)	L..1..Q	28)	T..A..O	28)	C..C..L	28)	L..F..J
29)	W..Y..Y			29)	M..2..R	29)	U..B..P	29)	D..D..M	29)	M..G..K
30)	X..Z..Z			30)	N..3..S	30)	V..C..Q	30)	E..E..N	30)	N..H..L
31)	A..1..1			31)	O..A..T			31)	F..F..O		

1907 1907

....JULY....		...AUGUST...		..SEPTEMBER.		..OCTOBER...		..NOVEMBER..		...DECEMBER..	
	P--E--I		P--E--I		P--E--I		P--E--I		P--E--I		P--E--I
1)	O..J..M	1)	W..M..K	1)	G..P..H	1)	O..R..E	1)	W..U..C	1)	F..W..8
2)	P..K..N	2)	X..N..L	2)	H..Q..J	2)	P..S..F	2)	X..V..D	2)	G..X..A
3)	Q..L..O	3)	A..O..M	3)	J..R..K	3)	Q..T..G	3)	A..W..E	3)	H..Y..B
4)	R..M..P	4)	B..P..N	4)	K..S..L	4)	R..U..H	4)	B..X..F	4)	J..Z..C
5)	S..N..Q	5)	C..Q..O	5)	L..T..M	5)	S..V..J	5)	C..Y..G	5)	K..1..D
6)	T..O..R	6)	D..R..P	6)	M..U..N	6)	T..W..K	6)	D..Z..H	6)	L..2..E
7)	U..P..S	7)	E..S..Q	7)	N..V..O	7)	U..X..L	7)	E..1..J	7)	M..3..F
8)	V..Q..T	8)	F..T..R	8)	O..W..P	8)	V..Y..M	8)	F..2..K	8)	N..A..G
9)	W..R..U	9)	G..U..S	9)	P..X..Q	9)	W..Z..N	9)	G..3..L	9)	O..B..H
10)	X..S..V	10)	H..V..T	10)	Q..Y..R	10)	X..1..O	10)	H..A..M	10)	P..C..J
11)	A..T..W	11)	J..W..U	11)	R..Z..S	11)	A..2..P	11)	J..B..N	11)	Q..D..K
12)	B..U..X	12)	K..X..V	12)	S..1..T	12)	B..3..Q	12)	K..C..O	12)	R..E..L
13)	C..V..Y	13)	L..Y..W	13)	T..2..U	13)	C..A..R	13)	L..D..P	13)	S..F..M
14)	D..W..Z	14)	M..Z..X	14)	U..3..V	14)	D..B..S	14)	M..E..Q	14)	T..G..N
15)	E..X..1	15)	N..1..Y	15)	V..A..W	15)	E..C..T	15)	N..F..R	15)	U..H..O
16)	F..Y..2	16)	O..2..Z	16)	W..B..X	16)	F..D..U	16)	O..G..S	16)	V..J..P
17)	G..Z..3	17)	P..3..1	17)	X..C..Y	17)	G..E..V	17)	P..H..T	17)	W..K..Q
18)	H..1..4	18)	Q..A..2	18)	A..D..Z	18)	H..F..W	18)	Q..J..U	18)	X..L..R
19)	J..2..5	19)	R..B..3	19)	B..E..1	19)	J..G..X	19)	R..K..V	19)	A..M..S
20)	K..3..6	20)	S..C..4	20)	C..F..2	20)	K..H..Y	20)	S..L..W	20)	B..N..T
21)	L..A..7	21)	T..D..5	21)	D..G..3	21)	L..J..Z	21)	T..M..X	21)	C..O..U
22)	M..B..8	22)	U..E..6	22)	E..H..4	22)	M..K..1	22)	U..N..Y	22)	D..P..V
23)	N..C..A	23)	V..F..7	23)	F..J..5	23)	N..L..2	23)	V..O..Z	23)	E..Q..W
24)	O..D..B	24)	W..G..8	24)	G..K..6	24)	O..M..3	24)	W..P..1	24)	F..R..X
25)	P..E..C	25)	X..H..A	25)	H..L..7	25)	P..N..4	25)	X..Q..2	25)	G..S..Y
26)	Q..F..D	26)	A..J..B	26)	J..M..8	26)	Q..O..5	26)	A..R..3	26)	H..T..Z
27)	R..G..E	27)	B..K..C	27)	K..N..A	27)	R..P..6	27)	B..S..4	27)	J..U..1
28)	S..H..F	28)	C..L..D	28)	L..O..B	28)	S..Q..7	28)	C..T..5	28)	K..V..2
29)	T..J..G	29)	D..M..E	29)	M..P..C	29)	T..R..8	29)	D..U..6	29)	L..W..3
30)	U..K..H	30)	E..N..F	30)	N..Q..D	30)	U..S..A	30)	E..V..7	30)	M..X..4
31)	V..L..J	31)	F..O..G			31)	V..T..B			31)	N..Y..5

CODES: P-PHYSICAL BIORHYTHM CURVE,E-EMOTIONAL BIORHYTHM CURVE,I-INTELLECTUAL BIORHYTHM CURVE

1908						1908
	...JANUARY..	..FEBRUARY..	....MARCH...	....APRIL...	....MAY....	....JUNE....
	P--E--I	P--E--I	P--E--I	P--E--I	P--E--I	P--E--I
1)	O..Z..6	W..3..4	E..A..Z	N..D..X	U..F..U	E..J..S
2)	P..1..7	X..A..5	F..B..1	O..E..Y	V..G..V	F..K..T
3)	Q..2..8	A..B..6	G..C..2	P..F..Z	W..H..W	G..L..U
4)	R..3..A	B..C..7	H..D..3	Q..G..1	X..J..X	H..M..V
5)	S..A..B	C..D..8	J..E..4	R..H..2	A..K..Y	J..N..W
6)	T..B..C	D..E..A	K..F..5	S..J..3	B..L..Z	K..O..K
7)	U..C..D	E..F..B	L..G..6	T..K..4	C..M..1	L..P..Y
8)	V..D..E	F..G..C	M..H..7	U..L..5	D..N..2	M..Q..Z
9)	W..E..F	G..H..D	N..J..8	V..M..6	E..O..3	N..R..1
10)	X..F..G	H..J..E	O..K..A	W..N..7	F..P..4	O..S..2
11)	A..G..H	J..K..F	P..L..B	X..O..8	G..Q..5	P..T..3
12)	B..H..J	K..L..G	Q..M..C	A..P..A	H..R..6	Q..U..4
13)	C..J..K	L..M..H	R..N..D	B..Q..B	J..S..7	R..V..5
14)	D..K..L	M..N..J	S..O..E	C..R..C	K..T..8	S..W..6
15)	E..L..M	N..O..K	T..P..F	D..S..D	L..U..A	T..X..7
16)	F..M..N	O..P..L	U..Q..G	E..T..E	M..V..B	U..Y..8
17)	G..N..O	P..Q..M	V..R..H	F..U..F	N..W..C	V..Z..A
18)	H..O..P	Q..R..N	W..S..J	G..V..G	O..X..D	W..1..B
19)	J..P..Q	R..S..O	X..T..K	H..W..H	P..Y..E	X..2..C
20)	K..Q..R	S..T..P	A..U..L	J..X..J	Q..Z..F	A..3..D
21)	L..R..S	T..U..Q	B..V..M	K..Y..K	R..1..G	B..A..E
22)	M..S..T	U..V..R	C..W..N	L..Z..L	S..2..H	C..B..F
23)	N..T..U	V..W..S	D..X..O	M..1..M	T..3..J	D..C..G
24)	O..U..V	W..X..T	E..Y..P	N..2..N	U..A..K	E..D..H
25)	P..V..W	X..Y..U	F..Z..Q	O..3..O	V..B..L	F..E..J
26)	Q..W..X	A..Z..V	G..1..R	P..A..P	W..C..M	G..F..K
27)	R..X..Y	B..1..W	H..2..S	Q..B..Q	X..D..N	H..G..L
28)	S..Y..Z	C..2..X	J..3..T	R..C..R	A..E..O	J..H..M
29)	T..Z..1	D..3..Y	K..A..U	S..D..S	B..F..P	K..J..N
30)	U..1..2		L..B..V	T..E..T	C..G..Q	L..K..Q
31)	V..2..3		M..C..W		D..H..R	

	....JULY....	...AUGUST...	..SEPTEMBER.	..OCTOBER...	..NOVEMBER..	..DECEMBER..
	P--E--I	P--E--I	P--E--I	P--E--I	P--E--I	P--E--I
1)	M..L..P	U..O..N	E..R..L	M..T..H	U..W..F	D..Y..C
2)	N..M..Q	V..P..O	F..S..M	N..U..J	V..X..G	E..Z..D
3)	O..N..R	W..Q..P	G..T..N	O..V..K	W..Y..H	F..1..E
4)	P..O..S	X..R..Q	H..U..O	P..W..L	X..Z..J	G..2..F
5)	Q..P..T	A..S..R	J..V..P	Q..X..M	A..1..K	H..3..G
6)	R..Q..U	B..T..S	K..W..Q	R..Y..N	B..2..L	J..A..H
7)	S..R..V	C..U..T	L..X..R	S..Z..O	C..3..M	K..B..J
8)	T..S..W	D..V..U	M..Y..S	T..1..P	D..A..N	L..C..K
9)	U..T..X	E..W..V	N..Z..T	U..2..Q	E..B..O	M..D..L
10)	V..U..Y	F..X..W	O..1..U	V..3..R	F..C..P	N..E..M
11)	W..V..Z	G..Y..X	P..2..V	W..A..S	G..D..Q	O..F..N
12)	X..W..1	H..Z..Y	Q..3..W	X..B..T	H..E..R	P..G..O
13)	A..X..2	J..1..Z	R..A..X	A..C..U	J..F..S	Q..H..P
14)	B..Y..3	K..2..1	S..B..Y	B..D..V	K..G..T	R..J..Q
15)	C..Z..4	L..3..2	T..C..Z	C..E..W	L..H..U	S..K..R
16)	D..1..5	M..A..3	U..D..1	D..F..X	M..J..V	T..L..S
17)	E..2..6	N..B..4	V..E..2	E..G..Y	N..K..W	U..M..T
18)	F..3..7	O..C..5	W..F..3	F..H..Z	O..L..X	V..N..U
19)	G..A..8	P..D..6	X..G..4	G..J..1	P..M..Y	W..O..V
20)	H..B..A	Q..E..7	A..H..5	H..K..2	Q..N..Z	X..P..W
21)	J..C..B	R..F..8	B..J..6	J..L..3	R..O..1	A..Q..X
22)	K..D..C	S..G..A	C..K..7	K..M..4	S..P..2	B..R..Y
23)	L..E..D	T..H..B	D..L..8	L..N..5	T..Q..3	C..S..Z
24)	M..F..E	U..J..C	E..M..A	M..O..6	U..R..4	D..T..1
25)	N..G..F	V..K..D	F..N..B	N..P..7	V..S..5	E..U..2
26)	O..H..G	W..L..E	G..O..C	O..Q..8	W..T..6	F..V..3
27)	P..J..H	X..M..F	H..P..D	P..R..A	X..U..7	G..W..4
28)	Q..K..J	A..N..G	J..Q..E	Q..S..B	A..V..8	H..X..5
29)	R..L..K	B..O..H	K..R..F	R..T..C	B..W..A	J..Y..6
30)	S..M..L	C..P..J	L..S..G	S..U..D	C..X..B	K..Z..7
31)	T..N..M	D..Q..K		T..V..E		L..1..8

CODES: P-PHYSICAL BIORHYTHM CURVE,E-EMOTIONAL BIORHYTHM CURVE,I-INTELLECTUAL BIORHYTHM CURVE

1909 | | | | | | 1909

	...JANUARY..	..FEBRUARY..	....MARCH...	....APRIL...	.....MAY....	....JUNE....
	P--E--I	P--E--I	P--E--I	P--E--I	P--E--I	P--E--I
1)	M..2..A	U..B..7	B..B..2	K..E..Z	R..G..W	B..K..U
2)	N..3..B	V..C..8	C..C..3	L..F..1	S..H..X	C..L..V
3)	O..A..C	W..D..A	D..D..4	M..G..2	T..J..Y	D..M..W
4)	P..B..D	X..E..B	E..E..5	N..H..3	U..K..Z	E..N..X
5)	Q..C..E	A..F..C	F..F..6	O..J..4	V..L..1	F..O..Y
6)	R..D..F	B..G..D	G..G..7	P..K..5	W..M..2	G..P..Z
7)	S..E..G	C..H..E	H..H..8	Q..L..6	X..N..3	H..Q..1
8)	T..F..H	D..J..F	J..J..A	R..M..7	A..O..4	J..R..2
9)	U..G..J	E..K..G	K..K..B	S..N..8	B..P..5	K..S..3
10)	V..H..K	F..L..H	L..L..C	T..O..A	C..Q..6	L..T..4
11)	W..J..L	G..M..J	M..M..D	U..P..B	D..R..7	M..U..5
12)	X..K..M	H..N..K	N..N..E	V..Q..C	E..S..8	N..V..6
13)	A..L..N	J..O..L	O..O..F	W..R..D	F..T..A	O..W..7
14)	B..M..O	K..P..M	P..P..G	X..S..E	G..U..B	P..X..8
15)	C..N..P	L..Q..N	Q..Q..H	A..T..F	H..V..C	Q..Y..A
16)	D..O..Q	M..R..O	R..R..J	B..U..G	J..W..D	R..Z..B
17)	E..P..R	N..S..P	S..S..K	C..V..H	K..X..E	S..1..C
18)	F..Q..S	O..T..Q	T..T..L	D..W..J	L..Y..F	T..2..D
19)	G..R..T	P..U..R	U..U..M	E..X..K	M..Z..G	U..3..E
20)	H..S..U	Q..V..S	V..V..N	F..Y..L	N..1..H	V..A..F
21)	J..T..V	R..W..T	W..W..O	G..Z..M	O..2..J	W..B..G
22)	K..U..W	S..X..U	X..X..P	H..1..N	P..3..K	X..C..H
23)	L..V..X	T..Y..V	A..Y..Q	J..2..O	Q..A..L	A..D..J
24)	M..W..Y	U..Z..W	B..Z..R	K..3..P	R..B..M	B..E..K
25)	N..X..Z	V..1..X	C..1..S	L..A..Q	S..C..N	C..F..L
26)	O..Y..1	W..2..Y	D..2..T	M..B..R	T..D..O	D..G..M
27)	P..Z..2	X..3..Z	E..3..U	N..C..S	U..E..P	E..H..N
28)	Q..1..3	A..A..1	F..A..V	O..D..T	V..F..Q	F..J..O
29)	R..2..4		G..B..W	P..E..U	W..G..R	G..K..P
30)	S..3..5		H..C..X	Q..F..V	X..H..S	H..L..Q
31)	T..A..6		J..D..Y		A..J..T	

1909 | 1909

	....JULY....	...AUGUST...	..SEPTEMBER.	..OCTOBER...	..NOVEMBER..	..DECEMBER..
	P--E--I	P--E--I	P--E--I	P--E--I	P--E--I	P--E--I
1)	J..M..R	R..P..P	B..S..N	J..U..K	R..X..H	A..Z..E
2)	K..N..S	S..Q..Q	C..T..O	K..V..L	S..Y..J	B..1..F
3)	L..O..T	T..R..R	D..U..P	L..W..M	T..Z..K	C..2..G
4)	M..P..U	U..S..S	E..V..Q	M..X..N	U..1..L	D..3..H
5)	N..Q..V	V..T..T	F..W..R	N..Y..O	V..2..M	E..A..J
6)	O..R..W	W..U..U	G..X..S	O..Z..P	W..3..N	F..B..K
7)	P..S..X	X..V..V	H..Y..T	P..1..Q	X..A..O	G..C..L
8)	Q..T..Y	A..W..W	J..Z..U	Q..2..R	A..B..P	H..D..M
9)	R..U..Z	B..X..X	K..1..V	R..3..S	B..C..Q	J..E..N
10)	S..V..1	C..Y..Y	L..2..W	S..A..T	C..D..R	K..F..O
11)	T..W..2	D..Z..Z	M..3..X	T..B..U	D..E..S	L..G..P
12)	U..X..3	E..1..1	N..A..Y	U..C..V	E..F..T	M..H..Q
13)	V..Y..4	F..2..2	O..B..Z	V..D..W	F..G..U	N..J..R
14)	W..Z..5	G..3..3	P..C..1	W..E..X	G..H..V	O..K..S
15)	X..1..6	H..A..4	Q..D..2	X..F..Y	H..J..W	P..L..T
16)	A..2..7	J..B..5	R..E..3	A..G..Z	J..K..X	Q..M..U
17)	B..3..8	K..C..6	S..F..4	B..H..1	K..L..Y	R..N..V
18)	C..A..A	L..D..7	T..G..5	C..J..2	L..M..Z	S..O..W
19)	D..B..B	M..E..8	U..H..6	D..K..3	M..N..1	T..P..X
20)	E..C..C	N..F..A	V..J..7	E..L..4	N..O..2	U..Q..Y
21)	F..D..D	O..G..B	W..K..8	F..M..5	O..P..3	V..R..Z
22)	G..E..E	P..H..C	X..L..A	G..N..6	P..Q..4	W..S..1
23)	H..F..F	Q..J..D	A..M..B	H..O..7	Q..R..5	X..T..2
24)	J..G..G	R..K..E	B..N..C	J..P..8	R..S..6	A..U..3
25)	K..H..H	S..L..F	C..O..D	K..Q..A	S..T..7	B..V..4
26)	L..J..J	T..M..G	D..P..E	L..R..B	T..U..8	C..W..5
27)	M..K..K	U..N..H	E..Q..F	M..S..C	U..V..A	D..X..6
28)	N..L..L	V..O..J	F..R..G	N..T..D	V..W..B	E..Y..7
29)	O..M..M	W..P..K	G..S..H	O..U..E	W..X..C	F..Z..8
30)	P..N..N	X..Q..L	H..T..J	P..V..F	X..Y..D	G..1..A
31)	Q..O..O	A..R..M		Q..W..G		H..2..B

CODES: P-PHYSICAL BIORHYTHM CURVE,E-EMOTIONAL BIORHYTHM CURVE,I-INTELLECTUAL BIORHYTHM CURVE

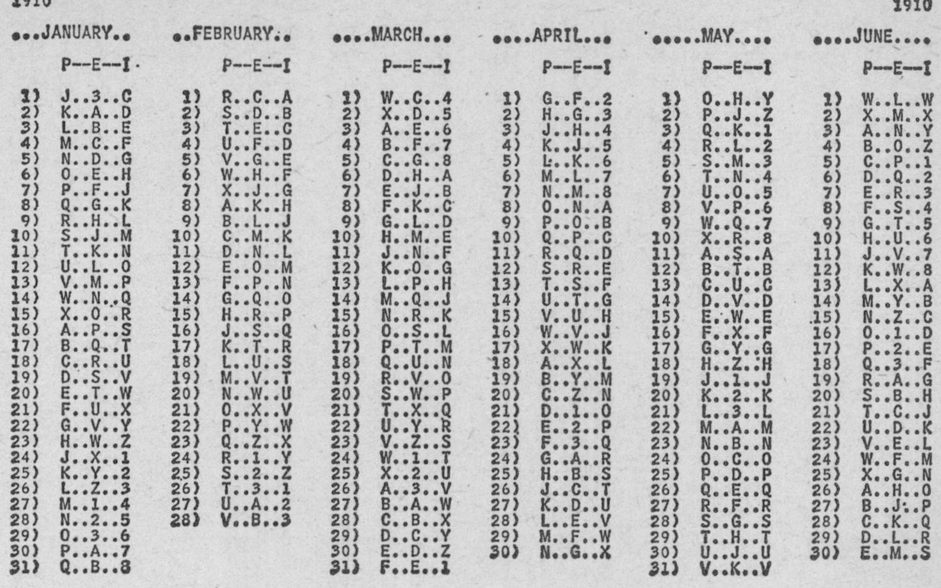

1910 | 1910

Day	...JANUARY.. P--E--I	..FEBRUARY.. P--E--I	....MARCH... P--E--I	....APRIL... P--E--I	.....MAY.... P--E--I	....JUNE.... P--E--I
1)	J..3..C	R..C..A	W..C..4	G..F..2	O..H..Y	W..L..W
2)	K..A..D	S..D..B	X..D..5	H..G..3	P..J..Z	X..M..X
3)	L..B..E	T..E..C	A..E..6	J..H..4	Q..K..1	A..N..Y
4)	M..C..F	U..F..D	B..F..7	K..J..5	R..L..2	B..O..Z
5)	N..D..G	V..G..E	C..G..8	L..K..6	S..M..3	C..P..1
6)	O..E..H	W..H..F	D..H..A	M..L..7	T..N..4	D..Q..2
7)	P..F..J	X..J..G	E..J..B	N..M..8	U..O..5	E..R..3
8)	Q..G..K	A..K..H	F..K..C	O..N..A	V..P..6	F..S..4
9)	R..H..L	B..L..J	G..L..D	P..O..B	W..Q..7	G..T..5
10)	S..J..M	C..M..K	H..M..E	Q..P..C	X..R..8	H..U..6
11)	T..K..N	D..N..L	J..N..F	R..Q..D	A..S..A	J..V..7
12)	U..L..O	E..O..M	K..O..G	S..R..E	B..T..B	K..W..8
13)	V..M..P	F..P..N	L..P..H	T..S..F	C..U..C	L..X..A
14)	W..N..Q	G..Q..O	M..Q..J	U..T..G	D..V..D	M..Y..B
15)	X..O..R	H..R..P	N..R..K	V..U..H	E..W..E	N..Z..C
16)	A..P..S	J..S..Q	O..S..L	W..V..J	F..X..F	O..1..D
17)	B..Q..T	K..T..R	P..T..M	X..W..K	G..Y..G	P..2..E
18)	C..R..U	L..U..S	Q..U..N	A..X..L	H..Z..H	Q..3..F
19)	D..S..V	M..V..T	R..V..O	B..Y..M	J..1..J	R..A..G
20)	E..T..W	N..W..U	S..W..P	C..Z..N	K..2..K	S..B..H
21)	F..U..X	O..X..V	T..X..Q	D..1..O	L..3..L	T..C..J
22)	G..V..Y	P..Y..W	U..Y..R	E..2..P	M..A..M	U..D..K
23)	H..W..Z	Q..Z..X	V..Z..S	F..3..Q	N..B..N	V..E..L
24)	J..X..1	R..1..Y	W..1..T	G..A..R	O..C..O	W..F..M
25)	K..Y..2	S..2..Z	X..2..U	H..B..S	P..D..P	X..G..N
26)	L..Z..3	T..3..1	A..3..V	J..C..T	Q..E..Q	A..H..O
27)	M..1..4	U..A..2	B..A..W	K..D..U	R..F..R	B..J..P
28)	N..2..5	V..B..3	C..B..X	L..E..V	S..G..S	C..K..Q
29)	O..3..6		D..C..Y	M..F..W	T..H..T	D..L..R
30)	P..A..7		E..D..Z	N..G..X	U..J..U	E..M..S
31)	Q..B..8		F..E..1		V..K..V	

1910 | 1910

	....JULY....	...AUGUST...	..SEPTEMBER.	..OCTOBER...	..NOVEMBER..	..DECEMBER..
	P--E--I	P--E--I	P--E--I	P--E--I	P--E--I	P--E--I
1)	F..N..T	O..Q..R	W..T..P	F..V..M	O..Y..K	V..1..G
2)	G..O..U	P..R..S	X..U..Q	G..W..N	P..Z..L	W..2..H
3)	H..P..V	Q..S..T	A..V..R	H..X..O	Q..1..M	X..3..J
4)	J..Q..W	R..T..U	B..W..S	J..Y..P	R..2..N	A..A..K
5)	K..R..X	S..U..V	C..X..T	K..Z..Q	S..3..O	B..B..L
6)	L..S..Y	T..V..W	D..Y..U	L..1..R	T..A..P	C..C..M
7)	M..T..Z	U..W..X	E..Z..V	M..2..S	U..B..Q	D..D..N
8)	N..U..1	V..X..Y	F..1..W	N..3..T	V..C..R	E..E..O
9)	O..V..2	W..Y..Z	G..2..X	O..A..U	W..D..S	F..F..P
10)	P..W..3	X..Z..1	H..3..Y	P..B..V	X..E..T	G..G..Q
11)	Q..X..4	A..1..2	J..A..Z	Q..C..W	A..F..U	H..H..R
12)	R..Y..5	B..2..3	K..B..1	R..D..X	B..G..V	J..J..S
13)	S..Z..6	C..3..4	L..C..2	S..E..Y	C..H..W	K..K..T
14)	T..1..7	D..A..5	M..D..3	T..F..Z	D..J..X	L..L..U
15)	U..2..8	E..B..6	N..E..4	U..G..1	E..K..Y	M..M..V
16)	V..3..A	F..C..7	O..F..5	V..H..2	F..L..Z	N..N..W
17)	W..A..B	G..D..8	P..G..6	W..J..3	G..M..1	O..O..X
18)	X..B..C	H..E..A	Q..H..7	X..K..4	H..N..2	P..P..Y
19)	A..C..D	J..F..B	R..J..8	A..L..5	J..O..3	Q..Q..Z
20)	B..D..E	K..G..C	S..K..A	B..M..6	K..P..4	R..R..1
21)	C..E..F	L..H..D	T..L..B	C..N..7	L..Q..5	S..S..2
22)	D..F..G	M..J..E	U..M..C	D..O..8	M..R..6	T..T..3
23)	E..G..H	N..K..F	V..N..D	E..P..A	N..S..7	U..U..4
24)	F..H..J	O..L..G	W..O..E	F..Q..B	O..T..8	V..V..5
25)	G..J..K	P..M..H	X..P..F	G..R..C	P..U..A	W..W..6
26)	H..K..L	Q..N..J	A..Q..G	H..S..D	Q..V..B	X..X..7
27)	J..L..M	R..O..K	B..R..H	J..T..E	R..W..C	A..Y..8
28)	K..M..N	S..P..L	C..S..J	K..U..F	S..X..D	B..Z..A
29)	L..N..O	T..Q..M	D..T..K	L..V..G	T..Y..E	C..1..B
30)	M..O..P	U..R..N	E..U..L	M..W..H	U..Z..F	D..2..C
31)	N..P..Q	V..S..O		N..X..J		E..3..D

CODES: P-PHYSICAL BIORHYTHM CURVE,E-EMOTIONAL BIORHYTHM CURVE,I-INTELLECTUAL BIORHYTHM CURVE

1911 | | 1911

	...JANUARY..	..FEBRUARY..	....MARCH...	....APRIL...	....MAY....	....JUNE...
	P--E--I	P--E--I	P--E--I	P--E--I	P--E--I	P--E--I
1)	F..A..E	O..D..C	T..D..6	D..G..4	L..J..1	T..M..Y
2)	G..B..F	P..E..D	U..E..7	E..H..5	M..K..2	U..N..Z
3)	H..C..G	Q..F..E	V..F..8	F..J..6	N..L..3	V..O..1
4)	J..D..H	R..G..F	W..G..A	G..K..7	O..M..4	W..P..2
5)	K..E..J	S..H..G	X..H..B	H..L..8	P..N..5	X..Q..3
6)	L..F..K	T..J..H	A..J..C	J..M..A	Q..O..6	A..R..4
7)	M..G..L	U..K..J	B..K..D	K..N..B	R..P..7	B..S..5
8)	N..H..M	V..L..K	C..L..E	L..O..C	S..Q..8	C..T..6
9)	O..J..N	W..M..L	D..M..F	M..P..D	T..R..A	D..U..7
10)	P..K..O	X..N..M	E..N..G	N..Q..E	U..S..B	E..V..8
11)	Q..L..P	A..O..N	F..O..H	O..R..F	V..T..C	F..W..A
12)	R..M..Q	B..P..O	G..P..J	P..S..G	W..U..D	G..X..B
13)	S..N..R	C..Q..P	H..Q..K	Q..T..H	X..V..E	H..Y..C
14)	T..O..S	D..R..Q	J..R..L	R..U..J	A..W..F	J..Z..D
15)	U..P..T	E..S..R	K..S..M	S..V..K	B..X..G	K..1..E
16)	V..Q..U	F..T..S	L..T..N	T..W..L	C..Y..H	L..2..F
17)	W..R..V	G..U..T	M..U..O	U..X..M	D..Z..J	M..3..G
18)	X..S..W	H..V..U	N..V..P	V..Y..N	E..1..K	N..A..H
19)	A..T..X	J..W..V	O..W..Q	W..Z..O	F..2..L	O..B..J
20)	B..U..Y	K..X..W	P..X..R	X..1..P	G..3..M	P..C..K
21)	C..V..Z	L..Y..X	Q..Y..S	A..2..Q	H..A..N	Q..D..L
22)	D..W..1	M..Z..Y	R..Z..T	B..3..R	J..B..O	R..E..M
23)	E..X..2	N..1..Z	S..1..U	C..A..S	K..C..P	S..F..N
24)	F..Y..3	O..2..1	T..2..V	D..B..T	L..D..Q	T..G..O
25)	G..Z..4	P..3..2	U..3..W	E..C..U	M..E..R	U..H..P
26)	H..1..5	Q..A..3	V..A..X	F..D..V	N..F..S	V..J..Q
27)	J..2..6	R..B..4	W..B..Y	G..E..W	O..G..T	W..K..R
28)	K..3..7	S..C..5	X..C..Z	H..F..X	P..H..U	X..L..S
29)	L..A..8		A..D..1	J..G..Y	Q..J..V	A..M..T
30)	M..B..A		B..E..2	K..H..Z	R..K..W	B..N..U
31)	N..C..B		C..F..3		S..L..X	

1911						1911
Day	....JULY....	...AUGUST...	..SEPTEMBER.	..OCTOBER...	..NOVEMBER..	..DECEMBER..
	P--E--I	P--E--I	P--E--I	P--E--I	P--E--I	P--E--I
1)	C..O..V	L..R..T	T..U..R	C..W..O	L..Z..M	S..2..J
2)	D..P..W	M..S..U	U..V..S	D..X..P	M..1..N	T..3..K
3)	E..Q..X	N..T..V	V..W..T	E..Y..Q	N..2..O	U..A..L
4)	F..R..Y	O..U..W	W..X..U	F..Z..R	O..3..P	V..B..M
5)	G..S..Z	P..V..X	X..Y..V	G..1..S	P..A..Q	W..C..N
6)	H..T..1	Q..W..Y	A..Z..W	H..2..T	Q..B..R	X..D..O
7)	J..U..2	R..X..Z	B..1..X	J..3..U	R..C..S	A..E..P
8)	K..V..3	S..Y..1	C..2..Y	K..A..V	S..D..T	B..F..Q
9)	L..W..4	T..Z..2	D..3..Z	L..B..W	T..E..U	C..G..R
10)	M..X..5	U..1..3	E..A..1	M..C..X	U..F..V	D..H..S
11)	N..Y..6	V..2..4	F..B..2	N..D..Y	V..G..W	E..J..T
12)	O..Z..7	W..3..5	G..C..3	O..E..Z	W..H..X	F..K..U
13)	P..1..8	X..A..6	H..D..4	P..F..1	X..J..Y	G..L..V
14)	Q..2..A	A..B..7	J..E..5	Q..G..2	A..K..Z	H..M..W
15)	R..3..B	B..C..8	K..F..6	R..H..3	B..L..1	J..N..X
16)	S..A..C	C..D..A	L..G..7	S..J..4	C..M..2	K..O..Y
17)	T..B..D	D..E..B	M..H..8	T..K..5	D..N..3	L..P..Z
18)	U..C..E	E..F..C	N..J..A	U..L..6	E..O..4	M..Q..1
19)	V..D..F	F..G..D	O..K..B	V..M..7	F..P..5	N..R..2
20)	W..E..G	G..H..E	P..L..C	W..N..8	G..Q..6	O..S..3
21)	X..F..H	H..J..F	Q..M..D	X..O..A	H..R..7	P..T..4
22)	A..G..J	J..K..G	R..N..E	A..P..B	J..S..8	Q..U..5
23)	B..H..K	K..L..H	S..O..F	B..Q..C	K..T..A	R..V..6
24)	C..J..L	L..M..J	T..P..G	C..R..D	L..U..B	S..W..7
25)	D..K..M	M..N..K	U..Q..H	D..S..E	M..V..C	T..X..8
26)	E..L..N	N..O..L	V..R..J	E..T..F	N..W..D	U..Y..A
27)	F..M..O	O..P..M	W..S..K	F..U..G	O..X..E	V..Z..B
28)	G..N..P	P..Q..N	X..T..L	G..V..H	P..Y..F	W..1..C
29)	H..O..Q	Q..R..O	A..U..M	H..W..J	Q..Z..G	X..2..D
30)	J..P..R	R..S..P	E..V..N	J..X..K	R..1..H	A..3..E
31)	K..Q..S	S..T..Q		K..Y..L		B..A..F

CODES: P-PHYSICAL BIORHYTHM CURVE, E-EMOTIONAL BIORHYTHM CURVE, I-INTELLECTUAL BIORHYTHM CURVE

1912 1912

	•...JANUARY..	..FEBRUARY:.	•...MARCH...	•...APRIL...	•....MAY....	•...JUNE....•
	P--E--I	P--E--I	P--E--I	P--E--I	P--E--I	P--E--I
1)	C..B..G	L..E..E	R..F..A	B..J..7	J..L..4	R..O..2
2)	D..C..H	M..F..F	S..G..B	C..K..8	K..M..5	S..P..3
3)	E..D..J	N..G..G	T..H..C	D..L..A	L..N..6	T..Q..4
4)	F..E..K	O..H..H	U..J..D	E..M..B	M..O..7	U..R..5
5)	G..F..L	P..J..J	V..K..E	F..N..C	N..P..8	V..S..6
6)	H..G..M	Q..K..K	W..L..F	G..O..D	O..Q..A	W..T..7
7)	J..H..N	R..L..L	X..M..G	H..P..E	P..R..B	X..U..8
8)	K..J..O	S..M..M	A..N..H	J..Q..F	Q..S..C	A..V..A
9)	L..K..P	T..N..N	B..O..J	K..R..G	R..T..D	B..W..B
10)	M..L..Q	U..O..O	C..P..K	L..S..H	S..U..E	C..X..C
11)	N..M..R	V..P..P	D..Q..L	M..T..J	T..V..F	D..Y..D
12)	O..N..S	W..Q..Q	E..R..M	N..U..K	U..W..G	E..Z..E
13)	P..O..T	X..R..R	F..S..N	O..V..L	V..X..H	F..1..F
14)	Q..P..U	A..S..S	G..T..O	P..W..M	W..Y..J	G..2..G
15)	R..Q..V	B..T..T	H..U..P	Q..X..N	X..Z..K	H..3..H
16)	S..R..W	C..U..U	J..V..Q	R..Y..O	A..1..L	J..A..J
17)	T..S..X	D..V..V	K..W..R	S..Z..P	B..2..M	K..B..K
18)	U..T..Y	E..W..W	L..X..S	T..1..Q	C..3..N	L..C..L
19)	V..U..Z	F..X..X	M..Y..T	U..2..R	D..A..O	M..D..M
20)	W..V..1	G..Y..Y	N..Z..U	V..3..S	E..B..P	N..E..N
21)	X..W..2	H..Z..Z	O..1..V	W..A..T	F..C..Q	O..F..O
22)	A..X..3	J..1..1	P..2..W	X..B..U	G..D..R	P..G..P
23)	B..Y..4	K..2..2	Q..3..X	A..C..V	H..E..S	Q..H..Q
24)	C..Z..5	L..3..3	R..A..Y	B..D..W	J..F..T	R..J..R
25)	D..1..6	M..A..4	S..B..Z	C..E..X	K..G..U	S..K..S
26)	E..2..7	N..B..5	T..C..1	D..F..Y	L..H..V	T..L..T
27)	F..3..8	O..C..6	U..D..2	E..G..Z	M..J..W	U..M..U
28)	G..A..A	P..D..7	V..E..3	F..H..1	N..K..X	V..N..V
29)	H..B..B	Q..E..8	W..F..4	G..J..2	O..L..Y	W..O..W
30)	J..C..C		X..G..5	H..K..3	P..M..Z	X..P..X
31)	K..D..D		A..H..6		Q..N..1	

1912 | 1912

	....JULY....	...AUGUST...	..SEPTEMBER.	..OCTOBER...	..NOVEMBER..	..DECEMBER..
	P--E--I	P--E--I	P--E--I	P--E--I	P--E--I	P--E--I
1)	A..Q..Y	J..T..W	R..W..U	A..Y..R	J..2..P	Q..A..M
2)	B..R..Z	K..U..X	S..X..V	B..Z..S	K..3..Q	R..B..N
3)	C..S..1	L..V..Y	T..Y..W	C..1..T	L..A..R	S..C..O
4)	D..T..2	M..W..Z	U..Z..X	D..2..U	M..B..S	T..D..P
5)	E..U..3	N..X..1	V..1..Y	E..3..V	N..C..T	U..E..Q
6)	F..V..4	O..Y..2	W..2..Z	F..A..W	O..D..U	V..F..R
7)	G..W..5	P..Z..3	X..3..1	G..B..X	P..E..V	W..G..S
8)	H..X..6	Q..1..4	A..A..2	H..C..Y	Q..F..W	X..H..T
9)	J..Y..7	R..2..5	B..B..3	J..D..Z	R..G..X	A..J..U
10)	K..Z..8	S..3..6	C..C..4	K..E..1	S..H..Y	B..K..V
11)	L..1..A	T..A..7	D..D..5	L..F..2	T..J..Z	C..L..W
12)	M..2..B	U..B..8	E..E..6	M..G..3	U..K..1	D..M..X
13)	N..3..C	V..C..A	F..F..7	N..H..4	V..L..2	E..N..Y
14)	O..A..D	W..D..B	G..G..8	O..J..5	W..M..3	F..O..Z
15)	P..B..E	X..E..C	H..H..A	P..K..6	X..N..4	G..P..1
16)	Q..C..F	A..F..D	J..J..B	Q..L..7	A..O..5	H..Q..2
17)	R..D..G	B..G..E	K..K..C	R..M..8	B..P..6	J..R..3
18)	S..E..H	C..H..F	L..L..D	S..N..A	C..Q..7	K..S..4
19)	T..F..J	D..J..G	M..M..E	T..O..B	D..R..8	L..T..5
20)	U..G..K	E..K..H	N..N..F	U..P..C	E..S..A	M..U..6
21)	V..H..L	F..L..J	O..O..G	V..Q..D	F..T..B	N..V..7
22)	W..J..M	G..M..K	P..P..H	W..R..E	G..U..C	O..W..8
23)	X..K..N	H..N..L	Q..Q..J	X..S..F	H..V..D	P..X..A
24)	A..L..O	J..O..M	R..R..K	A..T..G	J..W..E	Q..Y..B
25)	B..M..P	K..P..N	S..S..L	B..U..H	K..X..F	R..Z..C
26)	C..N..Q	L..Q..O	T..T..M	C..V..J	L..Y..G	S..1..D
27)	D..O..R	M..R..P	U..U..N	D..W..K	M..Z..H	T..2..E
28)	E..P..S	N..S..Q	V..V..O	E..X..L	N..1..J	U..3..F
29)	F..Q..T	O..T..R	W..W..P	F..Y..M	O..2..K	V..A..G
30)	G..R..U	P..U..S	X..X..Q	G..Z..N	P..3..L	W..B..H
31)	H..S..V	Q..V..T		H..1..O		X..C..J

CODES: P-PHYSICAL BIORHYTHM CURVE, E-EMOTIONAL BIORHYTHM CURVE, I-INTELLECTUAL BIORHYTHM CURVE

1913 1913

	...JANUARY..	..FEBRUARY..	....MARCH....	....APRIL...	....MAY....	....JUNE....
	P--E--I	P--E--I	P--E--I	P--E--I	P--E--I	P--E--I
1)	A..D..K	J..G..H	O..G..C	W..K..A	F..M..6	O..P..4
2)	B..E..L	K..H..J	P..H..D	X..L..B	G..N..7	P..Q..5
3)	C..F..M	L..J..K	Q..J..E	A..M..C	H..O..8	Q..R..6
4)	D..G..N	M..K..L	R..K..F	B..N..D	J..P..A	R..S..7
5)	E..H..O	N..L..M	S..L..G	C..O..E	K..Q..B	S..T..8
6)	F..J..P	O..M..N	T..M..H	D..P..F	L..R..C	T..U..A
7)	G..K..Q	P..N..O	U..N..J	E..Q..G	M..S..D	U..V..B
8)	H..L..R	Q..O..P	V..O..K	F..R..H	N..T..E	V..W..C
9)	J..M..S	R..P..Q	W..P..L	G..S..J	O..U..F	W..X..D
10)	K..N..T	S..Q..R	X..Q..M	H..T..K	P..V..G	X..Y..E
11)	L..O..U	T..R..S	A..R..N	J..U..L	Q..W..H	A..Z..F
12)	M..P..V	U..S..T	B..S..O	K..V..M	R..X..J	B..1..G
13)	N..Q..W	V..T..U	C..T..P	L..W..N	S..Y..K	C..2..H
14)	O..R..X	W..U..V	D..U..Q	M..X..O	T..Z..L	D..3..J
15)	P..S..Y	X..V..W	E..V..R	N..Y..P	U..1..M	E..A..K
16)	Q..T..Z	A..W..X	F..W..S	O..Z..Q	V..2..N	F..B..L
17)	R..U..1	B..X..Y	G..X..T	P..1..R	W..3..O	G..C..M
18)	S..V..2	C..Y..Z	H..Y..U	Q..2..S	X..A..P	H..D..N
19)	T..W..3	D..Z..1	J..Z..V	R..3..T	A..B..Q	J..E..O
20)	U..X..4	E..1..2	K..1..W	S..A..U	B..C..R	K..F..P
21)	V..Y..5	F..2..3	L..2..X	T..B..V	C..D..S	L..G..Q
22)	W..Z..6	G..3..4	M..3..Y	U..C..W	D..E..T	M..H..R
23)	X..1..7	H..A..5	N..A..Z	V..D..X	E..F..U	N..J..S
24)	A..2..8	J..B..6	O..B..1	W..E..Y	F..G..V	O..K..T
25)	B..3..A	K..C..7	P..C..2	X..F..Z	G..H..W	P..L..U
26)	C..A..B	L..D..8	Q..D..3	A..G..1	H..J..X	Q..M..V
27)	D..B..C	M..E..A	R..E..4	B..H..2	J..K..Y	R..N..W
28)	E..C..D	N..F..B	S..F..5	C..J..3	K..L..Z	S..O..X
29)	F..D..E		T..G..6	D..K..4	L..M..1	T..P..Y
30)	G..E..F		U..H..7	E..L..5	M..N..2	U..Q..Z
31)	H..F..G		V..J..8		N..O..3	

1913 1913

....JULY....		...AUGUST...		..SEPTEMBER.		..OCTOBER...		..NOVEMBER..		..DECEMBER..	
	P--E--I		P--E--I		P--E--I		P--E--I		P--E--I		P--E--I
1)	V..R..1	1)	F..U..Y	1)	O..X..W	1)	V..Z..T	1)	F..3..R	1)	N..B..O
2)	W..S..2	2)	G..V..Z	2)	P..Y..X	2)	W..1..U	2)	G..A..S	2)	O..C..P
3)	X..T..3	3)	H..W..1	3)	Q..Z..Y	3)	X..2..V	3)	H..B..T	3)	P..D..Q
4)	A..U..4	4)	J..X..2	4)	R..1..Z	4)	A..3..W	4)	J..C..U	4)	Q..E..R
5)	B..V..5	5)	K..Y..3	5)	S..2..1	5)	B..A..X	5)	K..D..V	5)	R..F..S
6)	C..W..6	6)	L..Z..4	6)	T..3..2	6)	C..B..Y	6)	L..E..W	6)	S..G..T
7)	D..X..7	7)	M..1..5	7)	U..A..3	7)	D..C..Z	7)	M..F..X	7)	T..H..U
8)	E..Y..8	8)	N..2..6	8)	V..B..4	8)	E..D..1	8)	N..G..Y	8)	U..J..V
9)	F..Z..A	9)	O..3..7	9)	W..C..5	9)	F..E..2	9)	O..H..Z	9)	V..K..W
10)	G..1..B	10)	P..A..8	10)	X..D..6	10)	G..F..3	10)	P..J..1	10)	W..L..X
11)	H..2..C	11)	Q..B..A	11)	A..E..7	11)	H..G..4	11)	Q..K..2	11)	X..M..Y
12)	J..3..D	12)	R..C..B	12)	B..F..8	12)	J..H..5	12)	R..L..3	12)	A..N..Z
13)	K..A..E	13)	S..D..C	13)	C..G..A	13)	K..J..6	13)	S..M..4	13)	B..O..1
14)	L..B..F	14)	T..E..D	14)	D..H..B	14)	L..K..7	14)	T..N..5	14)	C..P..2
15)	M..C..G	15)	U..F..E	15)	E..J..C	15)	M..L..8	15)	U..O..6	15)	D..Q..3
16)	N..D..H	16)	V..G..F	16)	F..K..D	16)	N..M..A	16)	V..P..7	16)	E..R..4
17)	O..E..J	17)	W..H..G	17)	G..L..E	17)	O..N..B	17)	W..Q..8	17)	F..S..5
18)	P..F..K	18)	X..J..H	18)	H..M..F	18)	P..O..C	18)	X..R..A	18)	G..T..6
19)	Q..G..L	19)	A..K..J	19)	J..N..G	19)	Q..P..D	19)	A..S..B	19)	H..U..7
20)	R..H..M	20)	B..L..K	20)	K..O..H	20)	R..Q..E	20)	B..T..C	20)	J..V..8
21)	S..J..N	21)	C..M..L	21)	L..P..J	21)	S..R..F	21)	C..U..D	21)	K..W..A
22)	T..K..O	22)	D..N..M	22)	M..Q..K	22)	T..S..G	22)	D..V..E	22)	L..X..B
23)	U..L..P	23)	E..O..N	23)	N..R..L	23)	U..T..H	23)	E..W..F	23)	M..Y..C
24)	V..M..Q	24)	F..P..O	24)	O..S..M	24)	V..U..J	24)	F..X..G	24)	N..Z..D
25)	W..N..R	25)	G..Q..P	25)	P..T..N	25)	W..V..K	25)	G..Y..H	25)	O..1..E
26)	X..O..S	26)	H..R..Q	26)	Q..U..O	26)	X..W..L	26)	H..Z..J	26)	P..2..F
27)	A..P..T	27)	J..S..R	27)	R..V..P	27)	A..X..M	27)	J..1..K	27)	Q..3..G
28)	B..Q..U	28)	K..T..S	28)	S..W..Q	28)	B..Y..N	28)	K..2..L	28)	R..A..H
29)	C..R..V	29)	L..U..T	29)	T..X..R	29)	C..Z..O	29)	L..3..M	29)	S..B..J
30)	D..S..W	30)	M..V..U	30)	U..Y..S	30)	D..1..P	30)	M..A..N	30)	T..C..K
31)	E..T..X	31)	N..W..V			31)	E..2..Q			31)	U..D..L

CODES: P-PHYSICAL BIORHYTHM CURVE,E-EMOTIONAL BIORHYTHM CURVE,I-INTELLECTUAL BIORHYTHM CURVE

1914 1914

	...JANUARY..	..FEBRUARY..	....MARCH...	....APRIL...	....MAY....	....JUNE....
	P--E--I	P--E--I	P--E--I	P--E--I	P--E--I	P--E--I
1)	V..E..M	F..H..K	L..H..É	T..L..C	C..N..8	L..Q..6
2)	W..F..N	G..J..L	M..J..F	U..M..D	D..O..A	M..R..7
3)	X..G..O	H..K..M	N..K..G	V..N..E	E..P..B	N..S..8
4)	A..H..P	J..L..N	O..L..H	W..O..F	F..Q..C	O..T..A
5)	B..J..Q	K..M..O	P..M..J	X..P..G	G..R..D	P..U..B
6)	C..K..R	L..N..P	Q..N..K	A..Q..H	H..S..E	Q..V..C
7)	D..L..S	M..O..Q	R..O..L	B..R..J	J..T..F	R..W..D
8)	E..M..T	N..P..R	S..P..M	C..S..K	K..U..G	S..X..E
9)	F..N..U	O..Q..S	T..Q..N	D..T..L	L..V..H	T..Y..F
10)	G..O..V	P..R..T	U..R..O	E..U..M	M..W..J	U..Z..G
11)	H..P..W	Q..S..U	V..S..P	F..V..N	N..X..K	V..1..H
12)	J..Q..X	R..T..V	W..T..Q	G..W..O	O..Y..L	W..2..J
13)	K..R..Y	S..U..W	X..U..R	H..X..P	P..Z..M	X..3..K
14)	L..S..Z	T..V..X	A..V..S	J..Y..Q	Q..1..N	A..A..L
15)	M..T..1	U..W..Y	B..W..T	K..Z..R	R..2..O	B..B..M
16)	N..U..2	V..X..Z	C..X..U	L..1..S	S..3..P	C..C..N
17)	O..V..3	W..Y..1	D..Y..V	M..2..T	T..A..Q	D..D..O
18)	P..W..4	X..Z..2	E..Z..W	N..3..U	U..B..R	E..E..P
19)	Q..X..5	A..1..3	F..1..X	O..A..V	V..C..S	F..F..Q
20)	R..Y..6	B..2..4	G..2..Y	P..B..W	W..D..T	G..G..R
21)	S..Z..7	C..3..5	H..3..Z	Q..C..X	X..E..U	H..H..S
22)	T..1..8	D..A..6	J..A..1	R..D..Y	A..F..V	J..J..T
23)	U..2..A	E..B..7	K..B..2	S..E..Z	B..G..W	K..K..U
24)	V..3..B	F..C..8	L..C..3	T..F..1	C..H..X	L..L..V
25)	W..A..C	G..D..A	M..D..4	U..G..2	D..J..Y	M..M..W
26)	X..B..D	H..E..B	N..E..5	V..H..3	E..K..Z	N..N..X
27)	A..C..E	J..F..C	O..F..6	W..J..4	F..L..1	O..O..Y
28)	B..D..F	K..G..D	P..G..7	X..K..5	G..M..2	P..P..Z
29)	C..E..G		Q..H..8	A..L..6	H..N..3	Q..Q..1
30)	D..F..H		R..J..A	B..M..7	J..O..4	R..R..2
31)	E..G..J		S..K..B		K..P..5	

1914 1914

	....JULY....		...AUGUST...		..SEPTEMBER.		..OCTOBER...		..NOVEMBER..		..DECEMBER..
	P--E--I		P--E--I		P--E--I		P--E--I		P--E--I		P--E--I
1)	S..S..3	1)	C..V..1	1)	L..Y..Y	1)	S..1..V	1)	C..A..T	1)	K..C..Q
2)	T..T..4	2)	D..W..2	2)	M..Z..Z	2)	T..2..W	2)	D..B..U	2)	L..D..R
3)	U..U..5	3)	E..X..3	3)	N..1..1	3)	U..3..X	3)	E..C..V	3)	M..E..S
4)	V..V..6	4)	F..Y..4	4)	O..2..2	4)	V..A..Y	4)	F..D..W	4)	N..F..T
5)	W..W..7	5)	G..Z..5	5)	P..3..3	5)	W..B..Z	5)	G..E..X	5)	O..G..U
6)	X..X..8	6)	H..1..6	6)	Q..A..4	6)	X..C..1	6)	H..F..Y	6)	P..H..V
7)	A..Y..A	7)	J..2..7	7)	R..B..5	7)	A..D..2	7)	J..G..Z	7)	Q..J..W
8)	B..Z..B	8)	K..3..8	8)	S..C..6	8)	B..E..3	8)	K..H..1	8)	R..K..X
9)	C..1..C	9)	L..A..A	9)	T..D..7	9)	C..F..4	9)	L..J..2	9)	S..L..Y
10)	D..2..D	10)	M..B..B	10)	U..E..8	10)	D..G..5	10)	M..K..3	10)	T..M..Z
11)	E..3..E	11)	N..C..C	11)	V..F..A	11)	E..H..6	11)	N..L..4	11)	U..N..1
12)	F..A..F	12)	O..D..D	12)	W..G..B	12)	F..J..7	12)	O..M..5	12)	V..O..2
13)	G..B..G	13)	P..E..E	13)	X..H..C	13)	G..K..8	13)	P..N..6	13)	W..P..3
14)	H..C..H	14)	Q..F..F	14)	A..J..D	14)	H..L..A	14)	Q..O..7	14)	X..Q..4
15)	J..D..J	15)	R..G..G	15)	B..K..E	15)	J..M..B	15)	R..P..8	15)	A..R..5
16)	K..E..K	16)	S..H..H	16)	C..L..F	16)	K..N..C	16)	S..Q..A	16)	B..S..6
17)	L..F..L	17)	T..J..J	17)	D..M..G	17)	L..O..D	17)	T..R..B	17)	C..T..7
18)	M..G..M	18)	U..K..K	18)	E..N..H	18)	M..P..E	18)	U..S..C	18)	D..U..8
19)	N..H..N	19)	V..L..L	19)	F..O..J	19)	N..Q..F	19)	V..T..D	19)	E..V..A
20)	O..J..O	20)	W..M..M	20)	G..P..K	20)	O..R..G	20)	W..U..E	20)	F..W..B
21)	P..K..P	21)	X..N..N	21)	H..Q..L	21)	P..S..H	21)	X..V..F	21)	G..X..C
22)	Q..L..Q	22)	A..O..O	22)	J..R..M	22)	Q..T..J	22)	A..W..G	22)	H..Y..D
23)	R..M..R	23)	B..P..P	23)	K..S..N	23)	R..U..K	23)	B..X..H	23)	J..Z..E
24)	S..N..S	24)	C..Q..Q	24)	L..T..O	24)	S..V..L	24)	C..Y..J	24)	K..1..F
25)	T..O..T	25)	D..R..R	25)	M..U..P	25)	T..W..M	25)	D..Z..K	25)	L..2..G
26)	U..P..U	26)	E..S..S	26)	N..V..Q	26)	U..X..N	26)	E..1..L	26)	M..3..H
27)	V..Q..V	27)	F..T..T	27)	O..W..R	27)	V..Y..O	27)	F..2..M	27)	N..A..J
28)	W..R..W	28)	G..U..U	28)	P..X..S	28)	W..Z..P	28)	G..3..N	28)	O..B..K
29)	X..S..X	29)	H..V..V	29)	Q..Y..T	29)	X..1..Q	29)	H..A..O	29)	P..C..L
30)	A..T..Y	30)	J..W..W	30)	R..Z..U	30)	A..2..R	30)	J..B..P	30)	Q..D..M
31)	B..U..Z	31)	K..X..X			31)	B..3..S			31)	R..E..N

CODES: P-PHYSICAL BIORHYTHM CURVE,E-EMOTIONAL BIORHYTHM CURVE,I-INTELLECTUAL BIORHYTHM CURVE

1915 1915

Day	...JANUARY..	..FEBRUARY..	....MARCH....	....APRIL...	.....MAY....	....JUNE....
	P--E--I	P--E--I	P--E--I	P--E--I	P--E--I	P--E--I
1)	S..F..O	C..J..M	H..J..G	Q..M..E	X..O..B	H..R..8
2)	T..G..P	D..K..N	J..K..H	R..N..F	A..P..C	J..S..A
3)	U..H..Q	E..L..O	K..L..J	S..O..G	B..Q..D	K..T..B
4)	V..J..R	F..M..P	L..M..K	T..P..H	C..R..E	L..U..C
5)	W..K..S	G..N..Q	M..N..L	U..Q..J	D..S..F	M..V..D
6)	X..L..T	H..O..R	N..O..M	V..R..K	E..T..G	N..W..E
7)	A..M..U	J..P..S	O..P..N	W..S..L	F..U..H	O..X..F
8)	B..N..V	K..Q..T	P..Q..O	X..T..M	G..V..J	P..Y..G
9)	C..O..W	L..R..U	Q..R..P	A..U..N	H..W..K	Q..Z..H
10)	D..P..X	M..S..V	R..S..Q	B..V..O	J..X..L	R..1..J
11)	E..Q..Y	N..T..W	S..T..R	C..W..P	K..Y..M	S..2..K
12)	F..R..Z	O..U..X	T..U..S	D..X..Q	L..Z..N	T..3..L
13)	G..S..1	P..V..Y	U..V..T	E..Y..R	M..1..O	U..A..M
14)	H..T..2	Q..W..Z	V..W..U	F..Z..S	N..2..P	V..B..N
15)	J..U..3	R..X..1	W..X..V	G..1..T	O..3..Q	W..C..O
16)	K..V..4	S..Y..2	X..Y..W	H..2..U	P..A..R	X..D..P
17)	L..W..5	T..Z..3	A..Z..X	J..3..V	Q..B..S	A..E..Q
18)	M..X..6	U..1..4	B..1..Y	K..A..W	R..C..T	B..F..R
19)	N..Y..7	V..2..5	C..2..Z	L..B..X	S..D..U	C..G..S
20)	O..Z..8	W..3..6	D..3..1	M..C..Y	T..E..V	D..H..T
21)	P..1..A	X..A..7	E..A..2	N..D..Z	U..F..W	E..J..U
22)	Q..2..B	A..B..8	F..B..3	O..E..1	V..G..X	F..K..V
23)	R..3..C	B..C..A	G..C..4	P..F..2	W..H..Y	G..L..W
24)	S..A..D	C..D..B	H..D..5	Q..G..3	X..J..Z	H..M..X
25)	T..B..E	D..E..C	J..E..6	R..H..4	A..K..1	J..N..Y
26)	U..C..F	E..F..D	K..F..7	S..J..5	B..L..2	K..O..Z
27)	V..D..G	F..G..E	L..G..8	T..K..6	C..M..3	L..P..1
28)	W..E..H	G..H..F	M..H..A	U..L..7	D..N..4	M..Q..2
29)	X..F..J		N..J..B	V..M..8	E..O..5	N..R..3
30)	A..G..K		O..K..C	W..N..A	F..P..6	O..S..4
31)	B..H..L		P..L..D		G..Q..7	

1915 1915

Day	....JULY.... P--E--I	Day	...AUGUST... P--E--I	Day	..SEPTEMBER. P--E--I	Day	..OCTOBER... P--E--I	Day	..NOVEMBER.. P--E--I	Day	..DECEMBER.. P--E--I
1)	P..T..5	1)	X..W..3	1)	H..Z..1	1)	P..2..X	1)	X..B..V	1)	G..D..S
2)	Q..U..6	2)	A..X..4	2)	J..1..2	2)	Q..3..Y	2)	A..C..W	2)	H..E..T
3)	R..V..7	3)	B..Y..5	3)	K..2..3	3)	R..A..Z	3)	B..D..X	3)	J..F..U
4)	S..W..8	4)	C..Z..6	4)	L..3..4	4)	S..B..1	4)	C..E..Y	4)	K..G..V
5)	T..X..A	5)	D..1..7	5)	M..A..5	5)	T..C..2	5)	D..F..Z	5)	L..H..W
6)	U..Y..B	6)	E..2..8	6)	N..B..6	6)	U..D..3	6)	E..G..1	6)	M..J..X
7)	V..Z..C	7)	F..3..A	7)	O..C..7	7)	V..E..4	7)	F..H..2	7)	N..K..Y
8)	W..1..D	8)	G..A..B	8)	P..D..8	8)	W..F..5	8)	G..J..3	8)	O..L..Z
9)	X..2..E	9)	H..B..C	9)	Q..E..A	9)	X..G..6	9)	H..K..4	9)	P..M..1
10)	A..3..F	10)	J..C..D	10)	R..F..B	10)	A..H..7	10)	J..L..5	10)	Q..N..2
11)	B..A..G	11)	K..D..E	11)	S..G..C	11)	B..J..8	11)	K..M..6	11)	R..O..3
12)	C..B..H	12)	L..E..F	12)	T..H..D	12)	C..K..A	12)	L..N..7	12)	S..P..4
13)	D..C..J	13)	M..F..G	13)	U..J..E	13)	D..L..B	13)	M..O..8	13)	T..Q..5
14)	E..D..K	14)	N..G..H	14)	V..K..F	14)	E..M..C	14)	N..P..A	14)	U..R..6
15)	F..E..L	15)	O..H..J	15)	W..L..G	15)	F..N..D	15)	O..Q..B	15)	V..S..7
16)	G..F..M	16)	P..J..K	16)	X..M..H	16)	G..O..E	16)	P..R..C	16)	W..T..8
17)	H..G..N	17)	Q..K..L	17)	A..N..J	17)	H..P..F	17)	Q..S..D	17)	X..U..A
18)	J..H..O	18)	R..L..M	18)	B..O..K	18)	J..Q..G	18)	R..T..E	18)	A..V..B
19)	K..J..P	19)	S..M..N	19)	C..P..L	19)	K..R..H	19)	S..U..F	19)	B..W..C
20)	L..K..Q	20)	T..N..O	20)	D..Q..M	20)	L..S..J	20)	T..V..G	20)	C..X..D
21)	M..L..R	21)	U..O..P	21)	E..R..N	21)	M..T..K	21)	U..W..H	21)	D..Y..E
22)	N..M..S	22)	V..P..Q	22)	F..S..O	22)	N..U..L	22)	V..X..J	22)	E..Z..F
23)	O..N..T	23)	W..Q..R	23)	G..T..P	23)	O..V..M	23)	W..Y..K	23)	F..1..G
24)	P..O..U	24)	X..R..S	24)	H..U..Q	24)	P..W..N	24)	X..Z..L	24)	G..2..H
25)	Q..P..V	25)	A..S..T	25)	J..V..R	25)	Q..X..O	25)	A..1..M	25)	H..3..J
26)	R..Q..W	26)	B..T..U	26)	K..W..S	26)	R..Y..P	26)	B..2..N	26)	J..A..K
27)	S..R..X	27)	C..U..V	27)	L..X..T	27)	S..Z..Q	27)	C..3..O	27)	K..B..L
28)	T..S..Y	28)	D..V..W	28)	M..Y..U	28)	T..1..R	28)	D..A..P	28)	L..C..M
29)	U..T..Z	29)	E..W..X	29)	N..Z..V	29)	U..2..S	29)	E..B..Q	29)	M..D..N
30)	V..U..1	30)	F..X..Y	30)	O..1..W	30)	V..3..T	30)	F..C..R	30)	N..E..O
31)	W..V..2	31)	G..Y..Z			31)	W..A..U			31)	O..F..P

CODES: P-PHYSICAL BIORHYTHM CURVE,E-EMOTIONAL BIORHYTHM CURVE,I-INTELLECTUAL BIORHYTHM CURVE

1916 1916

	...JANUARY..	..FEBRUARY..	....MARCH...	....APRIL...	.....MAY....	....JUNE....
	P--E--I	P--E--I	P--E--I	P--E--I	P--E--I	P--E--I
1)	P..G..Q	X..K..O	F..L..K	O..O..H	V..Q..E	F..T..C
2)	Q..H..R	A..L..P	G..M..L	P..P..J	W..R..F	G..U..D
3)	R..J..S	B..M..Q	H..N..M	Q..Q..K	X..S..G	H..V..E
4)	S..K..T	C..N..R	J..O..N	R..R..L	A..T..H	J..W..F
5)	T..L..U	D..O..S	K..P..O	S..S..M	B..U..J	K..X..G
6)	U..M..V	E..P..T	L..Q..P	T..T..N	C..V..K	L..Y..H
7)	V..N..W	F..Q..U	M..R..Q	U..U..O	D..W..L	M..Z..J
8)	W..O..X	G..R..V	N..S..R	V..V..P	E..X..M	N..1..K
9)	X..P..Y	H..S..W	O..T..S	W..W..Q	F..Y..N	O..2..L
10)	A..Q..Z	J..T..X	P..U..T	X..X..R	G..Z..O	P..3..M
11)	B..R..1	K..U..Y	Q..V..U	A..Y..S	H..1..P	Q..A..N
12)	C..S..2	L..V..Z	R..W..V	B..Z..T	J..2..Q	R..B..O
13)	D..T..3	M..W..1	S..X..W	C..1..U	K..3..R	S..C..P
14)	E..U..4	N..X..2	T..Y..X	D..2..V	L..A..S	T..D..Q
15)	F..V..5	O..Y..3	U..Z..Y	E..3..W	M..B..T	U..E..R
16)	G..W..6	P..Z..4	V..1..Z	F..A..X	N..C..U	V..F..S
17)	H..X..7	Q..1..5	W..2..1	G..B..Y	O..D..V	W..G..T
18)	J..Y..8	R..2..6	X..3..2	H..C..Z	P..E..W	X..H..U
19)	K..Z..A	S..3..7	A..A..3	J..D..1	Q..F..X	A..J..V
20)	L..1..B	T..A..8	B..B..4	K..E..2	R..G..Y	B..K..W
21)	M..2..C	U..B..A	C..C..5	L..F..3	S..H..Z	C..L..X
22)	N..3..D	V..C..B	D..D..6	M..G..4	T..J..1	D..M..Y
23)	O..A..E	W..D..C	E..E..7	N..H..5	U..K..2	E..N..Z
24)	P..B..F	X..E..D	F..F..8	O..J..6	V..L..3	F..O..1
25)	Q..C..G	A..F..E	G..G..A	P..K..7	W..M..4	G..P..2
26)	R..D..H	B..G..F	H..H..B	Q..L..8	X..N..5	H..Q..3
27)	S..E..J	C..H..G	J..J..C	R..M..A	A..O..6	J..R..4
28)	T..F..K	D..J..H	K..K..D	S..N..B	B..P..7	K..S..5
29)	U..G..L	E..K..J	L..L..E	T..O..C	C..Q..8	L..T..6
30)	V..H..M		M..M..F	U..P..D	D..R..A	M..U..7
31)	W..J..N		N..N..G		E..S..B	

1916											1916
....JULY....		...AUGUST...		..SEPTEMBER.		..OCTOBER...		..NOVEMBER..		..DECEMBER..	
	P--E--I		P--E--I		P--E--I		P--E--I		P--E--I		P--E--I
1)	N..V..8	1)	V..Y..6	1)	F..2..4	1)	N..A..1	1)	V..D..Y	1)	E..F..V
2)	O..W..A	2)	W..Z..7	2)	G..3..5	2)	O..B..2	2)	W..E..Z	2)	F..G..W
3)	P..X..B	3)	X..1..8	3)	H..A..6	3)	P..C..3	3)	X..F..1	3)	G..H..X
4)	Q..Y..C	4)	A..2..A	4)	J..B..7	4)	Q..D..4	4)	A..G..2	4)	H..J..Y
5)	R..Z..D	5)	B..3..B	5)	K..C..8	5)	R..E..5	5)	B..H..3	5)	J..K..Z
6)	S..1..E	6)	C..A..C	6)	L..D..A	6)	S..F..6	6)	C..J..4	6)	K..L..1
7)	T..2..F	7)	D..B..D	7)	M..E..B	7)	T..G..7	7)	D..K..5	7)	L..M..2
8)	U..3..G	8)	E..C..E	8)	N..F..C	8)	U..H..8	8)	E..L..6	8)	M..N..3
9)	V..A..H	9)	F..D..F	9)	O..G..D	9)	V..J..A	9)	F..M..7	9)	N..O..4
10)	W..B..J	10)	G..E..G	10)	P..H..E	10)	W..K..B	10)	G..N..8	10)	O..P..5
11)	X..C..K	11)	H..F..H	11)	Q..J..F	11)	X..L..C	11)	H..O..A	11)	P..Q..6
12)	A..D..L	12)	J..G..J	12)	R..K..G	12)	A..M..D	12)	J..P..B	12)	Q..R..7
13)	B..E..M	13)	K..H..K	13)	S..L..H	13)	B..N..E	13)	K..Q..C	13)	R..S..8
14)	C..F..N	14)	L..J..L	14)	T..M..J	14)	C..O..F	14)	L..R..D	14)	S..T..A
15)	D..G..O	15)	M..K..M	15)	U..N..K	15)	D..P..G	15)	M..S..E	15)	T..U..B
16)	E..H..P	16)	N..L..N	16)	V..O..L	16)	E..Q..H	16)	N..T..F	16)	U..V..C
17)	F..J..Q	17)	O..M..O	17)	W..P..M	17)	F..R..J	17)	O..U..G	17)	V..W..D
18)	G..K..R	18)	P..N..P	18)	X..Q..N	18)	G..S..K	18)	P..V..H	18)	W..X..E
19)	H..L..S	19)	Q..O..Q	19)	A..R..O	19)	H..T..L	19)	Q..W..J	19)	X..Y..F
20)	J..M..T	20)	R..P..R	20)	B..S..P	20)	J..U..M	20)	R..X..K	20)	A..Z..G
21)	K..N..U	21)	S..Q..S	21)	C..T..Q	21)	K..V..N	21)	S..Y..L	21)	B..1..H
22)	L..O..V	22)	T..R..T	22)	D..U..R	22)	L..W..O	22)	T..Z..M	22)	C..2..J
23)	M..P..W	23)	U..S..U	23)	E..V..S	23)	M..X..P	23)	U..1..N	23)	D..3..K
24)	N..Q..X	24)	V..T..V	24)	F..W..T	24)	N..Y..Q	24)	V..2..O	24)	E..A..L
25)	O..R..Y	25)	W..U..W	25)	G..X..U	25)	O..Z..R	25)	W..3..P	25)	F..B..M
26)	P..S..Z	26)	X..V..X	26)	H..Y..V	26)	P..1..S	26)	X..A..Q	26)	G..C..N
27)	Q..T..1	27)	A..W..Y	27)	J..Z..W	27)	Q..2..T	27)	A..B..R	27)	H..D..O
28)	R..U..2	28)	B..X..Z	28)	K..1..X	28)	R..3..U	28)	B..C..S	28)	J..E..P
29)	S..V..3	29)	C..Y..1	29)	L..2..Y	29)	S..A..V	29)	C..D..T	29)	K..F..Q
30)	T..W..4	30)	D..Z..2	30)	M..3..Z	30)	T..B..W	30)	D..E..U	30)	L..G..R
31)	U..X..5	31)	E..1..3			31)	U..C..X			31)	M..H..S

CODES: P-PHYSICAL BIORHYTHM CURVE,E-EMOTIONAL BIORHYTHM CURVE,I-INTELLECTUAL BIORHYTHM CURVE

1917 1917

	...JANUARY..	..FEBRUARY..	....MARCH...	....APRIL...	.....MAY....	....JUNE....
	P--E--I	P--E--I	P--E--I	P--E--I	P--E--I	P--E--I
1)	N..J..T	V..M..R	C..M..M	L..P..K	S..R..G	C..U..E
2)	O..K..U	W..N..S	D..N..N	M..Q..L	T..S..H	D..V..F
3)	P..L..V	X..O..T	E..O..O	N..R..M	U..T..J	E..W..G
4)	Q..M..W	A..P..U	F..P..P	O..S..N	V..U..K	F..X..H
5)	R..N..X	B..Q..V	G..Q..Q	P..T..O	W..V..L	G..Y..J
6)	S..O..Y	C..R..W	H..R..R	Q..U..P	X..W..M	H..Z..K
7)	T..P..Z	D..S..X	J..S..S	R..V..Q	A..X..N	J..1..L
8)	U..Q..1	E..T..Y	K..T..T	S..W..R	B..Y..O	K..2..M
9)	V..R..2	F..U..Z	L..U..U	T..X..S	C..Z..P	L..3..N
10)	W..S..3	G..V..1	M..V..V	U..Y..T	D..1..Q	M..A..O
11)	X..T..4	H..W..2	N..W..W	V..Z..U	E..2..R	N..B..P
12)	A..U..5	J..X..3	O..X..X	W..1..V	F..3..S	O..C..Q
13)	B..V..6	K..Y..4	P..Y..Y	X..2..W	G..A..T	P..D..R
14)	C..W..7	L..Z..5	Q..Z..Z	A..3..X	H..B..U	Q..E..S
15)	D..X..8	M..1..6	R..1..1	B..A..Y	J..C..V	R..F..T
16)	E..Y..A	N..2..7	S..2..2	C..B..Z	K..D..W	S..G..U
17)	F..Z..B	O..3..8	T..3..3	D..C..1	L..E..X	T..H..V
18)	G..1..C	P..A..A	U..A..4	E..D..2	M..F..Y	U..J..W
19)	H..2..D	Q..B..B	V..B..5	F..E..3	N..G..Z	V..K..X
20)	J..3..E	R..C..C	W..C..6	G..F..4	O..H..1	W..L..Y
21)	K..A..F	S..D..D	X..D..7	H..G..5	P..J..2	X..M..Z
22)	L..B..G	T..E..E	A..E..8	J..H..6	Q..K..3	A..N..1
23)	M..C..H	U..F..F	B..F..A	K..J..7	R..L..4	B..O..2
24)	N..D..J	V..G..G	C..G..B	L..K..8	S..M..5	C..P..3
25)	O..E..K	W..H..H	D..H..C	M..L..A	T..N..6	D..Q..4
26)	P..F..L	X..J..J	E..J..D	N..M..B	U..O..7	E..R..5
27)	Q..G..M	A..K..K	F..K..E	O..N..C	V..P..8	F..S..6
28)	R..H..N	B..L..L	G..L..F	P..O..D	W..Q..A	G..T..7
29)	S..J..O		H..M..G	Q..P..E	X..R..B	H..U..8
30)	T..K..P		J..N..H	R..Q..F	A..S..C	J..V..A
31)	U..L..Q		K..O..J		B..T..D	

1917 | 1917

	....JULY....		...AUGUST...		..SEPTEMBER.		..OCTOBER...		..NOVEMBER..		..DECEMBER..
	P--E--I		P--E--I		P--E--I		P--E--I		P--E--I		P--E--I
1)	K..W..B	1)	S..Z..8	1)	C..3..6	1)	K..B..3	1)	S..E..1	1)	B..G..X
2)	L..X..C	2)	T..1..A	2)	D..A..7	2)	L..C..4	2)	T..F..2	2)	C..H..Y
3)	M..Y..D	3)	U..2..B	3)	E..B..8	3)	M..D..5	3)	U..G..3	3)	D..J..Z
4)	N..Z..E	4)	V..3..C	4)	F..C..A	4)	N..E..6	4)	V..H..4	4)	E..K..1
5)	O..1..F	5)	W..A..D	5)	G..D..B	5)	O..F..7	5)	W..J..5	5)	F..L..2
6)	P..2..G	6)	X..B..E	6)	H..E..C	6)	P..G..8	6)	X..K..6	6)	G..M..3
7)	Q..3..H	7)	A..C..F	7)	J..F..D	7)	Q..H..A	7)	A..L..7	7)	H..N..4
8)	R..A..J	8)	B..D..G	8)	K..G..E	8)	R..J..B	8)	B..M..8	8)	J..O..5
9)	S..B..K	9)	C..E..H	9)	L..H..F	9)	S..K..C	9)	C..N..A	9)	K..P..6
10)	T..C..L	10)	D..F..J	10)	M..J..G	10)	T..L..D	10)	D..O..B	10)	L..Q..7
11)	U..D..M	11)	E..G..K	11)	N..K..H	11)	U..M..E	11)	E..P..C	11)	M..R..8
12)	V..E..N	12)	F..H..L	12)	O..L..J	12)	V..N..F	12)	F..Q..D	12)	N..S..A
13)	W..F..O	13)	G..J..M	13)	P..M..K	13)	W..O..G	13)	G..R..E	13)	O..T..B
14)	X..G..P	14)	H..K..N	14)	Q..N..L	14)	X..P..H	14)	H..S..F	14)	P..U..C
15)	A..H..Q	15)	J..L..O	15)	R..O..M	15)	A..Q..J	15)	J..T..G	15)	Q..V..D
16)	B..J..R	16)	K..M..P	16)	S..P..N	16)	B..R..K	16)	K..U..H	16)	R..W..E
17)	C..K..S	17)	L..N..Q	17)	T..Q..O	17)	C..S..L	17)	L..V..J	17)	S..X..F
18)	D..L..T	18)	M..O..R	18)	U..R..P	18)	D..T..M	18)	M..W..K	18)	T..Y..G
19)	E..M..U	19)	N..P..S	19)	V..S..Q	19)	E..U..N	19)	N..X..L	19)	U..Z..H
20)	F..N..V	20)	O..Q..T	20)	W..T..R	20)	F..V..O	20)	O..Y..M	20)	V..1..J
21)	G..O..W	21)	P..R..U	21)	X..U..S	21)	G..W..P	21)	P..Z..N	21)	W..2..K
22)	H..P..X	22)	Q..S..V	22)	A..V..T	22)	H..X..Q	22)	Q..1..O	22)	X..3..L
23)	J..Q..Y	23)	R..T..W	23)	B..W..U	23)	J..Y..R	23)	R..2..P	23)	A..A..M
24)	K..R..Z	24)	S..U..X	24)	C..X..V	24)	K..Z..S	24)	S..3..Q	24)	B..B..N
25)	L..S..1	25)	T..V..Y	25)	D..Y..W	25)	L..1..T	25)	T..A..R	25)	C..C..O
26)	M..T..2	26)	U..W..Z	26)	E..Z..X	26)	M..2..U	26)	U..B..S	26)	D..D..P
27)	N..U..3	27)	V..X..1	27)	F..1..Y	27)	N..3..V	27)	V..C..T	27)	E..E..Q
28)	O..V..4	28)	W..Y..2	28)	G..2..Z	28)	O..A..W	28)	W..D..U	28)	F..F..R
29)	P..W..5	29)	X..Z..3	29)	H..3..1	29)	P..B..X	29)	X..E..V	29)	G..G..S
30)	Q..X..6	30)	A..1..4	30)	J..A..2	30)	Q..C..Y	30)	A..F..W	30)	H..H..T
31)	R..Y..7	31)	B..2..5			31)	R..D..Z			31)	J..J..U

CODES: P-PHYSICAL BIORHYTHM CURVE, E-EMOTIONAL BIORHYTHM CURVE, I-INTELLECTUAL BIORHYTHM CURVE

1918											1918
...JANUARY..		..FEBRUARY..		....MARCH...		....APRIL...		....MAY....		....JUNE....	
	P--E--I		P--E--I		P--E--I		P--E--I		P--E--I		P--E--I
1)	K..K..V	1)	S..N..T	1)	X..N..O	1)	H..Q..M	1)	P..S..J	1)	X..V..G
2)	L..L..W	2)	T..O..U	2)	A..O..P	2)	J..R..N	2)	Q..T..K	2)	A..W..H
3)	M..M..X	3)	U..P..V	3)	B..P..Q	3)	K..S..O	3)	R..U..L	3)	B..X..J
4)	N..N..Y	4)	V..Q..W	4)	C..Q..R	4)	L..T..P	4)	S..V..M	4)	C..Y..K
5)	O..O..Z	5)	W..R..X	5)	D..R..S	5)	M..U..Q	5)	T..W..N	5)	D..Z..L
6)	P..P..1	6)	X..S..Y	6)	E..S..T	6)	N..V..R	6)	U..X..O	6)	E..1..M
7)	Q..Q..2	7)	A..T..Z	7)	F..T..U	7)	O..W..S	7)	V..Y..P	7)	F..2..N
8)	R..R..3	8)	B..U..1	8)	G..U..V	8)	P..X..T	8)	W..Z..Q	8)	G..3..O
9)	S..S..4	9)	C..V..2	9)	H..V..W	9)	Q..Y..U	9)	X..1..R	9)	H..A..P
10)	T..T..5	10)	D..W..3	10)	J..W..X	10)	R..Z..V	10)	A..2..S	10)	J..B..Q
11)	U..U..6	11)	E..X..4	11)	K..X..Y	11)	S..1..W	11)	B..3..T	11)	K..C..R
12)	V..V..7	12)	F..Y..5	12)	L..Y..Z	12)	T..2..X	12)	C..A..U	12)	L..D..S
13)	W..W..8	13)	G..Z..6	13)	M..Z..1	13)	U..3..Y	13)	D..B..V	13)	M..E..T
14)	X..X..A	14)	H..1..7	14)	N..1..2	14)	V..A..Z	14)	E..C..W	14)	N..F..U
15)	A..Y..B	15)	J..2..8	15)	O..2..3	15)	W..B..1	15)	F..D..X	15)	O..G..V
16)	B..Z..C	16)	K..3..A	16)	P..3..4	16)	X..C..2	16)	G..E..Y	16)	P..H..W
17)	C..1..D	17)	L..A..B	17)	Q..A..5	17)	A..D..3	17)	H..F..Z	17)	Q..J..X
18)	D..2..E	18)	M..B..C	18)	R..B..6	18)	B..E..4	18)	J..G..1	18)	R..K..Y
19)	E..3..F	19)	N..C..D	19)	S..C..7	19)	C..F..5	19)	K..H..2	19)	S..L..Z
20)	F..A..G	20)	O..D..E	20)	T..D..8	20)	D..G..6	20)	L..J..3	20)	T..M..1
21)	G..B..H	21)	P..E..F	21)	U..E..A	21)	E..H..7	21)	M..K..4	21)	U..N..2
22)	H..C..J	22)	Q..F..G	22)	V..F..B	22)	F..J..8	22)	N..L..5	22)	V..O..3
23)	J..D..K	23)	R..G..H	23)	W..G..C	23)	G..K..A	23)	O..M..6	23)	W..P..4
24)	K..E..L	24)	S..H..J	24)	X..H..D	24)	H..L..B	24)	P..N..7	24)	X..Q..5
25)	L..F..M	25)	T..J..K	25)	A..J..E	25)	J..M..C	25)	Q..O..8	25)	A..R..6
26)	M..G..N	26)	U..K..L	26)	B..K..F	26)	K..N..D	26)	R..P..A	26)	B..S..7
27)	N..H..O	27)	V..L..M	27)	C..L..G	27)	L..O..E	27)	S..Q..B	27)	C..T..8
28)	O..J..P	28)	W..M..N	28)	D..M..H	28)	M..P..F	28)	T..R..C	28)	D..U..A
29)	P..K..Q			29)	E..N..J	29)	N..Q..G	29)	U..S..D	29)	E..V..B
30)	Q..L..R			30)	F..O..K	30)	O..R..H	30)	V..T..E	30)	F..W..C
31)	R..M..S			31)	G..P..L			31)	W..U..F		

1918 1918

	....JULY....	...AUGUST...	..SEPTEMBER.	..OCTOBER...	..NOVEMBER..	..DECEMBER..
	P--E--I	P--E--I	P--E--I	P--E--I	P--E--I	P--E--I
1)	G..X..D	P..1..B	X..A..8	G..C..5	P..F..3	W..H..Z
2)	H..Y..E	Q..2..C	A..B..A	H..D..6	Q..G..4	X..J..1
3)	J..Z..F	R..3..D	B..C..B	J..E..7	R..H..5	A..K..2
4)	K..1..G	S..A..E	C..D..C	K..F..8	S..J..6	B..L..3
5)	L..2..H	T..B..F	D..E..D	L..G..A	T..K..7	C..M..4
6)	M..3..J	U..C..G	E..F..E	M..H..B	U..L..8	D..N..5
7)	N..A..K	V..D..H	F..G..F	N..J..C	V..M..A	E..O..6
8)	O..B..L	W..E..J	G..H..G	O..K..D	W..N..B	F..P..7
9)	P..C..M	X..F..K	H..J..H	P..L..E	X..O..C	G..Q..8
10)	Q..D..N	A..G..L	J..K..J	Q..M..F	A..P..D	H..R..A
11)	R..E..O	B..H..M	K..L..K	R..N..G	B..Q..E	J..S..B
12)	S..F..P	C..J..N	L..M..L	S..O..H	C..R..F	K..T..C
13)	T..G..Q	D..K..O	M..N..M	T..P..J	D..S..G	L..U..D
14)	U..H..R	E..L..P	N..O..N	U..Q..K	E..T..H	M..V..E
15)	V..J..S	F..M..Q	O..P..O	V..R..L	F..U..J	N..W..F
16)	W..K..T	G..N..R	P..Q..P	W..S..M	G..V..K	O..X..G
17)	X..L..U	H..O..S	Q..R..Q	X..T..N	H..W..L	P..Y..H
18)	A..M..V	J..P..T	R..S..R	A..U..O	J..X..M	Q..Z..J
19)	B..N..W	K..Q..U	S..T..S	B..V..P	K..Y..N	R..1..K
20)	C..O..X	L..R..V	T..U..T	C..W..Q	L..Z..O	S..2..L
21)	D..P..Y	M..S..W	U..V..U	D..X..R	M..1..P	T..3..M
22)	E..Q..Z	N..T..X	V..W..V	E..Y..S	N..2..Q	U..A..N
23)	F..R..1	O..U..Y	W..X..W	F..Z..T	O..3..R	V..B..O
24)	G..S..2	P..V..Z	X..Y..X	G..1..U	P..A..S	W..C..P
25)	H..T..3	Q..W..1	A..Z..Y	H..2..V	Q..B..T	X..D..Q
26)	J..U..4	R..X..2	B..1..Z	J..3..W	R..C..U	A..E..R
27)	K..V..5	S..Y..3	C..2..1	K..A..X	S..D..V	B..F..S
28)	L..W..6	T..Z..4	D..3..2	L..B..Y	T..E..W	C..G..T
29)	M..X..7	U..1..5	E..A..3	M..C..Z	U..F..X	D..H..U
30)	N..Y..8	V..2..6	F..B..4	N..D..1	V..G..Y	E..J..V
31)	O..Z..A	W..3..7		O..E..2		F..K..W

CODES: P-PHYSICAL BIORHYTHM CURVE,E-EMOTIONAL BIORHYTHM CURVE,I-INTELLECTUAL BIORHYTHM CURVE

1919 1919

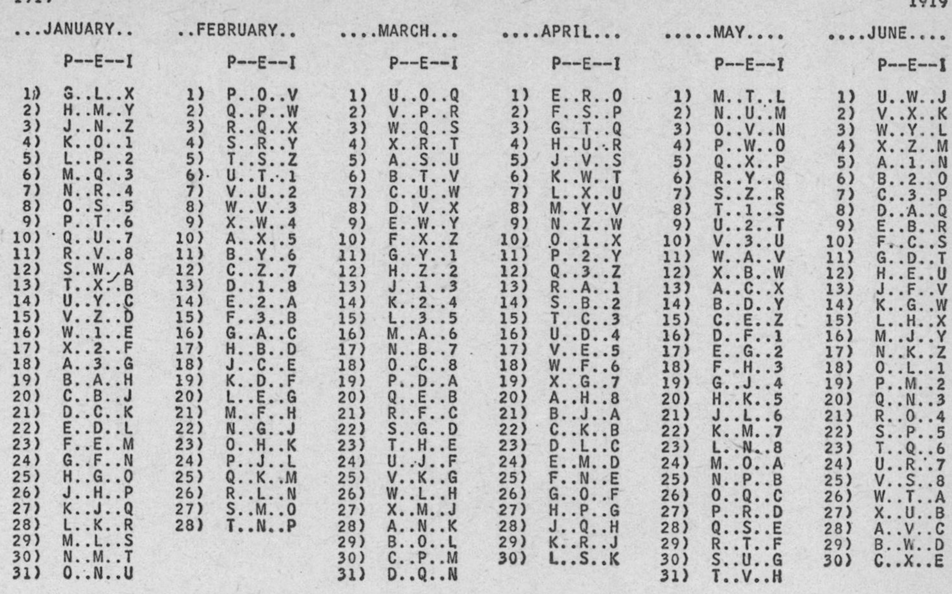

	...JANUARY..	..FEBRUARY..	....MARCH...	....APRIL...	.....MAY....	....JUNE....
	P--E--I	P--E--I	P--E--I	P--E--I	P--E--I	P--E--I
1)	G..L..X	P..O..V	U..O..Q	E..R..O	M..T..L	U..W..J
2)	H..M..Y	Q..P..W	V..P..R	F..S..P	N..U..M	V..X..K
3)	J..N..Z	R..Q..X	W..Q..S	G..T..Q	O..V..N	W..Y..L
4)	K..O..1	S..R..Y	X..R..T	H..U..R	P..W..O	X..Z..M
5)	L..P..2	T..S..Z	A..S..U	J..V..S	Q..X..P	A..1..N
6)	M..Q..3	U..T..1	B..T..V	K..W..T	R..Y..Q	B..2..O
7)	N..R..4	V..U..2	C..U..W	L..X..U	S..Z..R	C..3..P
8)	O..S..5	W..V..3	D..V..X	M..Y..V	T..1..S	D..A..Q
9)	P..T..6	X..W..4	E..W..Y	N..Z..W	U..2..T	E..B..R
10)	Q..U..7	A..X..5	F..X..Z	O..1..X	V..3..U	F..C..S
11)	R..V..8	B..Y..6	G..Y..1	P..2..Y	W..A..V	G..D..T
12)	S..W..A	C..Z..7	H..Z..2	Q..3..Z	X..B..W	H..E..U
13)	T..X..B	D..1..8	J..1..3	R..A..1	A..C..X	J..F..V
14)	U..Y..C	E..2..A	K..2..4	S..B..2	B..D..Y	K..G..W
15)	V..Z..D	F..3..B	L..3..5	T..C..3	C..E..Z	L..H..X
16)	W..1..E	G..A..C	M..A..6	U..D..4	D..F..1	M..J..Y
17)	X..2..F	H..B..D	N..B..7	V..E..5	E..G..2	N..K..Z
18)	A..3..G	J..C..E	O..C..8	W..F..6	F..H..3	O..L..1
19)	B..A..H	K..D..F	P..D..A	X..G..7	G..J..4	P..M..2
20)	C..B..J	L..E..G	Q..E..B	A..H..8	H..K..5	Q..N..3
21)	D..C..K	M..F..H	R..F..C	B..J..A	J..L..6	R..O..4
22)	E..D..L	N..G..J	S..G..D	C..K..B	K..M..7	S..P..5
23)	F..E..M	O..H..K	T..H..E	D..L..C	L..N..8	T..Q..6
24)	G..F..N	P..J..L	U..J..F	E..M..D	M..O..A	U..R..7
25)	H..G..O	Q..K..M	V..K..G	F..N..E	N..P..B	V..S..8
26)	J..H..P	R..L..N	W..L..H	G..O..F	O..Q..C	W..T..A
27)	K..J..Q	S..M..O	X..M..J	H..P..G	P..R..D	X..U..B
28)	L..K..R	T..N..P	A..N..K	J..Q..H	Q..S..E	A..V..C
29)	M..L..S		B..O..L	K..R..J	R..T..F	B..W..D
30)	N..M..T		C..P..M	L..S..K	S..U..G	C..X..E
31)	O..N..U		D..Q..N		T..V..H	

1919 | 1919

Day	....JULY.... P--E--I	...AUGUST... P--E--I	..SEPTEMBER. P--E--I	..OCTOBER... P--E--I	..NOVEMBER.. P--E--I	..DECEMBER.. P--E--I
1)	D..Y..F	M..2..D	U..B..B	D..D..7	M..G..5	T..J..2
2)	E..Z..G	N..3..E	V..C..C	E..E..8	N..H..6	U..K..3
3)	F..1..H	O..A..F	W..D..D	F..F..A	O..J..7	V..L..4
4)	G..2..J	P..B..G	X..E..E	G..G..B	P..K..8	W..M..5
5)	H..3..K	Q..C..H	A..F..F	H..H..C	Q..L..A	X..N..6
6)	J..A..L	R..D..J	B..G..G	J..J..D	R..M..B	A..O..7
7)	K..B..M	S..E..K	C..H..H	K..K..E	S..N..C	B..P..8
8)	L..C..N	T..F..L	D..J..J	L..L..F	T..O..D	C..Q..A
9)	M..D..O	U..G..M	E..K..K	M..M..G	U..P..E	D..R..B
10)	N..E..P	V..H..N	F..L..L	N..N..H	V..Q..F	E..S..C
11)	O..F..Q	W..J..O	G..M..M	O..O..J	W..R..G	F..T..D
12)	P..G..R	X..K..P	H..N..N	P..P..K	X..S..H	G..U..E
13)	Q..H..S	A..L..Q	J..O..O	Q..Q..L	A..T..J	H..V..F
14)	R..J..T	B..M..R	K..P..P	R..R..M	B..U..K	J..W..G
15)	S..K..U	C..N..S	L..Q..Q	S..S..N	C..V..L	K..X..H
16)	T..L..V	D..O..T	M..R..R	T..T..O	D..W..M	L..Y..J
17)	U..M..W	E..P..U	N..S..S	U..U..P	E..X..N	M..Z..K
18)	V..N..X	F..Q..V	O..T..T	V..V..Q	F..Y..O	N..1..L
19)	W..O..Y	G..R..W	P..U..U	W..W..R	G..Z..P	O..2..M
20)	X..P..Z	H..S..X	Q..V..V	X..X..S	H..1..Q	P..3..N
21)	A..Q..1	J..T..Y	R..W..W	A..Y..T	J..2..R	Q..A..O
22)	B..R..2	K..U..Z	S..X..X	B..Z..U	K..3..S	R..B..P
23)	C..S..3	L..V..1	T..Y..Y	C..1..V	L..A..T	S..C..Q
24)	D..T..4	M..W..2	U..Z..Z	D..2..W	M..B..U	T..D..R
25)	E..U..5	N..X..3	V..1..1	E..3..X	N..C..V	U..E..S
26)	F..V..6	O..Y..4	W..2..2	F..A..Y	O..D..W	V..F..T
27)	G..W..7	P..Z..5	X..3..3	G..B..Z	P..E..X	W..G..U
28)	H..X..8	Q..1..6	A..A..4	H..C..1	Q..F..Y	X..H..V
29)	J..Y..A	R..2..7	B..B..5	J..D..2	R..G..Z	A..J..W
30)	K..Z..B	S..3..8	C..C..6	K..E..3	S..H..1	B..K..X
31)	L..1..C	T..A..A		L..F..4		C..L..Y

CODES: P-PHYSICAL BIORHYTHM CURVE,E-EMOTIONAL BIORHYTHM CURVE,I-INTELLECTUAL BIORHYTHM CURVE

	...JANUARY..	..FEBRUARY..	....MARCH...	....APRIL...	.....MAY....	....JUNE....
	P--E--I	P--E--I	P--E--I	P--E--I	P--E--I	P--E--I
1)	D..M..Z	M..P..X	S..Q..T	C..T..R	K..V..O	S..Y..M
2)	E..N..1	N..Q..Y	T..R..U	D..U..S	L..W..P	T..Z..N
3)	F..O..2	O..R..Z	U..S..V	E..V..T	M..X..Q	U..1..O
4)	G..P..3	P..S..1	V..T..W	F..W..U	N..Y..R	V..2..P
5)	H..Q..4	Q..T..2	W..U..X	G..X..V	O..Z..S	W..3..Q
6)	J..R..5	R..U..3	X..V..Y	H..Y..W	P..1..T	X..A..R
7)	K..S..6	S..V..4	A..W..Z	J..Z..X	Q..2..U	A..B..S
8)	L..T..7	T..W..5	B..X..1	K..1..Y	R..3..V	B..C..T
9)	M..U..8	U..X..6	C..Y..2	L..2..Z	S..A..W	C..D..U
10)	N..V..A	V..Y..7	D..Z..3	M..3..1	T..B..X	D..E..V
11)	O..W..B	W..Z..8	E..1..4	N..A..2	U..C..Y	E..F..W
12)	P..X..C	X..1..A	F..2..5	O..B..3	V..D..Z	F..G..X
13)	Q..Y..D	A..2..B	G..3..6	P..C..4	W..E..1	G..H..Y
14)	R..Z..E	B..3..C	H..A..7	Q..D..5	X..F..2	H..J..Z
15)	S..1..F	C..A..D	J..B..8	R..E..6	A..G..3	J..K..1
16)	T..2..G	D..B..E	K..C..A	S..F..7	B..H..4	K..L..2
17)	U..3..H	E..C..F	L..D..B	T..G..8	C..J..5	L..M..3
18)	V..A..J	F..D..G	M..E..C	U..H..A	D..K..6	M..N..4
19)	W..B..K	G..E..H	N..F..D	V..J..B	E..L..7	N..O..5
20)	X..C..L	H..F..J	O..G..E	W..K..C	F..M..8	O..P..6
21)	A..D..M	J..G..K	P..H..F	X..L..D	G..N..A	P..Q..7
22)	B..E..N	K..H..L	Q..J..G	A..M..E	H..O..B	Q..R..8
23)	C..F..O	L..J..M	R..K..H	B..N..F	J..P..C	R..S..A
24)	D..G..P	M..K..N	S..L..J	C..O..G	K..Q..D	S..T..B
25)	E..H..Q	N..L..O	T..M..K	D..P..H	L..R..E	T..U..C
26)	F..J..R	O..M..P	U..N..L	E..Q..J	M..S..F	U..V..D
27)	G..K..S	P..N..Q	V..O..M	F..R..K	N..T..G	V..W..E
28)	H..L..T	Q..O..R	W..P..N	G..S..L	O..U..H	W..X..F
29)	J..M..U	R..P..S	X..Q..O	H..T..M	P..V..J	X..Y..G
30)	K..N..V		A..R..P	J..U..N	Q..W..K	A..Z..H
31)	L..O..W		B..S..Q		R..X..L	

1920 | | | | | 1920

	....JULY....		...AUGUST...		..SEPTEMBER.		..OCTOBER...		..NOVEMBER..		..DECEMBER..
	P--E--I		P--E--I		P--E--I		P--E--I		P--E--I		P--E--I
1)	B..1..J	1)	K..A..G	1)	S..D..E	1)	B..F..B	1)	K..J..8	1)	R..L..5
2)	C..2..K	2)	L..B..H	2)	T..E..F	2)	C..G..C	2)	L..K..A	2)	S..M..6
3)	D..3..L	3)	M..C..J	3)	U..F..G	3)	D..H..D	3)	M..L..B	3)	T..N..7
4)	E..A..M	4)	N..D..K	4)	V..G..H	4)	E..J..E	4)	N..M..C	4)	U..O..8
5)	F..B..N	5)	O..E..L	5)	W..H..J	5)	F..K..F	5)	O..N..D	5)	V..P..A
6)	G..C..O	6)	P..F..M	6)	X..J..K	6)	G..L..G	6)	P..O..E	6)	W..Q..B
7)	H..D..P	7)	Q..G..N	7)	A..K..L	7)	H..M..H	7)	Q..P..F	7)	X..R..C
8)	J..E..Q	8)	R..H..O	8)	B..L..M	8)	J..N..J	8)	R..Q..G	8)	A..S..D
9)	K..F..R	9)	S..J..P	9)	C..M..N	9)	K..O..K	9)	S..R..H	9)	B..T..E
10)	L..G..S	10)	T..K..Q	10)	D..N..O	10)	L..P..L	10)	T..S..J	10)	C..U..F
11)	M..H..T	11)	U..L..R	11)	E..O..P	11)	M..Q..M	11)	U..T..K	11)	D..V..G
12)	N..J..U	12)	V..M..S	12)	F..P..Q	12)	N..R..N	12)	V..U..L	12)	E..W..H
13)	O..K..V	13)	W..N..T	13)	G..Q..R	13)	O..S..O	13)	W..V..M	13)	F..X..J
14)	P..L..W	14)	X..O..U	14)	H..R..S	14)	P..T..P	14)	X..W..N	14)	G..Y..K
15)	Q..M..X	15)	A..P..V	15)	J..S..T	15)	Q..U..Q	15)	A..X..O	15)	H..Z..L
16)	R..N..Y	16)	B..Q..W	16)	K..T..U	16)	R..V..R	16)	B..Y..P	16)	J..1..M
17)	S..O..Z	17)	C..R..X	17)	L..U..V	17)	S..W..S	17)	C..Z..Q	17)	K..2..N
18)	T..P..1	18)	D..S..Y	18)	M..V..W	18)	T..X..T	18)	D..1..R	18)	L..3..O
19)	U..Q..2	19)	E..T..Z	19)	N..W..X	19)	U..Y..U	19)	E..2..S	19)	M..A..P
20)	V..R..3	20)	F..U..1	20)	O..X..Y	20)	V..Z..V	20)	F..3..T	20)	N..B..Q
21)	W..S..4	21)	G..V..2	21)	P..Y..Z	21)	W..1..W	21)	G..A..U	21)	O..C..R
22)	X..T..5	22)	H..W..3	22)	Q..Z..1	22)	X..2..X	22)	H..B..V	22)	P..D..S
23)	A..U..6	23)	J..X..4	23)	R..1..2	23)	A..3..Y	23)	J..C..W	23)	Q..E..T
24)	B..V..7	24)	K..Y..5	24)	S..2..3	24)	B..A..Z	24)	K..D..X	24)	R..F..U
25)	C..W..8	25)	L..Z..6	25)	T..3..4	25)	C..B..1	25)	L..E..Y	25)	S..G..V
26)	D..X..A	26)	M..1..7	26)	U..A..5	26)	D..C..2	26)	M..F..Z	26)	T..H..W
27)	E..Y..B	27)	N..2..8	27)	V..B..6	27)	E..D..3	27)	N..G..1	27)	U..J..X
28)	F..Z..C	28)	O..3..A	28)	W..C..7	28)	F..E..4	28)	O..H..2	28)	V..K..Y
29)	G..1..D	29)	P..A..B	29)	X..D..8	29)	G..F..5	29)	P..J..3	29)	W..L..Z
30)	H..2..E	30)	Q..B..C	30)	A..E..A	30)	H..G..6	30)	Q..K..4	30)	X..M..1
31)	J..3..F	31)	R..C..D			31)	J..H..7			31)	A..N..2

CODES: P-PHYSICAL BIORHYTHM CURVE, E-EMOTIONAL BIORHYTHM CURVE, I-INTELLECTUAL BIORHYTHM CURVE

1921 1921

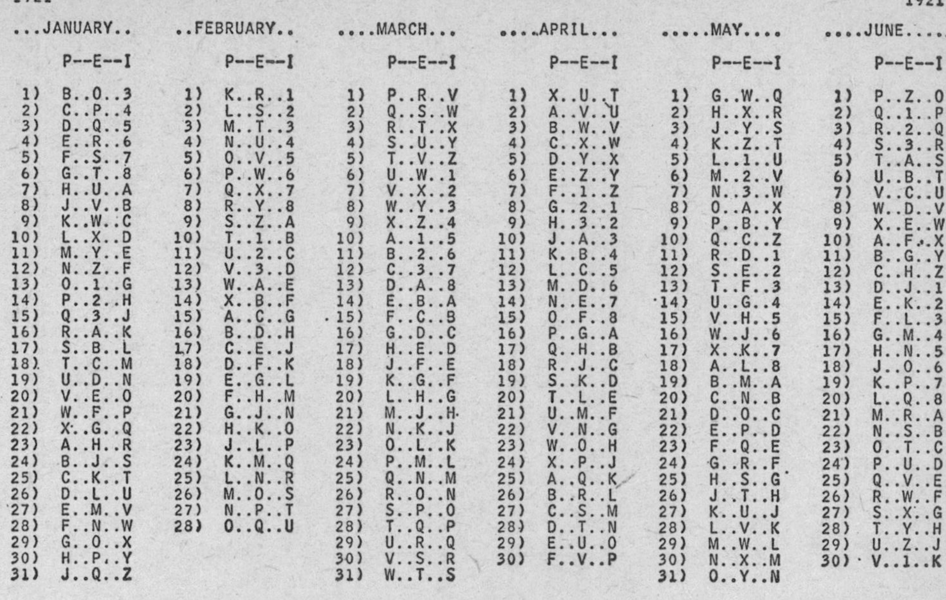

	...JANUARY..	..FEBRUARY..	....MARCH...	....APRIL...	.....MAY....	....JUNE.....
	P--E--I	P--E--I	P--E--I	P--E--I	P--E--I	P--E--I
1)	B..O..3	K..R..1	P..R..V	X..U..T	G..W..Q	P..Z..O
2)	C..P..4	L..S..2	Q..S..W	A..V..U	H..X..R	Q..1..P
3)	D..Q..5	M..T..3	R..T..X	B..W..V	J..Y..S	R..2..Q
4)	E..R..6	N..U..4	S..U..Y	C..X..W	K..Z..T	S..3..R
5)	F..S..7	O..V..5	T..V..Z	D..Y..X	L..1..U	T..A..S
6)	G..T..8	P..W..6	U..W..1	E..Z..Y	M..2..V	U..B..T
7)	H..U..A	Q..X..7	V..X..2	F..1..Z	N..3..W	V..C..U
8)	J..V..B	R..Y..8	W..Y..3	G..2..1	O..A..X	W..D..V
9)	K..W..C	S..Z..A	X..Z..4	H..3..2	P..B..Y	X..E..W
10)	L..X..D	T..1..B	A..1..5	J..A..3	Q..C..Z	A..F..X
11)	M..Y..E	U..2..C	B..2..6	K..B..4	R..D..1	B..G..Y
12)	N..Z..F	V..3..D	C..3..7	L..C..5	S..E..2	C..H..Z
13)	O..1..G	W..A..E	D..A..8	M..D..6	T..F..3	D..J..1
14)	P..2..H	X..B..F	E..B..A	N..E..7	U..G..4	E..K..2
15)	Q..3..J	A..C..G	F..C..B	O..F..8	V..H..5	F..L..3
16)	R..A..K	B..D..H	G..D..C	P..G..A	W..J..6	G..M..4
17)	S..B..L	C..E..J	H..E..D	Q..H..B	X..K..7	H..N..5
18)	T..C..M	D..F..K	J..F..E	R..J..C	A..L..8	J..O..6
19)	U..D..N	E..G..L	K..G..F	S..K..D	B..M..A	K..P..7
20)	V..E..O	F..H..M	L..H..G	T..L..E	C..N..B	L..Q..8
21)	W..F..P	G..J..N	M..J..H	U..M..F	D..O..C	M..R..A
22)	X..G..Q	H..K..O	N..K..J	V..N..G	E..P..D	N..S..B
23)	A..H..R	J..L..P	O..L..K	W..O..H	F..Q..E	O..T..C
24)	B..J..S	K..M..Q	P..M..L	X..P..J	G..R..F	P..U..D
25)	C..K..T	L..N..R	Q..N..M	A..Q..K	H..S..G	Q..V..E
26)	D..L..U	M..O..S	R..O..N	B..R..L	J..T..H	R..W..F
27)	E..M..V	N..P..T	S..P..O	C..S..M	K..U..J	S..X..G
28)	F..N..W	O..Q..U	T..Q..P	D..T..N	L..V..K	T..Y..H
29)	G..O..X		U..R..Q	E..U..O	M..W..L	U..Z..J
30)	H..P..Y		V..S..R	F..V..P	N..X..M	V..1..K
31)	J..Q..Z		W..T..S		O..Y..N	

1921 1921

Day	....JULY.... P--E--I.	...AUGUST... P--E--I	..SEPTEMBER. P--E--I	..OCTOBER... P--E--I	..NOVEMBER.. P--E--I	..DECEMBER.. P--E--I
1)	W..2..L	G..B..J	P..E..G	W..G..D	G..K..B	O..M..7
2)	X..3..M	H..C..K	Q..F..H	X..H..E	H..L..C	P..N..8
3)	A..A..N	J..D..L	R..G..J	A..J..F	J..M..D	Q..O..A
4)	B..B..O	K..E..M	S..H..K	B..K..G	K..N..E	R..P..B
5)	C..C..P	L..F..N	T..J..L	C..L..H	L..O..F	S..Q..C
6)	D..D..Q	M..G..O	U..K..M	D..M..J	M..P..G	T..R..D
7)	E..E..R	N..H..P	V..L..N	E..N..K	N..Q..H	U..S..E
8)	F..F..S	O..J..Q	W..M..O	F..O..L	O..R..J	V..T..F
9)	G..G..T	P..K..R	X..N..P	G..P..M	P..S..K	W..U..G
10)	H..H..U	Q..L..S	A..O..Q	H..Q..N	Q..T..L	X..V..H
11)	J..J..V	R..M..T	B..P..R	J..R..O	R..U..M	A..W..J
12)	K..K..W	S..N..U	C..Q..S	K..S..P	S..V..N	B..X..K
13)	L..L..X	T..O..V	D..R..T	L..T..Q	T..W..O	C..Y..L
14)	M..M..Y	U..P..W	E..S..U	M..U..R	U..X..P	D..Z..M
15)	N..N..Z	V..Q..X	F..T..V	N..V..S	V..Y..Q	E..1..N
16)	O..O..1	W..R..Y	G..U..W	O..W..T	W..Z..R	F..2..O
17)	P..P..2	X..S..Z	H..V..X	P..X..U	X..1..S	G..3..P
18)	Q..Q..3	A..T..1	J..W..Y	Q..Y..V	A..2..T	H..A..Q
19)	R..R..4	B..U..2	K..X..Z	R..Z..W	B..3..U	J..B..R
20)	S..S..5	C..V..3	L..Y..1	S..1..X	C..A..V	K..C..S
21)	T..T..6	D..W..4	M..Z..2	T..2..Y	D..B..W	L..D..T
22)	U..U..7	E..X..5	N..1..3	U..3..Z	E..C..X	M..E..U
23)	V..V..8	F..Y..6	O..2..4	V..A..1	F..D..Y	N..F..V
24)	W..W..A	G..Z..7	P..3..5	W..B..2	G..E..Z	O..G..W
25)	X..X..B	H..1..8	Q..A..6	X..C..3	H..F..1	P..H..X
26)	A..Y..C	J..2..A	R..B..7	A..D..4	J..G..2	Q..J..Y
27)	B..Z..D	K..3..B	S..C..8	B..E..5	K..H..3	R..K..Z
28)	C..1..E	L..A..C	T..D..A	C..F..6	L..J..4	S..L..1
29)	D..2..F	M..B..D	U..E..B	D..G..7	M..K..5	T..M..2
30)	E..3..G	N..C..E	V..F..C	E..H..8	N..L..6	U..N..3
31)	F..A..H	O..D..F		F..J..A		V..O..4

CODES: P-PHYSICAL BIORHYTHM CURVE, E-EMOTIONAL BIORHYTHM CURVE. I-INTELLECTUAL BIORHYTHM CURVE

1922 | 1922

...JANUARY..	..FEBRUARY..	....MARCH...	....APRIL...	.....MAY....	....JUNE....
P--E--I	P--E--I	P--E--I	P--E--I	P--E--I	P--E--I
1) W..P..5	1) G..S..3	1) M..S..X	1) U..V..V	1) D..X..S	1) M..1..Q
2) X..Q..6	2) H..T..4	2) N..T..Y	2) V..W..W	2) E..Y..T	2) N..2..R
3) A..R..7	3) J..U..5	3) O..U..Z	3) W..X..X	3) F..Z..U	3) O..3..S
4) B..S..8	4) K..V..6	4) P..V..1	4) X..Y..Y	4) G..1..V	4) P..A..T
5) C..T..A	5) L..W..7	5) Q..W..2	5) A..Z..Z	5) H..2..W	5) Q..B..U
6) D..U..B	6) M..X..8	6) R..X..3	6) B..1..1	6) J..3..X	6) R..C..V
7) E..V..C	7) N..Y..A	7) S..Y..4	7) C..2..2	7) K..A..Y	7) S..D..W
8) F..W..D	8) O..Z..B	8) T..Z..5	8) D..3..3	8) L..B..Z	8) T..E..X
9) G..X..E	9) P..1..C	9) U..1..6	9) E..A..4	9) M..C..1	9) U..F..Y
10) H..Y..F	10) Q..2..D	10) V..2..7	10) F..B..5	10) N..D..2	10) V..G..Z
11) J..Z..G	11) R..3..E	11) W..3..8	11) G..C..6	11) O..E..3	11) W..H..1
12) K..1..H	12) S..A..F	12) X..A..A	12) H..D..7	12) P..F..4	12) X..J..2
13) L..2..J	13) T..B..G	13) A..B..B	13) J..E..8	13) Q..G..5	13) A..K..3
14) M..3..K	14) U..C..H	14) B..C..C	14) K..F..A	14) R..H..6	14) B..L..4
15) N..A..L	15) V..D..J	15) C..D..D	15) L..G..B	15) S..J..7	15) C..M..5
16) O..B..M	16) W..E..K	16) D..E..E	16) M..H..C	16) T..K..8	16) D..N..6
17) P..C..N	17) X..F..L	17) E..F..F	17) N..J..D	17) U..L..A	17) E..O..7
18) Q..D..O	18) A..G..M	18) F..G..G	18) O..K..E	18) V..M..B	18) F..P..8
19) R..E..P	19) B..H..N	19) G..H..H	19) P..L..F	19) W..N..C	19) G..Q..A
20) S..F..Q	20) C..J..O	20) H..J..J	20) Q..M..G	20) X..O..D	20) H..R..B
21) T..G..R	21) D..K..P	21) J..K..K	21) R..N..H	21) A..P..E	21) J..S..C
22) U..H..S	22) E..L..Q	22) K..L..L	22) S..O..J	22) B..Q..F	22) K..T..D
23) V..J..T	23) F..M..R	23) L..M..M	23) T..P..K	23) C..R..G	23) L..U..E
24) W..K..U	24) G..N..S	24) M..N..N	24) U..Q..L	24) D..S..H	24) M..V..F
25) X..L..V	25) H..O..T	25) N..O..O	25) V..R..M	25) E..T..J	25) N..W..G
26) A..M..W	26) J..P..U	26) O..P..P	26) W..S..N	26) F..U..K	26) O..X..H
27) B..N..X	27) K..Q..V	27) P..Q..Q	27) X..T..O	27) G..V..L	27) P..Y..J
28) C..O..Y	28) L..R..W	28) Q..R..R	28) A..U..P	28) H..W..M	28) Q..Z..K
29) D..P..Z		29) R..S..S	29) B..V..Q	29) J..X..N	29) R..1..L
30) E..Q..1		30) S..T..T	30) C..W..R	30) K..Y..O	30) S..2..M
31) F..R..2		31) T..U..U		31) L..Z..P	

1922 1922

	....JULY....	...AUGUST...	..SEPTEMBER.	..OCTOBER...	..NOVEMBER..	..DECEMBER..
	P--E--I	P--E--I	P--E--I	P--E--I	P--E--I	P--E--I
1)	T..3..N	D..C..L	M..F..J	T..H..F	D..L..D	L..N..A
2)	U..A..O	E..D..M	N..G..K	U..J..G	E..M..E	M..O..B
3)	V..B..P	F..E..N	O..H..L	V..K..H	F..N..F	N..P..C
4)	W..C..Q	G..F..O	P..J..M	W..L..J	G..O..G	O..Q..D
5)	X..D..R	H..G..P	Q..K..N	X..M..K	H..P..H	P..R..E
6)	A..E..S	J..H..Q	R..L..O	A..N..L	J..Q..J	Q..S..F
7)	B..F..T	K..J..R	S..M..P	B..O..M	K..R..K	R..T..G
8)	C..G..U	L..K..S	T..N..Q	C..P..N	L..S..L	S..U..H
9)	D..H..V	M..L..T	U..O..R	D..Q..O	M..T..M	T..V..J
10)	E..J..W	N..M..U	V..P..S	E..R..P	N..U..N	U..W..K
11)	F..K..X	O..N..V	W..Q..T	F..S..Q	O..V..O	V..X..L
12)	G..L..Y	P..O..W	X..R..U	G..T..R	P..W..P	W..Y..M
13)	H..M..Z	Q..P..X	A..S..V	H..U..S	Q..X..Q	X..Z..N
14)	J..N..1	R..Q..Y	B..T..W	J..V..T	R..Y..R	A..1..O
15)	K..O..2	S..R..Z	C..U..X	K..W..U	S..Z..S	B..2..P
16)	L..P..3	T..S..1	D..V..Y	L..X..V	T..1..T	C..3..Q
17)	M..Q..4	U..T..2	E..W..Z	M..Y..W	U..2..U	D..A..R
18)	N..R..5	V..U..3	F..X..1	N..Z..X	V..3..V	E..B..S
19)	O..S..6	W..V..4	G..Y..2	O..1..Y	W..A..W	F..C..T
20)	P..T..7	X..W..5	H..Z..3	P..2..Z	X..B..X	G..D..U
21)	Q..U..8	A..X..6	J..1..4	Q..3..1	A..C..Y	H..E..V
22)	R..V..A	B..Y..7	K..2..5	R..A..2	B..D..Z	J..F..W
23)	S..W..B	C..Z..8	L..3..6	S..B..3	C..E..1	K..G..X
24)	T..X..C	D..1..A	M..A..7	T..C..4	D..F..2	L..H..Y
25)	U..Y..D	E..2..B	N..B..8	U..D..5	E..G..3	M..J..Z
26)	V..Z..E	F..3..C	O..C..A	V..E..6	F..H..4	N..K..1
27)	W..1..F	G..A..D	P..D..B	W..F..7	G..J..5	O..L..2
28)	X..2..G	H..B..E	Q..E..C	X..G..8	H..K..6	P..M..3
29)	A..3..H	J..C..F	R..F..D	A..H..A	J..L..7	Q..N..4
30)	B..A..J	K..D..G	S..G..E	B..J..B	K..M..8	R..O..5
31)	C..B..K	L..E..H		C..K..C		S..P..6

CODES: P-PHYSICAL BIORHYTHM CURVE,E-EMOTIONAL BIORHYTHM CURVE,I-INTELLECTUAL BIORHYTHM CURVE

1923 1923

	...JANUARY..	..FEBRUARY..	....MARCH...	....APRIL...	.....MAY....	....JUNE....
	P--E--I	P--E--I	P--E--I	P--E--I	P--E--I	P--E--I
1)	T..Q..7	D..T..5	J..T..Z	R..W..X	A..Y..U	J..2..S
2)	U..R..8	E..U..6	K..U..1	S..X..Y	B..Z..V	K..3..T
3)	V..S..A	F..V..7	L..V..2	T..Y..Z	C..1..W	L..A..U
4)	W..T..B	G..W..8	M..W..3	U..Z..1	D..2..X	M..B..V
5)	X..U..C	H..X..A	N..X..4	V..1..2	E..3..Y	N..C..W
6)	A..V..D	J..Y..B	O..Y..5	W..2..3	F..A..Z	O..D..X
7)	B..W..E	K..Z..C	P..Z..6	X..3..4	G..B..1	P..E..Y
8)	C..X..F	L..1..D	Q..1..7	A..A..5	H..C..2	Q..F..Z
9)	D..Y..G	M..2..E	R..2..8	B..B..6	J..D..3	R..G..1
10)	E..Z..H	N..3..F	S..3..A	C..C..7	K..E..4	S..H..2
11)	F..1..J	O..A..G	T..A..B	D..D..8	L..F..5	T..J..3
12)	G..2..K	P..B..H	U..B..C	E..E..A	M..G..6	U..K..4
13)	H..3..L	Q..C..J	V..C..D	F..F..B	N..H..7	V..L..5
14)	J..A..M	R..D..K	W..D..E	G..G..C	O..J..8	W..M..6
15)	K..B..N	S..E..L	X..E..F	H..H..D	P..K..A	X..N..7
16)	L..C..O	T..F..M	A..F..G	J..J..E	Q..L..B	A..O..8
17)	M..D..P	U..G..N	B..G..H	K..K..F	R..M..C	B..P..A
18)	N..E..Q	V..H..O	C..H..J	L..L..G	S..N..D	C..Q..B
19)	O..F..R	W..J..P	D..J..K	M..M..H	T..O..E	D..R..C
20)	P..G..S	X..K..Q	E..K..L	N..N..J	U..P..F	E..S..D
21)	Q..H..T	A..L..R	F..L..M	O..O..K	V..Q..G	F..T..E
22)	R..J..U	B..M..S	G..M..N	P..P..L	W..R..H	G..U..F
23)	S..K..V	C..N..T	H..N..O	Q..Q..M	X..S..J	H..V..G
24)	T..L..W	D..O..U	J..O..P	R..R..N	A..T..K	J..W..H
25)	U..M..X	E..P..V	K..P..Q	S..S..O	B..U..L	K..X..J
26)	V..N..Y	F..Q..W	L..Q..R	T..T..P	C..V..M	L..Y..K
27)	W..O..Z	G..R..X	M..R..S	U..U..Q	D..W..N	M..Z..L
28)	X..P..1	H..S..Y	N..S..T	V..V..R	E..X..O	N..1..M
29)	A..Q..2		O..T..U	W..W..S	F..Y..P	O..2..N
30)	B..R..3		P..U..V	X..X..T	G..Z..Q	P..3..O
31)	C..S..4		Q..V..W		H..1..R	

1923						1923
	....JULY....	...AUGUST...	..SEPTEMBER.	..OCTOBER...	..NOVEMBER..	..DECEMBER..
	P--E--I	P--E--I	P--E--I	P--E--I	P--E--I	P--E--I
1)	Q..A..P	A..D..N	J..G..L	Q..J..H	A..M..F	H..O..G
2)	R..B..Q	B..E..O	K..H..M	R..K..J	B..N..G	J..P..D
3)	S..C..R	C..F..P	L..J..N	S..L..K	C..O..H	K..Q..E
4)	T..D..S	D..G..Q	M..K..O	T..M..L	D..P..J	L..R..F
5)	U..E..T	E..H..R	N..L..P	U..N..M	E..Q..K	M..S..G
6)	V..F..U	F..J..S	O..M..Q	V..O..N	F..R..L	N..T..H
7)	W..G..V	G..K..T	P..N..R	W..P..O	G..S..M	O..U..J
8)	X..H..W	H..L..U	Q..O..S	X..Q..P	H..T..N	P..V..K
9)	A..J..X	J..M..V	R..P..T	A..R..Q	J..U..O	Q..W..L
10)	B..K..Y	K..N..W	S..Q..U	B..S..R	K..V..P	R..X..M
11)	C..L..Z	L..O..X	T..R..V	C..T..S	L..W..Q	S..Y..N
12)	D..M..1	M..P..Y	U..S..W	D..U..T	M..X..R	T..Z..O
13)	E..N..2	N..Q..Z	V..T..X	E..V..U	N..Y..S	U..1..P
14)	F..O..3	O..R..1	W..U..Y	F..W..V	O..Z..T	V..2..Q
15)	G..P..4	P..S..2	X..V..Z	G..X..W	P..1..U	W..3..R
16)	H..Q..5	Q..T..3	A..W..1	H..Y..X	Q..2..V	X..A..S
17)	J..R..6	R..U..4	B..X..2	J..Z..Y	R..3..W	A..B..T
18)	K..S..7	S..V..5	C..Y..3	K..1..Z	S..A..X	B..C..U
19)	L..T..8	T..W..6	D..Z..4	L..2..1	T..B..Y	C..D..V
20)	M..U..A	U..X..7	E..1..5	M..3..2	U..C..Z	D..E..W
21)	N..V..B	V..Y..8	F..2..6	N..A..3	V..D..1	E..F..X
22)	O..W..C	W..Z..A	G..3..7	O..B..4	W..E..2	F..G..Y
23)	P..X..D	X..1..B	H..A..8	P..C..5	X..F..3	G..H..Z
24)	Q..Y..E	A..2..C	J..B..A	Q..D..6	A..G..4	H..J..1
25)	R..Z..F	B..3..D	K..C..B	R..E..7	B..H..5	J..K..2
26)	S..1..G	C..A..E	L..D..C	S..F..8	C..J..6	K..L..3
27)	T..2..H	D..B..F	M..E..D	T..G..A	D..K..7	L..M..4
28)	U..3..J	E..C..G	N..F..E	U..H..B	E..L..8	M..N..5
29)	V..A..K	F..D..H	O..G..F	V..J..C	F..M..A	N..O..6
30)	W..B..L	G..E..J	P..H..G	W..K..D	G..N..B	O..P..7
31)	X..C..M	H..F..K		X..L..E		P..Q..8

CODES: P-PHYSICAL BIORHYTHM CURVE, E-EMOTIONAL BIORHYTHM CURVE, I-INTELLECTUAL BIORHYTHM CURVE

	..JANUARY..	..FEBRUARY..	....MARCH...	....APRIL...	.....MAY....	....JUNE....
	P--E--I	P--E--I	P--E--I	P--E--I	P--E--I	P--E--I
1)	Q..R..A	A..U..7	G..V..3	P..Y..1	W..1..X	G..A..V
2)	R..S..B	B..V..8	H..W..4	Q..Z..2	X..2..Y	H..B..W
3)	S..T..C	C..W..A	J..X..5	R..1..3	A..3..Z	J..C..X
4)	T..U..D	D..X..B	K..Y..6	S..2..4	B..A..1	K..D..Y
5)	U..V..E	E..Y..C	L..Z..7	T..3..5	C..B..2	L..E..Z
6)	V..W..F	F..Z..D	M..1..8	U..A..6	D..C..3	M..F..1
7)	W..X..G	G..1..E	N..2..A	V..B..7	E..D..4	N..G..2
8)	X..Y..H	H..2..F	O..3..B	W..C..8	F..E..5	O..H..3
9)	A..Z..J	J..3..G	P..A..C	X..D..A	G..F..6	P..J..4
10)	B..1..K	K..A..H	Q..B..D	A..E..B	H..G..7	Q..K..5
11)	C..2..L	L..B..J	R..C..E	B..F..C	J..H..8	R..L..6
12)	D..3..M	M..C..K	S..D..F	C..G..D	K..J..A	S..M..7
13)	E..A..N	N..D..L	T..E..G	D..H..E	L..K..B	T..N..8
14)	F..B..O	O..E..M	U..F..H	E..J..F	M..L..C	U..O..A
15)	G..C..P	P..F..N	V..G..J	F..K..G	N..M..D	V..P..B
16)	H..D..Q	Q..G..O	W..H..K	G..L..H	O..N..E	W..Q..C
17)	J..E..R	R..H..P	X..J..L	H..M..J	P..O..F	X..R..D
18)	K..F..S	S..J..Q	A..K..M	J..N..K	Q..P..G	A..S..E
19)	L..G..T	T..K..R	B..L..N	K..O..L	R..Q..H	B..T..F
20)	M..H..U	U..L..S	C..M..O	L..P..M	S..R..J	C..U..G
21)	N..J..V	V..M..T	D..N..P	M..Q..N	T..S..K	D..V..H
22)	O..K..W	W..N..U	E..O..Q	N..R..O	U..T..L	E..W..J
23)	P..L..X	X..O..V	F..P..R	O..S..P	V..U..M	F..X..K
24)	Q..M..Y	A..P..W	G..Q..S	P..T..Q	W..V..N	G..Y..L
25)	R..N..Z	B..Q..X	H..R..T	Q..U..R	X..W..O	H..Z..M
26)	S..O..1	C..R..Y	J..S..U	R..V..S	A..X..P	J..1..N
27)	T..P..2	D..S..Z	K..T..V	S..W..T	B..Y..Q	K..2..O
28)	U..Q..3	E..T..1	L..U..W	T..X..U	C..Z..R	L..3..P
29)	V..R..4	F..U..2	M..V..X	U..Y..V	D..1..S	M..A..Q
30)	W..S..5		N..W..Y	V..Z..W	E..2..T	N..B..R
31)	X..T..6		O..X..Z		F..3..U	

1924 1924

Day	....JULY.... P--E--I	...AUGUST... P--E--I	..SEPTEMBER. P--E--I	..OCTOBER... P--E--I	..NOVEMBER.. P--E--I	..DECEMBER.. P--E--I
1)	O..C..S	W..F..Q	G..J..O	O..L..L	W..?..J	F..Q..F
2)	P..D..T	X..G..R	H..K..P	P..M..M	X..P..K	G..R..G
3)	Q..E..U	A..H..S	J..L..Q	Q..N..N	A..Q..L	H..S..H
4)	R..F..V	B..J..T	K..M..R	R..O..O	B..R..M	J..T..J
5)	S..G..W	C..K..U	L..N..S	S..P..P	C..S..N	K..U..K
6)	T..H..X	D..L..V	M..O..T	T..Q..Q	D..T..O	L..V..L
7)	U..J..Y	E..M..W	N..P..U	U..R..R	E..U..P	M..W..M
8)	V..K..Z	F..N..X	O..Q..V	V..S..S	F..V..Q	N..X..N
9)	W..L..1	G..O..Y	P..R..W	W..T..T	G..W..R	O..Y..O
10)	X..M..2	H..P..Z	Q..S..X	X..U..U	H..X..S	P..Z..P
11)	A..N..3	J..Q..1	R..T..Y	A..V..V	J..Y..T	Q..1..Q
12)	3..O..4	K..R..2	S..U..Z	B..W..W	K..Z..U	R..2..R
13)	C..P..5	L..S..3	T..V..1	C..X..X	L..1..V	S..3..S
14)	D..Q..6	M..T..4	U..W..2	D..Y..Y	M..2..W	T..A..T
15)	E..R..7	N..U..5	V..X..3	E..Z..Z	N..3..X	U..B..U
16)	F..S..8	O..V..6	W..Y..4	F..1..1	O..A..Y	V..C..V
17)	G..T..A	P..W..7	X..Z..5	G..2..2	P..B..Z	W..D..W
18)	H..U..B	Q..X..8	A..1..6	H..3..3	Q..C..1	X..E..X
19)	J..V..C	R..Y..A	B..2..7	J..A..4	R..D..2	A..F..Y
20)	K..W..D	S..Z..B	C..3..8	K..B..5	S..E..3	B..G..Z
21)	L..X..E	T..1..C	D..A..A	L..C..6	T..F..4	C..H..1
22)	M..Y..F	U..2..D	E..B..B	M..D..7	U..G..5	D..J..2
23)	N..Z..G	V..3..E	F..C..C	N..E..8	V..H..6	E..K..3
24)	O..1..H	W..A..F	G..D..D	O..F..A	W..J..7	F..L..4
25)	P..2..J	X..B..G	H..E..E	P..G..B	X..K..8	G..M..5
26)	Q..3..K	A..C..H	J..F..F	Q..H..C	A..L..A	H..N..6
27)	R..A..L	B..D..J	K..G..G	R..J..D	B..M..B	J..O..7
28)	S..B..M	C..E..K	L..H..H	S..K..E	C..N..C	K..P..8
29)	T..C..N	D..F..L	M..J..J	T..L..F	D..O..D	L..Q..A
30)	U..D..O	E..G..M	N..K..K	U..M..G	E..P..E	M..R..B
31)	V..E..P	F..H..N		V..N..H		N..S..C

CODES: P-PHYSICAL BIORHYTHM CURVE,E-EMOTIONAL BIORHYTHM CURVE,I-INTELLECTUAL BIORHYTHM CURVE

1925 1925

	...JANUARY..	..FEBRUARY..	....MARCH...	....APRIL...	.....MAY....	....JUNE....
	P--E--I	P--E--I	P--E--I	P--E--I	P--E--I	P--E--I
1)	O..T..D	W..W..B	D..W..5	M..Z..3	T..2..Z	D..B..X
2)	P..U..E	X..X..C	E..X..6	N..1..4	U..3..1	E..C..Y
3)	Q..V..F	A..Y..D	F..Y..7	O..2..5	V..A..2	F..D..Z
4)	R..W..G	B..Z..E	G..Z..8	P..3..6	W..B..3	G..E..1
5)	S..X..H	C..1..F	H..1..A	Q..A..7	X..C..4	H..F..2
6)	T..Y..J	D..2..G	J..2..B	R..B..8	A..D..5	J..G..3
7)	U..Z..K	E..3..H	K..3..C	S..C..A	B..E..6	K..H..4
8)	V..1..L	F..A..J	L..A..D	T..D..B	C..F..7	L..J..5
9)	W..2..M	G..B..K	M..B..E	U..E..C	D..G..8	M..K..6
10)	X..3..N	H..C..L	N..C..F	V..F..D	E..H..A	N..L..7
11)	A..A..O	J..D..M	O..D..G	W..G..E	F..J..B	O..M..8
12)	B..B..P	K..E..N	P..E..H	X..H..F	G..K..C	P..N..A
13)	C..C..Q	L..F..O	Q..F..J	A..J..G	H..L..D	Q..O..B
14)	D..D..R	M..G..P	R..G..K	B..K..H	J..M..E	R..P..C
15)	E..E..S	N..H..Q	S..H..L	C..L..J	K..N..F	S..Q..D
16)	F..F..T	O..J..R	T..J..M	D..M..K	L..O..G	T..R..E
17)	G..G..U	P..K..S	U..K..N	E..N..L	M..P..H	U..S..F
18)	H..H..V	Q..L..T	V..L..O	F..O..M	N..Q..J	V..T..G
19)	J..J..W	R..M..U	W..M..P	G..P..N	O..R..K	W..U..H
20)	K..K..X	S..N..V	X..N..Q	H..Q..O	P..S..L	X..V..J
21)	L..L..Y	T..O..W	A..O..R	J..R..P	Q..T..M	A..W..K
22)	M..M..Z	U..P..X	B..P..S	K..S..Q	R..U..N	B..X..L
23)	N..N..1	V..Q..Y	C..Q..T	L..T..R	S..V..O	C..Y..M
24)	O..O..2	W..R..Z	D..R..U	M..U..S	T..W..P	D..Z..N
25)	P..P..3	X..S..1	E..S..V	N..V..T	U..X..Q	E..1..O
26)	Q..Q..4	A..T..2	F..T..W	O..W..U	V..Y..R	F..2..P
27)	R..R..5	B..U..3	G..U..X	P..X..V	W..Z..S	G..3..Q
28)	S..S..6	C..V..4	H..V..Y	Q..Y..W	X..1..T	H..A..R
29)	T..T..7		J..W..Z	R..Z..X	A..2..U	J..B..S
30)	U..U..8		K..X..1	S..1..Y	B..3..V	K..C..T
31)	V..V..A		L..Y..2		C..A..W	

1925						1925
	....JULY....	...AUGUST...	..SEPTEMBER.	..OCTOBER...	..NOVEMBER..	..DECEMBER:.
	P--E--I	P--E--I	P--E--I	P--E--I	P--E--I	P--E--I
1)	L..D..U	T..G..S	D..K..Q	L..M..N	T..P..L	C..R..H
2)	M..E..V	U..H..T	E..L..R	M..N..O	U..Q..M	D..S..J
3)	N..F..W	V..J..U	F..M..S	N..O..P	V..R..N	E..T..K
4)	O..G..X	W..K..V	G..N..T	O..P..Q	W..S..O	F..U..L
5)	P..H..Y	X..L..W	H..O..U	P..Q..R	X..T..P	G..V..M
6)	Q..J..Z	A..M..X	J..P..V	Q..R..S	A..U..Q	H..W..N
7)	R..K..1	B..N..Y	K..Q..W	R..S..T	B..V..R	J..X..O
8)	S..L..2	C..O..Z	L..R..X	S..T..U	C..W..S	K..Y..P
9)	T..M..3	D..P..1	M..S..Y	T..U..V	D..X..T	L..Z..Q
10)	U..N..4	E..Q..2	N..T..Z	U..V..W	E..Y..U	M..1..R
11)	V..O..5	F..R..3	O..U..1	V..W..X	F..Z..V	N..2..S
12)	W..P..6	G..S..4	P..V..2	W..X..Y	G..1..W	O..3..T
13)	X..Q..7	H..T..5	Q..W..3	X..Y..Z	H..2..X	P..A..U
14)	A..R..8	J..U..6	R..X..4	A..Z..1	J..3..Y	Q..B..V
15)	B..S..A	K..V..7	S..Y..5	B..1..2	K..A..Z	R..C..W
16)	C..T..B	L..W..8	T..Z..6	C..2..3	L..B..1	S..D..X
17)	D..U..C	M..X..A	U..1..7	D..3..4	M..C..2	T..E..Y
18)	E..V..D	N..Y..B	V..2..8	E..A..5	N..D..3	U..F..Z
19)	F..W..E	O..Z..C	W..3..A	F..B..6	O..E..4	V..G..1
20)	G..X..F	P..1..D	X..A..B	G..C..7	P..F..5	W..H..2
21)	H..Y..G	Q..2..E	A..B..C	H..D..8	Q..G..6	X..J..3
22)	J..Z..H	R..3..F	B..C..D	J..E..A	R..H..7	A..K..4
23)	K..1..J	S..A..G	C..D..E	K..F..B	S..J..8	B..L..5
24)	L..2..K	T..B..H	D..E..F	L..G..C	T..K..A	C..M..6
25)	M..3..L	U..C..J	E..F..G	M..H..D	U..L..B	D..N..7
26)	N..A..M	V..D..K	F..G..H	N..J..E	V..M..C	E..O..8
27)	O..B..N	W..E..L	G..H..J	O..K..F	W..N..D	F..P..A
28)	P..C..O	X..F..M	H..J..K	P..L..G	X..O..E	G..Q..B
29)	Q..D..P	A..G..N	J..K..L	Q..M..H	A..P..F	H..R..C
30)	R..E..Q	B..H..O	K..L..M	R..N..J	B..Q..G	J..S..D
31)	S..F..R	C..J..P		S..O..K		K..T..E

CODES: P-PHYSICAL BIORHYTHM CURVE, E-EMOTIONAL BIORHYTHM CURVE, I-INTELLECTUAL BIORHYTHM CURVE

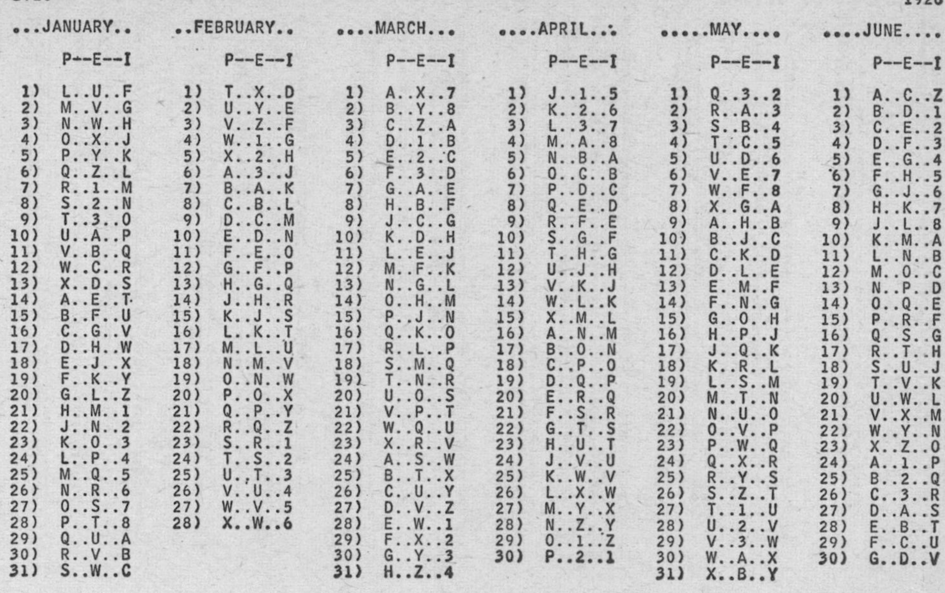

	...JANUARY.. P--E--I	..FEBRUARY.. P--E--I	....MARCH... P--E--I	....APRIL... P--E--I	.....MAY.... P--E--I	....JUNE.... P--E--I
1)	L..U..F	T..X..D	A..X..7	J..1..5	Q..3..2	A..C..Z
2)	M..V..G	U..Y..E	B..Y..8	K..2..6	R..A..3	B..D..1
3)	N..W..H	V..Z..F	C..Z..A	L..3..7	S..B..4	C..E..2
4)	O..X..J	W..1..G	D..1..B	M..A..8	T..C..5	D..F..3
5)	P..Y..K	X..2..H	E..2..C	N..B..A	U..D..6	E..G..4
6)	Q..Z..L	A..3..J	F..3..D	O..C..B	V..E..7	F..H..5
7)	R..1..M	B..A..K	G..A..E	P..D..C	W..F..8	G..J..6
8)	S..2..N	C..B..L	H..B..F	Q..E..D	X..G..A	H..K..7
9)	T..3..O	D..C..M	J..C..G	R..F..E	A..H..B	J..L..8
10)	U..A..P	E..D..N	K..D..H	S..G..F	B..J..C	K..M..A
11)	V..B..Q	F..E..O	L..E..J	T..H..G	C..K..D	L..N..B
12)	W..C..R	G..F..P	M..F..K	U..J..H	D..L..E	M..O..C
13)	X..D..S	H..G..Q	N..G..L	V..K..J	E..M..F	N..P..D
14)	A..E..T	J..H..R	O..H..M	W..L..K	F..N..G	O..Q..E
15)	B..F..U	K..J..S	P..J..N	X..M..L	G..O..H	P..R..F
16)	C..G..V	L..K..T	Q..K..O	A..N..M	H..P..J	Q..S..G
17)	D..H..W	M..L..U	R..L..P	B..O..N	J..Q..K	R..T..H
18)	E..J..X	N..M..V	S..M..Q	C..P..O	K..R..L	S..U..J
19)	F..K..Y	O..N..W	T..N..R	D..Q..P	L..S..M	T..V..K
20)	G..L..Z	P..O..X	U..O..S	E..R..Q	M..T..N	U..W..L
21)	H..M..1	Q..P..Y	V..P..T	F..S..R	N..U..O	V..X..M
22)	J..N..2	R..Q..Z	W..Q..U	G..T..S	O..V..P	W..Y..N
23)	K..O..3	S..R..1	X..R..V	H..U..T	P..W..Q	X..Z..O
24)	L..P..4	T..S..2	A..S..W	J..V..U	Q..X..R	A..1..P
25)	M..Q..5	U..T..3	B..T..X	K..W..V	R..Y..S	B..2..Q
26)	N..R..6	V..U..4	C..U..Y	L..X..W	S..Z..T	C..3..R
27)	O..S..7	W..V..5	D..V..Z	M..Y..X	T..1..U	D..A..S
28)	P..T..8	X..W..6	E..W..1	N..Z..Y	U..2..V	E..B..T
29)	Q..U..A		F..X..2	O..1..Z	V..3..W	F..C..U
30)	R..V..B		G..Y..3	P..2..1	W..A..X	G..D..V
31)	S..W..C		H..Z..4		X..B..Y	

1926 | 1926

Day	....JULY.... P--E--I	...AUGUST... P--E--I	..SEPTEMBER P--E--I	..OCTOBER... P--E--I	..NOVEMBER.. P--E--I	..DECEMBER.. P--E--I
1)	H..E..W	Q..H..U	A..L..S	H..N..P	Q..Q..N	X..S..K
2)	J..F..X	R..J..V	B..M..T	J..O..Q	R..R..O	A..T..L
3)	K..G..Y	S..K..W	C..N..U	K..P..R	S..S..P	B..U..M
4)	L..H..Z	T..L..X	D..O..V	L..Q..S	T..T..Q	C..V..N
5)	M..J..1	U..M..Y	E..P..W	M..R..T	U..U..R	D..W..O
6)	N..K..2	V..N..Z	F..Q..X	N..S..U	V..V..S	E..X..P
7)	O..L..3	W..O..1	G..R..Y	O..T..V	W..W..T	F..Y..Q
8)	P..M..4	X..P..2	H..S..Z	P..U..W	X..X..U	G..Z..R
9)	Q..N..5	A..Q..3	J..T..1	Q..V..X	A..Y..V	H..1..S
10)	R..O..6	B..R..4	K..U..2	R..W..Y	B..Z..W	J..2..T
11)	S..P..7	C..S..5	L..V..3	S..X..Z	C..1..X	K..3..U
12)	T..Q..8	D..T..6	M..W..4	T..Y..1	D..2..Y	L..A..V
13)	U..R..A	E..U..7	N..X..5	U..Z..2	E..3..Z	M..B..W
14)	V..S..B	F..V..8	O..Y..6	V..1..3	F..A..1	N..C..X
15)	W..T..C	G..W..A	P..Z..7	W..2..4	G..B..2	O..D..Y
16)	X..U..D	H..X..B	Q..1..8	X..3..5	H..C..3	P..E..Z
17)	A..V..E	J..Y..C	R..2..A	A..A..6	J..D..4	Q..F..1
18)	B..W..F	K..Z..D	S..3..B	B..B..7	K..E..5	R..G..2
19)	C..X..G	L..1..E	T..A..C	C..C..8	L..F..6	S..H..3
20)	D..Y..H	M..2..F	U..B..D	D..D..A	M..G..7	T..J..4
21)	E..Z..J	N..3..G	V..C..E	E..E..B	N..H..8	U..K..5
22)	F..1..K	O..A..H	W..D..F	F..F..C	O..J..A	V..L..6
23)	G..2..L	P..B..J	X..E..G	G..G..D	P..K..B	W..M..7
24)	H..3..M	Q..C..K	A..F..H	H..H..E	Q..L..C	X..N..8
25)	J..A..N	R..D..L	B..G..J	J..J..F	R..M..D	A..O..A
26)	K..B..O	S..E..M	C..H..K	K..K..G	S..N..E	B..P..B
27)	L..C..P	T..F..N	D..J..L	L..L..H	T..O..F	C..Q..C
28)	M..D..Q	U..G..O	E..K..M	M..M..J	U..P..G	D..R..D
29)	N..E..R	V..H..P	F..L..N	N..N..K	V..Q..H	E..S..E
30)	O..F..S	W..J..Q	G..M..O	O..O..L	W..R..J	F..T..F
31)	P..G..T	X..K..R		P..P..M		G..U..G

CODES: P-PHYSICAL BIORHYTHM CURVE, E-EMOTIONAL BIORHYTHM CURVE, I-INTELLECTUAL BIORHYTHM CURVE

1927 1927

Day	...JANUARY..	..FEBRUARY..	....MARCH...	....APRIL...	.....MAY....	....JUNE....
	P--E--I	P--E--I	P--E--I	P--E--I	P--E--I	P--E--I
1)	H..V..H	Q..Y..F	V..Y..A	F..2..7	N..A..4	V..D..2
2)	J..W..J	R..Z..G	W..Z..B	G..3..8	O..B..5	W..E..3
3)	K..X..K	S..1..H	X..1..C	H..A..A	P..C..6	X..F..4
4)	L..Y..L	T..2..J	A..2..D	J..B..B	Q..D..7	A..G..5
5)	M..Z..M	U..3..K	B..3..E	K..C..C	R..E..8	B..H..6
6)	N..1..N	V..A..L	C..A..F	L..D..D	S..F..A	C..J..7
7)	O..2..O	W..B..M	D..B..G	M..E..E	T..G..B	D..K..8
8)	P..3..P	X..C..N	E..C..H	N..F..F	U..H..C	E..L..A
9)	Q..A..Q	A..D..O	F..D..J	O..G..G	V..J..D	F..M..B
10)	R..B..R	B..E..P	G..E..K	P..H..H	W..K..E	G..N..C
11)	S..C..S	C..F..Q	H..F..L	Q..J..J	X..L..F	H..O..D
12)	T..D..T	D..G..R	J..G..M	R..K..K	A..M..G	J..P..E
13)	U..E..U	E..H..S	K..H..N	S..L..L	B..N..H	K..Q..F
14)	V..F..V	F..J..T	L..J..O	T..M..M	C..O..J	L..R..G
15)	W..G..W	G..K..U	M..K..P	U..N..N	D..P..K	M..S..H
16)	X..H..X	H..L..V	N..L..Q	V..O..O	E..Q..L	N..T..J
17)	A..J..Y	J..M..W	O..M..R	W..P..P	F..R..M	O..U..K
18)	B..K..Z	K..N..X	P..N..S	X..Q..Q	G..S..N	P..V..L
19)	C..L..1	L..O..Y	Q..O..T	A..R..R	H..T..O	Q..W..M
20)	D..M..2	M..P..Z	R..P..U	B..S..S	J..U..P	R..X..N
21)	E..N..3	N..Q..1	S..Q..V	C..T..T	K..V..Q	S..Y..O
22)	F..O..4	O..R..2	T..R..W	D..U..U	L..W..R	T..Z..P
23)	G..P..5	P..S..3	U..S..X	E..V..V	M..X..S	U..1..Q
24)	H..Q..6	Q..T..4	V..T..Y	F..W..W	N..Y..T	V..2..R
25)	J..R..7	R..U..5	W..U..Z	G..X..X	O..Z..U	W..3..S
26)	K..S..8	S..V..6	X..V..1	H..Y..Y	P..1..V	X..A..T
27)	L..T..A	T..W..7	A..W..2	J..Z..Z	Q..2..W	A..B..U
28)	M..U..B	U..X..8	B..X..3	K..1..1	R..3..X	B..C..V
29)	N..V..C		C..Y..4	L..2..2	S..A..Y	C..D..W
30)	O..W..D		D..Z..5	M..3..3	T..B..Z	D..E..X
31)	P..X..E		E..1..6		U..C..1	

1927						1927
	....JULY....	...AUGUST...	..SEPTEMBER.	..OCTOBER...	..NOVEMBER..	..DECEMBER..
	P--E--I	P--E--I	P--E--I	P--E--I	P--E--I	P--E--I
1)	E..F..Y	N..J..W	V..M..U	E..O..R	N..R..P	U..T..M
2)	F..G..Z	O..K..X	W..N..V	F..P..S	O..S..Q	V..U..N
3)	G..H..1	P..L..Y	X..O..W	G..Q..T	P..T..R	W..V..O
4)	H..J..2	Q..M..Z	A..P..X	H..R..U	Q..U..S	X..W..P
5)	J..K..3	R..N..1	B..Q..Y	J..S..V	R..V..T	A..X..Q
6)	K..L..4	S..O..2	C..R..Z	K..T..W	S..W..U	B..Y..R
7)	L..M..5	T..P..3	D..S..1	L..U..X	T..X..V	C..Z..S
8)	M..N..6	U..Q..4	E..T..2	M..V..Y	U..Y..W	D..1..T
9)	N..O..7	V..R..5	F..U..3	N..W..Z	V..Z..X	E..2..U
10)	O..P..8	W..S..6	G..V..4	O..X..1	W..1..Y	F..3..V
11)	P..Q..A	X..T..7	H..W..5	P..Y..2	X..2..Z	G..A..W
12)	Q..R..B	A..U..8	J..X..6	Q..Z..3	A..3..1	H..B..X
13)	R..S..C	B..V..A	K..Y..7	R..1..4	B..A..2	J..C..Y
14)	S..T..D	C..W..B	L..Z..8	S..2..5	C..B..3	K..D..Z
15)	T..U..E	D..X..C	M..1..A	T..3..6	D..C..4	L..E..1
16)	U..V..F	E..Y..D	N..2..B	U..A..7	E..D..5	M..F..2
17)	V..W..G	F..Z..E	O..3..C	V..B..8	F..E..6	N..G..3
18)	W..X..H	G..1..F	P..A..D	W..C..A	G..F..7	O..H..4
19)	X..Y..J	H..2..G	Q..B..E	X..D..B	H..G..8	P..J..5
20)	A..Z..K	J..3..H	R..C..F	A..E..C	J..H..A	Q..K..6
21)	B..1..L	K..A..J	S..D..G	B..F..D	K..J..B	R..L..7
22)	C..2..M	L..B..K	T..E..H	C..G..E	L..K..C	S..M..8
23)	D..3..N	M..C..L	U..F..J	D..H..F	M..L..D	T..N..A
24)	E..A..O	N..D..M	V..G..K	E..J..G	N..M..E	U..O..B
25)	F..B..P	O..E..N	W..H..L	F..K..H	O..N..F	V..P..C
26)	G..C..Q	P..F..O	X..J..M	G..L..J	P..O..G	W..Q..D
27)	H..D..R	Q..G..P	A..K..N	H..M..K	Q..P..H	X..R..E
28)	J..E..S	R..H..Q	B..L..O	J..N..L	R..Q..J	A..S..F
29)	K..F..T	S..J..R	C..M..P	K..O..M	S..R..K	B..T..G
30)	L..G..U	T..K..S	D..N..Q	L..P..N	T..S..L	C..U..H
31)	M..H..V	U..L..T		M..Q..O		D..V..J

CODES: P-PHYSICAL BIORHYTHM CURVE, E-EMOTIONAL BIORHYTHM CURVE, I-INTELLECTUAL BIORHYTHM CURVE

1928 | | | | | | 1928

	...JANUARY..	..FEBRUARY..	....MARCH...	....APRIL...	.....MAY....	....JUNE....
	P--E--I	P--E--I	P--E--I	P--E--I	P--E--I	P--E--I
1)	E..W..K	N..Z..H	T..1..D	D..A..B	L..C..7	T..F..5
2)	F..X..L	O..1..J	U..2..E	E..B..C	M..D..8	U..G..6
3)	G..Y..M	P..2..K	V..3..F	F..C..D	N..E..A	V..H..7
4)	H..Z..N	Q..3..L	W..A..G	G..D..E	O..F..B	W..J..8
5)	J..1..O	R..A..M	X..B..H	H..E..F	P..G..C	X..K..A
6)	K..2..P	S..B..N	A..C..J	J..F..G	Q..H..D	A..L..B
7)	L..3..Q	T..C..O	B..D..K	K..G..H	R..J..E	B..M..C
8)	M..A..R	U..D..P	C..E..L	L..H..J	S..K..F	C..N..D
9)	N..B..S	V..E..Q	D..F..M	M..J..K	T..L..G	D..O..E
10)	O..C..T	W..F..R	E..G..N	N..K..L	U..M..H	E..P..F
11)	P..D..U	X..G..S	F..H..O	O..L..M	V..N..J	F..Q..G
12)	Q..E..V	A..H..T	G..J..P	P..M..N	W..O..K	G..R..H
13)	R..F..W	B..J..U	H..K..Q	Q..N..O	X..P..L	H..S..J
14)	S..G..X	C..K..V	J..L..R	R..O..P	A..Q..M	J..T..K
15)	T..H..Y	D..L..W	K..M..S	S..P..Q	B..R..N	K..U..L
16)	U..J..Z	E..M..X	L..N..T	T..Q..R	C..S..O	L..V..M
17)	V..K..1	F..N..Y	M..O..U	U..R..S	D..T..P	M..W..N
18)	W..L..2	G..O..Z	N..P..V	V..S..T	E..U..Q	N..X..O
19)	X..M..3	H..P..1	O..Q..W	W..T..U	F..V..R	O..Y..P
20)	A..N..4	J..Q..2	P..R..X	X..U..V	G..W..S	P..Z..Q
21)	B..O..5	K..R..3	Q..S..Y	A..V..W	H..X..T	Q..1..R
22)	C..P..6	L..S..4	R..T..Z	B..W..X	J..Y..U	R..2..S
23)	D..Q..7	M..T..5	S..U..1	C..X..Y	K..Z..V	S..3..T
24)	E..R..8	N..U..6	T..V..2	D..Y..Z	L..1..W	T..A..U
25)	F..S..A	O..V..7	U..W..3	E..Z..1	M..2..X	U..B..V
26)	G..T..B	P..W..8	V..X..4	F..1..2	N..3..Y	V..C..W
27)	H..U..C	Q..X..A	W..Y..5	G..2..3	O..A..Z	W..D..X
28)	J..V..D	R..Y..B	X..Z..6	H..3..4	P..B..1	X..E..Y
29)	K..W..E	S..Z..C	A..1..7	J..A..5	Q..C..2	A..F..Z
30)	L..X..F		B..2..8	K..B..6	R..D..3	B..G..1
31)	M..Y..G		C..3..A		S..E..4	

1928 | 1928

Day	....JULY.... P--E--I	...AUGUST... P--E--I	..SEPTEMBER. P--E--I	..OCTOBER... P--E--I	..NOVEMBER.. P--E--I	..DECEMBER.. P--E--I
1)	C..H..2	L..L..Z	T..O..X	C..Q..U	L..T..S	S..V..P
2)	D..J..3	M..M..1	U..P..Y	D..R..V	M..U..T	T..W..Q
3)	E..K..4	N..N..2	V..Q..Z	E..S..W	N..V..U	U..X..R
4)	F..L..5	O..O..3	W..R..1	F..T..X	O..W..V	V..Y..S
5)	G..M..6	P..P..4	X..S..2	G..U..Y	P..X..W	W..Z..T
6)	H..N..7	Q..Q..5	A..T..3	H..V..Z	Q..Y..X	X..1..U
7)	J..O..8	R..R..6	B..U..4	J..W..1	R..Z..Y	A..2..V
8)	K..P..A	S..S..7	C..V..5	K..X..2	S..1..Z	B..3..W
9)	L..Q..B	T..T..8	D..W..6	L..Y..3	T..2..1	C..A..X
10)	M..R..C	U..U..A	E..X..7	M..Z..4	U..3..2	D..B..Y
11)	N..S..D	V..V..B	F..Y..8	N..1..5	V..A..3	E..C..Z
12)	O..T..E	W..W..C	G..Z..A	O..2..6	W..B..4	F..D..1
13)	P..U..F	X..X..D	H..1..B	P..3..7	X..C..5	G..E..2
14)	Q..V..G	A..Y..E	J..2..C	Q..A..8	A..D..6	H..F..3
15)	R..W..H	B..Z..F	K..3..D	R..B..A	B..E..7	J..G..4
16)	S..X..J	C..1..G	L..A..E	S..C..B	C..F..8	K..H..5
17)	T..Y..K	D..2..H	M..B..F	T..D..C	D..G..A	L..J..6
18)	U..Z..L	E..3..J	N..C..G	U..E..D	E..H..B	M..K..7
19)	V..1..M	F..A..K	O..D..H	V..F..E	F..J..C	N..L..8
20)	W..2..N	G..B..L	P..E..J	W..G..F	G..K..D	O..M..A
21)	X..3..O	H..C..M	Q..F..K	X..H..G	H..L..E	P..N..B
22)	A..A..P	J..D..N	R..G..L	A..J..H	J..M..F	Q..O..C
23)	B..B..Q	K..E..O	S..H..M	B..K..J	K..N..G	R..P..D
24)	C..C..R	L..F..P	T..J..N	C..L..K	L..O..H	S..Q..E
25)	D..D..S	M..G..Q	U..K..O	D..M..L	M..P..J	T..R..F
26)	E..E..T	N..H..R	V..L..P	E..N..M	N..Q..K	U..S..G
27)	F..F..U	O..J..S	W..M..Q	F..O..N	O..R..L	V..T..H
28)	G..G..V	P..K..T	X..N..R	G..P..O	P..S..M	W..U..J
29)	H..H..W	Q..L..U	A..O..S	H..Q..P	Q..T..N	X..V..K
30)	J..J..X	R..M..V	B..P..T	J..R..Q	R..U..O	A..W..L
31)	K..K..Y	S..N..W		K..S..R		B..X..M

CODES: P-PHYSICAL BIORHYTHM CURVE, E-EMOTIONAL BIORHYTHM CURVE, I-INTELLECTUAL BIORHYTHM CURVE

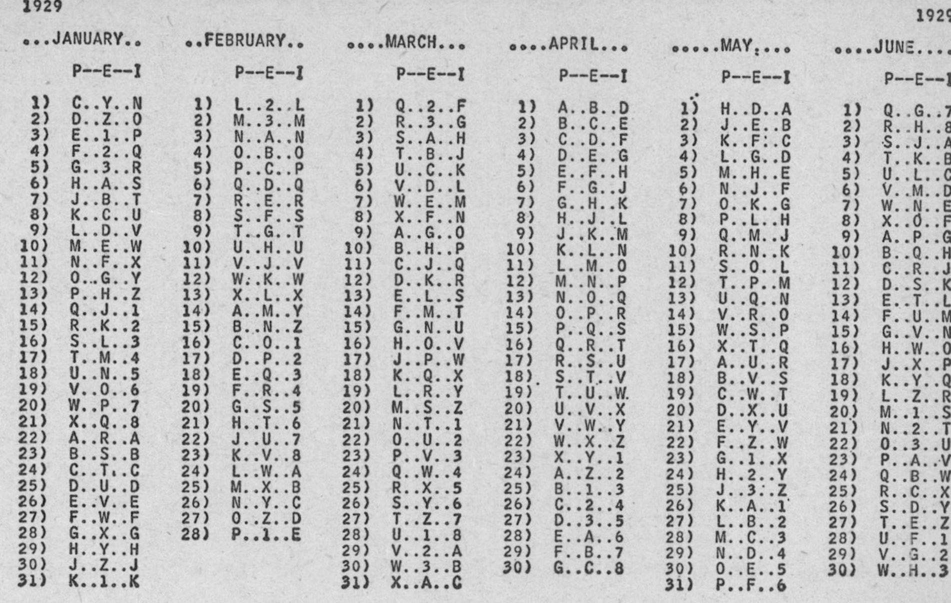

1929 1929

	...JANUARY..	..FEBRUARY..	....MARCH...	....APRIL...	.....MAY....	....JUNE....
	P--E--I	P--E--I	P--E--I	P--E--I	P--E--I	P--E--I
1)	C..Y..N	L..2..L	Q..2..F	A..B..D	H..D..A	Q..G..7
2)	D..Z..O	M..3..M	R..3..G	B..C..E	J..E..B	R..H..8
3)	E..1..P	N..A..N	S..A..H	C..D..F	K..F..C	S..J..A
4)	F..2..Q	O..B..O	T..B..J	D..E..G	L..G..D	T..K..B
5)	G..3..R	P..C..P	U..C..K	E..F..H	M..H..E	U..L..C
6)	H..A..S	Q..D..Q	V..D..L	F..G..J	N..J..F	V..M..D
7)	J..B..T	R..E..R	W..E..M	G..H..K	O..K..G	W..N..E
8)	K..C..U	S..F..S	X..F..N	H..J..L	P..L..H	X..O..F
9)	L..D..V	T..G..T	A..G..O	J..K..M	Q..M..J	A..P..G
10)	M..E..W	U..H..U	B..H..P	K..L..N	R..N..K	B..Q..H
11)	N..F..X	V..J..V	C..J..Q	L..M..O	S..O..L	C..R..J
12)	O..G..Y	W..K..W	D..K..R	M..N..P	T..P..M	D..S..K
13)	P..H..Z	X..L..X	E..L..S	N..O..Q	U..Q..N	E..T..L
14)	Q..J..1	A..M..Y	F..M..T	O..P..R	V..R..O	F..U..M
15)	R..K..2	B..N..Z	G..N..U	P..Q..S	W..S..P	G..V..N
16)	S..L..3	C..O..1	H..O..V	Q..R..T	X..T..Q	H..W..O
17)	T..M..4	D..P..2	J..P..W	R..S..U	A..U..R	J..X..P
18)	U..N..5	E..Q..3	K..Q..X	S..T..V	B..V..S	K..Y..Q
19)	V..O..6	F..R..4	L..R..Y	T..U..W	C..W..T	L..Z..R
20)	W..P..7	G..S..5	M..S..Z	U..V..X	D..X..U	M..1..S
21)	X..Q..8	H..T..6	N..T..1	V..W..Y	E..Y..V	N..2..T
22)	A..R..A	J..U..7	O..U..2	W..X..Z	F..Z..W	O..3..U
23)	B..S..B	K..V..8	P..V..3	X..Y..1	G..1..X	P..A..V
24)	C..T..C	L..W..A	Q..W..4	A..Z..2	H..2..Y	Q..B..W
25)	D..U..D	M..X..B	R..X..5	B..1..3	J..3..Z	R..C..X
26)	E..V..E	N..Y..C	S..Y..6	C..2..4	K..A..1	S..D..Y
27)	F..W..F	O..Z..D	T..Z..7	D..3..5	L..B..2	T..E..Z
28)	G..X..G	P..1..E	U..1..8	E..A..6	M..C..3	U..F..1
29)	H..Y..H		V..2..A	F..B..7	N..D..4	V..G..2
30)	J..Z..J		W..3..B	G..C..8	O..E..5	W..H..3
31)	K..1..K		X..A..C		P..F..6	

1929 1929

Day	....JULY....	....AUGUST....	..SEPTEMBER.	..OCTOBER...	..NOVEMBER..	..DECEMBER..
	P--E--I	P--E--I	P--E--I	P--E--I	P--E--I	P--E--I
1)	X..J..4	H..M..2	Q..P..Z	X..R..W	H..U..U	P..W..R
2)	A..K..5	J..N..3	R..Q..1	A..S..X	J..V..V	Q..X..S
3)	B..L..6	K..O..4	S..R..2	B..T..Y	K..W..W	R..Y..T
4)	C..M..7	L..P..5	T..S..3	C..U..Z	L..X..X	S..Z..U
5)	D..N..8	M..Q..6	U..T..4	D..V..1	M..Y..Y	T..1..V
6)	E..O..A	N..R..7	V..U..5	E..W..2	N..Z..Z	U..2..W
7)	F..P..B	O..S..8	W..V..6	F..X..3	O..1..1	V..3..X
8)	G..Q..C	P..T..A	X..W..7	G..Y..4	P..2..2	W..A..Y
9)	H..R..D	Q..U..B	A..X..8	H..Z..5	Q..3..3	X..B..Z
10)	J..S..E	R..V..C	B..Y..A	J..1..6	R..A..4	A..C..1
11)	K..T..F	S..W..D	C..Z..B	K..2..7	S..B..5	B..D..2
12)	L..U..G	T..X..E	D..1..C	L..3..8	T..C..6	C..E..3
13)	M..V..H	U..Y..F	E..2..D	M..A..A	U..D..7	D..F..4
14)	N..W..J	V..Z..G	F..3..E	N..B..B	V..E..8	E..G..5
15)	O..X..K	W..1..H	G..A..F	O..C..C	W..F..A	F..H..6
16)	P..Y..L	X..2..J	H..B..G	P..D..D	X..G..B	G..J..7
17)	Q..Z..M	A..3..K	J..C..H	Q..E..E	A..H..C	H..K..8
18)	R..1..N	B..A..L	K..D..J	R..F..F	B..J..D	J..L..A
19)	S..2..O	C..B..M	L..E..K	S..G..G	C..K..E	K..M..B
20)	T..3..P	D..C..N	M..F..L	T..H..H	D..L..F	L..N..C
21)	U..A..Q	E..D..O	N..G..M	U..J..J	E..M..G	M..O..D
22)	V..B..R	F..E..P	O..H..N	V..K..K	F..N..H	N..P..E
23)	W..C..S	G..F..Q	P..J..O	W..L..L	G..O..J	O..Q..F
24)	X..D..T	H..G..R	Q..K..P	X..M..M	H..P..K	P..R..G
25)	A..E..U	J..H..S	R..L..Q	A..N..N	J..Q..L	Q..S..H
26)	B..F..V	K..J..T	S..M..R	B..O..O	K..R..M	R..T..J
27)	C..G..W	L..K..U	T..N..S	C..P..P	L..S..N	S..U..K
28)	D..H..X	M..L..V	U..O..T	D..Q..Q	M..T..O	T..V..L
29)	E..J..Y	N..M..W	V..P..U	E..R..R	N..U..P	U..W..M
30)	F..K..Z	O..N..X	W..Q..V	F..S..S	O..V..Q	V..X..N
31)	G..L..1	P..O..Y		G..T..T		W..Y..O

CODES: P-PHYSICAL BIORHYTHM CURVE, E-EMOTIONAL BIORHYTHM CURVE, I-INTELLECTUAL BIORHYTHM CURVE

1930						1930
	...JANUARY..	..FEBRUARY..	....MARCH...	....APRIL...	.....MAY....	....JUNE....
	P--E--I	P--E--I	P--E--I	P--E--I	P--E--I	P--E--I
1)	X..Z..P	H..3..N	N..3..H	V..C..F	E..E..C	N..H..A
2)	A..1..Q	J..A..O	O..A..J	W..D..G	F..F..D	O..J..B
3)	B..2..R	K..B..P	P..B..K	X..E..H	G..G..E	P..K..C
4)	C..3..S	L..C..Q	Q..C..L	A..F..J	H..H..F	Q..L..D
5)	D..A..T	M..D..R	R..D..M	B..G..K	J..J..G	R..M..E
6)	E..B..U	N..E..S	S..E..N	C..H..L	K..K..H	S..N..F
7)	F..C..V	O..F..T	T..F..O	D..J..M	L..L..J	T..O..G
8)	G..D..W	P..G..U	U..G..P	E..K..N	M..M..K	U..P..H
9)	H..E..X	Q..H..V	V..H..Q	F..L..O	N..N..L	V..Q..J
10)	J..F..Y	R..J..W	W..J..R	G..M..P	O..O..M	W..R..K
11)	K..G..Z	S..K..X	X..K..S	H..N..Q	P..P..N	X..S..L
12)	L..H..1	T..L..Y	A..L..T	J..O..R	Q..Q..O	A..T..M
13)	M..J..2	U..M..Z	B..M..U	K..P..S	R..R..P	B..U..N
14)	N..K..3	V..N..1	C..N..V	L..Q..T	S..S..Q	C..V..O
15)	O..L..4	W..O..2	D..O..W	M..R..U	T..T..R	D..W..P
16)	P..M..5	X..P..3	E..P..X	N..S..V	U..U..S	E..X..Q
17)	Q..N..6	A..Q..4	F..Q..Y	O..T..W	V..V..T	F..Y..R
18)	R..O..7	B..R..5	G..R..Z	P..U..X	W..W..U	G..Z..S
19)	S..P..8	C..S..6	H..S..1	Q..V..Y	X..X..V	H..1..T
20)	T..Q..A	D..T..7	J..T..2	R..W..Z	A..Y..W	J..2..U
21)	U..R..B	E..U..8	K..U..3	S..X..1	B..Z..X	K..3..V
22)	V..S..C	F..V..A	L..V..4	T..Y..2	C..1..Y	L..A..W
23)	W..T..D	G..W..B	M..W..5	U..Z..3	D..2..Z	M..B..X
24)	X..U..E	H..X..C	N..X..6	V..1..4	E..3..1	N..C..Y
25)	A..V..F	J..Y..D	O..Y..7	W..2..5	F..A..2	O..D..Z
26)	B..W..G	K..Z..E	P..Z..8	X..3..6	G..B..3	P..E..1
27)	C..X..H	L..1..F	Q..1..A	A..A..7	H..C..4	Q..F..2
28)	D..Y..J	M..2..G	R..2..B	B..B..8	J..D..5	R..G..3
29)	E..Z..K		S..3..C	C..C..A	K..E..6	S..H..4
30)	F..1..L		T..A..D	D..D..B	L..F..7	T..J..5
31)	G..2..M		U..B..E		M..G..8	

1930 | 1930

Day	....JULY.... P--E--I	...AUGUST... P--E--I	..SEPTEMBER. P--E--I	..OCTOBER... P--E--I	..NOVEMBER.. P--E--I	..DECEMBER.. P--E--I
1)	U..K..6	E..N..4	N..Q..2	U..S..Y	E..V..W	M..X..T
2)	V..L..7	F..O..5	O..R..3	V..T..Z	F..W..X	N..Y..U
3)	W..M..8	G..P..6	P..S..4	W..U..1	G..X..Y	O..Z..V
4)	X..N..A	H..Q..7	Q..T..5	X..V..2	H..Y..Z	P..1..W
5)	A..O..B	J..R..8	R..U..6	A..W..3	J..Z..1	Q..2..X
6)	B..P..C	K..S..A	S..V..7	B..X..4	K..1..2	R..3..Y
7)	C..Q..D	L..T..B	T..W..8	C..Y..5	L..2..3	S..A..Z
8)	D..R..E	M..U..C	U..X..A	D..Z..6	M..3..4	T..B..1
9)	E..S..F	N..V..D	V..Y..B	E..1..7	N..A..5	U..C..2
10)	F..T..G	O..W..E	W..Z..C	F..2..8	O..B..6	V..D..3
11)	G..U..H	P..X..F	X..1..D	G..3..A	P..C..7	W..E..4
12)	H..V..J	Q..Y..G	A..2..E	H..A..B	Q..D..8	X..F..5
13)	J..W..K	R..Z..H	B..3..F	J..B..C	R..E..A	A..G..6
14)	K..X..L	S..1..J	C..A..G	K..C..D	S..F..B	B..H..7
15)	L..Y..M	T..2..K	D..B..H	L..D..E	T..G..C	C..J..8
16)	M..Z..N	U..3..L	E..C..J	M..E..F	U..H..D	D..K..A
17)	N..1..O	V..A..M	F..D..K	N..F..G	V..J..E	E..L..B
18)	O..2..P	W..B..N	G..E..L	O..G..H	W..K..F	F..M..C
19)	P..3..Q	X..C..O	H..F..M	P..H..J	X..L..G	G..N..D
20)	Q..A..R	A..D..P	J..G..N	Q..J..K	A..M..H	H..O..E
21)	R..B..S	B..E..Q	K..H..O	R..K..L	B..N..J	J..P..F
22)	S..C..T	C..F..R	L..J..P	S..L..M	C..O..K	K..Q..G
23)	T..D..U	D..G..S	M..K..Q	T..M..N	D..P..L	L..R..H
24)	U..E..V	E..H..T	N..L..R	U..N..O	E..Q..M	M..S..J
25)	V..F..W	F..J..U	O..M..S	V..O..P	F..R..N	N..T..K
26)	W..G..X	G..K..V	P..N..T	W..P..Q	G..S..O	O..U..L
27)	X..H..Y	H..L..W	Q..O..U	X..Q..R	H..T..P	P..V..M
28)	A..J..Z	J..M..X	R..P..V	A..R..S	J..U..Q	Q..W..N
29)	B..K..1	K..N..Y	S..Q..W	B..S..T	K..V..R	R..X..O
30)	C..L..2	L..O..Z	T..R..X	C..T..U	L..W..S	S..Y..P
31)	D..M..3	M..P..1		D..U..V		T..Z..Q

CODES: P-PHYSICAL BIORHYTHM CURVE, E-EMOTIONAL BIORHYTHM CURVE, I-INTELLECTUAL BIORHYTHM CURVE

	...JANUARY..	..FEBRUARY..	....MARCH...	....APRIL...	....MAY....	....JUNE....
	P--E--I	P--E--I	P--E--I	P--E--I	P--E--I	P--E--I
1)	U..1..R	E..A..P	K..A..K	S..D..H	B..F..E	K..J..C
2)	V..2..S	F..B..Q	L..B..L	T..E..J	C..G..F	L..K..D
3)	W..3..T	G..C..R	M..C..M	U..F..K	D..H..G	M..L..E
4)	X..A..U	H..D..S	N..D..N	V..G..L	E..J..H	N..M..F
5)	A..B..V	J..E..T	O..E..O	W..H..M	F..K..J	O..N..G
6)	B..C..W	K..F..U	P..F..P	X..J..N	G..L..K	P..O..H
7)	C..D..X	L..G..V	Q..G..Q	A..K..O	H..M..L	Q..P..J
8)	D..E..Y	M..H..W	R..H..R	B..L..P	J..N..M	R..Q..K
9)	E..F..Z	N..J..X	S..J..S	C..M..Q	K..O..N	S..R..L
10)	F..G..1	O..K..Y	T..K..T	D..N..R	L..P..O	T..S..M
11)	G..H..2	P..L..Z	U..L..U	E..O..S	M..Q..P	U..T..N
12)	H..J..3	Q..M..1	V..M..V	F..P..T	N..R..Q	V..U..O
13)	J..K..4	R..N..2	W..N..W	G..Q..U	O..S..R	W..V..P
14)	K..L..5	S..O..3	X..O..X	H..R..V	P..T..S	X..W..Q
15)	L..M..6	T..P..4	A..P..Y	J..S..W	Q..U..T	A..X..R
16)	M..N..7	U..Q..5	B..Q..Z	K..T..X	R..V..U	B..Y..S
17)	N..O..8	V..R..6	C..R..1	L..U..Y	S..W..V	C..Z..T
18)	O..P..A	W..S..7	D..S..2	M..V..Z	T..X..W	D..1..U
19)	P..Q..B	X..T..8	E..T..3	N..W..1	U..Y..X	E..2..V
20)	Q..R..C	A..U..A	F..U..4	O..X..2	V..Z..Y	F..3..W
21)	R..S..D	B..V..B	G..V..5	P..Y..3	W..1..Z	G..A..X
22)	S..T..E	C..W..C	H..W..6	Q..Z..4	X..2..1	H..B..Y
23)	T..U..F	D..X..D	J..X..7	R..1..5	A..3..2	J..C..Z
24)	U..V..G	E..Y..E	K..Y..8	S..2..6	B..A..3	K..D..1
25)	V..W..H	F..Z..F	L..Z..A	T..3..7	C..B..4	L..E..2
26)	W..X..J	G..1..G	M..1..B	U..A..8	D..C..5	M..F..3
27)	X..Y..K	H..2..H	N..2..C	V..B..A	E..D..6	N..G..4
28)	A..Z..L	J..3..J	O..3..D	W..C..B	F..E..7	O..H..5
29)	B..1..M		P..A..E	X..D..C	G..F..8	P..J..6
30)	C..2..N		Q..B..F	A..E..D	H..G..A	Q..K..7
31)	D..3..O		R..C..G		J..H..B	

1931						1931
	....JULY....	...AUGUST...	..SEPTEMBER.	..OCTOBER...	..NOVEMBER..	..DECEMBER..
	P--E--I	P--E--I	P--E--I	P--E--I	P--E--I	P--E--I
1)	R..L..8	B..O..6	K..R..4	R..T..1	B..W..Y	J..Y..V
2)	S..M..A	C..P..7	L..S..5	S..U..2	C..X..Z	K..Z..W
3)	T..N..B	D..Q..8	M..T..6	T..V..3	D..Y..1	L..1..X
4)	U..O..C	E..R..A	N..U..7	U..W..4	E..Z..2	M..2..Y
5)	V..P..D	F..S..B	O..V..8	V..X..5	F..1..3	N..3..Z
6)	W..Q..E	G..T..C	P..W..A	W..Y..6	G..2..4	O..A..1
7)	X..R..F	H..U..D	Q..X..B	X..Z..7	H..3..5	P..B..2
8)	A..S..G	J..V..E	R..Y..C	A..1..8	J..A..6	Q..C..3
9)	B..T..H	K..W..F	S..Z..D	B..2..A	K..B..7	R..D..4
10)	C..U..J	L..X..G	T..1..E	C..3..B	L..C..8	S..E..5
11)	D..V..K	M..Y..H	U..2..F	D..A..C	M..D..A	T..F..6
12)	E..W..L	N..Z..J	V..3..G	E..B..D	N..E..B	U..G..7
13)	F..X..M	O..1..K	W..A..H	F..C..E	O..F..C	V..H..8
14)	G..Y..N	P..2..L	X..B..J	G..D..F	P..G..D	W..J..A
15)	H..Z..O	Q..3..M	A..C..K	H..E..G	Q..H..E	X..K..B
16)	J..1..P	R..A..N	B..D..L	J..F..H	R..J..F	A..L..C
17)	K..2..Q	S..B..O	C..E..M	K..G..J	S..K..G	B..M..D
18)	L..3..R	T..C..P	D..F..N	L..H..K	T..L..H	C..N..E
19)	M..A..S	U..D..Q	E..G..O	M..J..L	U..M..J	D..O..F
20)	N..B..T	V..E..R	F..H..P	N..K..M	V..N..K	E..P..G
21)	O..C..U	W..F..S	G..J..Q	O..L..N	W..O..L	F..Q..H
22)	P..D..V	X..G..T	H..K..R	P..M..O	X..P..M	G..R..J
23)	Q..E..W	A..H..U	J..L..S	Q..N..P	A..Q..N	H..S..K
24)	R..F..X	B..J..V	K..M..T	R..O..Q	B..R..O	J..T..L
25)	S..G..Y	C..K..W	L..N..U	S..P..R	C..S..P	K..U..M
26)	T..H..Z	D..L..X	M..O..V	T..Q..S	D..T..Q	L..V..N
27)	U..J..1	E..M..Y	N..P..W	U..R..T	E..U..R	M..W..O
28)	V..K..2	F..N..Z	O..Q..X	V..S..U	F..V..S	N..X..P
29)	W..L..3	G..O..1	P..R..Y	W..T..V	G..W..T	O..Y..Q
30)	X..M..4	H..P..2	Q..S..Z	X..U..W	H..X..U	P..Z..R
31)	A..N..5	J..Q..3		A..V..X		Q..1..S

CODES: P-PHYSICAL BIORHYTHM CURVE,E-EMOTIONAL BIORHYTHM CURVE,I-INTELLECTUAL BIORHYTHM CURVE

1932 1932

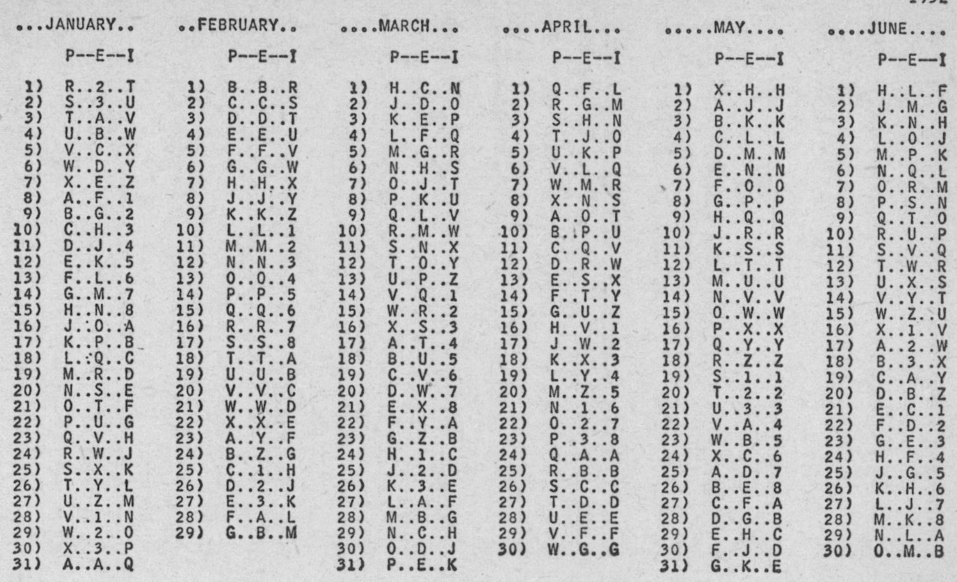

	...JANUARY.. P--E--I		..FEBRUARY.. P--E--I		....MARCH... P--E--I		....APRIL... P--E--I		.....MAY.... P--E--I		....JUNE.... P--E--I
1)	R..2..T	1)	B..B..R	1)	H..C..N	1)	Q..F..L	1)	X..H..H	1)	H..L..F
2)	S..3..U	2)	C..C..S	2)	J..D..O	2)	R..G..M	2)	A..J..J	2)	J..M..G
3)	T..A..V	3)	D..D..T	3)	K..E..P	3)	S..H..N	3)	B..K..K	3)	K..N..H
4)	U..B..W	4)	E..E..U	4)	L..F..Q	4)	T..J..O	4)	C..L..L	4)	L..O..J
5)	V..C..X	5)	F..F..V	5)	M..G..R	5)	U..K..P	5)	D..M..M	5)	M..P..K
6)	W..D..Y	6)	G..G..W	6)	N..H..S	6)	V..L..Q	6)	E..N..N	6)	N..Q..L
7)	X..E..Z	7)	H..H..X	7)	O..J..T	7)	W..M..R	7)	F..O..O	7)	O..R..M
8)	A..F..1	8)	J..J..Y	8)	P..K..U	8)	X..N..S	8)	G..P..P	8)	P..S..N
9)	B..G..2	9)	K..K..Z	9)	Q..L..V	9)	A..O..T	9)	H..Q..Q	9)	Q..T..O
10)	C..H..3	10)	L..L..1	10)	R..M..W	10)	B..P..U	10)	J..R..R	10)	R..U..P
11)	D..J..4	11)	M..M..2	11)	S..N..X	11)	C..Q..V	11)	K..S..S	11)	S..V..Q
12)	E..K..5	12)	N..N..3	12)	T..O..Y	12)	D..R..W	12)	L..T..T	12)	T..W..R
13)	F..L..6	13)	O..O..4	13)	U..P..Z	13)	E..S..X	13)	M..U..U	13)	U..X..S
14)	G..M..7	14)	P..P..5	14)	V..Q..1	14)	F..T..Y	14)	N..V..V	14)	V..Y..T
15)	H..N..8	15)	Q..Q..6	15)	W..R..2	15)	G..U..Z	15)	O..W..W	15)	W..Z..U
16)	J..O..A	16)	R..R..7	16)	X..S..3	16)	H..V..1	16)	P..X..X	16)	X..1..V
17)	K..P..B	17)	S..S..8	17)	A..T..4	17)	J..W..2	17)	Q..Y..Y	17)	A..2..W
18)	L..Q..C	18)	T..T..A	18)	B..U..5	18)	K..X..3	18)	R..Z..Z	18)	B..3..X
19)	M..R..D	19)	U..U..B	19)	C..V..6	19)	L..Y..4	19)	S..1..1	19)	C..A..Y
20)	N..S..E	20)	V..V..C	20)	D..W..7	20)	M..Z..5	20)	T..2..2	20)	D..B..Z
21)	O..T..F	21)	W..W..D	21)	E..X..8	21)	N..1..6	21)	U..3..3	21)	E..C..1
22)	P..U..G	22)	X..X..E	22)	F..Y..A	22)	O..2..7	22)	V..A..4	22)	F..D..2
23)	Q..V..H	23)	A..Y..F	23)	G..Z..B	23)	P..3..8	23)	W..B..5	23)	G..E..3
24)	R..W..J	24)	B..Z..G	24)	H..1..C	24)	Q..A..A	24)	X..C..6	24)	H..F..4
25)	S..X..K	25)	C..1..H	25)	J..2..D	25)	R..B..B	25)	A..D..7	25)	J..G..5
26)	T..Y..L	26)	D..2..J	26)	K..3..E	26)	S..C..C	26)	B..E..8	26)	K..H..6
27)	U..Z..M	27)	E..3..K	27)	L..A..F	27)	T..D..D	27)	C..F..A	27)	L..J..7
28)	V..1..N	28)	F..A..L	28)	M..B..G	28)	U..E..E	28)	D..G..B	28)	M..K..8
29)	W..2..O	29)	G..B..M	29)	N..C..H	29)	V..F..F	29)	E..H..C	29)	N..L..A
30)	X..3..P			30)	O..D..J	30)	W..G..G	30)	F..J..D	30)	O..M..B
31)	A..A..Q			31)	P..E..K			31)	G..K..E		

1932 | | | | | 1932

Day	....JULY.... P--E--I	...AUGUST... P--E--I	..SEPTEMBER. P--E--I	..OCTOBER... P--E--I	..NOVEMBER.. P--E--I	..DECEMBER.. P--E--I
1)	P..N..C	X..Q..A	H..T..7	P..V..4	X..Y..2	G..1..Y
2)	Q..O..D	A..R..B	J..U..8	Q..W..5	A..Z..3	H..2..Z
3)	R..P..E	B..S..C	K..V..A	R..X..6	B..1..4	J..3..1
4)	S..Q..F	C..T..D	L..W..B	S..Y..7	C..2..5	K..A..2
5)	T..R..G	D..U..E	M..X..C	T..Z..8	D..3..6	L..B..3
6)	U..S..H	E..V..F	N..Y..D	U..1..A	E..A..7	M..C..4
7)	V..T..J	F..W..G	O..Z..E	V..2..B	F..B..8	N..D..5
8)	W..U..K	G..X..H	P..1..F	W..3..C	G..C..A	O..E..6
9)	X..V..L	H..Y..J	Q..2..G	X..A..D	H..D..B	P..F..7
10)	A..W..M	J..Z..K	R..3..H	A..B..E	J..E..C	Q..G..8
11)	B..X..N	K..1..L	S..A..J	B..C..F	K..F..D	R..H..A
12)	C..Y..O	L..2..M	T..B..K	C..D..G	L..G..E	S..J..B
13)	D..Z..P	M..3..N	U..C..L	D..E..H	M..H..F	T..K..C
14)	E..1..Q	N..A..O	V..D..M	E..F..J	N..J..G	U..L..D
15)	F..2..R	O..B..P	W..E..N	F..G..K	O..K..H	V..M..E
16)	G..3..S	P..C..Q	X..F..O	G..H..L	P..L..J	W..N..F
17)	H..A..T	Q..D..R	A..G..P	H..J..M	Q..M..K	X..O..G
18)	J..B..U	R..E..S	B..H..Q	J..K..N	R..N..L	A..P..H
19)	K..C..V	S..F..T	C..J..R	K..L..O	S..O..M	B..Q..J
20)	L..D..W	T..G..U	D..K..S	L..M..P	T..P..N	C..R..K
21)	M..E..X	U..H..V	E..L..T	M..N..Q	U..Q..O	D..S..L
22)	N..F..Y	V..J..W	F..M..U	N..O..R	V..R..P	E..T..M
23)	O..G..Z	W..K..X	G..N..V	O..P..S	W..S..Q	F..U..N
24)	P..H..1	X..L..Y	H..O..W	P..Q..T	X..T..R	G..V..O
25)	Q..J..2	A..M..Z	J..P..X	Q..R..U	A..U..S	H..W..P
26)	R..K..3	B..N..1	K..Q..Y	R..S..V	B..V..T	J..X..Q
27)	S..L..4	C..O..2	L..R..Z	S..T..W	C..W..U	K..Y..R
28)	T..M..5	D..P..3	M..S..1	T..U..X	D..X..V	L..Z..S
29)	U..N..6	E..Q..4	N..T..2	U..V..Y	E..Y..W	M..1..T
30)	V..O..7	F..R..5	O..U..3	V..W..Z	F..Z..X	N..2..U
31)	W..P..8	G..S..6		W..X..1		O..3..V

CODES: P-PHYSICAL BIORHYTHM CURVE,E-EMOTIONAL BIORHYTHM CURVE,I-INTELLECTUAL BIORHYTHM CURVE

1933 1933

	...JANUARY..	..FEBRUARY..	....MARCH...	....APRIL...	.....MAY....	....JUNE....
	P--E--I	P--E--I	P--E--I	P--E--I	P--E--I	P--E--I
1)	P..A..W	X..D..U	E..D..P	N..G..N	U..J..K	E..M..H
2)	Q..B..X	A..E..V	F..E..Q	O..H..O	V..K..L	F..N..J
3)	R..C..Y	B..F..W	G..F..R	P..J..P	W..L..M	G..O..K
4)	S..D..Z	C..G..X	H..G..S	Q..K..Q	X..M..N	H..P..L
5)	T..E..1	D..H..Y	J..H..T	R..L..R	A..N..O	J..Q..M
6)	U..F..2	E..J..Z	K..J..U	S..M..S	B..O..P	K..R..N
7)	V..G..3	F..K..1	L..K..V	T..N..T	C..P..Q	L..S..O
8)	W..H..4	G..L..2	M..L..W	U..O..U	D..Q..R	M..T..P
9)	X..J..5	H..M..3	N..M..X	V..P..V	E..R..S	N..U..Q
10)	A..K..6	J..N..4	O..N..Y	W..Q..W	F..S..T	O..V..R
11)	B..L..7	K..O..5	P..O..Z	X..R..X	G..T..U	P..W..S
12)	C..M..8	L..P..6	Q..P..1	A..S..Y	H..U..V	Q..X..T
13)	D..N..A	M..Q..7	R..Q..2	B..T..Z	J..V..W	R..Y..U
14)	E..O..B	N..R..8	S..R..3	C..U..1	K..W..X	S..Z..V
15)	F..P..C	O..S..A	T..S..4	D..V..2	L..X..Y	T..1..W
16)	G..Q..D	P..T..B	U..T..5	E..W..3	M..Y..Z	U..2..X
17)	H..R..E	Q..U..C	V..U..6	F..X..4	N..Z..1	V..3..Y
18)	J..S..F	R..V..D	W..V..7	G..Y..5	O..1..2	W..A..Z
19)	K..T..G	S..W..E	X..W..8	H..Z..6	P..2..3	X..B..1
20)	L..U..H	T..X..F	A..X..A	J..1..7	Q..3..4	A..C..2
21)	M..V..J	U..Y..G	B..Y..B	K..2..8	R..A..5	B..D..3
22)	N..W..K	V..Z..H	C..Z..C	L..3..A	S..B..6	C..E..4
23)	O..X..L	W..1..J	D..1..D	M..A..B	T..C..7	D..F..5
24)	P..Y..M	X..2..K	E..2..E	N..B..C	U..D..8	E..G..6
25)	Q..Z..N	A..3..L	F..3..F	O..C..D	V..E..A	F..H..7
26)	R..1..O	B..A..M	G..A..G	P..D..E	W..F..B	G..J..8
27)	S..2..P	C..B..N	H..B..H	Q..E..F	X..G..C	H..K..A
28)	T..3..Q	D..C..O	J..C..J	R..F..G	A..H..D	J..L..B
29)	U..A..R		K..D..K	S..G..H	B..J..E	K..M..C
30)	V..B..S		L..E..L	T..H..J	C..K..F	L..N..D
31)	W..C..T		M..F..M		D..L..G	

1933 | | | | | | | | | | | 1933

	....JULY....		...AUGUST...		..SEPTEMBER.		..OCTOBER...		..NOVEMBER..		..DECEMBER..
	P--E--I		P--E--I		P--E--I		P--E--I		P--E--I		P--E--I
1)	M..O..E	1)	U..R..C	1)	E..U..A	1)	M..W..6	1)	U..Z..4	1)	D..2..1
2)	N..P..F	2)	V..S..D	2)	F..V..B	2)	N..X..7	2)	V..1..5	2)	E..3..2
3)	O..Q..G	3)	W..T..E	3)	G..W..C	3)	O..Y..8	3)	W..2..6	3)	F..A..3
4)	P..R..H	4)	X..U..F	4)	H..X..D	4)	P..Z..A	4)	X..3..7	4)	G..B..4
5)	Q..S..J	5)	A..V..G	5)	J..Y..E	5)	Q..1..B	5)	A..A..8	5)	H..C..5
6)	R..T..K	6)	B..W..H	6)	K..Z..F	6)	R..2..C	6)	B..B..A	6)	J..D..6
7)	S..U..L	7)	C..X..J	7)	L..1..G	7)	S..3..D	7)	C..C..B	7)	K..E..7
8)	T..V..M	8)	D..Y..K	8)	M..2..H	8)	T..A..E	8)	D..D..C	8)	L..F..8
9)	U..W..N	9)	E..Z..L	9)	N..3..J	9)	U..B..F	9)	E..E..D	9)	M..G..A
10)	V..X..O	10)	F..1..M	10)	O..A..K	10)	V..C..G	10)	F..F..E	10)	N..H..B
11)	W..Y..P	11)	G..2..N	11)	P..B..L	11)	W..D..H	11)	G..G..F	11)	O..J..C
12)	X..Z..Q	12)	H..3..O	12)	Q..C..M	12)	X..E..J	12)	H..H..G	12)	P..K..D
13)	A..1..R	13)	J..A..P	13)	R..D..N	13)	A..F..K	13)	J..J..H	13)	Q..L..E
14)	B..2..S	14)	K..B..Q	14)	S..E..O	14)	B..G..L	14)	K..K..J	14)	R..M..F
15)	C..3..T	15)	L..C..R	15)	T..F..P	15)	C..H..M	15)	L..L..K	15)	S..N..G
16)	D..A..U	16)	M..D..S	16)	U..G..Q	16)	D..J..N	16)	M..M..L	16)	T..O..H
17)	E..B..V	17)	N..E..T	17)	V..H..R	17)	E..K..O	17)	N..N..M	17)	U..P..J
18)	F..C..W	18)	O..F..U	18)	W..J..S	18)	F..L..P	18)	O..O..N	18)	V..Q..K
19)	G..D..X	19)	P..G..V	19)	X..K..T	19)	G..M..Q	19)	P..P..O	19)	W..R..L
20)	H..E..Y	20)	Q..H..W	20)	A..L..U	20)	H..N..R	20)	Q..Q..P	20)	X..S..M
21)	J..F..Z	21)	R..J..X	21)	B..M..V	21)	J..O..S	21)	R..R..Q	21)	A..T..N
22)	K..G..1	22)	S..K..Y	22)	C..N..W	22)	K..P..T	22)	S..S..R	22)	B..U..O
23)	L..H..2	23)	T..L..Z	23)	D..O..X	23)	L..Q..U	23)	T..T..S	23)	C..V..P
24)	M..J..3	24)	U..M..1	24)	E..P..Y	24)	M..R..V	24)	U..U..T	24)	D..W..Q
25)	N..K..4	25)	V..N..2	25)	F..Q..Z	25)	N..S..W	25)	V..V..U	25)	E..X..R
26)	O..L..5	26)	W..O..3	26)	G..R..1	26)	O..T..X	26)	W..W..V	26)	F..Y..S
27)	P..M..6	27)	X..P..4	27)	H..S..2	27)	P..U..Y	27)	X..X..W	27)	G..Z..T
28)	Q..N..7	28)	A..Q..5	28)	J..T..3	28)	Q..V..Z	28)	A..Y..X	28)	H..1..U
29)	R..O..8	29)	B..R..6	29)	K..U..4	29)	R..W..1	29)	B..Z..Y	29)	J..2..V
30)	S..P..A	30)	C..S..7	30)	L..V..5	30)	S..X..2	30)	C..1..Z	30)	K..3..W
31)	T..Q..B	31)	D..T..8			31)	T..Y..3			31)	L..A..X

CODES: P-PHYSICAL BIORHYTHM CURVE,E-EMOTIONAL BIORHYTHM CURVE,I-INTELLECTUAL BIORHYTHM CURVE

	...JANUARY..	..FEBRUARY..	....MARCH...	....APRIL...	.....MAY....	....JUNE....
	P--E--I	P--E--I	P--E--I	P--E--I	P--E--I	P--E--I
1)	M..B..Y	U..E..W	B..E..R	K..H..P	R..K..M	B..N..K
2)	N..C..Z	V..F..X	C..F..S	L..J..Q	S..L..N	C..O..L
3)	O..D..1	W..G..Y	D..G..T	M..K..R	T..M..O	D..P..M
4)	P..E..2	X..H..Z	E..H..U	N..L..S	U..N..P	E..Q..N
5)	Q..F..3	A..J..1	F..J..V	O..M..T	V..O..Q	F..R..O
6)	R..G..4	B..K..2	G..K..W	P..N..U	W..P..R	G..S..P
7)	S..H..5	C..L..3	H..L..X	Q..O..V	X..Q..S	H..T..Q
8)	T..J..6	D..M..4	J..M..Y	R..P..W	A..R..T	J..U..R
9)	U..K..7	E..N..5	K..N..Z	S..Q..X	B..S..U	K..V..S
10)	V..L..8	F..O..6	L..O..1	T..R..Y	C..T..V	L..W..T
11)	W..M..A	G..P..7	M..P..2	U..S..Z	D..U..W	M..X..U
12)	X..N..B	H..Q..8	N..Q..3	V..T..1	E..V..X	N..Y..V
13)	A..O..C	J..R..A	O..R..4	W..U..2	F..W..Y	O..Z..W
14)	B..P..D	K..S..B	P..S..5	X..V..3	G..X..Z	P..1..X
15)	C..Q..E	L..T..C	Q..T..6	A..W..4	H..Y..1	Q..2..Y
16)	D..R..F	M..U..D	R..U..7	B..X..5	J..Z..2	R..3..Z
17)	E..S..G	N..V..E	S..V..8	C..Y..6	K..1..3	S..A..1
18)	F..T..H	O..W..F	T..W..A	D..Z..7	L..2..4	T..B..2
19)	G..U..J	P..X..G	U..X..B	E..1..8	M..3..5	U..C..3
20)	H..V..K	Q..Y..H	V..Y..C	F..2..A	N..A..6	V..D..4
21)	J..W..L	R..Z..J	W..Z..D	G..3..B	O..B..7	W..E..5
22)	K..X..M	S..1..K	X..1..E	H..A..C	P..C..8	X..F..6
23)	L..Y..N	T..2..L	A..2..F	J..B..D	Q..D..A	A..G..7
24)	M..Z..O	U..3..M	B..3..G	K..C..E	R..E..B	B..H..8
25)	N..1..P	V..A..N	C..A..H	L..D..F	S..F..C	C..J..A
26)	O..2..Q	W..B..O	D..B..J	M..E..G	T..G..D	D..K..B
27)	P..3..R	X..C..P	E..C..K	N..F..H	U..H..E	E..L..C
28)	Q..A..S	A..D..Q	F..D..L	O..G..J	V..J..F	F..M..D
29)	R..B..T		G..E..M	P..H..K	W..K..G	G..N..E
30)	S..C..U		H..F..N	Q..J..L	X..L..H	H..O..F
31)	T..D..V		J..G..O		A..M..J	

1934 1934

	....JULY....	...AUGUST...	..SEPTEMBER.	..OCTOBER...	..NOVEMBER..	..DECEMBER..
	P--E--I	P--E--I	P--E--I	P--E--I	P--E--I	P--E--I
1)	J..P..G	R..S..E	B..V..C	J..X..8	R..1..6	A..3..3
2)	K..Q..H	S..T..F	C..W..D	K..Y..A	S..2..7	B..A..4
3)	L..R..J	T..U..G	D..X..E	L..Z..B	T..3..8	C..B..5
4)	M..S..K	U..V..H	E..Y..F	M..1..C	U..A..A	D..C..6
5)	N..T..L	V..W..J	F..Z..G	N..2..D	V..B..B	E..D..7
6)	O..U..M	W..X..K	G..1..H	O..3..E	W..C..C	F..E..8
7)	P..V..N	X..Y..L	H..2..J	P..A..F	X..D..D	G..F..A
8)	Q..W..O	A..Z..M	J..3..K	Q..B..G	A..E..E	H..G..B
9)	R..X..P	B..1..N	K..A..L	R..C..H	B..F..F	J..H..C
10)	S..Y..Q	C..2..O	L..B..M	S..D..J	C..G..G	K..J..D
11)	T..Z..R	D..3..P	M..C..N	T..E..K	D..H..H	L..K..E
12)	U..1..S	E..A..Q	N..D..O	U..F..L	E..J..J	M..L..F
13)	V..2..T	F..B..R	O..E..P	V..G..M	F..K..K	N..M..G
14)	W..3..U	G..C..S	P..F..Q	W..H..N	G..L..L	O..N..H
15)	X..A..V	H..D..T	Q..G..R	X..J..O	H..M..M	P..O..J
16)	A..B..W	J..E..U	R..H..S	A..K..P	J..N..N	Q..P..K
17)	B..C..X	K..F..V	S..J..T	B..L..Q	K..O..O	R..Q..L
18)	C..D..Y	L..G..W	T..K..U	C..M..R	L..P..P	S..R..M
19)	D..E..Z	M..H..X	U..L..V	D..N..S	M..Q..Q	T..S..N
20)	E..F..1	N..J..Y	V..M..W	E..O..T	N..R..R	U..T..O
21)	F..G..2	O..K..Z	W..N..X	F..P..U	O..S..S	V..U..P
22)	G..H..3	P..L..1	X..O..Y	G..Q..V	P..T..T	W..V..Q
23)	H..J..4	Q..M..2	A..P..Z	H..R..W	Q..U..U	X..W..R
24)	J..K..5	R..N..3	B..Q..1	J..S..X	R..V..V	A..X..S
25)	K..L..6	S..O..4	C..R..2	K..T..Y	S..W..W	B..Y..T
26)	L..M..7	T..P..5	D..S..3	L..U..Z	T..X..X	C..Z..U
27)	M..N..8	U..Q..6	E..T..4	M..V..1	U..Y..Y	D..1..V
28)	N..O..A	V..R..7	F..U..5	N..W..2	V..Z..Z	E..2..W
29)	O..P..B	W..S..8	G..V..6	O..X..3	W..1..1	F..3..X
30)	P..Q..C	X..T..A	H..W..7	P..Y..4	X..2..2	G..A..Y
31)	Q..R..D	A..U..B		Q..Z..5		H..B..Z

CODES: P-PHYSICAL BIORHYTHM CURVE, E-EMOTIONAL BIORHYTHM CURVE, I-INTELLECTUAL BIORHYTHM CURVE

	...JANUARY..	..FEBRUARY..	....MARCH...	....APRIL...	.....MAY....	....JUNE....
	P--E--I	P--E--I	P--E--I	P--E--I	P--E--I	P--E--I
1)	J..C..1	R..F..Y	W..F..T	G..J..R	O..L..O	W..O..M
2)	K..D..2	S..G..Z	X..G..U	H..K..S	P..M..P	X..P..N
3)	L..E..3	T..H..1	A..H..V	J..L..T	Q..N..Q	A..Q..O
4)	M..F..4	U..J..2	B..J..W	K..M..U	R..O..R	B..R..P
5)	N..G..5	V..K..3	C..K..X	L..N..V	S..P..S	C..S..Q
6)	O..H..6	W..L..4	D..L..Y	M..O..W	T..Q..T	D..T..R
7)	P..J..7	X..M..5	E..M..Z	N..P..X	U..R..U	E..U..S
8)	Q..K..8	A..N..6	F..N..1	O..Q..Y	V..S..V	F..V..T
9)	R..L..A	B..O..7	G..O..2	P..R..Z	W..T..W	G..W..U
10)	S..M..B	C..P..8	H..P..3	Q..S..1	X..U..X	H..X..V
11)	T..N..C	D..Q..A	J..Q..4	R..T..2	A..V..Y	J..Y..W
12)	U..O..D	E..R..B	K..R..5	S..U..3	B..W..Z	K..Z..X
13)	V..P..E	F..S..C	L..S..6	T..V..4	C..X..1	L..1..Y
14)	W..Q..F	G..T..D	M..T..7	U..W..5	D..Y..2	M..2..Z
15)	X..R..G	H..U..E	N..U..8	V..X..6	E..Z..3	N..3..1
16)	A..S..H	J..V..F	O..V..A	W..Y..7	F..1..4	O..A..2
17)	B..T..J	K..W..G	P..W..B	X..Z..8	G..2..5	P..B..3
18)	C..U..K	L..X..H	Q..X..C	A..1..A	H..3..6	Q..C..4
19)	D..V..L	M..Y..J	R..Y..D	B..2..B	J..A..7	R..D..5
20)	E..W..M	N..Z..K	S..Z..E	C..3..C	K..B..8	S..E..6
21)	F..X..N	O..1..L	T..1..F	D..A..D	L..C..A	T..F..7
22)	G..Y..O	P..2..M	U..2..G	E..B..E	M..D..B	U..G..8
23)	H..Z..P	Q..3..N	V..3..H	F..C..F	N..E..C	V..H..A
24)	J..1..Q	R..A..O	W..A..J	G..D..G	O..F..D	W..J..B
25)	K..2..R	S..B..P	X..B..K	H..E..H	P..G..E	X..K..C
26)	L..3..S	T..C..Q	A..C..L	J..F..J	Q..H..F	A..L..D
27)	M..A..T	U..D..R	B..D..M	K..G..K	R..J..G	B..M..E
28)	N..B..U	V..E..S	C..E..N	L..H..L	S..K..H	C..N..F
29)	O..C..V		D..F..O	M..J..M	T..L..J	D..O..G
30)	P..D..W		E..G..P	N..K..N	U..M..K	E..P..H
31)	Q..E..X		F..H..Q		V..N..L	

1935 1935

	....JULY....	...AUGUST...	..SEPTEMBER.	..OCTOBER...	..NOVEMBER..	..DECEMBER..
	P--E--I	P--E--I	P--E--I	P--E--I	P--E--I	P--E--I
1)	F..Q..J	O..T..G	W..W..E	F..Y..B	O..2..8	V..A..5
2)	G..R..K	P..U..H	X..X..F	G..Z..C	P..3..A	W..B..6
3)	H..S..L	Q..V..J	A..Y..G	H..1..D	Q..A..B	X..C..7
4)	J..T..M	R..W..K	B..Z..H	J..2..E	R..B..C	A..D..8
5)	K..U..N	S..X..L	C..1..J	K..3..F	S..C..D	B..E..A
6)	L..V..O	T..Y..M	D..2..K	L..A..G	T..D..E	C..F..B
7)	M..W..P	U..Z..N	E..3..L	M..B..H	U..E..F	D..G..C
8)	N..X..Q	V..1..O	F..A..M	N..C..J	V..F..G	E..H..D
9)	O..Y..R	W..2..P	G..B..N	O..D..K	W..G..H	F..J..E
10)	P..Z..S	X..3..Q	H..C..O	P..E..L	X..H..J	G..K..F
11)	Q..1..T	A..A..R	J..D..P	Q..F..M	A..J..K	H..L..G
12)	R..2..U	B..B..S	K..E..Q	R..G..N	B..K..L	J..M..H
13)	S..3..V	C..C..T	L..F..R	S..H..O	C..L..M	K..N..J
14)	T..A..W	D..D..U	M..G..S	T..J..P	D..M..N	L..O..K
15)	U..B..X	E..E..V	N..H..T	U..K..Q	E..N..O	M..P..L
16)	V..C..Y	F..F..W	O..J..U	V..L..R	F..O..P	N..Q..M
17)	W..D..Z	G..G..X	P..K..V	W..M..S	G..P..Q	O..R..N
18)	X..E..1	H..H..Y	Q..L..W	X..N..T	H..Q..R	P..S..O
19)	A..F..2	J..J..Z	R..M..X	A..O..U	J..R..S	Q..T..P
20)	B..G..3	K..K..1	S..N..Y	B..P..V	K..S..T	R..U..Q
21)	C..H..4	L..L..2	T..O..Z	C..Q..W	L..T..U	S..V..R
22)	D..J..5	M..M..3	U..P..1	D..R..X	M..U..V	T..W..S
23)	E..K..6	N..N..4	V..Q..2	E..S..Y	N..V..W	U..X..T
24)	F..L..7	O..O..5	W..R..3	F..T..Z	O..W..X	V..Y..U
25)	G..M..8	P..P..6	X..S..4	G..U..1	P..X..Y	W..Z..V
26)	H..N..A	Q..Q..7	A..T..5	H..V..2	Q..Y..Z	X..1..W
27)	J..O..B	R..R..8	B..U..6	J..W..3	R..Z..1	A..2..X
28)	K..P..C	S..S..A	C..V..7	K..X..4	S..1..2	B..3..Y
29)	L..Q..D	T..T..B	D..W..8	L..Y..5	T..2..3	C..A..Z
30)	M..R..E	U..U..C	E..X..A	M..Z..6	U..3..4	D..B..1
31)	N..S..F	V..V..D		N..1..7		E..C..2

CODES: P-PHYSICAL BIORHYTHM CURVE, E-EMOTIONAL BIORHYTHM CURVE, I-INTELLECTUAL BIORHYTHM CURVE

1936 1936

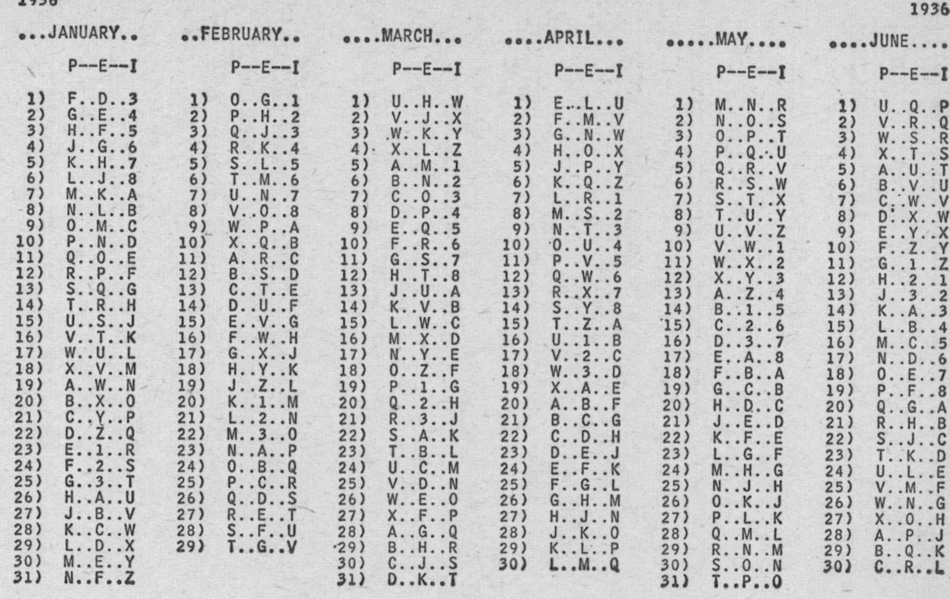

Day	...JANUARY.. P--E--I	..FEBRUARY.. P--E--I	....MARCH... P--E--I	....APRIL... P--E--I	.....MAY.... P--E--I	....JUNE.... P--E--I
1)	F..D..3	O..G..1	U..H..W	E..L..U	M..N..R	U..Q..P
2)	G..E..4	P..H..2	V..J..X	F..M..V	N..O..S	V..R..Q
3)	H..F..5	Q..J..3	W..K..Y	G..N..W	O..P..T	W..S..R
4)	J..G..6	R..K..4	X..L..Z	H..O..X	P..Q..U	X..T..S
5)	K..H..7	S..L..5	A..M..1	J..P..Y	Q..R..V	A..U..T
6)	L..J..8	T..M..6	B..N..2	K..Q..Z	R..S..W	B..V..U
7)	M..K..A	U..N..7	C..O..3	L..R..1	S..T..X	C..W..V
8)	N..L..B	V..O..8	D..P..4	M..S..2	T..U..Y	D..X..W
9)	O..M..C	W..P..A	E..Q..5	N..T..3	U..V..Z	E..Y..X
10)	P..N..D	X..Q..B	F..R..6	O..U..4	V..W..1	F..Z..Y
11)	Q..O..E	A..R..C	G..S..7	P..V..5	W..X..2	G..1..Z
12)	R..P..F	B..S..D	H..T..8	Q..W..6	X..Y..3	H..2..1
13)	S..Q..G	C..T..E	J..U..A	R..X..7	A..Z..4	J..3..2
14)	T..R..H	D..U..F	K..V..B	S..Y..8	B..1..5	K..A..3
15)	U..S..J	E..V..G	L..W..C	T..Z..A	C..2..6	L..B..4
16)	V..T..K	F..W..H	M..X..D	U..1..B	D..3..7	M..C..5
17)	W..U..L	G..X..J	N..Y..E	V..2..C	E..A..8	N..D..6
18)	X..V..M	H..Y..K	O..Z..F	W..3..D	F..B..A	O..E..7
19)	A..W..N	J..Z..L	P..1..G	X..A..E	G..C..B	P..F..8
20)	B..X..O	K..1..M	Q..2..H	A..B..F	H..D..C	Q..G..A
21)	C..Y..P	L..2..N	R..3..J	B..C..G	J..E..D	R..H..B
22)	D..Z..Q	M..3..O	S..A..K	C..D..H	K..F..E	S..J..C
23)	E..1..R	N..A..P	T..B..L	D..E..J	L..G..F	T..K..D
24)	F..2..S	O..B..Q	U..C..M	E..F..K	M..H..G	U..L..E
25)	G..3..T	P..C..R	V..D..N	F..G..L	N..J..H	V..M..F
26)	H..A..U	Q..D..S	W..E..O	G..H..M	O..K..J	W..N..G
27)	J..B..V	R..E..T	X..F..P	H..J..N	P..L..K	X..O..H
28)	K..C..W	S..F..U	A..G..Q	J..K..O	Q..M..L	A..P..J
29)	L..D..X	T..G..V	B..H..R	K..L..P	R..N..M	B..Q..K
30)	M..E..Y		C..J..S	L..M..Q	S..O..N	C..R..L
31)	N..F..Z		D..K..T		T..P..O	

1936 | 1936

Day	....JULY.... P--E--I	...AUGUST... P--E--I	..SEPTEMBER. P--E--I	..OCTOBER... P--E--I	..NOVEMBER.. P--E--I	..DECEMBER.. P--E--I
1)	D..S..M	M..V..K	U..Y..H	D..1..E	M..A..C	T..C..8
2)	E..T..N	N..W..L	V..Z..J	E..2..F	N..B..D	U..D..A
3)	F..U..O	O..X..M	W..1..K	F..3..G	O..C..E	V..E..B
4)	G..V..P	P..Y..N	X..2..L	G..A..H	P..D..F	W..F..C
5)	H..W..Q	Q..Z..O	A..3..M	H..B..J	Q..E..G	X..G..D
6)	J..X..R	R..1..P	B..A..N	J..C..K	R..F..H	A..H..E
7)	K..Y..S	S..2..Q	C..B..O	K..D..L	S..G..J	B..J..F
8)	L..Z..T	T..3..R	D..C..P	L..E..M	T..H..K	C..K..G
9)	M..1..U	U..A..S	E..D..Q	M..F..N	U..J..L	D..L..H
10)	N..2..V	V..B..T	F..E..R	N..G..O	V..K..M	E..M..J
11)	O..3..W	W..C..U	G..F..S	O..H..P	W..L..N	F..N..K
12)	P..A..X	X..D..V	H..G..T	P..J..Q	X..M..O	G..O..L
13)	Q..B..Y	A..E..W	J..H..U	Q..K..R	A..N..P	H..P..M
14)	R..C..Z	B..F..X	K..J..V	R..L..S	B..O..Q	J..Q..N
15)	S..D..1	C..G..Y	L..K..W	S..M..T	C..P..R	K..R..O
16)	T..E..2	D..H..Z	M..L..X	T..N..U	D..Q..S	L..S..P
17)	U..F..3	E..J..1	N..M..Y	U..O..V	E..R..T	M..T..Q
18)	V..G..4	F..K..2	O..N..Z	V..P..W	F..S..U	N..U..R
19)	W..H..5	G..L..3	P..O..1	W..Q..X	G..T..V	O..V..S
20)	X..J..6	H..M..4	Q..P..2	X..R..Y	H..U..W	P..W..T
21)	A..K..7	J..N..5	R..Q..3	A..S..Z	J..V..X	Q..X..U
22)	B..L..8	K..O..6	S..R..4	B..T..1	K..W..Y	R..Y..V
23)	C..M..A	L..P..7	T..S..5	C..U..2	L..X..Z	S..Z..W
24)	D..N..B	M..Q..8	U..T..6	D..V..3	M..Y..1	T..1..X
25)	E..O..C	N..R..A	V..U..7	E..W..4	N..Z..2	U..2..Y
26)	F..P..D	O..S..B	W..V..8	F..X..5	O..1..3	V..3..Z
27)	G..Q..E	P..T..C	X..W..A	G..Y..6	P..2..4	W..A..1
28)	H..R..F	Q..U..D	A..X..B	H..Z..7	Q..3..5	X..B..2
29)	J..S..G	R..V..E	B..Y..C	J..1..8	R..A..6	A..C..3
30)	K..T..H	S..W..F	C..Z..D	K..2..A	S..B..7	B..D..4
31)	L..U..J	T..X..G		L..3..B		C..E..5

CODES: P-PHYSICAL BIORHYTHM CURVE, E-EMOTIONAL BIORHYTHM CURVE, I-INTELLECTUAL BIORHYTHM CURVE

1937 1937

	...JANUARY..	..FEBRUARY..	....MARCH...	....APRIL...	.....MAY....	....JUNE....
	P--E--I	P--E--I	P--E--I	P--E--I	P--E--I	P--E--I
1)	D..F..6	M..J..4	R..J..Y	B..M..W	J..O..T	R..R..R
2)	E..G..7	N..K..5	S..K..Z	C..N..X	K..P..U	S..S..S
3)	F..H..8	O..L..6	T..L..1	D..O..Y	L..Q..V	T..T..T
4)	G..J..A	P..M..7	U..M..2	E..P..Z	M..R..W	U..U..U
5)	H..K..B	Q..N..8	V..N..3	F..Q..1	N..S..X	V..V..V
6)	J..L..C	R..O..A	W..O..4	G..R..2	O..T..Y	W..W..W
7)	K..M..D	S..P..B	X..P..5	H..S..3	P..U..Z	X..X..X
8)	L..N..E	T..Q..C	A..Q..6	J..T..4	Q..V..1	A..Y..Y
9)	M..O..F	U..R..D	B..R..7	K..U..5	R..W..2	B..Z..Z
10)	N..P..G	V..S..E	C..S..8	L..V..6	S..X..3	C..1..1
11)	O..Q..H	W..T..F	D..T..A	M..W..7	T..Y..4	D..2..2
12)	P..R..J	X..U..G	E..U..B	N..X..8	U..Z..5	E..3..3
13)	Q..S..K	A..V..H	F..V..C	O..Y..A	V..1..6	F..A..4
14)	R..T..L	B..W..J	G..W..D	P..Z..B	W..2..7	G..B..5
15)	S..U..M	C..X..K	H..X..E	Q..1..C	X..3..8	H..C..6
16)	T..V..N	D..Y..L	J..Y..F	R..2..D	A..A..A	J..D..7
17)	U..W..O	E..Z..M	K..Z..G	S..3..E	B..B..B	K..E..8
18)	V..X..P	F..1..N	L..1..H	T..A..F	C..C..C	L..F..A
19)	W..Y..Q	G..2..O	M..2..J	U..B..G	D..D..D	M..G..B
20)	X..Z..R	H..3..P	N..3..K	V..C..H	E..E..E	N..H..C
21)	A..1..S	J..A..Q	O..A..L	W..D..J	F..F..F	O..J..D
22)	B..2..T	K..B..R	P..B..M	X..E..K	G..G..G	P..K..E
23)	C..3..U	L..C..S	Q..C..N	A..F..L	H..H..H	Q..L..F
24)	D..A..V	M..D..T	R..D..O	B..G..M	J..J..J	R..M..G
25)	E..B..W	N..E..U	S..E..P	C..H..N	K..K..K	S..N..H
26)	F..C..X	O..F..V	T..F..Q	D..J..O	L..L..L	T..O..J
27)	G..D..Y	P..G..W	U..G..R	E..K..P	M..M..M	U..P..K
28)	H..E..Z	Q..H..X	V..H..S	F..L..Q	N..N..N	V..Q..L
29)	J..F..1		W..J..T	G..M..R	O..O..O	W..R..M
30)	K..G..2		X..K..U	H..N..S	P..P..P	X..S..N
31)	L..H..3		A..L..V		Q..Q..Q	

1937 | 1937

Day	....JULY.... P--E--I	...AUGUST... P--E--I	..SEPTEMEER. P--E--I	..OCTOBER... P--E--I	..NOVEMBER.. P--E--I	..DECEMBER.. P--E--I
1)	A..T..O	J..W..M	R..Z..K	A..2..G	J..B..E	Q..D..B
2)	B..U..P	K..X..N	S..1..L	B..3..H	K..C..F	R..E..C
3)	C..V..Q	L..Y..O	T..2..M	C..A..J	L..D..G	S..F..D
4)	D..W..R	M..Z..P	U..3..N	D..B..K	M..E..H	T..G..E
5)	E..X..S	N..1..Q	V..A..O	E..C..L	N..F..J	U..H..F
6)	F..Y..T	O..2..R	W..B..P	F..D..M	O..G..K	V..J..G
7)	G..Z..U	P..3..S	X..C..Q	G..E..N	P..H..L	W..K..H
8)	H..1..V	Q..A..T	A..D..R	H..F..O	Q..J..M	X..L..J
9)	J..2..W	R..B..U	B..E..S	J..G..P	R..K..N	A..M..K
10)	K..3..X	S..C..V	C..F..T	K..H..Q	S..L..O	B..N..L
11)	L..A..Y	T..D..W	D..G..U	L..J..R	T..M..P	C..O..M
12)	M..B..Z	U..E..X	E..H..V	M..K..S	U..N..Q	D..P..N
13)	N..C..1	V..F..Y	F..J..W	N..L..T	V..O..R	E..Q..O
14)	O..D..2	W..G..Z	G..K..X	O..M..U	W..P..S	F..R..P
15)	P..E..3	X..H..1	H..L..Y	P..N..V	X..Q..T	G..S..Q
16)	Q..F..4	A..J..2	J..M..Z	Q..O..W	A..R..U	H..T..R
17)	R..G..5	B..K..3	K..N..1	R..P..X	B..S..V	J..U..S
18)	S..H..6	C..L..4	L..O..2	S..Q..Y	C..T..W	K..V..T
19)	T..J..7	D..M..5	M..P..3	T..R..Z	D..U..X	L..W..U
20)	U..K..8	E..N..6	N..Q..4	U..S..1	E..V..Y	M..X..V
21)	V..L..A	F..O..7	O..R..5	V..T..2	F..W..Z	N..Y..W
22)	W..M..B	G..P..8	P..S..6	W..U..3	G..X..1	O..Z..X
23)	X..N..C	H..Q..A	Q..T..7	X..V..4	H..Y..2	P..1..Y
24)	A..O..D	J..R..B	R..U..8	A..W..5	J..Z..3	Q..2..Z
25)	B..P..E	K..S..C	S..V..A	B..X..6	K..1..4	R..3..1
26)	C..Q..F	L..T..D	T..W..B	C..Y..7	L..2..5	S..A..2
27)	D..R..G	M..U..E	U..X..C	D..Z..8	M..3..6	T..B..3
28)	E..S..H	N..V..F	V..Y..D	E..1..A	N..A..7	U..C..4
29)	F..T..J	O..W..G	W..Z..E	F..2..B	O..B..8	V..D..5
30)	G..U..K	P..X..H	X..1..F	G..3..C	P..C..A	W..E..6
31)	H..V..L	Q..Y..J		H..A..D		X..F..7

CODES: P-PHYSICAL BIORHYTHM CURVE,E-EMOTIONAL BIORHYTHM CURVE,I-INTELLECTUAL BIORHYTHM CURVE

1938 1938

Day	...JANUARY.. P--E--I	..FEBRUARY.. P--E--I	....MARCH... P--E--I	....APRIL... P--E--I	.....MAY.... P--E--I	....JUNE.... P--E--I
1)	A..G..8	J..K..6	O..K..1	W..N..Y	F..P..V	O..S..T
2)	B..H..A	K..L..7	P..L..2	X..O..Z	G..Q..W	P..T..U
3)	C..J..B	L..M..8	Q..M..3	A..P..1	H..R..X	Q..U..V
4)	D..K..C	M..N..A	R..N..4	B..Q..2	J..S..Y	R..V..W
5)	E..L..D	N..O..B	S..O..5	C..R..3	K..T..Z	S..W..X
6)	F..M..E	O..P..C	T..P..6	D..S..4	L..U..1	T..X..Y
7)	G..N..F	P..Q..D	U..Q..7	E..T..5	M..V..2	U..Y..Z
8)	H..O..G	Q..R..E	V..R..8	F..U..6	N..W..3	V..Z..1
9)	J..P..H	R..S..F	W..S..A	G..V..7	O..X..4	W..1..2
10)	K..Q..J	S..T..G	X..T..B	H..W..8	P..Y..5	X..2..3
11)	L..R..K	T..U..H	A..U..C	J..X..A	Q..Z..6	A..3..4
12)	M..S..L	U..V..J	B..V..D	K..Y..B	R..1..7	B..A..5
13)	N..T..M	V..W..K	C..W..E	L..Z..C	S..2..8	C..B..6
14)	O..U..N	W..X..L	D..X..F	M..1..D	T..3..A	D..C..7
15)	P..V..O	X..Y..M	E..Y..G	N..2..E	U..A..B	E..D..8
16)	Q..W..P	A..Z..N	F..Z..H	O..3..F	V..B..C	F..E..A
17)	R..X..Q	B..1..O	G..1..J	P..A..G	W..C..D	G..F..B
18)	S..Y..R	C..2..P	H..2..K	Q..B..H	X..D..E	H..G..C
19)	T..Z..S	D..3..Q	J..3..L	R..C..J	A..E..F	J..H..D
20)	U..1..T	E..A..R	K..A..M	S..D..K	B..F..G	K..J..E
21)	V..2..U	F..B..S	L..B..N	T..E..L	C..G..H	L..K..F
22)	W..3..V	G..C..T	M..C..O	U..F..M	D..H..J	M..L..G
23)	X..A..W	H..D..U	N..D..P	V..G..N	E..J..K	N..M..H
24)	A..B..X	J..E..V	O..E..Q	W..H..O	F..K..L	O..N..J
25)	B..C..Y	K..F..W	P..F..R	X..J..P	G..L..M	P..O..K
26)	C..D..Z	L..G..X	Q..G..S	A..K..Q	H..M..N	Q..P..L
27)	D..E..1	M..H..Y	R..H..T	B..L..R	J..N..O	R..Q..M
28)	E..F..2	N..J..Z	S..J..U	C..M..S	K..O..P	S..R..N
29)	F..G..3		T..K..V	D..N..T	L..P..Q	T..S..O
30)	G..H..4		U..L..W	E..O..U	M..Q..R	U..T..P
31)	H..J..5		V..M..X		N..R..S	

1938						1938
	....JULY....	...AUGUST...	..SEPTEMBER.	..OCTOBER...	..NOVEMBER..	..DECEMBER..
	P--E--I	P--E--I	P--E--I	P--E--I	P--E--I	P--E--I
1)	V..U..Q	F..X..O	O..1..M	V..3..J	F..C..G	N..E..D
2)	W..V..R	G..Y..P	P..2..N	W..A..K	G..D..H	O..F..E
3)	X..W..S	H..Z..Q	Q..3..O	X..B..L	H..E..J	P..G..F
4)	A..X..T	J..1..R	R..A..P	A..C..M	J..F..K	Q..H..G
5)	B..Y..U	K..2..S	S..B..Q	B..D..N	K..G..L	R..J..H
6)	C..Z..V	L..3..T	T..C..R	C..E..O	L..H..M	S..K..J
7)	D..1..W	M..A..U	U..D..S	D..F..P	M..J..N	T..L..K
8)	E..2..X	N..B..V	V..E..T	E..G..Q	N..K..O	U..M..L
9)	F..3..Y	O..C..W	W..F..U	F..H..R	O..L..P	V..N..M
10)	G..A..Z	P..D..X	X..G..V	G..J..S	P..M..Q	W..O..N
11)	H..B..1	Q..E..Y	A..H..W	H..K..T	Q..N..R	X..P..O
12)	J..C..2	R..F..Z	B..J..X	J..L..U	R..O..S	A..Q..P
13)	K..D..3	S..G..1	C..K..Y	K..M..V	S..P..T	B..R..Q
14)	L..E..4	T..H..2	D..L..Z	L..N..W	T..Q..U	C..S..R
15)	M..F..5	U..J..3	E..M..1	M..O..X	U..R..V	D..T..S
16)	N..G..6	V..K..4	F..N..2	N..P..Y	V..S..W	E..U..T
17)	O..H..7	W..L..5	G..O..3	O..Q..Z	W..T..X	F..V..U
18)	P..J..8	X..M..6	H..P..4	P..R..1	X..U..Y	G..W..V
19)	Q..K..A	A..N..7	J..Q..5	Q..S..2	A..V..Z	H..X..W
20)	R..L..B	B..O..8	K..R..6	R..T..3	B..W..1	J..Y..X
21)	S..M..C	C..P..A	L..S..7	S..U..4	C..X..2	K..Z..Y
22)	T..N..D	D..Q..B	M..T..8	T..V..5	D..Y..3	L..1..Z
23)	U..O..E	E..R..C	N..U..A	U..W..6	E..Z..4	M..2..1
24)	V..P..F	F..S..D	O..V..B	V..X..7	F..1..5	N..3..2
25)	W..Q..G	G..T..E	P..W..C	W..Y..8	G..2..6	O..A..3
26)	X..R..H	H..U..F	Q..X..D	X..Z..A	H..3..7	P..B..4
27)	A..S..J	J..V..G	R..Y..E	A..1..B	J..A..8	Q..C..5
28)	B..T..K	K..W..H	S..Z..F	B..2..C	K..B..A	R..D..6
29)	C..U..L	L..X..J	T..1..G	C..3..D	L..C..B	S..E..7
30)	D..V..M	M..Y..K	U..2..H	D..A..E	M..D..C	T..F..8
31)	E..W..N	N..Z..L		E..B..F		U..G..A

CODES: P-PHYSICAL BIORHYTHM CURVE, E-EMOTIONAL BIORHYTHM CURVE, I-INTELLECTUAL BIORHYTHM CURVE

1939 | 1939

	...JANUARY..	..FEBRUARY..	....MARCH...	....APRIL...	.....MAY....	....JUNE....
	P--E--I	P--E--I	P--E--I	P--E--I	P--E--I	P--E--I
1)	V..H..B	F..L..8	L..L..3	T..O..1	C..Q..X	L..T..V
2)	W..J..C	G..M..A	M..M..4	U..P..2	D..R..Y	M..U..W
3)	X..K..D	H..N..B	N..N..5	V..Q..3	E..S..Z	N..V..X
4)	A..L..E	J..O..C	O..O..6	W..R..4	F..T..1	O..W..Y
5)	B..M..F	K..P..D	P..P..7	X..S..5	G..U..2	P..X..Z
6)	C..N..G	L..Q..E	Q..Q..8	A..T..6	H..V..3	Q..Y..1
7)	D..O..H	M..R..F	R..R..A	B..U..7	J..W..4	R..Z..2
8)	E..P..J	N..S..G	S..S..B	C..V..8	K..X..5	S..1..3
9)	F..Q..K	O..T..H	T..T..C	D..W..A	L..Y..6	T..2..4
10)	G..R..L	P..U..J	U..U..D	E..X..B	M..Z..7	U..3..5
11)	H..S..M	Q..V..K	V..V..E	F..Y..C	N..1..8	V..A..6
12)	J..T..N	R..W..L	W..W..F	G..Z..D	O..2..A	W..B..7
13)	K..U..O	S..X..M	X..X..G	H..1..E	P..3..B	X..C..8
14)	L..V..P	T..Y..N	A..Y..H	J..2..F	Q..A..C	A..D..A
15)	M..W..Q	U..Z..O	B..Z..J	K..3..G	R..B..D	B..E..B
16)	N..X..R	V..1..P	C..1..K	L..A..H	S..C..E	C..F..C
17)	O..Y..S	W..2..Q	D..2..L	M..B..J	T..D..F	D..G..D
18)	P..Z..T	X..3..R	E..3..M	N..C..K	U..E..G	E..H..E
19)	Q..1..U	A..A..S	F..A..N	O..D..L	V..F..H	F..J..F
20)	R..2..V	B..B..T	G..B..O	P..E..M	W..G..J	G..K..G
21)	S..3..W	C..C..U	H..C..P	Q..F..N	X..H..K	H..L..H
22)	T..A..X	D..D..V	J..D..Q	R..G..O	A..J..L	J..M..J
23)	U..B..Y	E..E..W	K..E..R	S..H..P	B..K..M	K..N..K
24)	V..C..Z	F..F..X	L..F..S	T..J..Q	C..L..N	L..O..L
25)	W..D..1	G..G..Y	M..G..T	U..K..R	D..M..O	M..P..M
26)	X..E..2	H..H..Z	N..H..U	V..L..S	E..N..P	N..Q..N
27)	A..F..3	J..J..1	O..J..V	W..M..T	F..O..Q	O..R..O
28)	B..G..4	K..K..2	P..K..W	X..N..U	G..P..R	P..S..P
29)	C..H..5		Q..L..X	A..O..V	H..Q..S	Q..T..Q
30)	D..J..6		R..M..Y	B..P..W	J..R..T	R..U..R
31)	E..K..7		S..N..Z		K..S..U	

1939 1939

	....JULY....	...AUGUST...	..SEPTEMBER.	..OCTOBER...	..NOVEMBER..	..DECEMBER..
	P--E--I	P--E--I	P--E--I	P--E--I	P--E--I	P--E--I
1)	S..V..S	C..Y..Q	L..2..O	S..A..L	C..D..J	K..F..F
2)	T..W..T	D..Z..R	M..3..P	T..B..M	D..E..K	L..G..G
3)	U..X..U	E..1..S	N..A..Q	U..C..N	E..F..L	M..H..H
4)	V..Y..V	F..2..T	O..B..R	V..D..O	F..G..M	N..J..J
5)	W..Z..W	G..3..U	P..C..S	W..E..P	G..H..N	O..K..K
6)	X..1..X	H..A..V	Q..D..T	X..F..Q	H..J..O	P..L..L
7)	A..2..Y	J..B..W	R..E..U	A..G..R	J..K..P	Q..M..M
8)	B..3..Z	K..C..X	S..F..V	B..H..S	K..L..Q	R..N..N
9)	C..A..1	L..D..Y	T..G..W	C..J..T	L..M..R	S..O..O
10)	D..B..2	M..E..Z	U..H..X	D..K..U	M..N..S	T..P..P
11)	E..C..3	N..F..1	V..J..Y	E..L..V	N..O..T	U..Q..Q
12)	F..D..4	O..G..2	W..K..Z	F..M..W	O..P..U	V..R..R
13)	G..E..5	P..H..3	X..L..1	G..N..X	P..Q..V	W..S..S
14)	H..F..6	Q..J..4	A..M..2	H..O..Y	Q..R..W	X..T..T
15)	J..G..7	R..K..5	B..N..3	J..P..Z	R..S..X	A..U..U
16)	K..H..8	S..L..6	C..O..4	K..Q..1	S..T..Y	B..V..V
17)	L..J..A	T..M..7	D..P..5	L..R..2	T..U..Z	C..W..W
18)	M..K..B	U..N..8	E..Q..6	M..S..3	U..V..1	D..X..X
19)	N..L..C	V..O..A	F..R..7	N..T..4	V..W..2	E..Y..Y
20)	O..M..D	W..P..B	G..S..8	O..U..5	W..X..3	F..Z..Z
21)	P..N..E	X..Q..C	H..T..A	P..V..6	X..Y..4	G..1..1
22)	Q..O..F	A..R..D	J..U..B	Q..W..7	A..Z..5	H..2..2
23)	R..P..G	B..S..E	K..V..C	R..X..8	B..1..6	J..3..3
24)	S..Q..H	C..T..F	L..W..D	S..Y..A	C..2..7	K..A..4
25)	T..R..J	D..U..G	M..X..E	T..Z..B	D..3..8	L..B..5
26)	U..S..K	E..V..H	N..Y..F	U..1..C	E..A..A	M..C..6
27)	V..T..L	F..W..J	O..Z..G	V..2..D	F..B..B	N..D..7
28)	W..U..M	G..X..K	P..1..H	W..3..E	G..C..C	O..E..8
29)	X..V..N	H..Y..L	Q..2..J	X..A..F	H..D..D	P..F..A
30)	A..W..O	J..Z..M	R..3..K	A..B..G	J..E..E	Q..G..B
31)	B..X..P	K..1..N		B..C..H		R..H..C

CODES: P-PHYSICAL BIORHYTHM CURVE,E-EMOTIONAL BIORHYTHM CURVE,I-INTELLECTUAL BIORHYTHM CURVE

1940 | | | | | | 1940

Day	...JANUARY.. P--E--I	..FEBRUARY.. P--E--I	....MARCH... P--E--I	....APRIL... P--E--I	.....MAY.... P--E--I	....JUNE.... P--E--I
1)	S..J..D	C..M..B	J..N..6	R..Q..4	A..S..1	J..V..Y
2)	T..K..E	D..N..C	K..O..7	S..R..5	B..T..2	K..W..Z
3)	U..L..F	E..O..D	L..P..8	T..S..6	C..U..3	L..X..1
4)	V..M..G	F..P..E	M..Q..A	U..T..7	D..V..4	M..Y..2
5)	W..N..H	G..Q..F	N..R..B	V..U..8	E..W..5	N..Z..3
6)	X..O..J	H..R..G	O..S..C	W..V..A	F..X..6	O..1..4
7)	A..P..K	J..S..H	P..T..D	X..W..B	G..Y..7	P..2..5
8)	B..Q..L	K..T..J	Q..U..E	A..X..C	H..Z..8	Q..3..6
9)	C..R..M	L..U..K	R..V..F	B..Y..D	J..1..A	R..A..7
10)	D..S..N	M..V..L	S..W..G	C..Z..E	K..2..B	S..B..8
11)	E..T..O	N..W..M	T..X..H	D..1..F	L..3..C	T..C..A
12)	F..U..P	O..X..N	U..Y..J	E..2..G	M..A..D	U..D..B
13)	G..V..Q	P..Y..O	V..Z..K	F..3..H	N..B..E	V..E..C
14)	H..W..R	Q..Z..P	W..1..L	G..A..J	O..C..F	W..F..D
15)	J..X..S	R..1..Q	X..2..M	H..B..K	P..D..G	X..G..E
16)	K..Y..T	S..2..R	A..3..N	J..C..L	Q..E..H	A..H..F
17)	L. Z..U	T..3..S	B..A..O	K..D..M	R..F..J	B..J..G
18)	M .1..V	U..A..T	C..B..P	L..E..N	S..G..K	C..K..H
19)	N..2..W	V..B..U	D..C..Q	M..F..O	T..H..L	D..L..J
20)	O..3..X	W..C..V	E..D..R	N..G..P	U..J..M	E..M..K
21)	P..A..Y	X..D..W	F..E..S	O..H..Q	V..K..N	F..N..L
22)	Q..B..Z	A..E..X	G..F..T	P..J..R	W..L..O	G..O..M
23)	R..C..1	B..F..Y	H..G..U	Q..K..S	X..M..P	H..P..N
24)	S..D..2	C..G..Z	J..H..V	R..L..T	A..N..Q	J..Q..O
25)	T..E..3	D..H..1	K..J..W	S..M..U	B..O..R	K..R..P
26)	U..F..4	E..J..2	L..K..X	T..N..V	C..P..S	L..S..Q
27)	V..G..5	F..K..3	M..L..Y	U..O..W	D..Q..T	M..T..R
28)	W..H..6	G..L..4	N..M..Z	V..P..X	E..R..U	N..U..S
29)	X..J..7	H..M..5	O..N..1	W..Q..Y	F..S..V	O..V..T
30)	A..K..8		P..O..2	X..R..Z	G..T..W	P..W..U
31)	B..L..A		Q..P..3		H..U..X	

1940 1940

	....JULY....	...AUGUST...	..SEPTEMBER.	..OCTOBER...	..NOVEMBER..	..DECEMBER..
	P--E--I	P--E--I	P--E--I	P--E--I	P--E--I	P--E--I
1)	Q..X..V	A..1..T	J..A..R	Q..C..O	A..F..M	H..H..J
2)	R..Y..W	B..2..U	K..B..S	R..D..P	B..G..N	J..J..K
3)	S..Z..X	C..3..V	L..C..T	S..E..Q	C..H..O	K..K..L
4)	T..1..Y	D..A..W	M..D..U	T..F..R	D..J..P	L..L..M
5)	U..2..Z	E..B..X	N..E..V	U..G..S	E..K..Q	M..M..N
6)	V..3..1	F..C..Y	O..F..W	V..H..T	F..L..R	N..N..O
7)	W..A..2	G..D..Z	P..G..X	W..J..U	G..M..S	O..O..P
8)	X..B..3	H..E..1	Q..H..Y	X..K..V	H..N..T	P..P..Q
9)	A..C..4	J..F..2	R..J..Z	A..L..W	J..O..U	Q..Q..R
10)	B..D..5	K..G..3	S..K..1	B..M..X	K..P..V	R..R..S
11)	C..E..6	L..H..4	T..L..2	C..N..Y	L..Q..W	S..S..T
12)	D..F..7	M..J..5	U..M..3	D..O..Z	M..R..X	T..T..U
13)	E..G..8	N..K..6	V..N..4	E..P..1	N..S..Y	U..U..V
14)	F..H..A	O..L..7	W..O..5	F..Q..2	O..T..Z	V..V..W
15)	G..J..B	P..M..8	X..P..6	G..R..3	P..U..1	W..W..X
16)	H..K..C	Q..N..A	A..Q..7	H..S..4	Q..V..2	X..X..Y
17)	J..L..D	R..O..B	B..R..8	J..T..5	R..W..3	A..Y..Z
18)	K..M..E	S..P..C	C..S..A	K..U..6	S..X..4	B..Z..1
19)	L..N..F	T..Q..D	D..T..B	L..V..7	T..Y..5	C..1..2
20)	M..O..G	U..R..E	E..U..C	M..W..8	U..Z..6	D..2..3
21)	N..P..H	V..S..F	F..V..D	N..X..A	V..1..7	E..3..4
22)	O..Q..J	W..T..G	G..W..E	O..Y..B	W..2..8	F..A..5
23)	P..R..K	X..U..H	H..X..F	P..Z..C	X..3..A	G..B..6
24)	Q..S..L	A..V..J	J..Y..G	Q..1..D	A..A..B	H..C..7
25)	R..T..M	B..W..K	K..Z..H	R..2..E	B..B..C	J..D..8
26)	S..U..N	C..X..L	L..1..J	S..3..F	C..C..D	K..E..A
27)	T..V..O	D..Y..M	M..2..K	T..A..G	D..D..E	L..F..B
28)	U..W..P	E..Z..N	N..3..L	U..B..H	E..E..F	M..G..C
29)	V..X..Q	F..1..O	O..A..M	V..C..J	F..F..G	N..H..D
30)	W..Y..R	G..2..P	P..B..N	W..D..K	G..G..H	O..J..E
31)	X..Z..S	H..3..Q		X..E..L		P..K..F

CODES: P-PHYSICAL BIORHYTHM CURVE,E-EMOTIONAL BIORHYTHM CURVE,I-INTELLECTUAL BIORHYTHM CURVE

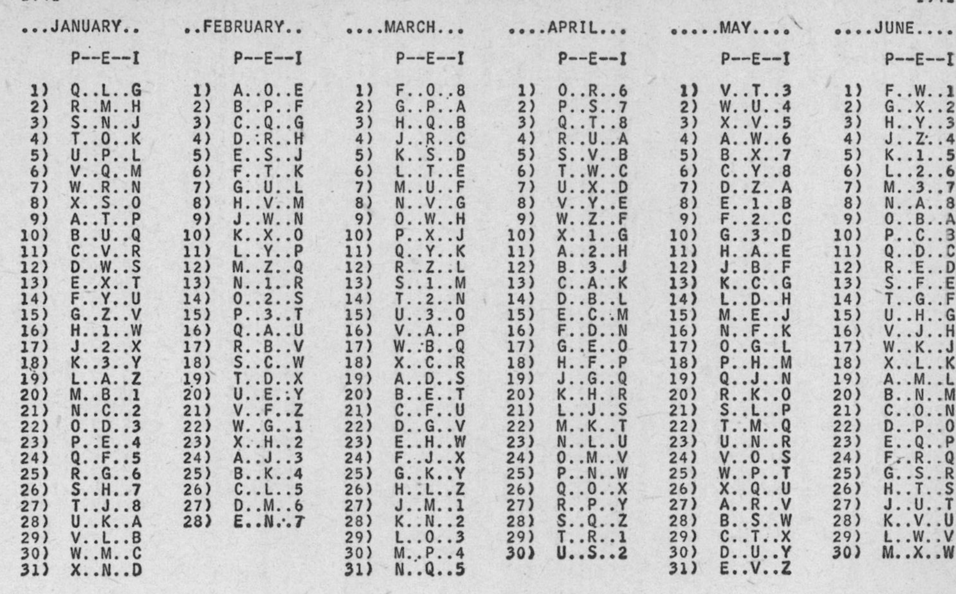

	1941					1941
	...JANUARY..	..FEBRUARY..	....MARCH...	....APRIL...	.....MAY....	....JUNE....
	P--E--I	P--E--I	P--E--I	P--E--I	P--E--I	P--E--I
1)	Q..L..G	A..O..E	F..O..8	O..R..6	V..T..3	F..W..1
2)	R..M..H	B..P..F	G..P..A	P..S..7	W..U..4	G..X..2
3)	S..N..J	C..Q..G	H..Q..B	Q..T..8	X..V..5	H..Y..3
4)	T..O..K	D..R..H	J..R..C	R..U..A	A..W..6	J..Z..4
5)	U..P..L	E..S..J	K..S..D	S..V..B	B..X..7	K..1..5
6)	V..Q..M	F..T..K	L..T..E	T..W..C	C..Y..8	L..2..6
7)	W..R..N	G..U..L	M..U..F	U..X..D	D..Z..A	M..3..7
8)	X..S..O	H..V..M	N..V..G	V..Y..E	E..1..B	N..A..8
9)	A..T..P	J..W..N	O..W..H	W..Z..F	F..2..C	O..B..A
10)	B..U..Q	K..X..O	P..X..J	X..1..G	G..3..D	P..C..B
11)	C..V..R	L..Y..P	Q..Y..K	A..2..H	H..A..E	Q..D..C
12)	D..W..S	M..Z..Q	R..Z..L	B..3..J	J..B..F	R..E..D
13)	E..X..T	N..1..R	S..1..M	C..A..K	K..C..G	S..F..E
14)	F..Y..U	O..2..S	T..2..N	D..B..L	L..D..H	T..G..F
15)	G..Z..V	P..3..T	U..3..O	E..C..M	M..E..J	U..H..G
16)	H..1..W	Q..A..U	V..A..P	F..D..N	N..F..K	V..J..H
17)	J..2..X	R..B..V	W..B..Q	G..E..O	O..G..L	W..K..J
18)	K..3..Y	S..C..W	X..C..R	H..F..P	P..H..M	X..L..K
19)	L..A..Z	T..D..X	A..D..S	J..G..Q	Q..J..N	A..M..L
20)	M..B..1	U..E..Y	B..E..T	K..H..R	R..K..O	B..N..M
21)	N..C..2	V..F..Z	C..F..U	L..J..S	S..L..P	C..O..N
22)	O..D..3	W..G..1	D..G..V	M..K..T	T..M..Q	D..P..O
23)	P..E..4	X..H..2	E..H..W	N..L..U	U..N..R	E..Q..P
24)	Q..F..5	A..J..3	F..J..X	O..M..V	V..O..S	F..R..Q
25)	R..G..6	B..K..4	G..K..Y	P..N..W	W..P..T	G..S..R
26)	S..H..7	C..L..5	H..L..Z	Q..O..X	X..Q..U	H..T..S
27)	T..J..8	D..M..6	J..M..1	R..P..Y	A..R..V	J..U..T
28)	U..K..A	E..N..7	K..N..2	S..Q..Z	B..S..W	K..V..U
29)	V..L..B		L..O..3	T..R..1	C..T..X	L..W..V
30)	W..M..C		M..P..4	U..S..2	D..U..Y	M..X..W
31)	X..N..D		N..Q..5		E..V..Z	

1941 | 1941

	....JULY....	...AUGUST...	..SEPTEMBER.	..OCTOBER...	..NOVEMBER..	..DECEMBER..
	P--E--I	P--E--I	P--E--I	P--E--I	P--E--I	P--E--I
1)	N..Y..X	V..2..V	F..B..T	N..D..Q	V..G..O	E..J..L
2)	O..Z..Y	W..3..W	G..C..U	O..E..R	W..H..P	F..K..M
3)	P..1..Z	X..A..X	H..D..V	P..F..S	X..J..Q	G..L..N
4)	Q..2..1	A..B..Y	J..E..W	Q..G..T	A..K..R	H..M..O
5)	R..3..2	B..C..Z	K..F..X	R..H..U	B..L..S	J..N..P
6)	S..A..3	C..D..1	L..G..Y	S..J..V	C..M..T	K..O..Q
7)	T..B..4	D..E..2	M..H..Z	T..K..W	D..N..U	L..P..R
8)	U..C..5	E..F..3	N..J..1	U..L..X	E..O..V	M..Q..S
9)	V..D..6	F..G..4	O..K..2	V..M..Y	F..P..W	N..R..T
10)	W..E..7	G..H..5	P..L..3	W..N..Z	G..Q..X	O..S..U
11)	X..F..8	H..J..6	Q..M..4	X..O..1	H..R..Y	P..T..V
12)	A..G..A	J..K..7	R..N..5	A..P..2	J..S..Z	Q..U..W
13)	B..H..B	K..L..8	S..O..6	B..Q..3	K..T..1	R..V..X
14)	C..J..C	L..M..A	T..P..7	C..R..4	L..U..2	S..W..Y
15)	D..K..D	M..N..B	U..Q..8	D..S..5	M..V..3	T..X..Z
16)	E..L..E	N..O..C	V..R..A	E..T..6	N..W..4	U..Y..1
17)	F..M..F	O..P..D	W..S..B	F..U..7	O..X..5	V..Z..2
18)	G..N..G	P..Q..E	X..T..C	G..V..8	P..Y..6	W..1..3
19)	H..O..H	Q..R..E	A..U..D	H..W..A	Q..Z..7	X..2..4
20)	J..P..J	R..S..G	B..V..E	J..X..B	R..1..8	A..3..5
21)	K..Q..K	S..T..H	C..W..F	K..Y..C	S..2..A	B..A..6
22)	L..R..L	T..U..J	D..X..G	L..Z..D	T..3..B	C..B..7
23)	M..S..M	U..V..K	E..Y..H	M..1..E	U..A..C	D..C..8
24)	N..T..N	V..W..L	F..Z..J	N..2..F	V..B..D	E..D..A
25)	O..U..O	W..X..M	G..1..K	O..3..G	W..C..E	F..E..B
26)	P..V..P	X..Y..N	H..2..L	P..A..H	X..D..F	G..F..C
27)	Q..W..Q	A..Z..O	J..3..M	Q..B..J	A..E..G	H..G..D
28)	R..X..R	B..1..P	K..A..N	R..C..K	B..F..H	J..H..E
29)	S..Y..S	C..2..Q	L..B..O	S..D..L	C..G..J	K..J..F
30)	T..Z..T	D..3..R	M..C..P	T..E..M	D..H..K	L..K..G
31)	U..1..U	E..A..S		U..F..N		M..L..H

CODES: P-PHYSICAL BIORHYTHM CURVE, E-EMOTIONAL BIORHYTHM CURVE, I-INTELLECTUAL BIORHYTHM CURVE

1942 | 1942

	...JANUARY..	..FEBRUARY..	....MARCH...	....APRIL...	.....MAY....	....JUNE....
	P--E--I	P--E--I	P--E--I	P--E--I	P--E--I	P--E--I
1)	N..M..J	V..P..G	C..P..B	L..S..8	S..U..5	C..X..3
2)	O..N..K	W..Q..H	D..Q..C	M..T..A	T..V..6	D..Y..4
3)	P..O..L	X..R..J	E..R..D	N..U..B	U..W..7	E..Z..5
4)	Q..P..M	A..S..K	F..S..E	O..V..C	V..X..8	F..1..6
5)	R..Q..N	B..T..L	G..T..F	P..W..D	W..Y..A	G..2..7
6)	S..R..O	C..U..M	H..U..G	Q..X..E	X..Z..B	H..3..8
7)	T..S..P	D..V..N	J..V..H	R..Y..F	A..1..C	J..A..A
8)	U..T..Q	E..W..O	K..W..J	S..Z..G	B..2..D	K..B..B
9)	V..U..R	F..X..P	L..X..K	T..1..H	C..3..E	L..C..C
10)	W..V..S	G..Y..Q	M..Y..L	U..2..J	D..A..F	M..D..D
11)	X..W..T	H..Z..R	N..Z..M	V..3..K	E..B..G	N..E..E
12)	A..X..U	J..1..S	O..1..N	W..A..L	F..C..H	O..F..F
13)	B..Y..V	K..2..T	P..2..O	X..B..M	G..D..J	P..G..G
14)	C..Z..W	L..3..U	Q..3..P	A..C..N	H..E..K	Q..H..H
15)	D..1..X	M..A..V	R..A..Q	B..D..O	J..F..L	R..J..J
16)	E..2..Y	N..B..W	S..B..R	C..E..P	K..G..M	S..K..K
17)	F..3..Z	O..C..X	T..C..S	D..F..Q	L..H..N	T..L..L
18)	G..A..1	P..D..Y	U..D..T	E..G..R	M..J..O	U..M..M
19)	H..B..2	Q..E..Z	V..E..U	F..H..S	N..K..P	V..N..N
20)	J..C..3	R..F..1	W..F..V	G..J..T	O..L..Q	W..O..O
21)	K..D..4	S..G..2	X..G..W	H..K..U	P..M..R	X..P..P
22)	L..E..5	T..H..3	A..H..X	J..L..V	Q..N..S	A..Q..Q
23)	M..F..6	U..J..4	B..J..Y	K..M..W	R..O..T	B..R..R
24)	N..G..7	V..K..5	C..K..Z	L..N..X	S..P..U	C..S..S
25)	O..H..8	W..L..6	D..L..1	M..O..Y	T..Q..V	D..T..T
26)	P..J..A	X..M..7	E..M..2	N..P..Z	U..R..W	E..U..U
27)	Q..K..B	A..N..8	F..N..3	O..Q..1	V..S..X	F..V..V
28)	R..L..C	B..O..A	G..O..4	P..R..2	W..T..Y	G..W..W
29)	S..M..D		H..P..5	Q..S..3	X..U..Z	H..X..X
30)	T..N..E		J..Q..6	R..T..4	A..V..1	J..Y..Y
31)	U..O..F		K..R..7		B..W..2	

1942 | 1942

	....JULY....	...AUGUST...	..SEPTEMBER.	..OCTOBER...	..NOVEMBER..	..DECEMBER..
	P--E--I	P--E--I	P--E--I	P--E--I	P--E--I	P--E--I
1)	K..Z..Z	S..3..X	C..C..V	K..E..S	S..H..Q	B..K..N
2)	L..1..1	T..A..Y	D..D..W	L..F..T	T..J..R	C..L..O
3)	M..2..2	U..B..Z	E..E..X	M..G..U	U..K..S	D..M..P
4)	N..3..3	V..C..1	F..F..Y	N..H..V	V..L..T	E..N..Q
5)	O..A..4	W..D..2	G..G..Z	O..J..W	W..M..U	F..O..R
6)	P..B..5	X..E..3	H..H..1	P..K..X	X..N..V	G..P..S
7)	Q..C..6	A..F..4	J..J..2	Q..L..Y	A..O..W	H..Q..T
8)	R..D..7	B..G..5	K..K..3	R..M..Z	B..P..X	J..R..U
9)	S..E..8	C..H..6	L..L..4	S..N..1	C..Q..Y	K..S..V
10)	T..F..A	D..J..7	M..M..5	T..O..2	D..R..Z	L..T..W
11)	U..G..B	E..K..8	N..N..6	U..P..3	E..S..1	M..U..X
12)	V..H..C	F..L..A	O..O..7	V..Q..4	F..T..2	N..V..Y
13)	W..J..D	G..M..B	P..P..8	W..R..5	G..U..3	O..W..Z
14)	X..K..E	H..N..C	Q..Q..A	X..S..6	H..V..4	P..X..1
15)	A..L..F	J..O..D	R..R..B	A..T..7	J..W..5	Q..Y..2
16)	B..M..G	K..P..E	S..S..C	B..U..8	K..X..6	R..Z..3
17)	C..N..H	L..Q..F	T..T..D	C..V..A	L..Y..7	S..1..4
18)	D..O..J	M..R..G	U..U..E	D..W..B	M..Z..8	T..2..5
19)	E..P..K	N..S..H	V..V..F	E..X..C	N..1..A	U..3..6
20)	F..Q..L	O..T..J	W..W..G	F..Y..D	O..2..B	V..A..7
21)	G..R..M	P..U..K	X..X..H	G..Z..E	P..3..C	W..B..8
22)	H..S..N	Q..V..L	A..Y..J	H..1..F	Q..A..D	X..C..A
23)	J..T..O	R..W..M	B..Z..K	J..2..G	R..B..E	A..D..B
24)	K..U..P	S..X..N	C..1..L	K..3..H	S..C..F	B..E..C
25)	L..V..Q	T..Y..O	D..2..M	L..A..J	T..D..G	C..F..D
26)	M..W..R	U..Z..P	E..3..N	M..B..K	U..E..H	D..G..E
27)	N..X..S	V..1..Q	F..A..O	N..C..L	V..F..J	E..H..F
28)	O..Y..T	W..2..R	G..B..P	O..D..M	W..G..K	F..J..G
29)	P..Z..U	X..3..S	H..C..Q	P..E..N	X..H..L	G..K..H
30)	Q..1..V	A..A..T	J..D..R	Q..F..O	A..J..M	H..L..J
31)	R..2..W	B..B..U		R..G..P		J..M..K

CODES: P-PHYSICAL BIORHYTHM CURVE,E-EMOTIONAL BIORHYTHM CURVE,I-INTELLECTUAL BIORHYTHM CURVE

1943 1943

Day	...JANUARY.. P--E--I	..FEBRUARY.. P--E--I	....MARCH... P--E--I	....APRIL... P--E--I	.....MAY.... P--E--I	....JUNE.... P--E--I
1)	K..N..L	S..Q..J	X..Q..D	H..T..B	P..V..7	X..Y..5
2)	L..O..M	T..R..K	A..R..E	J..U..C	Q..W..8	A..Z..6
3)	M..P..N	U..S..L	B..S..F	K..V..D	R..X..A	B..1..7
4)	N..Q..O	V..T..M	C..T..G	L..W..E	S..Y..B	C..2..8
5)	O..R..P	W..U..N	D..U..H	M..X..F	T..Z..C	D..3..A
6)	P..S..Q	X..V..O	E..V..J	N..Y..G	U..1..D	E..A..B
7)	Q..T..R	A..W..P	F..W..K	O..Z..H	V..2..E	F..B..C
8)	R..U..S	B..X..Q	G..X..L	P..1..J	W..3..F	G..C..D
9)	S..V..T	C..Y..R	H..Y..M	Q..2..K	X..A..G	H..D..E
10)	T..W..U	D..Z..S	J..Z..N	R..3..L	A..B..H	J..E..F
11)	U..X..V	E..1..T	K..1..O	S..A..M	B..C..J	K..F..G
12)	V..Y..W	F..2..U	L..2..P	T..B..N	C..D..K	L..G..H
13)	W..Z..X	G..3..V	M..3..Q	U..C..O	D..E..L	M..H..J
14)	X..1..Y	H..A..W	N..A..R	V..D..P	E..F..M	N..J..K
15)	A..2..Z	J..B..X	O..B..S	W..E..Q	F..G..N	O..K..L
16)	B..3..1	K..C..Y	P..C..T	X..F..R	G..H..O	P..L..M
17)	C..A..2	L..D..Z	Q..D..U	A..G..S	H..J..P	Q..M..N
18)	D..B..3	M..E..1	R..E..V	B..H..T	J..K..Q	R..N..O
19)	E..C..4	N..F..2	S..F..W	C..J..U	K..L..R	S..O..P
20)	F..D..5	O..G..3	T..G..X	D..K..V	L..M..S	T..P..Q
21)	G..E..6	P..H..4	U..H..Y	E..L..W	M..N..T	U..Q..R
22)	H..F..7	Q..J..5	V..J..Z	F..M..X	N..O..U	V..R..S
23)	J..G..8	R..K..6	W..K..1	G..N..Y	O..P..V	W..S..T
24)	K..H..A	S..L..7	X..L..2	H..O..Z	P..Q..W	X..T..U
25)	L..J..B	T..M..8	A..M..3	J..P..1	Q..R..X	A..U..V
26)	M..K..C	U..N..A	B..N..4	K..Q..2	R..S..Y	B..V..W
27)	N..L..D	V..O..B	C..O..5	L..R..3	S..T..Z	C..W..X
28)	O..M..E	W..P..C	D..P..6	M..S..4	T..U..1	D..X..Y
29)	P..N..F		E..Q..7	N..T..5	U..V..2	E..Y..Z
30)	Q..O..G		F..R..8	O..U..6	V..W..3	F..Z..1
31)	R..P..H		G..S..A		W..X..4	

....JULY....	P--E--I	...AUGUST...	P--E--I	..SEPTEMBER.	P--E--I	..OCTOBER...	P--E--I	..NOVEMBER..	P--E--I	..DECEMBER..	P--E--I
1)	G..1..2	1)	P..A..Z	1)	X..D..X	1)	G..F..U	1)	P,.J..S	1)	W..L..P
2)	H..2..3	2)	Q..B..1	2)	A..E..Y	2)	H..G..V	2)	Q..K..T	2)	X..M..Q
3)	J..3..4	3)	R..C..2	3)	B..F..Z	3)	J..H..W	3)	R..L..U	3)	A..N..R
4)	K..A..5	4)	S..D..3	4)	C..G..1	4)	K..J..X	4)	S..M..V	4)	B..O..S
5)	L..B..6	5)	T..E..4	5)	D..H..2	5)	L..K..Y	5)	T..N..W	5)	C..P..T
6)	M..C..7	6)	U..F..5	6)	E..J..3	6)	M..L..Z	6)	U..O..X	6)	D..Q..U
7)	N..D..8	7)	V..G..6	7)	F..K..4	7)	N..M..1	7)	V..P..Y	7)	E..R..V
8)	O..E..A	8)	W..H..7	8)	G..L..5	8)	O..N..2	8)	W..Q..Z	8)	F..S..W
9)	P..F..B	9)	X..J..8	9)	H..M..6	9)	P..O..3	9)	X..R..1	9)	G..T..X
10)	Q..G..C	10)	A..K..A	10)	J..N..7	10)	Q..P..4	10)	A..S..2	10)	H..U..Y
11)	R..H..D	11)	B..L..B	11)	K..O..8	11)	R..Q..5	11)	B..T..3	11)	J..V..Z
12)	S..J..E	12)	C..M..C	12)	L..P..A	12)	S..R..6	12)	C..U..4	12)	K..W..1
13)	T..K..F	13)	D..N..D	13)	M..Q..B	13)	T..S..7	13)	D..V..5	13)	L..X..2
14)	U..L..G	14)	E..O..E	14)	N..R..C	14)	U..T..8	14)	E..W..6	14)	M..Y..3
15)	V..M..H	15)	F..P..F	15)	O..S..D	15)	V..U..A	15)	F..X..7	15)	N..Z..4
16)	W..N..J	16)	G..Q..G	16)	P..T..E	16)	W..V..B	16)	G..Y..8	16)	O..1..5
17)	X..O..K	17)	H..R..H	17)	Q..U..F	17)	X..W..C	17)	H..Z..A	17)	P..2..6
18)	A..P..L	18)	J..S..J	18)	R..V..G	18)	A..X..D	18)	J..1..B	18)	Q..3..7
19)	B..Q..M	19)	K..T..K	19)	S..W..H	19)	B..Y..E	19)	K..2..C	19)	R..A..8
20)	C..R..N	20)	L..U..L	20)	T..X..J	20)	C..Z..F	20)	L..3..D	20)	S..B..A
21)	D..S..O	21)	M..V..M	21)	U..Y..K	21)	D..1..G	21)	M..A..E	21)	T..C..B
22)	E..T..P	22)	N..W..N	22)	V..Z..L	22)	E..2..H	22)	N..B..F	22)	U..D..C
23)	F..U..Q	23)	O..X..O	23)	W..1..M	23)	F..3..J	23)	O..C..G	23)	V..E..D
24)	G..V..R	24)	P..Y..P	24)	X..2..N	24)	G..A..K	24)	P..D..H	24)	W..F..E
25)	H..W..S	25)	Q..Z..Q	25)	A..3..O	25)	H..B..L	25)	Q..E..J	25)	X..G..F
26)	J..X..T	26)	R..1..R	25)	B..A..P	26)	J..C..M	26)	R..F..K	26)	A..H..G
27)	K..Y..U	27)	S..2..S	27)	C..B..Q	27)	K..D..N	27)	S..G..L	27)	B..J..H
28)	L..Z..V	28)	T..3..T	23)	D..C..R	28)	L..E..O	28)	T..H..M	28)	C..K..J
29)	M..1..W	29)	U..A..U	29)	E..D..S	29)	M..F..P	29)	U..J..N	29)	D..L..K
30)	N..2..X	30)	V..B..V	30)	F..E..T	30)	N..G..Q	30)	V..K..O	30)	E..M..L
31)	O..3..Y	31)	W..C..W			31)	O..H..R			31)	F..N..M

CODES: P-PHYSICAL BIORHYTHM CURVE,E-EMOTIONAL BIORHYTHM CURVE,I-INTELLECTUAL BIORHYTHM CURVE

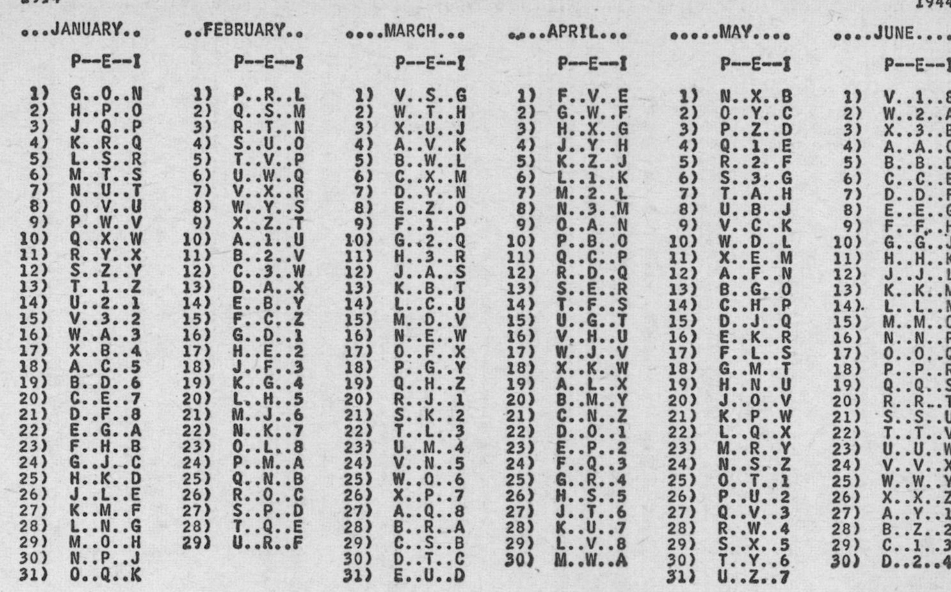

1944											1944
...JANUARY..		..FEBRUARY..		....MARCH...		....APRIL...		.....MAY....		....JUNE....	
	P--E--I		P--E--I		P--E--I		P--E--I		P--E--I		P--E--I
1)	G..O..N	1)	P..R..L	1)	V..S..G	1)	F..V..E	1)	N..X..B	1)	V..1..8
2)	H..P..O	2)	Q..S..M	2)	W..T..H	2)	G..W..F	2)	O..Y..C	2)	W..2..A
3)	J..Q..P	3)	R..T..N	3)	X..U..J	3)	H..X..G	3)	P..Z..D	3)	X..3..B
4)	K..R..Q	4)	S..U..O	4)	A..V..K	4)	J..Y..H	4)	Q..1..E	4)	A..A..C
5)	L..S..R	5)	T..V..P	5)	B..W..L	5)	K..Z..J	5)	R..2..F	5)	B..B..D
6)	M..T..S	6)	U..W..Q	6)	C..X..M	6)	L..1..K	6)	S..3..G	6)	C..C..E
7)	N..U..T	7)	V..X..R	7)	D..Y..N	7)	M..2..L	7)	T..A..H	7)	D..D..F
8)	O..V..U	8)	W..Y..S	8)	E..Z..O	8)	N..3..M	8)	U..B..J	8)	E..E..G
9)	P..W..V	9)	X..Z..T	9)	F..1..P	9)	O..A..N	9)	V..C..K	9)	F..F..H
10)	Q..X..W	10)	A..1..U	10)	G..2..Q	10)	P..B..O	10)	W..D..L	10)	G..G..J
11)	R..Y..X	11)	B..2..V	11)	H..3..R	11)	Q..C..P	11)	X..E..M	11)	H..H..K
12)	S..Z..Y	12)	C..3..W	12)	J..A..S	12)	R..D..Q	12)	A..F..N	12)	J..J..L
13)	T..1..Z	13)	D..A..X	13)	K..B..T	13)	S..E..R	13)	B..G..O	13)	K..K..M
14)	U..2..1	14)	E..B..Y	14)	L..C..U	14)	T..F..S	14)	C..H..P	14)	L..L..N
15)	V..3..2	15)	F..C..Z	15)	M..D..V	15)	U..G..T	15)	D..J..Q	15)	M..M..O
16)	W..A..3	16)	G..D..1	16)	N..E..W	16)	V..H..U	16)	E..K..R	16)	N..N..P
17)	X..B..4	17)	H..E..2	17)	O..F..X	17)	W..J..V	17)	F..L..S	17)	O..O..Q
18)	A..C..5	18)	J..F..3	18)	P..G..Y	18)	X..K..W	18)	G..M..T	18)	P..P..R
19)	B..D..6	19)	K..G..4	19)	Q..H..Z	19)	A..L..X	19)	H..N..U	19)	Q..Q..S
20)	C..E..7	20)	L..H..5	20)	R..J..1	20)	B..M..Y	20)	J..O..V	20)	R..R..T
21)	D..F..8	21)	M..J..6	21)	S..K..2	21)	C..N..Z	21)	K..P..W	21)	S..S..U
22)	E..G..A	22)	N..K..7	22)	T..L..3	22)	D..O..1	22)	L..Q..X	22)	T..T..V
23)	F..H..B	23)	O..L..8	23)	U..M..4	23)	E..P..2	23)	M..R..Y	23)	U..U..W
24)	G..J..C	24)	P..M..A	24)	V..N..5	24)	F..Q..3	24)	N..S..Z	24)	V..V..X
25)	H..K..D	25)	Q..N..B	25)	W..O..6	25)	G..R..4	25)	O..T..1	25)	W..W..Y
26)	J..L..E	26)	R..O..C	26)	X..P..7	26)	H..S..5	26)	P..U..2	26)	X..X..Z
27)	K..M..F	27)	S..P..D	27)	A..Q..8	27)	J..T..6	27)	Q..V..3	27)	A..Y..1
28)	L..N..G	28)	T..Q..E	28)	B..R..A	28)	K..U..7	28)	R..W..4	28)	B..Z..2
29)	M..O..H	29)	U..R..F	29)	C..S..B	29)	L..V..8	29)	S..X..5	29)	C..1..3
30)	N..P..J			30)	D..T..C	30)	M..W..A	30)	T..Y..6	30)	D..2..4
31)	O..Q..K			31)	E..U..D			31)	U..Z..7		

1944 1944

	....JULY....	...AUGUST...	..SEPTEMBER.	..OCTOBER...	..NOVEMBER..	..DECEMBER..
	P--E--I	P--E--I	P--E--I	P--E--I	P--E--I	P--E--I
1)	E..3..5	N..C..3	V..F..1	E..H..X	N..L..V	U..N..S
2)	F..A..6	O..D..4	W..G..2	F..J..Y	O..M..W	V..O..T
3)	G..B..7	P..E..5	X..H..3	G..K..Z	P..N..X	W..P..U
4)	H..C..8	Q..F..6	A..J..4	H..L..1	Q..O..Y	X..Q..V
5)	J..D..A	R..G..7	B..K..5	J..M..2	R..P..Z	A..R..W
6)	K..E..B	S..H..8	C..L..6	K..N..3	S..Q..1	B..S..X
7)	L..F..C	T..J..A	D..M..7	L..O..4	T..R..2	C..T..Y
8)	M..G..D	U..K..B	E..N..8	M..P..5	U..S..3	D..U..Z
9)	N..H..E	V..L..C	F..O..A	N..Q..6	V..T..4	E..V..1
10)	O..J..F	W..M..D	G..P..B	O..R..7	W..U..5	F..W..2
11)	P..K..G	X..N..E	H..Q..C	P..S..8	X..V..6	G..X..3
12)	Q..L..H	A..O..F	J..R..D	Q..T..A	A..W..7	H..Y..4
13)	R..M..J	B..P..G	K..S..E	R..U..B	B..X..8	J..Z..5
14)	S..N..K	C..Q..H	L..T..F	S..V..C	C..Y..A	K..1..6
15)	T..O..L	D..R..J	M..U..G	T..W..D	D..Z..B	L..2..7
16)	U..P..M	E..S..K	N..V..H	U..X..E	E..1..C	M..3..8
17)	V..Q..N	F..T..L	O..W..J	V..Y..F	F..2..D	N..A..A
18)	W..R..O	G..U..M	P..X..K	W..Z..G	G..3..E	O..B..B
19)	X..S..P	H..V..N	Q..Y..L	X..1..H	H..A..F	P..C..C
20)	A..T..Q	J..W..O	R..Z..M	A..2..J	J..B..G	Q..D..D
21)	B..U..R	K..X..P	S..1..N	B..3..K	K..C..H	R..E..E
22)	C..V..S	L..Y..Q	T..2..O	C..A..L	L..D..J	S..F..F
23)	D..W..T	M..Z..R	U..3..P	D..B..M	M..E..K	T..G..G
24)	E..X..U	N..1..S	V..A..Q	E..C..N	N..F..L	U..H..H
25)	F..Y..V	O..2..T	W..B..R	F..D..O	O..G..M	V..J..J
26)	G..Z..W	P..3..U	X..C..S	G..E..P	P..H..N	W..K..K
27)	H..1..X	Q..A..V	A..D..T	H..F..Q	Q..J..O	X..L..L
28)	J..2..Y	R..B..W	B..E..U	J..G..R	R..K..P	A..M..M
29)	K..3..Z	S..C..X	C..F..V	K..H..S	S..L..Q	B..N..N
30)	L..A..1	T..D..Y	D..G..W	L..J..T	T..M..R	C..O..O
31)	M..B..2	U..E..Z		M..K..U		D..P..P

CODES: P-PHYSICAL BIORHYTHM CURVE,E-EMOTIONAL BIORHYTHM CURVE,I-INTELLECTUAL BIORHYTHM CURVE

1945 1945

	...JANUARY..	..FEBRUARY..	....MARCH...	....APRIL...	.....MAY....	....JUNE....
	P--E--I	P--E--I	P--E--I	P--E--I	P--E--I	P--E--I
1)	E..Q..Q	N..T..O	S..T..J	C..W..G	K..Y..D	S..2..B
2)	F..R..R	O..U..P	T..U..K	D..X..H	L..Z..E	T..3..C
3)	G..S..S	P..V..Q	U..V..L	E..Y..J	M..1..F	U..A..D
4)	H..T..T	Q..W..R	V..W..M	F..Z..K	N..2..G	V..B..E
5)	J..U..U	R..X..S	W..X..N	G..1..L	O..3..H	W..C..F
6)	K..V..V	S..Y..T	X..Y..O	H..2..M	P..A..J	X..D..G
7)	L..W..W	T..Z..U	A..Z..P	J..3..N	Q..B..K	A..E..H
8)	M..X..X	U..1..V	B..1..Q	K..A..O	R..C..L	B..F..J
9)	N..Y..Y	V..2..W	C..2..R	L..B..P	S..D..M	C..G..K
10)	O..Z..Z	W..3..X	D..3..S	M..C..Q	T..E..N	D..H..L
11)	P..1..1	X..A..Y	E..A..T	N..D..R	U..F..O	E..J..M
12)	Q..2..2	A..B..Z	F..B..U	O..E..S	V..G..P	F..K..N
13)	R..3..3	B..C..1	G..C..V	P..F..T	W..H..Q	G..L..O
14)	S..A..4	C..D..2	H..D..W	Q..G..U	X..J..R	H..M..P
15)	T..B..5	D..E..3	J..E..X	R..H..V	A..K..S	J..N..Q
16)	U..C..6	E..F..4	K..F..Y	S..J..W	B..L..T	K..O..R
17)	V..D..7	F..G..5	L..G..Z	T..K..X	C..M..U	L..P..S
18)	W..E..8	G..H..6	M..H..1	U..L..Y	D..N..V	M..Q..T
19)	X..F..A	H..J..7	N..J..2	V..M..Z	E..O..W	N..R..U
20)	A..G..B	J..K..8	O..K..3	W..N..1	F..P..X	O..S..V
21)	B..H..C	K..L..A	P..L..4	X..O..2	G..Q..Y	P..T..W
22)	C..J..D	L..M..B	Q..M..5	A..P..3	H..R..Z	Q..U..X
23)	D..K..E	M..N..C	R..N..6	B..Q..4	J..S..1	R..V..Y
24)	E..L..F	N..O..D	S..O..7	C..R..5	K..T..2	S..W..Z
25)	F..M..G	O..P..E	T..P..8	D..S..6	L..U..3	T..X..1
26)	G..N..H	P..Q..F	U..Q..A	E..T..7	M..V..4	U..Y..2
27)	H..O..J	Q..R..G	V..R..B	F..U..8	N..W..5	V..Z..3
28)	J..P..K	R..S..H	W..S..C	G..V..A	O..X..6	W..1..4
29)	K..Q..L		X..T..D	H..W..B	P..Y..7	X..2..5
30)	L..R..M		A..U..E	J..X..C	Q..Z..8	A..3..6
31)	M..S..N		B..V..F		R..1..A	

....JULY....	...AUGUST...	..SEPTEMBER.	..OCTOBER...	..NOVEMBER..	..DECEMBER..
P--E--I	P--E--I	P--E--I	P--E--I	P--E--I	P--E--I
1) B..A..7	1) K..D..5	1) S..G..3	1) B..J..Z	1) K..M..X	1) R..O..U
2) C..B..8	2) L..E..6	2) T..H..4	2) C..K..1	2) L..N..Y	2) S..P..V
3) D..C..A	3) M..F..7	3) U..J..5	3) D..L..2	3) M..O..Z	3) T..Q..W
4) E..D..B	4) N..G..8	4) V..K..6	4) E..M..3	4) N..P..1	4) U..R..X
5) F..E..C	5) O..H..A	5) W..L..7	5) F..N..4	5) O..Q..2	5) V..S..Y
6) G..F..D	6) P..J..B	6) X..M..8	6) G..O..5	6) P..R..3	6) W..T..Z
7) H..G..E	7) Q..K..C	7) A..N..A	7) H..P..6	7) Q..S..4	7) X..U..1
8) J..H..F	8) R..L..D	8) B..O..B	8) J..Q..7	8) R..T..5	8) A..V..2
9) K..J..G	9) S..M..E	9) C..P..C	9) K..R..8	9) S..U..6	9) B..W..3
10) L..K..H	10) T..N..F	10) D..Q..D	10) L..S..A	10) T..V..7	10) C..X..4
11) M..L..J	11) U..O..G	11) E..R..E	11) M..T..B	11) U..W..8	11) D..Y..5
12) N..M..K	12) V..P..H	12) F..S..F	12) N..U..C	12) V..X..A	12) E..Z..6
13) O..N..L	13) W..Q..J	13) G..T..G	13) O..V..D	13) W..Y..B	13) F..1..7
14) P..O..M	14) X..R..K	14) H..U..H	14) P..W..E	14) X..Z..C	14) G..2..8
15) Q..P..N	15) A..S..L	15) J..V..J	15) Q..X..F	15) A..1..D	15) H..3..A
16) R..Q..O	16) B..T..M	16) K..W..K	16) R..Y..G	16) B..2..E	16) J..A..B
17) S..R..P	17) C..U..N	17) L..X..L	17) S..Z..H	17) C..3..F	17) K..B..C
18) T..S..Q	18) D..V..O	18) M..Y..M	18) T..1..J	18) D..A..G	18) L..C..D
19) U..T..R	19) E..W..P	19) N..Z..N	19) U..2..K	19) E..B..H	19) M..D..E
20) V..U..S	20) F..X..Q	20) O..1..O	20) V..3..L	20) F..C..J	20) N..E..F
21) W..V..T	21) G..Y..R	21) P..2..P	21) W..A..M	21) G..D..K	21) O..F..G
22) X..W..U	22) H..Z..S	22) Q..3..Q	22) X..B..N	22) H..E..L	22) P..G..H
23) A..X..V	23) J..1..T	23) R..A..R	23) A..C..O	23) J..F..M	23) Q..H..J
24) B..Y..W	24) K..2..U	24) S..B..S	24) B..D..P	24) K..G..N	24) R..J..K
25) C..Z..X	25) L..3..V	25) T..C..T	25) C..E..Q	25) L..H..O	25) S..K..L
26) D..1..Y	26) M..A..W	26) U..D..U	26) D..F..R	26) M..J..P	26) T..L..M
27) E..2..Z	27) N..B..X	27) V..E..V	27) E..G..S	27) N..K..Q	27) U..M..N
28) F..3..1	28) O..C..Y	28) W..F..W	28) F..H..T	28) O..L..R	28) V..N..O
29) G..A..2	29) P..D..Z	29) X..G..X	29) G..J..U	29) P..M..S	29) W..O..P
30) H..B..3	30) Q..E..1	30) A..H..Y	30) H..K..V	30) Q..N..T	30) X..P..Q
31) J..C..4	31) R..F..2		31) J..L..W		31) A..Q..R

CODES: P-PHYSICAL BIORHYTHM CURVE,E-EMOTIONAL BIORHYTHM CURVE,I-INTELLECTUAL BIORHYTHM CURVE

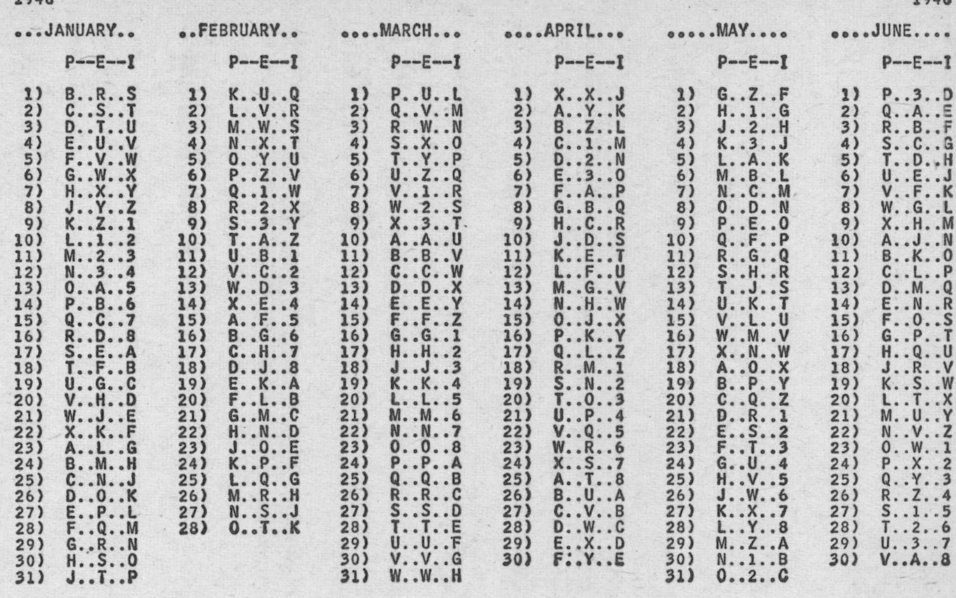

1946						1946
	...JANUARY..	..FEBRUARY..	....MARCH...	....APRIL...	.....MAY....	....JUNE....
	P--E--I	P--E--I	P--E--I	P--E--I	P--E--I	P--E--I
1)	B..R..S	K..U..Q	P..U..L	X..X..J	G..Z..F	P..3..D
2)	C..S..T	L..V..R	Q..V..M	A..Y..K	H..1..G	Q..A..E
3)	D..T..U	M..W..S	R..W..N	B..Z..L	J..2..H	R..B..F
4)	E..U..V	N..X..T	S..X..O	C..1..M	K..3..J	S..C..G
5)	F..V..W	O..Y..U	T..Y..P	D..2..N	L..A..K	T..D..H
6)	G..W..X	P..Z..V	U..Z..Q	E..3..O	M..B..L	U..E..J
7)	H..X..Y	Q..1..W	V..1..R	F..A..P	N..C..M	V..F..K
8)	J..Y..Z	R..2..X	W..2..S	G..B..Q	O..D..N	W..G..L
9)	K..Z..1	S..3..Y	X..3..T	H..C..R	P..E..O	X..H..M
10)	L..1..2	T..A..Z	A..A..U	J..D..S	Q..F..P	A..J..N
11)	M..2..3	U..B..1	B..B..V	K..E..T	R..G..Q	B..K..O
12)	N..3..4	V..C..2	C..C..W	L..F..U	S..H..R	C..L..P
13)	O..A..5	W..D..3	D..D..X	M..G..V	T..J..S	D..M..Q
14)	P..B..6	X..E..4	E..E..Y	N..H..W	U..K..T	E..N..R
15)	Q..C..7	A..F..5	F..F..Z	O..J..X	V..L..U	F..O..S
16)	R..D..8	B..G..6	G..G..1	P..K..Y	W..M..V	G..P..T
17)	S..E..A	C..H..7	H..H..2	Q..L..Z	X..N..W	H..Q..U
18)	T..F..B	D..J..8	J..J..3	R..M..1	A..O..X	J..R..V
19)	U..G..C	E..K..A	K..K..4	S..N..2	B..P..Y	K..S..W
20)	V..H..D	F..L..B	L..L..5	T..O..3	C..Q..Z	L..T..X
21)	W..J..E	G..M..C	M..M..6	U..P..4	D..R..1	M..U..Y
22)	X..K..F	H..N..D	N..N..7	V..Q..5	E..S..2	N..V..Z
23)	A..L..G	J..O..E	O..O..8	W..R..6	F..T..3	O..W..1
24)	B..M..H	K..P..F	P..P..A	X..S..7	G..U..4	P..X..2
25)	C..N..J	L..Q..G	Q..Q..B	A..T..8	H..V..5	Q..Y..3
26)	D..O..K	M..R..H	R..R..C	B..U..A	J..W..6	R..Z..4
27)	E..P..L	N..S..J	S..S..D	C..V..B	K..X..7	S..1..5
28)	F..Q..M	O..T..K	T..T..E	D..W..C	L..Y..8	T..2..6
29)	G..R..N		U..U..F	E..X..D	M..Z..A	U..3..7
30)	H..S..O		V..V..G	F..Y..E	N..1..B	V..A..8
31)	J..T..P		W..W..H		O..2..C	

1946						1946
	....JULY....	...AUGUST...	..SEPTEMBER.	..OCTOBER...	..NOVEMBER..	..DECEMBER..
	P--E--I	P--E--I	P--E--I	P--E--I	P--E--I	P--E--I
1)	W..B..A	G..E..7	P..H..5	W..K..2	G..N..Z	O..P..W
2)	X..C..B	H..F..8	Q..J..6	X..L..3	H..O..1	P..Q..X
3)	A..D..C	J..G..A	R..K..7	A..M..4	J..P..2	Q..R..Y
4)	B..E..D	K..H..B	S..L..8	B..N..5	K..Q..3	R..S..Z
5)	C..F..E	L..J..C	T..M..A	C..O..6	L..R..4	S..T..1
6)	D..G..F	M..K..D	U..N..B	D..P..7	M..S..5	T..U..2
7)	E..H..G	N..L..E	V..O..C	E..Q..8	N..T..6	U..V..3
8)	F..J..H	O..M..F	W..P..D	F..R..A	O..U..7	V..W..4
9)	G..K..J	P..N..G	X..Q..E	G..S..B	P..V..8	W..X..5
10)	H..L..K	Q..O..H	A..R..F	H..T..C	Q..W..A	X..Y..6
11)	J..M..L	R..P..J	B..S..G	J..U..D	R..X..B	A..Z..7
12)	K..N..M	S..Q..K	C..T..H	K..V..E	S..Y..C	B..1..8
13)	L..O..N	T..R..L	D..U..J	L..W..F	T..Z..D	C..2..A
14)	M..P..O	U..S..M	E..V..K	M..X..G	U..1..E	D..3..B
15)	N..Q..P	V..T..N	F..W..L	N..Y..H	V..2..F	E..A..C
16)	O..R..Q	W..U..O	G..X..M	O..Z..J	W..3..G	F..B..D
17)	P..S..R	X..V..P	H..Y..N	P..1..K	X..A..H	G..C..E
18)	Q..T..S	A..W..Q	J..Z..O	Q..2..L	A..B..J	H..D..F
19)	R..U..T	B..X..R	K..1..P	R..3..M	B..C..K	J..E..G
20)	S..V..U	C..Y..S	L..2..Q	S..A..N	C..D..L	K..F..H
21)	T..W..V	D..Z..T	M..3..R	T..B..O	D..E..M	L..G..J
22)	U..X..W	E..1..U	N..A..S	U..C..P	E..F..N	M..H..K
23)	V..Y..X	F..2..V	O..B..T	V..D..Q	F..G..O	N..J..L
24)	W..Z..Y	G..3..W	P..C..U	W..E..R	G..H..P	O..K..M
25)	X..1..Z	H..A..X	Q..D..V	X..F..S	H..J..Q	P..L..N
26)	A..2..1	J..B..Y	R..E..W	A..G..T	J..K..R	Q..M..O
27)	B..3..2	K..C..Z	S..F..X	B..H..U	K..L..S	R..N..P
28)	C..A..3	L..D..1	T..G..Y	C..J..V	L..M..T	S..O..Q
29)	D..B..4	M..E..2	U..H..Z	D..K..W	M..N..U	T..P..R
30)	E..C..5	N..F..3	V..J..1	E..L..X	N..O..V	U..Q..S
31)	F..D..6	O..G..4		F..M..Y		V..R..T

CODES: P-PHYSICAL BIORHYTHM CURVE,E-EMOTIONAL BIORHYTHM CURVE,I-INTELLECTUAL BIORHYTHM CURVE

1947											1947
...JANUARY..		..FEBRUARY..		....MARCH...		....APRIL...		.....MAY....		....JUNE....	
	P--E--I		P--E--I		P--E--I		P--E--I		P--E--I		P--E--I
1)	W..S..U	1)	G..V..S	1)	M..V..N	1)	U..Y..L	1)	D..1..H	1)	M..A..F
2)	X..T..V	2)	H..W..T	2)	N..W..O	2)	V..Z..M	2)	E..2..J	2)	N..B..G
3)	A..U..W	3)	J..X..U	3)	O..X..P	3)	W..1..N	3)	F..3..K	3)	O..C..H
4)	B..V..X	4)	K..Y..V	4)	P..Y..Q	4)	X..2..O	4)	G..A..L	4)	P..D..J
5)	C..W..Y	5)	L..Z..W	5)	Q..Z..R	5)	A..3..P	5)	H..B..M	5)	Q..E..K
6)	D..X..Z	6)	M..1..X	6)	R..1..S	6)	B..A..Q	6)	J..C..N	6)	R..F..L
7)	E..Y..1	7)	N..2..Y	7)	S..2..T	7)	C..B..R	7)	K..D..O	7)	S..G..M
8)	F..Z..2	8)	O..3..Z	8)	T..3..U	8)	D..C..S	8)	L..E..P	8)	T..H..N
9)	G..1..3	9)	P..A..1	9)	U..A..V	9)	E..D..T	9)	M..F..Q	9)	U..J..O
10)	H..2..4	10)	Q..B..2	10)	V..B..W	10)	F..E..U	10)	N..G..R	10)	V..K..P
11)	J..3..5	11)	R..C..3	11)	W..C..X	11)	G..F..V	11)	O..H..S	11)	W..L..Q
12)	K..A..6	12)	S..D..4	12)	X..D..Y	12)	H..G..W	12)	P..J..T	12)	X..M..R
13)	L..B..7	13)	T..E..5	13)	A..E..Z	13)	J..H..X	13)	Q..K..U	13)	A..N..S
14)	M..C..8	14)	U..F..6	14)	B..F..1	14)	K..J..Y	14)	R..L..V	14)	B..O..T
15)	N..D..A	15)	V..G..7	15)	C..G..2	15)	L..K..Z	15)	S..M..W	15)	C..P..U
16)	O..E..B	16)	W..H..8	16)	D..H..3	16)	M..L..1	16)	T..N..X	16)	D..Q..V
17)	P..F..C	17)	X..J..A	17)	E..J..4	17)	N..M..2	17)	U..O..Y	17)	E..R..W
18)	Q..G..D	18)	A..K..B	18)	F..K..5	18)	O..N..3	18)	V..P..Z	18)	F..S..X
19)	R..H..E	19)	B..L..C	19)	G..L..6	19)	P..O..4	19)	W..Q..1	19)	G..T..Y
20)	S..J..F	20)	C..M..D	20)	H..M..7	20)	Q..P..5	20)	X..R..2	20)	H..U..Z
21)	T..K..G	21)	D..N..E	21)	J..N..8	21)	R..Q..6	21)	A..S..3	21)	J..V..1
22)	U..L..H	22)	E..O..F	22)	K..O..A	22)	S..R..7	22)	B..T..4	22)	K..W..2
23)	V..M..J	23)	F..P..G	23)	L..P..B	23)	T..S..8	23)	C..U..5	23)	L..X..3
24)	W..N..K	24)	G..Q..H	24)	M..Q..C	24)	U..T..A	24)	D..V..6	24)	M..Y..4
25)	X..O..L	25)	H..R..J	25)	N..R..D	25)	V..U..B	25)	E..W..7	25)	N..Z..5
26)	A..P..M	26)	J..S..K	26)	O..S..E	26)	W..V..C	26)	F..X..8	26)	O..1..6
27)	B..Q..N	27)	K..T..L	27)	P..T..F	27)	X..W..D	27)	G..Y..A	27)	P..2..7
28)	C..R..O	28)	L..U..M	28)	Q..U..G	28)	A..X..E	28)	H..Z..B	28)	Q..3..8
29)	D..S..P			29)	R..V..H	29)	B..Y..F	29)	J..1..C	29)	R..A..A
30)	E..T..Q			30)	S..W..J	30)	C..Z..G	30)	K..2..D	30)	S..B..B
31)	F..U..R			31)	T..X..K			31)	L..3..E		

	....JULY.... P--E--I	...AUGUST... P--E--I	..SEPTEMBER. P--E--I	..OCTOBER... P--E--I	..NOVEMBER.. P--E--I	..DECEMBER.. P--E--I
1)	T..C..C	D..F..A	M..J..7	T..L..4	D..O..2	L..Q..Y
2)	U..D..D	E..G..B	N..K..8	U..M..5	E..P..3	M..R..Z
3)	V..E..E	F..H..C	O..L..A	V..N..6	F..Q..4	N..S..1
4)	W..F..F	G..J..D	P..M..B	W..O..7	G..R..5	O..T..2
5)	X..G..G	H..K..E	Q..N..C	X..P..8	H..S..6	P..U..3
6)	A..H..H	J..L..F	R..O..D	A..Q..A	J..T..7	Q..V..4
7)	B..J..J	K..M..G	S..P..E	B..R..B	K..U..8	R..W..5
8)	C..K..K	L..N..H	T..Q..F	C..S..C	L..V..A	S..X..6
9)	D..L..L	M..O..J	U..R..G	D..T..D	M..W..B	T..Y..7
10)	E..M..M	N..P..K	V..S..H	E..U..E	N..X..C	U..Z..8
11)	F..N..N	O..Q..L	W..T..J	F..V..F	O..Y..D	V..1..A
12)	G..O..O	P..R..M	X..U..K	G..W..G	P..Z..E	W..2..B
13)	H..P..P	Q..S..N	A..V..L	H..X..H	Q..1..F	X..3..C
14)	J..Q..Q	R..T..O	B..W..M	J..Y..J	R..2..G	A..A..D
15)	K..R..R	S..U..P	C..X..N	K..Z..K	S..3..H	B..B..E
16)	L..S..S	T..V..Q	D..Y..O	L..1..L	T..A..J	C..C..F
17)	M..T..T	U..W..R	E..Z..P	M..2..M	U..B..K	D..D..G
18)	N..U..U	V..X..S	F..1..Q	N..3..N	V..C..L	E..E..H
19)	O..V..V	W..Y..T	G..2..R	O..A..O	W..D..M	F..F..J
20)	P..W..W	X..Z..U	H..3..S	P..B..P	X..E..N	G..G..K
21)	Q..X..X	A..1..V	J..A..T	Q..C..Q	A..F..O	H..H..L
22)	R..Y..Y	B..2..W	K..B..U	R..D..R	B..G..P	J..J..M
23)	S..Z..Z	C..3..X	L..C..V	S..E..S	C..H..Q	K..K..N
24)	T..1..1	D..A..Y	M..D..W	T..F..T	D..J..R	L..L..O
25)	U..2..2	E..B..Z	N..E..X	U..G..U	E..K..S	M..M..P
26)	V..3..3	F..C..1	O..F..Y	V..H..V	F..L..T	N..N..Q
27)	W..A..4	G..D..2	P..G..Z	W..J..W	G..M..U	O..O..R
28)	X..B..5	H..E..3	Q..H..1	X..K..X	H..N..V	P..P..S
29)	A..C..6	J..F..4	R..J..2	A..L..Y	J..O..W	Q..Q..T
30)	B..D..7	K..G..5	S..K..3	B..M..Z	K..P..X	R..R..U
31)	C..E..8	L..H..6		C..N..1		S..S..V

CODES: P-PHYSICAL BIORHYTHM CURVE, E-EMOTIONAL BIORHYTHM CURVE, I-INTELLECTUAL BIORHYTHM CURVE

1948 | 1948

	...JANUARY..	..FEBRUARY..	....MARCH...	....APRIL...	.....MAY....	....JUNE....
	P--E--I	P--E--I	P--E--I	P--E--I	P--E--I	P--E--I
1)	T..T..W	D..W..U	K..X..Q	S..1..O	B..3..L	K..C..J
2)	U..U..X	E..X..V	L..Y..R	T..2..P	C..A..M	L..D..K
3)	V..V..Y	F..Y..W	M..Z..S	U..3..Q	D..B..N	M..E..L
4)	W..W..Z	G..Z..X	N..1..T	V..A..R	E..C..O	N..F..M
5)	X..X..1	H..1..Y	O..2..U	W..B..S	F..D..P	O..G..N
6)	A..Y..2	J..2..Z	P..3..V	X..C..T	G..E..Q	P..H..O
7)	B..Z..3	K..3..1	Q..A..W	A..D..U	H..F..R	Q..J..P
8)	C..1..4	L..A..2	R..B..X	B..E..V	J..G..S	R..K..Q
9)	D..2..5	M..B..3	S..C..Y	C..F..W	K..H..T	S..L..R
10)	E..3..6	N..C..4	T..D..Z	D..G..X	L..J..U	T..M..S
11)	F..A..7	O..D..5	U..E..1	E..H..Y	M..K..V	U..N..T
12)	G..B..8	P..E..6	V..F..2	F..J..Z	N..L..W	V..O..U
13)	H..C..A	Q..F..7	W..G..3	G..K..1	O..M..X	W..P..V
14)	J..D..B	R..G..8	X..H..4	H..L..2	P..N..Y	X..Q..W
15)	K..E..C	S..H..A	A..J..5	J..M..3	Q..O..Z	A..R..X
16)	L..F..D	T..J..B	B..K..6	K..N..4	R..P..1	B..S..Y
17)	M..G..E	U..K..C	C..L..7	L..O..5	S..Q..2	C..T..Z
18)	N..H..F	V..L..D	D..M..8	M..P..6	T..R..3	D..U..1
19)	O..J..G	W..M..E	E..N..A	N..Q..7	U..S..4	E..V..2
20)	P..K..H	X..N..F	F..O..B	O..R..8	V..T..5	F..W..3
21)	Q..L..J	A..O..G	G..P..C	P..S..A	W..U..6	G..X..4
22)	R..M..K	B..P..H	H..Q..D	Q..T..B	X..V..7	H..Y..5
23)	S..N..L	C..Q..J	J..R..E	R..U..C	A..W..8	J..Z..6
24)	T..O..M	D..R..K	K..S..F	S..V..D	B..X..A	K..1..7
25)	U..P..N	E..S..L	L..T..G	T..W..E	C..Y..B	L..2..8
26)	V..Q..O	F..T..M	M..U..H	U..X..F	D..Z..C	M..3..A
27)	W..R..P	G..U..N	N..V..J	V..Y..G	E..1..D	N..A..B
28)	X..S..Q	H..V..O	O..W..K	W..Z..H	F..2..E	O..B..C
29)	A..T..R	J..W..P	P..X..L	X..1..J	G..3..F	P..C..D
30)	B..U..S		Q..Y..M	A..2..K	H..A..G	Q..D..E
31)	C..V..T		R..Z..N		J..B..H	

1948 1948

	....JULY....	...AUGUST...	..SEPTEMBER.	..OCTOBER...	..NOVEMBER..	..DECEMBER..
	P--E--I	P--E--I	P--E--I	P--E--I	P--E--I	P--E--I
1)	R..E..F	B..H..D	K..L..B	R..N..7	B..Q..5	J..S..2
2)	S..F..G	C..J..E	L..M..C	S..O..8	C..R..6	K..T..3
3)	T..G..H	D..K..F	M..N..D	T..P..A	D..S..7	L..U..4
4)	U..H..J	E..L..G	N..O..E	U..Q..B	E..T..8	M..V..5
5)	V..J..K	F..M..H	O..P..F	V..R..C	F..U..A	N..W..6
6)	W..K..L	G..N..J	P..Q..G	W..S..D	G..V..B	O..X..7
7)	X..L..M	H..O..K	Q..R..H	X..T..E	H..W..C	P..Y..8
8)	A..M..N	J..P..L	R..S..J	A..U..F	J..X..D	Q..Z..A
9)	B..N..O	K..Q..M	S..T..K	B..V..G	K..Y..E	R..1..B
10)	C..O..P	L..R..N	T..U..L	C..W..H	L..Z..F	S..2..C
11)	D..P..Q	M..S..O	U..V..M	D..X..J	M..1..G	T..3..D
12)	E..Q..R	N..T..P	V..W..N	E..Y..K	N..2..H	U..A..E
13)	F..R..S	O..U..Q	W..X..O	F..Z..L	O..3..J	V..B..F
14)	G..S..T	P..V..R	X..Y..P	G..1..M	P..A..K	W..C..G
15)	H..T..U	Q..W..S	A..Z..Q	H..2..N	Q..B..L	X..D..H
16)	J..U..V	R..X..T	B..1..R	J..3..O	R..C..M	A..E..J
17)	K..V..W	S..Y..U	C..2..S	K..A..P	S..D..N	B..F..K
18)	L..W..X	T..Z..V	D..3..T	L..B..Q	T..E..O	C..G..L
19)	M..X..Y	U..1..W	E..A..U	M..C..R	U..F..P	D..H..M
20)	N..Y..Z	V..2..X	F..B..V	N..D..S	V..G..Q	E..J..N
21)	O..Z..1	W..3..Y	G..C..W	O..E..T	W..H..R	F..K..O
22)	P..1..2	X..A..Z	H..D..X	P..F..U	X..J..S	G..L..P
23)	Q..2..3	A..B..1	J..E..Y	Q..G..V	A..K..T	H..M..Q
24)	R..3..4	B..C..2	K..F..Z	R..H..W	B..L..U	J..N..R
25)	S..A..5	C..D..3	L..G..1	S..J..X	C..M..V	K..O..S
26)	T..B..6	D..E..4	M..H..2	T..K..Y	D..N..W	L..P..T
27)	U..C..7	E..F..5	N..J..3	U..L..Z	E..O..X	M..Q..U
28)	V..D..8	F..G..6	O..K..4	V..M..1	F..P..Y	N..R..V
29)	W..E..A	G..H..7	P..L..5	W..N..2	G..Q..Z	O..S..W
30)	X..F..B	H..J..8	Q..M..6	X..O..3	H..R..1	P..T..X
31)	A..G..C	J..K..A		A..P..4		Q..U..Y

CODES: P-PHYSICAL BIORHYTHM CURVE,E-EMOTIONAL BIORHYTHM CURVE,I-INTELLECTUAL BIORHYTHM CURVE

...JANUARY..	..FEBRUARY..	....MARCH...	....APRIL...	.....MAY....	....JUNE....
P--E--I	P--E--I	P--E--I	P--E--I	P--E--I	P--E--I
1) R..V..Z	1) B..Y..X	1) G..Y..S	1) P..2..Q	1) W..A..N	1) G..D..L
2) S..W..1	2) C..Z..Y	2) H..Z..T	2) Q..3..R	2) X..B..O	2) H..E..M
3) T..X..2	3) D..1..Z	3) J..1..U	3) R..A..S	3) A..C..P	3) J..F..N
4) U..Y..3	4) E..2..1	4) K..2..V	4) S..B..T	4) B..D..Q	4) K..G..O
5) V..Z..4	5) F..3..2	5) L..3..W	5) T..C..U	5) C..E..R	5) L..H..P
6) W..1..5	6) G..A..3	6) M..A..X	6) U..D..V	6) D..F..S	6) M..J..Q
7) X..2..6	7) H..B..4	7) N..B..Y	7) V..E..W	7) E..G..T	7) N..K..R
8) A..3..7	8) J..C..5	8) O..C..Z	8) W..F..X	8) F..H..U	8) O..L..S
9) B..A..8	9) K..D..6	9) P..D..1	9) X..G..Y	9) G..J..V	9) P..M..T
10) C..B..A	10) L..E..7	10) Q..E..2	10) A..H..Z	10) H..K..W	10) Q..N..U
11) D..C..B	11) M..F..8	11) R..F..3	11) B..J..1	11) J..L..X	11) R..O..V
12) E..D..C	12) N..G..A	12) S..G..4	12) C..K..2	12) K..M..Y	12) S..P..W
13) F..E..D	13) O..H..B	13) T..H..5	13) D..L..3	13) L..N..Z	13) T..Q..X
14) G..F..E	14) P..J..C	14) U..J..6	14) E..M..4	14) M..O..1	14) U..R..Y
15) H..G..F	15) Q..K..D	15) V..K..7	15) F..N..5	15) N..P..2	15) V..S..Z
16) J..H..G	16) R..L..E	16) W..L..8	16) G..O..6	16) O..Q..3	16) W..T..1
17) K..J..H	17) S..M..F	17) X..M..A	17) H..P..7	17) P..R..4	17) X..U..2
18) L..K..J	18) T..N..G	18) A..N..B	18) J..Q..8	18) Q..S..5	18) A..V..3
19) M..L..K	19) U..O..H	19) B..O..C	19) K..R..A	19) R..T..6	19) B..W..4
20) N..M..L	20) V..P..J	20) C..P..D	20) L..S..B	20) S..U..7	20) C..X..5
21) O..N..M	21) W..Q..K	21) D..Q..E	21) M..T..C	21) T..V..8	21) D..Y..6
22) P..O..N	22) X..R..L	22) E..R..F	22) N..U..D	22) U..W..A	22) E..Z..7
23) Q..P..O	23) A..S..M	23) F..S..G	23) O..V..E	23) V..X..B	23) F..1..8
24) R..Q..P	24) B..T..N	24) G..T..H	24) P..W..F	24) W..Y..C	24) G..2..A
25) S..R..Q	25) C..U..O	25) H..U..J	25) Q..X..G	25) X..Z..D	25) H..3..B
26) T..S..R	26) D..V..P	26) J..V..K	26) R..Y..H	26) A..1..E	26) J..A..C
27) U..T..S	27) E..W..Q	27) K..W..L	27) S..Z..J	27) B..2..F	27) K..B..D
28) V..U..T	28) F..X..R	28) L..X..M	28) T..1..K	28) C..3..G	28) L..C..E
29) W..V..U		29) M..Y..N	29) U..2..L	29) D..A..H	29) M..D..F
30) X..W..V		30) N..Z..O	30) V..3..M	30) E..B..J	30) N..E..G
31) A..X..W		31) O..1..P		31) F..C..K	

1949 | 1949

Day	....JULY.... P--E--I	...AUGUST... P--E--I	..SEPTEMBER. P--E--I	..OCTOBER... P--E--I	..NOVEMBER.. P--E--I	..DECEMBER.. P--E--I
1)	O..F..H	W..J..F	G..M..D	O..O..A	W..R..7	F..T..4
2)	P..G..J	X..K..G	H..N..E	P..P..B	X..S..8	G..U..5
3)	Q..H..K	A..L..H	J..O..F	Q..Q..C	A..T..A	H..V..6
4)	R..J..L	B..M..J	K..P..G	R..R..D	B..U..B	J..W..7
5)	S..K..M	C..N..K	L..Q..H	S..S..E	C..V..C	K..X..8
6)	T..L..N	D..O..L	M..R..J	T..T..F	D..W..D	L..Y..A
7)	U..M..O	E..P..M	N..S..K	U..U..G	E..X..E	M..Z..B
8)	V..N..P	F..Q..N	O..T..L	V..V..H	F..Y..F	N..1..C
9)	W..O..Q	G..R..O	P..U..M	W..W..J	G..Z..G	O..2..D
10)	X..P..R	H..S..P	Q..V..N	X..X..K	H..1..H	P..3..E
11)	A..Q..S	J..T..Q	R..W..O	A..Y..L	J..2..J	Q..A..F
12)	B..R..T	K..U..R	S..X..P	B..Z..M	K..3..K	R..B..G
13)	C..S..U	L..V..S	T..Y..Q	C..1..N	L..A..L	S..C..H
14)	D..T..V	M..W..T	U..Z..R	D..2..O	M..B..M	T..D..J
15)	E..U..W	N..X..U	V..1..S	E..3..P	N..C..N	U..E..K
16)	F..V..X	O..Y..V	W..2..T	F..A..Q	O..D..O	V..F..L
17)	G..W..Y	P..Z..W	X..3..U	G..B..R	P..E..P	W..G..M
18)	H..X..Z	Q..1..X	A..A..V	H..C..S	Q..F..Q	X..H..N
19)	J..Y..1	R..2..Y	B..B..W	J..D..T	R..G..R	A..J..O
20)	K..Z..2	S..3..Z	C..C..X	K..E..U	S..H..S	B..K..P
21)	L..1..3	T..A..1	D..D..Y	L..F..V	T..J..T	C..L..Q
22)	M..2..4	U..B..2	E..E..Z	M..G..W	U..K..U	D..M..R
23)	N..3..5	V..C..3	F..F..1	N..H..X	V..L..V	E..N..S
24)	O..A..6	W..D..4	G..G..2	O..J..Y	W..M..W	F..O..T
25)	P..B..7	X..E..5	H..H..3	P..K..Z	X..N..X	G..P..U
26)	Q..C..8	A..F..6	J..J..4	Q..L..1	A..O..Y	H..Q..V
27)	R..D..A	B..G..7	K..K..5	R..M..2	B..P..Z	J..R..W
28)	S..E..B	C..H..8	L..L..6	S..N..3	C..Q..1	K..S..X
29)	T..F..C	D..J..A	M..M..7	T..O..4	D..R..2	L..T..Y
30)	U..G..D	E..K..B	N..N..8	U..P..5	E..S..3	M..U..Z
31)	V..H..E	F..L..C		V..Q..6		N..V..1

CODES: P-PHYSICAL BIORHYTHM CURVE,E-EMOTIONAL BIORHYTHM CURVE,I-INTELLECTUAL BIORHYTHM CURVE

1950

1950

	...JANUARY.. P--E--I	..FEBRUARY.. P--E--I	....MARCH... P--E--I	....APRIL... P--E--I	.....MAY.... P--E--I	....JUNE.... P--E--I
1)	O..W..2	W..Z..Z	D..Z..U	M..3..S	T..B..P	D..E..N
2)	P..X..3	X..1..I	E..1..V	N..A..T	U..C..Q	E..F..O
3)	Q..Y..4	A..2..2	F..2..W	O..B..U	V..D..R	F..G..P
4)	R..Z..5	B..3..3	G..3..X	P..C..V	W..E..S	G..H..Q
5)	S..1..6	C..A..4	H..A..Y	Q..D..W	X..F..T	H..J..R
6)	T..2..7	D..B..5	J..B..Z	R..E..X	A..G..U	J..K..S
7)	U..3..8	E..C..6	K..C..1	S..F..Y	B..H..V	K..L..T
8)	V..A..A	F..D..7	L..D..2	T..G..Z	C..J..W	L..M..U
9)	W..B..B	G..E..8	M..E..3	U..H..1	D..K..X	M..N..V
10)	X..C..C	H..F..A	N..F..4	V..J..2	E..L..Y	N..O..W
11)	A..D..D	J..G..B	O..G..5	W..K..3	F..M..Z	O..P..X
12)	B..E..E	K..H..C	P..H..6	X..L..4	G..N..1	P..Q..Y
13)	C..F..F	L..J..D	Q..J..7	A..M..5	H..O..2	Q..R..Z
14)	D..G..G	M..K..E	R..K..8	B..N..6	J..P..3	R..S..1
15)	E..H..H	N..L..F	S..L..A	C..O..7	K..Q..4	S..T..2
16)	F..J..J	O..M..G	T..M..B	D..P..8	L..R..5	T..U..3
17)	G..K..K	P..N..H	U..N..C	E..Q..A	M..S..6	U..V..4
18)	H..L..L	Q..O..J	V..O..D	F..R..B	N..T..7	V..W..5
19)	J..M..M	R..P..K	W..P..E	G..S..C	O..U..8	W..X..6
20)	K..N..N	S..Q..L	X..Q..F	H..T..D	P..V..A	X..Y..7
21)	L..O..O	T..R..M	A..R..G	J..U..E	Q..W..B	A..Z..8
22)	M..P..P	U..S..N	B..S..H	K..V..F	R..X..C	B..1..A
23)	N..Q..Q	V..T..O	C..T..J	L..W..G	S..Y..D	C..2..B
24)	O..R..R	W..U..P	D..U..K	M..X..H	T..Z..E	D..3..C
25)	P..S..S	X..V..Q	E..V..L	N..Y..J	U..1..F	E..A..D
26)	Q..T..T	A..W..R	F..W..M	O..Z..K	V..2..G	F..B..E
27)	R..U..U	B..X..S	G..X..N	P..1..L	W..3..H	G..C..F
28)	S..V..V	C..Y..T	H..Y..O	Q..2..M	X..A..J	H..D..G
29)	T..W..W		J..Z..P	R..3..N	A..B..K	J..E..H
30)	U..X..X		K..1..Q	S..A..O	B..C..L	K..F..J
31)	V..Y..Y		L..2..R		C..D..M	

Day	....JULY.... P--E--I	...AUGUST... P--E--I	..SEPTEMBER. P--E--I	..OCTOBER... P--E--I	..NOVEMBER.. P--E--I	..DECEMBER.. P--E--I
1)	L..G..K	T..K..H	D..N..F	L..P..C	T..S..A	C..U..6
2)	M..H..L	U..L..J	E..O..G	M..Q..D	U..T..B	D..V..7
3)	N..J..M	V..M..K	F..P..H	N..R..E	V..U..C	E..W..8
4)	O..K..N	W..N..L	G..Q..J	O..S..F	W..V..D	F..X..A
5)	P..L..O	X..O..M	H..R..K	P..T..G	X..W..E	G..Y..B
6)	Q..M..P	A..P..N	J..S..L	Q..U..H	A..X..F	H..Z..C
7)	R..N..Q	B..Q..O	K..T..M	R..V..J	B..Y..G	J..1..D
8)	S..O..R	C..R..P	L..U..N	S..W..K	C..Z..H	K..2..E
9)	T..P..S	D..S..Q	M..V..O	T..X..L	D..1..J	L..3..F
10)	U..Q..T	E..T..R	N..W..P	U..Y..M	E..2..K	M..A..G
11)	V..R..U	F..U..S	O..X..Q	V..Z..N	F..3..L	N..B..H
12)	W..S..V	G..V..T	P..Y..R	W..1..O	G..A..M	O..C..J
13)	X..T..W	H..W..U	Q..Z..S	X..2..P	H..B..N	P..D..K
14)	A..U..X	J..X..V	R..1..T	A..3..Q	J..C..O	Q..E..L
15)	B..V..Y	K..Y..W	S..2..U	B..A..R	K..D..P	R..F..M
16)	C..W..Z	L..Z..X	T..3..V	C..B..S	L..E..Q	S..G..N
17)	D..X..1	M..1..Y	U..A..W	D..C..T	M..F..R	T..H..O
18)	E..Y..2	N..2..Z	V..B..X	E..D..U	N..G..S	U..J..P
19)	F..Z..3	O..3..1	W..C..Y	F..E..V	O..H..T	V..K..Q
20)	G..1..4	P..A..2	X..D..Z	G..F..W	P..J..U	W..L..R
21)	H..2..5	Q..B..3	A..E..1	H..G..X	Q..K..V	X..M..S
22)	J..3..6	R..C..4	B..F..2	J..H..Y	R..L..W	A..N..T
23)	K..A..7	S..D..5	C..G..3	K..J..Z	S..M..X	B..O..U
24)	L..B..8	T..E..6	D..H..4	L..K..1	T..N..Y	C..P..V
25)	M..C..A	U..F..7	E..J..5	M..L..2	U..O..Z	D..Q..W
26)	N..D..B	V..G..8	F..K..6	N..M..3	V..P..1	E..R..X
27)	O..E..C	W..H..A	G..L..7	O..N..4	W..Q..2	F..S..Y
28)	P..F..D	X..J..B	H..M..8	P..O..5	X..R..3	G..T..Z
29)	Q..G..E	A..K..C	J..N..A	Q..P..6	A..S..4	H..U..1
30)	R..H..F	B..L..D	K..O..B	R..Q..7	B..T..5	J..V..2
31)	S..J..G	C..M..E		S..R..8		K..W..3

CODES: P-PHYSICAL BIORHYTHM CURVE, E-EMOTIONAL BIORHYTHM CURVE, I-INTELLECTUAL BIORHYTHM CURVE

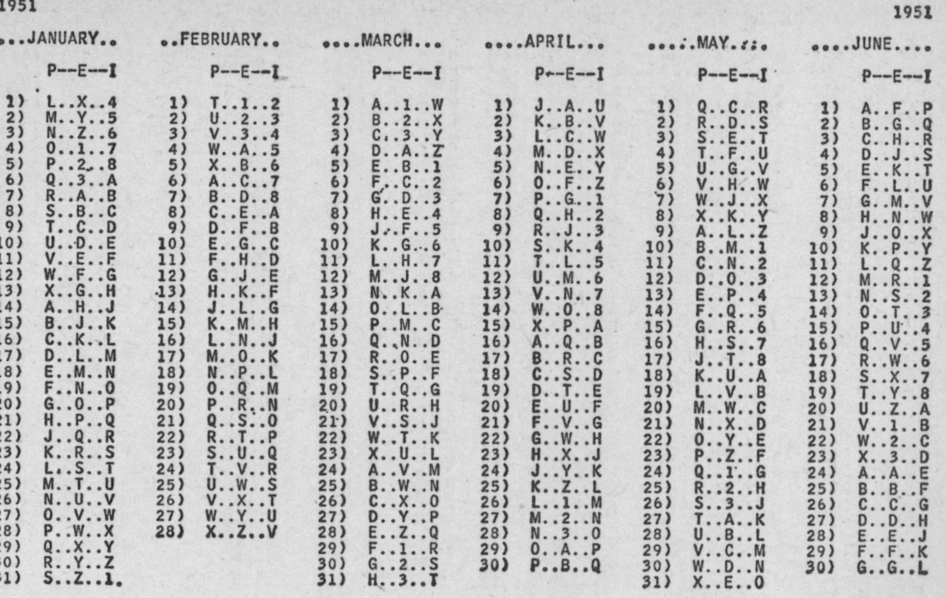

1951 1951

	...JANUARY..	..FEBRUARY..	....MARCH...	....APRIL...	....MAY....	....JUNE....
	P--E--I	P--E--I	P--E--I	P--E--I	P--E--I	P--E--I
1)	L..X..4	T..1..2	A..1..W	J..A..U	Q..C..R	A..F..P
2)	M..Y..5	U..2..3	B..2..X	K..B..V	R..D..S	B..G..Q
3)	N..Z..6	V..3..4	C..3..Y	L..C..W	S..E..T	C..H..R
4)	O..1..7	W..A..5	D..A..Z	M..D..X	T..F..U	D..J..S
5)	P..2..8	X..B..6	E..B..1	N..E..Y	U..G..V	E..K..T
6)	Q..3..A	A..C..7	F..C..2	O..F..Z	V..H..W	F..L..U
7)	R..A..B	B..D..8	G..D..3	P..G..1	W..J..X	G..M..V
8)	S..B..C	C..E..A	H..E..4	Q..H..2	X..K..Y	H..N..W
9)	T..C..D	D..F..B	J..F..5	R..J..3	A..L..Z	J..O..X
10)	U..D..E	E..G..C	K..G..6	S..K..4	B..M..1	K..P..Y
11)	V..E..F	F..H..D	L..H..7	T..L..5	C..N..2	L..Q..Z
12)	W..F..G	G..J..E	M..J..8	U..M..6	D..O..3	M..R..1
13)	X..G..H	H..K..F	N..K..A	V..N..7	E..P..4	N..S..2
14)	A..H..J	J..L..G	O..L..B	W..O..8	F..Q..5	O..T..3
15)	B..J..K	K..M..H	P..M..C	X..P..A	G..R..6	P..U..4
16)	C..K..L	L..N..J	Q..N..D	A..Q..B	H..S..7	Q..V..5
17)	D..L..M	M..O..K	R..O..E	B..R..C	J..T..8	R..W..6
18)	E..M..N	N..P..L	S..P..F	C..S..D	K..U..A	S..X..7
19)	F..N..O	O..Q..M	T..Q..G	D..T..E	L..V..B	T..Y..8
20)	G..O..P	P..R..N	U..R..H	E..U..F	M..W..C	U..Z..A
21)	H..P..Q	Q..S..O	V..S..J	F..V..G	N..X..D	V..1..B
22)	J..Q..R	R..T..P	W..T..K	G..W..H	O..Y..E	W..2..C
23)	K..R..S	S..U..Q	X..U..L	H..X..J	P..Z..F	X..3..D
24)	L..S..T	T..V..R	A..V..M	J..Y..K	Q..1..G	A..A..E
25)	M..T..U	U..W..S	B..W..N	K..Z..L	R..2..H	B..B..F
26)	N..U..V	V..X..T	C..X..O	L..1..M	S..3..J	C..C..G
27)	O..V..W	W..Y..U	D..Y..P	M..2..N	T..A..K	D..D..H
28)	P..W..X	X..Z..V	E..Z..Q	N..3..O	U..B..L	E..E..J
29)	Q..X..Y		F..1..R	O..A..P	V..C..M	F..F..K
30)	R..Y..Z		G..2..S	P..B..Q	W..D..N	G..G..L
31)	S..Z..1.		H..3..T		X..E..O	

1951 | 1951

Day	....JULY.... P--E--I	...AUGUST... P--E--I	..SEPTEMBER. P--E--I	..OCTOBER... P--E--I	..NOVEMBER.. P--E--I	..DECEMBER.. P--E--I
1)	H..H..M	Q..L..K	A..O..H	H..Q..E	Q..T..C	X..V..8
2)	J..J..N	R..M..L	B..P..J	J..R..F	R..U..D	A..W..A
3)	K..K..O	S..N..M	C..Q..K	K..S..G	S..V..E	B..X..B
4)	L..L..P	T..O..N	D..R..L	L..T..H	T..W..F	C..Y..C
5)	M..M..Q	U..P..O	E..S..M	M..U..J	U..X..G	D..Z..D
6)	N..N..R	V..Q..P	F..T..N	N..V..K	V..Y..H	E..1..E
7)	O..O..S	W..R..Q	G..U..O	O..W..L	W..Z..J	F..2..F
8)	P..P..T	X..S..R	H..V..P	P..X..M	X..1..K	G..3..G
9)	Q..Q..U	A..T..S	J..W..Q	Q..Y..N	A..2..L	H..A..H
10)	R..R..V	B..U..T	K..X..R	R..Z..O	B..3..M	J..B..J
11)	S..S..W	C..V..U	L..Y..S	S..1..P	C..A..N	K..C..K
12)	T..T..X	D..W..V	M..Z..T	T..2..Q	D..B..O	L..D..L
13)	U..U..Y	E..X..W	N..1..U	U..3..R	E..C..P	M..E..M
14)	V..V..Z	F..Y..X	O..2..V	V..A..S	F..D..Q	N..F..N
15)	W..W..1	G..Z..Y	P..3..W	W..B..T	G..E..R	O..G..O
16)	X..X..2	H..1..Z	Q..A..X	X..C..U	H..F..S	P..H..P
17)	A..Y..3	J..2..1	R..B..Y	A..D..V	J..G..T	Q..J..Q
18)	B..Z..4	K..3..2	S..C..Z	B..E..W	K..H..U	R..K..R
19)	C..1..5	L..A..3	T..D..1	C..F..X	L..J..V	S..L..S
20)	D..2..6	M..B..4	U..E..2	D..G..Y	M..K..W	T..M..T
21)	E..3..7	N..C..5	V..F..3	E..H..Z	N..L..X	U..N..U
22)	F..A..8	O..D..6	W..G..4	F..J..1	O..M..Y	V..O..V
23)	G..B..A	P..E..7	X..H..5	G..K..2	P..N..Z	W..P..W
24)	H..C..B	Q..F..8	A..J..6	H..L..3	Q..O..1	X..Q..X
25)	J..D..C	R..G..A	B..K..7	J..M..4	R..P..2	A..R..Y
26)	K..E..D	S..H..B	C..L..8	K..N..5	S..Q..3	B..S..Z
27)	L..F..E	T..J..C	D..M..A	L..O..6	T..R..4	C..T..1
28)	M..G..F	U..K..D	E..N..B	M..P..7	U..S..5	D..U..2
29)	N..H..G	V..L..E	F..O..C	N..Q..8	V..T..6	E..V..3
30)	O..J..H	W..M..F	G..P..D	O..R..A	W..U..7	F..W..4
31)	P..K..J	X..N..G		P..S..B		G..X..5

CODES: P-PHYSICAL BIORHYTHM CURVE, E-EMOTIONAL BIORHYTHM CURVE, I-INTELLECTUAL BIORHYTHM CURVE

1952 1952

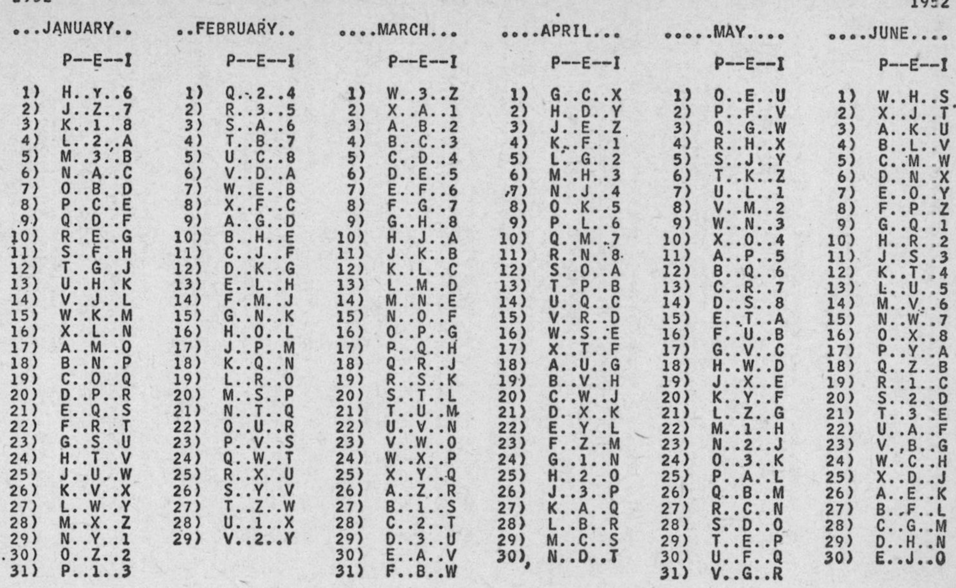

	...JANUARY..	..FEBRUARY..	....MARCH...	....APRIL...	.....MAY....	....JUNE....
	P--E--I	P--E--I	P--E--I	P--E--I	P--E--I	P--E--I
1)	H..Y..6	Q..2..4	W..3..Z	G..C..X	O..E..U	W..H..S
2)	J..Z..7	R..3..5	X..A..1	H..D..Y	P..F..V	X..J..T
3)	K..1..8	S..A..6	A..B..2	J..E..Z	Q..G..W	A..K..U
4)	L..2..A	T..B..7	B..C..3	K..F..1	R..H..X	B..L..V
5)	M..3..B	U..C..8	C..D..4	L..G..2	S..J..Y	C..M..W
6)	N..A..C	V..D..A	D..E..5	M..H..3	T..K..Z	D..N..X
7)	O..B..D	W..E..B	E..F..6	N..J..4	U..L..1	E..O..Y
8)	P..C..E	X..F..C	F..G..7	O..K..5	V..M..2	F..P..Z
9)	Q..D..F	A..G..D	G..H..8	P..L..6	W..N..3	G..Q..1
10)	R..E..G	B..H..E	H..J..A	Q..M..7	X..O..4	H..R..2
11)	S..F..H	C..J..F	J..K..B	R..N..8	A..P..5	J..S..3
12)	T..G..J	D..K..G	K..L..C	S..O..A	B..Q..6	K..T..4
13)	U..H..K	E..L..H	L..M..D	T..P..B	C..R..7	L..U..5
14)	V..J..L	F..M..J	M..N..E	U..Q..C	D..S..8	M..V..6
15)	W..K..M	G..N..K	N..O..F	V..R..D	E..T..A	N..W..7
16)	X..L..N	H..O..L	O..P..G	W..S..E	F..U..B	O..X..8
17)	A..M..O	J..P..M	P..Q..H	X..T..F	G..V..C	P..Y..A
18)	B..N..P	K..Q..N	Q..R..J	A..U..G	H..W..D	Q..Z..B
19)	C..O..Q	L..R..O	R..S..K	B..V..H	J..X..E	R..1..C
20)	D..P..R	M..S..P	S..T..L	C..W..J	K..Y..F	S..2..D
21)	E..Q..S	N..T..Q	T..U..M	D..X..K	L..Z..G	T..3..E
22)	F..R..T	O..U..R	U..V..N	E..Y..L	M..1..H	U..A..F
23)	G..S..U	P..V..S	V..W..O	F..Z..M	N..2..J	V..B..G
24)	H..T..V	Q..W..T	W..X..P	G..1..N	O..3..K	W..C..H
25)	J..U..W	R..X..U	X..Y..Q	H..2..O	P..A..L	X..D..J
26)	K..V..X	S..Y..V	A..Z..R	J..3..P	Q..B..M	A..E..K
27)	L..W..Y	T..Z..W	B..1..S	K..A..Q	R..C..N	B..F..L
28)	M..X..Z	U..1..X	C..2..T	L..B..R	S..D..O	C..G..M
29)	N..Y..1	V..2..Y	D..3..U	M..C..S	T..E..P	D..H..N
30)	O..Z..2		E..A..V	N..D..T	U..F..Q	E..J..O
31)	P..1..3		F..B..W		V..G..R	

1952 | 1952

Day	....JULY.... P--E--I	...AUGUST... P--E--I	..SEPTEMBER. P--E--I	..OCTOBER... P--E--I	..NOVEMBER.. P--E--I	..DECEMBER.. P--E--I
1)	F..K..P	O..N..N	W..Q..L	F..S..H	O..V..F	V..X..C
2)	G..L..Q	P..O..O	X..R..M	G..T..J	P..W..G	W..Y..D
3)	H..M..R	Q..P..P	A..S..N	H..U..K	Q..X..H	X..Z..E
4)	J..N..S	R..Q..Q	B..T..O	J..V..L	R..Y..J	A..1..F
5)	K..O..T	S..R..R	C..U..P	K..W..M	S..Z..K	B..2..G
6)	L..P..U	T..S..S	D..V..Q	L..X..N	T..1..L	C..3..H
7)	M..Q..V	U..T..T	E..W..R	M..Y..O	U..2..M	D..A..J
8)	N..R..W	V..U..U	F..X..S	N..Z..P	V..3..N	E..B..K
9)	O..S..X	W..V..V	G..Y..T	O..1..Q	W..A..O	F..C..L
10)	P..T..Y	X..W..W	H..Z..U	P..2..R	X..B..P	G..D..M
11)	Q..U..Z	A..X..X	J..1..V	Q..3..S	A..C..Q	H..E..N
12)	R..V..1	B..Y..Y	K..2..W	R..A..T	B..D..R	J..F..O
13)	S..W..2	C..Z..Z	L..3..X	S..B..U	C..E..S	K..G..P
14)	T..X..3	D..1..1	M..A..Y	T..C..V	D..F..T	L..H..Q
15)	U..Y..4	E..2..2	N..B..Z	U..D..W	E..G..U	M..J..R
16)	V..Z..5	F..3..3	O..C..1	V..E..X	F..H..V	N..K..S
17)	W..1..6	G..A..4	P..D..2	W..F..Y	G..J..W	O..L..T
18)	X..2..7	H..B..5	Q..E..3	X..G..Z	H..K..X	P..M..U
19)	A..3..8	J..C..6	R..F..4	A..H..1	J..L..Y	Q..N..V
20)	B..A..A	K..D..7	S..G..5	B..J..2	K..M..Z	R..O..W
21)	C..B..B	L..E..8	T..H..6	C..K..3	L..N..1	S..P..X
22)	D..C..C	M..F..A	U..J..7	D..L..4	M..O..2	T..Q..Y
23)	E..D..D	N..G..B	V..K..8	E..M..5	N..P..3	U..R..Z
24)	F..E..E	O..H..C	W..L..A	F..N..6	O..Q..4	V..S..1
25)	G..F..F	P..J..D	X..M..B	G..O..7	P..R..5	W..T..2
26)	H..G..G	Q..K..E	A..N..C	H..P..8	Q..S..6	X..U..3
27)	J..H..H	R..L..F	B..O..D	J..Q..A	R..T..7	A..V..4
28)	K..J..J	S..M..G	C..P..E	K..R..B	S..U..8	B..W..5
29)	L..K..K	T..N..H	D..Q..F	L..S..C	T..V..A	C..X..6
30)	M..L..L	U..O..J	E..R..G	M..T..D	U..W..B	D..Y..7
31)	N..M..M	V..P..K		N..U..E		E..Z..8

CODES: P-PHYSICAL BIORHYTHM CURVE,E-EMOTIONAL BIORHYTHM CURVE,I-INTELLECTUAL BIORHYTHM CURVE

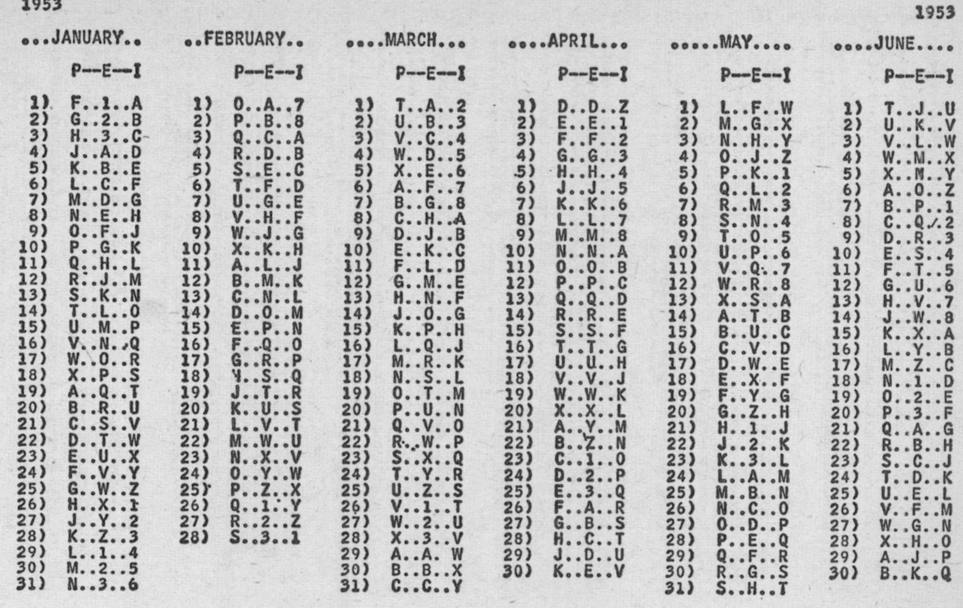

1953						1953
	...JANUARY..	..FEBRUARY..	....MARCH...	....APRIL...	....MAY....	....JUNE....
	P--E--I	P--E--I	P--E--I	P--E--I	P--E--I	P--E--I
1)	F..1..A	O..A..7	T..A..2	D..D..Z	L..F..W	T..J..U
2)	G..2..B	P..B..8	U..B..3	E..E..1	M..G..X	U..K..V
3)	H..3..C	Q..C..A	V..C..4	F..F..2	N..H..Y	V..L..W
4)	J..A..D	R..D..B	W..D..5	G..G..3	O..J..Z	W..M..X
5)	K..B..E	S..E..C	X..E..6	H..H..4	P..K..1	X..N..Y
6)	L..C..F	T..F..D	A..F..7	J..J..5	Q..L..2	A..O..Z
7)	M..D..G	U..G..E	B..G..8	K..K..6	R..M..3	B..P..1
8)	N..E..H	V..H..F	C..H..A	L..L..7	S..N..4	C..Q..2
9)	O..F..J	W..J..G	D..J..B	M..M..8	T..O..5	D..R..3
10)	P..G..K	X..K..H	E..K..C	N..N..A	U..P..6	E..S..4
11)	Q..H..L	A..L..J	F..L..D	O..O..B	V..Q..7	F..T..5
12)	R..J..M	B..M..K	G..M..E	P..P..C	W..R..8	G..U..6
13)	S..K..N	C..N..L	H..N..F	Q..Q..D	X..S..A	H..V..7
14)	T..L..O	D..O..M	J..O..G	R..R..E	A..T..B	J..W..8
15)	U..M..P	E..P..N	K..P..H	S..S..F	B..U..C	K..X..A
16)	V..N..Q	F..Q..O	L..Q..J	T..T..G	C..V..D	L..Y..B
17)	W..O..R	G..R..P	M..R..K	U..U..H	D..W..E	M..Z..C
18)	X..P..S	H..S..Q	N..S..L	V..V..J	E..X..F	N..1..D
19)	A..Q..T	J..T..R	O..T..M	W..W..K	F..Y..G	O..2..E
20)	B..R..U	K..U..S	P..U..N	X..X..L	G..Z..H	P..3..F
21)	C..S..V	L..V..T	Q..V..O	A..Y..M	H..1..J	Q..A..G
22)	D..T..W	M..W..U	R..W..P	B..Z..N	J..2..K	R..B..H
23)	E..U..X	N..X..V	S..X..Q	C..1..O	K..3..L	S..C..J
24)	F..V..Y	O..Y..W	T..Y..R	D..2..P	L..A..M	T..D..K
25)	G..W..Z	P..Z..X	U..Z..S	E..3..Q	M..B..N	U..E..L
26)	H..X..1	Q..1..Y	V..1..T	F..A..R	N..C..O	V..F..M
27)	J..Y..2	R..2..Z	W..2..U	G..B..S	O..D..P	W..G..N
28)	K..Z..3	S..3..1	X..3..V	H..C..T	P..E..Q	X..H..O
29)	L..1..4		A..A..W	J..D..U	Q..F..R	A..J..P
30)	M..2..5		B..B..X	K..E..V	R..G..S	B..K..Q
31)	N..3..6		C..C..Y		S..H..T	

1953 — 1953

Day	....JULY.... P--E--I	...AUGUST... P--E--I	..SEPTEMBER. P--E--I	..OCTOBER... P--E--I	..NOVEMBER.. P--E--I	..DECEMBER.. P--E--I
1)	C..L..R	L..O..P	T..R..N	C..T..K	L..W..H	S..Y..E
2)	D..M..S	M..P..Q	U..S..O	D..U..L	M..X..J	T..Z..F
3)	E..N..T	N..Q..R	V..T..P	E..V..M	N..Y..K	U..1..G
4)	F..O..U	O..R..S	W..U..Q	F..W..N	O..Z..L	V..2..H
5)	G..P..V	P..S..T	X..V..R	G..X..O	P..1..M	W..3..J
6)	H..Q..W	Q..T..U	A..W..S	H..Y..P	Q..2..N	X..A..K
7)	J..R..X	R..U..V	B..X..T	J..Z..Q	R..3..O	A..B..L
8)	K..S..Y	S..V..W	C..Y..U	K..1..R	S..A..P	B..C..M
9)	L..T..Z	T..W..X	D..Z..V	L..2..S	T..B..Q	C..D..N
10)	M..U..1	U..X..Y	E..1..W	M..3..T	U..C..R	D..E..O
11)	N..V..2	V..Y..Z	F..2..X	N..A..U	V..D..S	E..F..P
12)	O..W..3	W..Z..1	G..3..Y	O..B..V	W..E..T	F..G..Q
13)	P..X..4	X..1..2	H..A..Z	P..C..W	X..F..U	G..H..R
14)	Q..Y..5	A..2..3	J..B..1	Q..D..X	A..G..V	H..J..S
15)	R..Z..6	B..3..4	K..C..2	R..E..Y	B..H..W	J..K..T
16)	S..1..7	C..A..5	L..D..3	S..F..Z	C..J..X	K..L..U
17)	T..2..8	D..B..6	M..E..4	T..G..1	D..K..Y	L..M..V
18)	U..3..A	E..C..7	N..F..5	U..H..2	E..L..Z	M..N..W
19)	V..A..B	F..D..8	O..G..6	V..J..3	F..M..1	N..O..X
20)	W..B..C	G..E..A	P..H..7	W..K..4	G..N..2	O..P..Y
21)	X..C..D	H..F..B	Q..J..8	X..L..5	H..O..3	P..Q..Z
22)	A..D..E	J..G..C	R..K..A	A..M..6	J..P..4	Q..R..1
23)	B..E..F	K..H..D	S..L..B	B..N..7	K..Q..5	R..S..2
24)	C..F..G	L..J..E	T..M..C	C..O..8	L..R..6	S..T..3
25)	D..G..H	M..K..F	U..N..D	D..P..A	M..S..7	T..U..4
26)	E..H..J	N..L..G	V..O..E	E..Q..B	N..T..8	U..V..5
27)	F..J..K	O..M..H	W..P..F	F..R..C	O..U..A	V..W..6
28)	G..K..L	P..N..J	X..Q..G	G..S..D	P..V..B	W..X..7
29)	H..L..M	Q..O..K	A..R..H	H..T..E	Q..W..C	X..Y..8
30)	J..M..N	R..P..L	B..S..J	J..U..F	R..X..D	A..Z..A
31)	K..N..O	S..Q..M		K..V..G		B..1..B

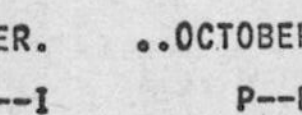
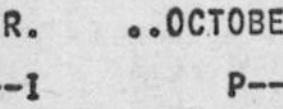

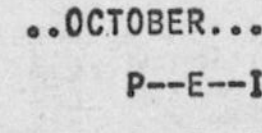
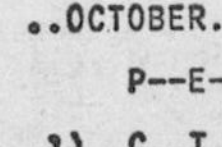
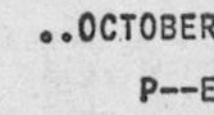
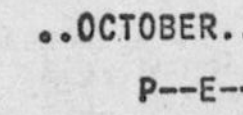
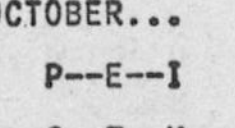
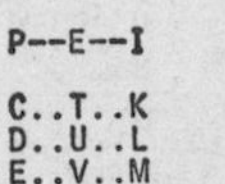
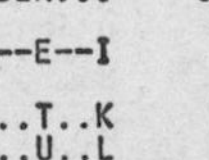
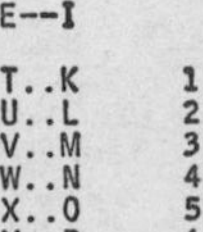
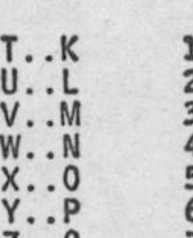
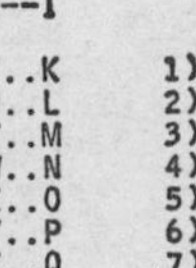
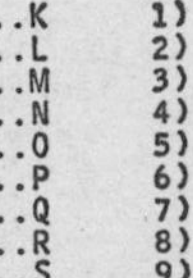
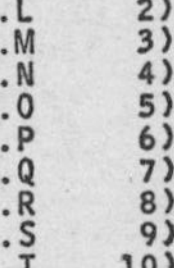
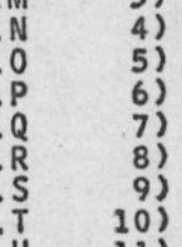
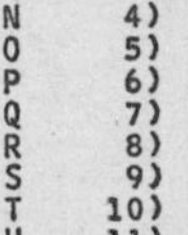
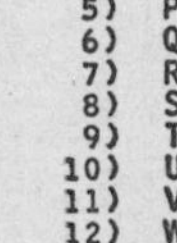
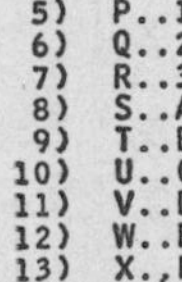

CODES: P-PHYSICAL BIORHYTHM CURVE, E-EMOTIONAL BIORHYTHM CURVE, I-INTELLECTUAL BIORHYTHM CURVE

1954 1954

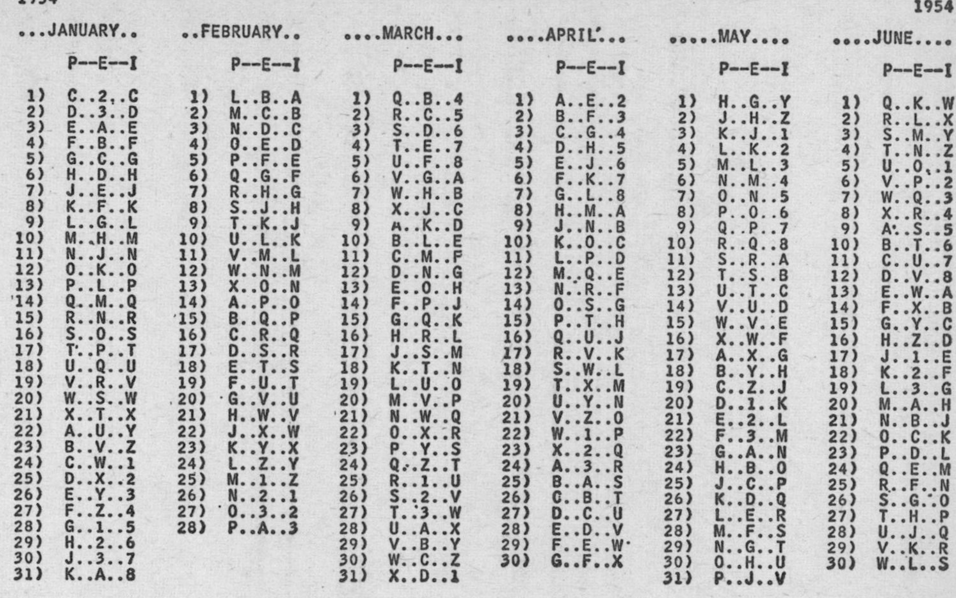

	...JANUARY..	..FEBRUARY..	....MARCH...	....APRIL...	.....MAY....	....JUNE....
	P--E--I	P--E--I	P--E--I	P--E--I	P--E--I	P--E--I
1)	C..2..C	L..B..A	Q..B..4	A..E..2	H..G..Y	Q..K..W
2)	D..3..D	M..C..B	R..C..5	B..F..3	J..H..Z	R..L..X
3)	E..A..E	N..D..C	S..D..6	C..G..4	K..J..1	S..M..Y
4)	F..B..F	O..E..D	T..E..7	D..H..5	L..K..2	T..N..Z
5)	G..C..G	P..F..E	U..F..8	E..J..6	M..L..3	U..O..1
6)	H..D..H	Q..G..F	V..G..A	F..K..7	N..M..4	V..P..2
7)	J..E..J	R..H..G	W..H..B	G..L..8	O..N..5	W..Q..3
8)	K..F..K	S..J..H	X..J..C	H..M..A	P..O..6	X..R..4
9)	L..G..L	T..K..J	A..K..D	J..N..B	Q..P..7	A..S..5
10)	M..H..M	U..L..K	B..L..E	K..O..C	R..Q..8	B..T..6
11)	N..J..N	V..M..L	C..M..F	L..P..D	S..R..A	C..U..7
12)	O..K..O	W..N..M	D..N..G	M..Q..E	T..S..B	D..V..8
13)	P..L..P	X..O..N	E..O..H	N..R..F	U..T..C	E..W..A
14)	Q..M..Q	A..P..O	F..P..J	O..S..G	V..U..D	F..X..B
15)	R..N..R	B..Q..P	G..Q..K	P..T..H	W..V..E	G..Y..C
16)	S..O..S	C..R..Q	H..R..L	Q..U..J	X..W..F	H..Z..D
17)	T..P..T	D..S..R	J..S..M	R..V..K	A..X..G	J..1..E
18)	U..Q..U	E..T..S	K..T..N	S..W..L	B..Y..H	K..2..F
19)	V..R..V	F..U..T	L..U..O	T..X..M	C..Z..J	L..3..G
20)	W..S..W	G..V..U	M..V..P	U..Y..N	D..1..K	M..A..H
21)	X..T..X	H..W..V	N..W..Q	V..Z..O	E..2..L	N..B..J
22)	A..U..Y	J..X..W	O..X..R	W..1..P	F..3..M	O..C..K
23)	B..V..Z	K..Y..X	P..Y..S	X..2..Q	G..A..N	P..D..L
24)	C..W..1	L..Z..Y	Q..Z..T	A..3..R	H..B..O	Q..E..M
25)	D..X..2	M..1..Z	R..1..U	B..A..S	J..C..P	R..F..N
26)	E..Y..3	N..2..1	S..2..V	C..B..T	K..D..Q	S..G..O
27)	F..Z..4	O..3..2	T..3..W	D..C..U	L..E..R	T..H..P
28)	G..1..5	P..A..3	U..A..X	E..D..V	M..F..S	U..J..Q
29)	H..2..6		V..B..Y	F..E..W	N..G..T	V..K..R
30)	J..3..7		W..C..Z	G..F..X	O..H..U	W..L..S
31)	K..A..8		X..D..1		P..J..V	

1954 | 1954

Day	....JULY.... P--E--I	...AUGUST.... P--E--I	..SEPTEMBER. P--E--I	..OCTOBER... P--E--I	..NOVEMBER.. P--E--I	..DECEMBER.. P--E--I
1)	X..M..T	H..P..R	Q..S..P	X..U..M	H..X..K	P..Z..G
2)	A..N..U	J..Q..S	R..T..Q	A..V..N	J..Y..L	Q..1..H
3)	B..O..V	K..R..T	S..U..R	B..W..O	K..Z..M	R..2..J
4)	C..P..W	L..S..U	T..V..S	C..X..P	L..1..N	S..3..K
5)	D..Q..X	M..T..V	U..W..T	D..Y..Q	M..2..O	T..A..L
6)	E..R..Y	N..U..W	V..X..U	E..Z..R	N..3..P	U..B..M
7)	F..S..Z	O..V..X	W..Y..V	F..1..S	O..A..Q	V..C..N
8)	G..T..1	P..W..Y	X..Z..W	G..2..T	P..B..R	W..D..O
9)	H..U..2	Q..X..Z	A..1..X	H..3..U	Q..C..S	X..E..P
10)	J..V..3	R..Y..1	B..2..Y	J..A..V	R..D..T	A..F..Q
11)	K..W..4	S..Z..2	C..3..Z	K..B..W	S..E..U	B..G..R
12)	L..X..5	T..1..3	D..A..1	L..C..X	T..F..V	C..H..S
13)	M..Y..6	U..2..4	E..B..2	M..D..Y	U..G..W	D..J..T
14)	N..Z..7	V..3..5	F..C..3	N..E..Z	V..H..X	E..K..U
15)	O..1..8	W..A..6	G..D..4	O..F..1	W..J..Y	F..L..V
16)	P..2..A	X..B..7	H..E..5	P..G..2	X..K..Z	G..M..W
17)	Q..3..B	A..C..8	J..F..6	Q..H..3	A..L..1	H..N..X
18)	R..A..C	B..D..A	K..G..7	R..J..4	B..M..2	J..O..Y
19)	S..B..D	C..E..B	L..H..8	S..K..5	C..N..3	K..P..Z
20)	T..C..E	D..F..C	M..J..A	T..L..6	D..O..4	L..Q..1
21)	U..D..F	E..G..D	N..K..B	U..M..7	E..P..5	M..R..2
22)	V..E..G	F..H..E	O..L..C	V..N..8	F..Q..6	N..S..3
23)	W..F..H	G..J..F	P..M..D	W..O..A	G..R..7	O..T..4
24)	X..G..J	H..K..G	Q..N..E	X..P..B	H..S..8	P..U..5
25)	A..H..K	J..L..H	R..O..F	A..Q..C	J..T..A	Q..V..6
26)	B..J..L	K..M..J	S..P..G	B..R..D	K..U..B	R..W..7
27)	C..K..M	L..N..K	T..Q..H	C..S..E	L..V..C	S..X..8
28)	D..L..N	M..O..L	U..R..J	D..T..F	M..W..D	T..Y..A
29)	E..M..O	N..P..M	V..S..K	E..U..G	N..X..E	U..Z..B
30)	F..N..P	O..Q..N	W..T..L	F..V..H	O..Y..F	V..1..C
31)	G..O..Q	P..R..O		G..W..J		W..2..D

CODES: P-PHYSICAL BIORHYTHM CURVE,E-EMOTIONAL BIORHYTHM CURVE,I-INTELLECTUAL BIORHYTHM CURVE

1955 | 1955

	...JANUARY..	..FEBRUARY..	....MARCH...	....APRIL...	.....MAY....	....JUNE....
	P--E--I	P--E--I	P--E--I	P--E--I	P--E--I	P--E--I
1)	X..3..E	H..C..C	N..C..6	V..F..4	E..H..1	N..L..Y
2)	A..A..F	J..D..D	O..D..7	W..G..5	F..J..2	O..M..Z
3)	B..B..G	K..E..E	P..E..8	X..H..6	G..K..3	P..N..1
4)	C..C..H	L..F..F	Q..F..A	A..J..7	H..L..4	Q..O..2
5)	D..D..J	M..G..G	R..G..B	B..K..8	J..M..5	R..P..3
6)	E..E..K	N..H..H	S..H..C	C..L..A	K..N..6	S..Q..4
7)	F..F..L	O..J..J	T..J..D	D..M..B	L..O..7	T..R..5
8)	G..G..M	P..K..K	U..K..E	E..N..C	M..P..8	U..S..6
9)	H..H..N	Q..L..L	V..L..F	F..O..D	N..Q..A	V..T..7
10)	J..J..O	R..M..M	W..M..G	G..P..E	O..R..B	W..U..8
11)	K..K..P	S..N..N	X..N..H	H..Q..F	P..S..C	X..V..A
12)	L..L..Q	T..O..O	A..O..J	J..R..G	Q..T..D	A..W..B
13)	M..M..R	U..P..P	B..P..K	K..S..H	R..U..E	B..X..C
14)	N..N..S	V..Q..Q	C..Q..L	L..T..J	S..V..F	C..Y..D
15)	O..O..T	W..R..R	D..R..M	M..U..K	T..W..G	D..Z..E
16)	P..P..U	X..S..S	E..S..N	N..V..L	U..X..H	E..1..F
17)	Q..Q..V	A..T..T	F..T..O	O..W..M	V..Y..J	F..2..G
18)	R..R..W	B..U..U	G..U..P	P..X..N	W..Z..K	G..3..H
19)	S..S..X	C..V..V	H..V..Q	Q..Y..O	X..1..L	H..A..J
20)	T..T..Y	D..W..W	J..W..R	R..Z..P	A..2..M	J..B..K
21)	U..U..Z	E..X..X	K..X..S	S..1..Q	B..3..N	K..C..L
22)	V..V..1	F..Y..Y	L..Y..T	T..2..R	C..A..O	L..D..M
23)	W..W..2	G..Z..Z	M..Z..U	U..3..S	D..B..P	M..E..N
24)	X..X..3	H..1..1	N..1..V	V..A..T	E..C..Q	N..F..O
25)	A..Y..4	J..2..2	O..2..W	W..B..U	F..D..R	O..G..P
26)	B..Z..5	K..3..3	P..3..X	X..C..V	G..E..S	P..H..Q
27)	C..1..6	L..A..4	Q..A..Y	A..D..W	H..F..T	Q..J..R
28)	D..2..7	M..B..5	R..B..Z	B..E..X	J..G..U	R..K..S
29)	E..3..8		S..C..1	C..F..Y	K..H..V	S..L..T
30)	F..A..A		T..D..2	D..G..Z	L..J..W	T..M..U
31)	G..B..B		U..E..3		M..K..X	

1955 | 1955

Day	....JULY.... P--E--I	...AUGUST... P--E--I	..SEPTEMBER. P--E--I	..OCTOBER... P--E--I	..NOVEMBER.. P--E--I	..DECEMBER.. P--E--I
1)	U..N..V	E..Q..T	N..T..R	U..V..O	E..Y..M	M..1..J
2)	V..O..W	F..R..U	O..U..S	V..W..P	F..Z..N	N..2..K
3)	W..P..X	G..S..V	P..V..T	W..X..Q	G..1..O	O..3..L
4)	X..Q..Y	H..T..W	Q..W..U	X..Y..R	H..2..P	P..A..M
5)	A..R..Z	J..U..X	R..X..V	A..Z..S	J..3..Q	Q..B..N
6)	B..S..1	K..V..Y	S..Y..W	B..1..T	K..A..R	R..C..O
7)	C..T..2	L..W..Z	T..Z..X	C..2..U	L..B..S	S..D..P
8)	D..U..3	M..X..1	U..1..Y	D..3..V	M..C..T	T..E..Q
9)	E..V..4	N..Y..2	V..2..Z	E..A..W	N..D..U	U..F..R
10)	F..W..5	O..Z..3	W..3..1	F..B..X	O..E..V	V..G..S
11)	G..X..6	P..1..4	X..A..2	G..C..Y	P..F..W	W..H..T
12)	H..Y..7	Q..2..5	A..B..3	H..D..Z	Q..G..X	X..J..U
13)	J..Z..8	R..3..6	B..C..4	J..E..1	R..H..Y	A..K..V
14)	K..1..A	S..A..7	C..D..5	K..F..2	S..J..Z	B..L..W
15)	L..2..B	T..B..8	D..E..6	L..G..3	T..K..1	C..M..X
16)	M..3..C	U..C..A	E..F..7	M..H..4	U..L..2	D..N..Y
17)	N..A..D	V..D..B	F..G..8	N..J..5	V..M..3	E..O..Z
18)	O..B..E	W..E..C	G..H..A	O..K..6	W..N..4	F..P..1
19)	P..C..F	X..F..D	H..J..B	P..L..7	X..O..5	G..Q..2
20)	Q..D..G	A..G..E	J..K..C	Q..M..8	A..P..6	H..R..3
21)	R..E..H	B..H..F	K..L..D	R..N..A	B..Q..7	J..S..4
22)	S..F..J	C..J..G	L..M..E	S..O..B	C..R..8	K..T..5
23)	T..G..K	D..K..H	M..N..F	T..P..C	D..S..A	L..U..6
24)	U..H..L	E..L..J	N..O..G	U..Q..D	E..T..B	M..V..7
25)	V..J..M	F..M..K	O..P..H	V..R..E	F..U..C	N..W..8
26)	W..K..N	G..N..L	P..Q..J	W..S..F	G..V..D	O..X..A
27)	X..L..O	H..O..M	Q..R..K	X..T..G	H..W..E	P..Y..B
28)	A..M..P	J..P..N	R..S..L	A..U..H	J..X..F	Q..Z..C
29)	B..N..Q	K..Q..O	S..T..M	B..V..J	K..Y..G	R..1..D
30)	C..O..R	L..R..P	T..U..N	C..W..K	L..Z..H	S..2..E
31)	D..P..S	M..S..Q		D..X..L		T..3..F

CODES: P-PHYSICAL BIORHYTHM CURVE,E-EMOTIONAL BIORHYTHM CURVE,I-INTELLECTUAL BIORHYTHM CURVE

1956 1956

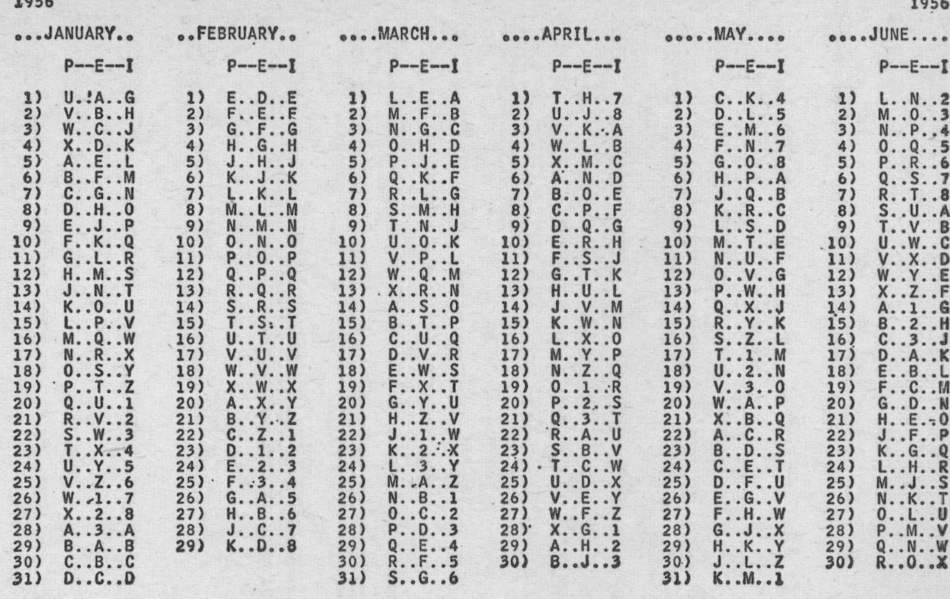

Day	...JANUARY.. P--E--I	..FEBRUARY.. P--E--I	....MARCH... P--E--I	....APRIL... P--E--I	.....MAY.... P--E--I	....JUNE.... P--E--I
1)	U..A..G	E..D..E	L..E..A	T..H..7	C..K..4	L..N..2
2)	V..B..H	F..E..F	M..F..B	U..J..8	D..L..5	M..O..3
3)	W..C..J	G..F..G	N..G..C	V..K..A	E..M..6	N..P..4
4)	X..D..K	H..G..H	O..H..D	W..L..B	F..N..7	O..Q..5
5)	A..E..L	J..H..J	P..J..E	X..M..C	G..O..8	P..R..6
6)	B..F..M	K..J..K	Q..K..F	A..N..D	H..P..A	Q..S..7
7)	C..G..N	L..K..L	R..L..G	B..O..E	J..Q..B	R..T..8
8)	D..H..O	M..L..M	S..M..H	C..P..F	K..R..C	S..U..A
9)	E..J..P	N..M..N	T..N..J	D..Q..G	L..S..D	T..V..B
10)	F..K..Q	O..N..O	U..O..K	E..R..H	M..T..E	U..W..C
11)	G..L..R	P..O..P	V..P..L	F..S..J	N..U..F	V..X..D
12)	H..M..S	Q..P..Q	W..Q..M	G..T..K	O..V..G	W..Y..E
13)	J..N..T	R..Q..R	X..R..N	H..U..L	P..W..H	X..Z..F
14)	K..O..U	S..R..S	A..S..O	J..V..M	Q..X..J	A..1..G
15)	L..P..V	T..S..T	B..T..P	K..W..N	R..Y..K	B..2..H
16)	M..Q..W	U..T..U	C..U..Q	L..X..O	S..Z..L	C..3..J
17)	N..R..X	V..U..V	D..V..R	M..Y..P	T..1..M	D..A..K
18)	O..S..Y	W..V..W	E..W..S	N..Z..Q	U..2..N	E..B..L
19)	P..T..Z	X..W..X	F..X..T	O..1..R	V..3..O	F..C..M
20)	Q..U..1	A..X..Y	G..Y..U	P..2..S	W..A..P	G..D..N
21)	R..V..2	B..Y..Z	H..Z..V	Q..3..T	X..B..Q	H..E..O
22)	S..W..3	C..Z..1	J..1..W	R..A..U	A..C..R	J..F..P
23)	T..X..4	D..1..2	K..2..X	S..B..V	B..D..S	K..G..Q
24)	U..Y..5	E..2..3	L..3..Y	T..C..W	C..E..T	L..H..R
25)	V..Z..6	F..3..4	M..A..Z	U..D..X	D..F..U	M..J..S
26)	W..1..7	G..A..5	N..B..1	V..E..Y	E..G..V	N..K..T
27)	X..2..8	H..B..6	O..C..2	W..F..Z	F..H..W	O..L..U
28)	A..3..A	J..C..7	P..D..3	X..G..1	G..J..X	P..M..V
29)	B..A..B	K..D..8	Q..E..4	A..H..2	H..K..Y	Q..N..W
30)	C..B..C		R..F..5	B..J..3	J..L..Z	R..O..X
31)	D..C..D		S..G..6		K..M..1	

1956 1956

	....JULY....	...AUGUST...	..SEPTEMBER.	..OCTOBER...	..NOVEMBER..	..DECEMBER..
	P--E--I	P--E--I	P--E--I	P--E--I	P--E--I	P--E--I
1)	S..P..Y	C..S..W	L..V..U	S..X..R	C..1..P	K..3..M
2)	T..Q..Z	D..T..X	M..W..V	T..Y..S	D..2..Q	L..A..N
3)	U..R..1	E..U..Y	N..X..W	U..Z..T	E..3..R	M..B..O
4)	V..S..2	F..V..Z	O..Y..X	V..1..U	F..A..S	N..C..P
5)	W..T..3	G..W..1	P..Z..Y	W..2..V	G..B..T	O..D..Q
6)	X..U..4	H..X..2	Q..1..Z	X..3..W	H..C..U	P..E..R
7)	A..V..5	J..Y..3	R..2..1	A..A..X	J..D..V	Q..F..S
8)	B..W..6	K..Z..4	S..3..2	B..B..Y	K..E..W	R..G..T
9)	C..X..7	L..1..5	T..A..3	C..C..Z	L..F..X	S..H..U
10)	D..Y..8	M..2..6	U..B..4	D..D..1	M..G..Y	T..J..V
11)	E..Z..A	N..3..7	V..C..5	E..E..2	N..H..Z	U..K..W
12)	F..1..B	O..A..8	W..D..6	F..F..3	O..J..1	V..L..X
13)	G..2..C	P..B..A	X..E..7	G..G..4	P..K..2	W..M..Y
14)	H..3..D	Q..C..B	A..F..8	H..H..5	Q..L..3	X..N..Z
15)	J..A..E	R..D..C	B..G..A	J..J..6	R..M..4	A..O..1
16)	K..B..F	S..E..D	C..H..B	K..K..7	S..N..5	B..P..2
17)	L..C..G	T..F..E	D..J..C	L..L..8	T..O..6	C..Q..3
18)	M..D..H	U..G..F	E..K..D	M..M..A	U..P..7	D..R..4
19)	N..E..J	V..H..G	F..L..E	N..N..B	V..Q..8	E..S..5
20)	O..F..K	W..J..H	G..M..F	O..O..C	W..R..A	F..T..6
21)	P..G..L	X..K..J	H..N..G	P..P..D	X..S..B	G..U..7
22)	Q..H..M	A..L..K	J..O..H	Q..Q..E	A..T..C	H..V..8
23)	R..J..N	B..M..L	K..P..J	R..R..F	B..U..D	J..W..A
24)	S..K..O	C..N..M	L..Q..K	S..S..G	C..V..E	K..X..B
25)	T..L..P	D..O..N	M..R..L	T..T..H	D..W..F	L..Y..C
26)	U..M..Q	E..P..O	N..S..M	U..U..J	E..X..G	M..Z..D
27)	V..N..R	F..Q..P	O..T..N	V..V..K	F..Y..H	N..1..E
28)	W..O..S	G..R..Q	P..U..O	W..W..L	G..Z..J	O..2..F
29)	X..P..T	H..S..R	Q..V..P	X..X..M	H..1..K	P..3..G
30)	A..Q..U	J..T..S	R..W..Q	A..Y..N	J..2..L	Q..A..H
31)	B..R..V	K..U..T		B..Z..O		R..B..J

CODES: P-PHYSICAL BIORHYTHM CURVE, E-EMOTIONAL BIORHYTHM CURVE, I-INTELLECTUAL BIORHYTHM CURVE

1957 1957

	...JANUARY..		..FEBRUARY..		....MARCH...		....APRIL...		.....MAY....		....JUNE....
	P--E--I		P--E--I		P--E--I		P--E--I		P--E--I		P--E--I
1)	S..C..K	1)	C..F..H	1)	H..F..C	1)	Q..J..A	1)	X..L..6	1)	H..O..4
2)	T..D..L	2)	D..G..J	2)	J..G..D	2)	R..K..B	2)	A..M..7	2)	J..P..5
3)	U..E..M	3)	E..H..K	3)	K..H..E	3)	S..L..C	3)	B..N..8	3)	K..Q..6
4)	V..F..N	4)	F..J..L	4)	L..J..F	4)	T..M..D	4)	C..O..A	4)	L..R..7
5)	W..G..O	5)	G..K..M	5)	M..K..G	5)	U..N..E	5)	D..P..B	5)	M..S..8
6)	X..H..P	6)	H..L..N	6)	N..L..H	6)	V..O..F	6)	E..Q..C	6)	N..T..A
7)	A..J..Q	7)	J..M..O	7)	O..M..J	7)	W..P..G	7)	F..R..D	7)	O..U..B
8)	B..K..R	8)	K..N..P	8)	P..N..K	8)	X..Q..H	8)	G..S..E	8)	P..V..C
9)	C..L..S	9)	L..O..Q	9)	Q..O..L	9)	A..R..J	9)	H..T..F	9)	Q..W..D
10)	D..M..T	10)	M..P..R	10)	R..P..M	10)	B..S..K	10)	J..U..G	10)	R..X..E
11)	E..N..U	11)	N..Q..S	11)	S..Q..N	11)	C..T..L	11)	K..V..H	11)	S..Y..F
12)	F..O..V	12)	O..R..T	12)	T..R..O	12)	D..U..M	12)	L..W..J	12)	T..Z..G
13)	G..P..W	13)	P..S..U	13)	U..S..P	13)	E..V..N	13)	M..X..K	13)	U..1..H
14)	H..Q..X	14)	Q..T..V	14)	V..T..Q	14)	F..W..O	14)	N..Y..L	14)	V..2..J
15)	J..R..Y	15)	R..U..W	15)	W..U..R	15)	G..X..P	15)	O..Z..M	15)	W..3..K
16)	K..S..Z	16)	S..V..X	16)	X..V..S	16)	H..Y..Q	16)	P..1..N	16)	X..A..L
17)	L..T..1	17)	T..W..Y	17)	A..W..T	17)	J..Z..R	17)	Q..2..O	17)	A..B..M
18)	M..U..2	18)	U..X..Z	18)	B..X..U	18)	K..1..S	18)	R..3..P	18)	B..C..N
19)	N..V..3	19)	V..Y..1	19)	C..Y..V	19)	L..2..T	19)	S..A..Q	19)	C..D..O
20)	O..W..4	20)	W..Z..2	20)	D..Z..W	20)	M..3..U	20)	T..B..R	20)	D..E..P
21)	P..X..5	21)	X..1..3	21)	E..1..X	21)	N..A..V	21)	U..C..S	21)	E..F..Q
22)	Q..Y..6	22)	A..2..4	22)	F..2..Y	22)	O..B..W	22)	V..D..T	22)	F..G..R
23)	R..Z..7	23)	B..3..5	23)	G..3..Z	23)	P..C..X	23)	W..E..U	23)	G..H..S
24)	S..1..8	24)	C..A..6	24)	H..A..1	24)	Q..D..Y	24)	X..F..V	24)	H..J..T
25)	T..2..A	25)	D..B..7	25)	J..B..2	25)	R..E..Z	25)	A..G..W	25)	J..K..U
26)	U..3..B	26)	E..C..8	26)	K..C..3	26)	S..F..1	26)	B..H..X	26)	K..L..V
27)	V..A..C	27)	F..D..A	27)	L..D..4	27)	T..G..2	27)	C..J..Y	27)	L..M..W
28)	W..B..D	28)	G..E..B	28)	M..E..5	28)	U..H..3	28)	D..K..Z	28)	M..N..X
29)	X..C..E			29)	N..F..6	29)	V..J..4	29)	E..L..1	29)	N..O..Y
30)	A..D..F			30)	O..G..7	30)	W..K..5	30)	F..M..2	30)	O..P..Z
31)	B..E..G			31)	P..H..8			31)	G..N..3		

1957 1957

	....JULY....	...AUGUST...	..SEPTEMBER.	..OCTOBER...	..NOVEMBER..	..DECEMBER..
	P--E--I	P--E--I	P--E--I	P--E--I	P--E--I	P--E--I
1)	P..Q..1	X..T..Y	H..W..W	P..Y..T	X..2..R	G..A..O
2)	Q..R..2	A..U..Z	J..X..X	Q..Z..U	A..3..S	H..B..P
3)	R..S..3	B..V..1	K..Y..Y	R..1..V	B..A..T	J..C..Q
4)	S..T..4	C..W..2	L..Z..Z	S..2..W	C..B..U	K..D..R
5)	T..U..5	D..X..3	M..1..1	T..3..X	D..C..V	L..E..S
6)	U..V..6	E..Y..4	N..2..2	U..A..Y	E..D..W	M..F..T
7)	V..W..7	F..Z..5	O..3..3	V..B..Z	F..E..X	N..G..U
8)	W..X..8	G..1..6	P..A..4	W..C..1	G..F..Y	O..H..V
9)	X..Y..A	H..2..7	Q..B..5	X..D..2	H..G..Z	P..J..W
10)	A..Z..B	J..3..8	R..C..6	A..E..3	J..H..1	Q..K..X
11)	B..1..C	K..A..A	S..D..7	B..F..4	K..J..2	R..L..Y
12)	C..2..D	L..B..B	T..E..8	C..G..5	L..K..3	S..M..Z
13)	D..3..E	M..C..C	U..F..A	D..H..6	M..L..4	T..N..1
14)	E..A..F	N..D..D	V..G..B	E..J..7	N..M..5	U..O..2
15)	F..B..G	O..E..E	W..H..C	F..K..8	O..N..6	V..P..3
16)	G..C..H	P..F..F	X..J..D	G..L..A	P..O..7	W..Q..4
17)	H..D..J	Q..G..G	A..K..E	H..M..B	Q..P..8	X..R..5
18)	J..E..K	R..H..H	B..L..F	J..N..C	R..Q..A	A..S..6
19)	K..F..L	S..J..J	C..M..G	K..O..D	S..R..B	B..T..7
20)	L..G..M	T..K..K	D..N..H	L..P..E	T..S..C	C..U..8
21)	M..H..N	U..L..L	E..O..J	M..Q..F	U..T..D	D..V..A
22)	N..J..O	V..M..M	F..P..K	N..R..G	V..U..E	E..W..B
23)	O..K..P	W..N..N	G..Q..L	O..S..H	W..V..F	F..X..C
24)	P..L..Q	X..O..O	H..R..M	P..T..J	X..W..G	G..Y..D
25)	Q..M..R	A..P..P	J..S..N	Q..U..K	A..X..H	H..Z..E
26)	R..N..S	B..Q..Q	K..T..O	R..V..L	B..Y..J	J..1..F
27)	S..O..T	C..R..R	L..U..P	S..W..M	C..Z..K	K..2..G
28)	T..P..U	D..S..S	M..V..Q	T..X..N	D..1..L	L..3..H
29)	U..Q..V	E..T..T	N..W..R	U..Y..O	E..2..M	M..A..J
30)	V..R..W	F..U..U	O..X..S	V..Z..P	F..3..N	N..B..K
31)	W..S..X	G..V..V		W..1..Q		O..C..L

CODES: P-PHYSICAL BIORHYTHM CURVE,E-EMOTIONAL BIORHYTHM CURVE,I-INTELLECTUAL BIORHYTHM CURVE

1958 | 1958

	...JANUARY..	..FEBRUARY..	....MARCH...	....APRIL...	.....MAY....	....JUNE....
	P--E--I	P--E--I	P--E--I	P--E--I	P--E--I	P--E--I
1)	P..D..M	X..G..K	E..G..E	N..K..C	U..M..8	E..P..6
2)	Q..E..N	A..H..L	F..H..F	O..L..D	V..N..A	F..Q..7
3)	R..F..O	B..J..M	G..J..G	P..M..E	W..O..B	G..R..8
4)	S..G..P	C..K..N	H..K..H	Q..N..F	X..P..C	H..S..A
5)	T..H..Q	D..L..O	J..L..J	R..O..G	A..Q..D	J..T..B
6)	U..J..R	E..M..P	K..M..K	S..P..H	B..R..E	K..U..C
7)	V..K..S	F..N..Q	L..N..L	T..Q..J	C..S..F	L..V..D
8)	W..L..T	G..O..R	M..O..M	U..R..K	D..T..G	M..W..E
9)	X..M..U	H..P..S	N..P..N	V..S..L	E..U..H	N..X..F
10)	A..N..V	J..Q..T	O..Q..O	W..T..M	F..V..J	O..Y..G
11)	B..O..W	K..R..U	P..R..P	X..U..N	G..W..K	P..Z..H
12)	C..P..X	L..S..V	Q..S..Q	A..V..O	H..X..L	Q..1..J
13)	D..Q..Y	M..T..W	R..T..R	B..W..P	J..Y..M	R..2..K
14)	E..R..Z	N..U..X	S..U..S	C..X..Q	K..Z..N	S..3..L
15)	F..S..1	O..V..Y	T..V..T	D..Y..R	L..1..O	T..A..M
16)	G..T..2	P..W..Z	U..W..U	E..Z..S	M..2..P	U..B..N
17)	H..U..3	Q..X..1	V..X..V	F..1..T	N..3..Q	V..C..O
18)	J..V..4	R..Y..2	W..Y..W	G..2..U	O..A..R	W..D..P
19)	K..W..5	S..Z..3	X..Z..X	H..3..V	P..B..S	X..E..Q
20)	L..X..6	T..1..4	A..1..Y	J..A..W	Q..C..T	A..F..R
21)	M..Y..7	U..2..5	B..2..Z	K..B..X	R..D..U	B..G..S
22)	N..Z..8	V..3..6	C..3..1	L..C..Y	S..E..V	C..H..T
23)	O..1..A	W..A..7	D..A..2	M..D..Z	T..F..W	D..J..U
24)	P..2..B	X..B..8	E..B..3	N..E..1	U..G..X	E..K..V
25)	Q..3..C	A..C..A	F..C..4	O..F..2	V..H..Y	F..L..W
26)	R..A..D	B..D..B	G..D..5	P..G..3	W..J..Z	G..M..X
27)	S..B..E	C..E..C	H..E..6	Q..H..4	X..K..1	H..N..Y
28)	T..C..F	D..F..D	J..F..7	R..J..5	A..L..2	J..O..Z
29)	U..D..G		K..G..8	S..K..6	B..M..3	K..P..1
30)	V..E..H		L..H..A	T..L..7	C..N..4	L..Q..2
31)	W..F..J		M..J..B		D..O..5	

1958											1958
	....JULY....		...AUGUST...		..SEPTEMBER.		..OCTOBER...		..NOVEMBER..		..DECEMBER..
	P--E--I		P--E--I		P--E--I		P--E--I		P--E--I		P--E--I
1)	M..R..3	1)	U..U..1	1)	E..X..Y	1)	M..Z..V	1)	U..3..T	1)	D..B..Q
2)	N..S..4	2)	V..V..2	2)	F..Y..Z	2)	N..1..W	2)	V..A..U	2)	E..C..R
3)	O..T..5	3)	W..W..3	3)	G..Z..1	3)	O..2..X	3)	W..B..V	3)	F..D..S
4)	P..U..6	4)	X..X..4	4)	H..1..2	4)	P..3..Y	4)	X..C..W	4)	G..E..T
5)	Q..V..7	5)	A..Y..5	5)	J..2..3	5)	Q..A..Z	5)	A..D..X	5)	H..F..U
6)	R..W..8	6)	B..Z..6	6)	K..3..4	6)	R..B..1	6)	B..E..Y	6)	J..G..V
7)	S..X..A	7)	C..1..7	7)	L..A..5	7)	S..C..2	7)	C..F..Z	7)	K..H..W
8)	T..Y..B	8)	D..2..8	8)	M..B..6	8)	T..D..3	8)	D..G..1	8)	L..J..X
9)	U..Z..C	9)	E..3..A	9)	N..C..7	9)	U..E..4	9)	E..H..2	9)	M..K..Y
10)	V..1..D	10)	F..A..B	10)	O..D..8	10)	V..F..5	10)	F..J..3	10)	N..L..Z
11)	W..2..E	11)	G..B .C	11)	P..E..A	11)	W..G..6	11)	G..K..4	11)	O..M..1
12)	X..3..F	12)	H..C..D	12)	Q..F..B	12)	X..H..7	12)	H..L..5	12)	P..N..2
13)	A..A..G	13)	J..D..E	13)	R..G..C	13)	A..J..8	13)	J..M..6	13)	Q..O..3
14)	B..B..H	14)	K..E..F	14)	S..H..D	14)	B..K..A	14)	K..N..7	14)	R..P..4
15)	C..C..J	15)	L..F..G	15)	T..J..E	15)	C..L..B	15)	L..O..8	15)	S..Q..5
16)	D..D..K	16)	M..G..H	16)	U..K..F	16)	D..M..C	16)	M..P..A	16)	T..R..6
17)	E..E..L	17)	N..H..J	17)	V..L..G	17)	E..N..D	17)	N..Q..B	17)	U..S..7
18)	F..F..M	18)	O..J..K	18)	W..M..H	18)	F..O..E	18)	O..R..C	18)	V..T..8
19)	G..G..N	19)	P..K..L	19)	X..N..J	19)	G..P..F	19)	P..S..D	19)	W..U..A
20)	H..H..O	20)	Q..L..M	20)	A..O..K	20)	H..Q..G	20)	Q..T..E	20)	X..V..B
21)	J..J..P	21)	R..M..N	21)	B..P..L	21)	J..R..H	21)	R..U..F	21)	A..W..C
22)	K..K..Q	22)	S..N..O	22)	C..Q..M	22)	K..S..J	22)	S..V..G	22)	B..X..D
23)	L..L..R	23)	T..O..P	23)	D..R..N	23)	L..T..K	23)	T..W..H	23)	C..Y..E
24)	M..M..S	24)	U..P..Q	24)	E..S..O	24)	M..U..L	24)	U..X..J	24)	D..Z..F
25)	N..N..T	25)	V..Q..R	25)	F..T..P	25)	N..V..M	25)	V..Y..K	25)	E..1..G
26)	O..O..U	26)	W..R..S	26)	G..U..Q	26)	O..W..N	26)	W..Z..L	26)	F..2..H
27)	P..P..V	27)	X..S..T	27)	H..V..R	27)	P..X..O	27)	X..1..M	27)	G..3..J
28)	Q..Q..W	28)	A..T..U	28)	J..W..S	28)	Q..Y..P	28)	A..2..N	28)	H..A..K
29)	R..R..X	29)	B..U..V	29)	K..X..T	29)	R..Z..Q	29)	B..3..O	29)	J..B..L
30)	S..S..Y	30)	C..V..W	30)	L..Y..U	30)	S..1..R	30)	C..A..P	30)	K..C..M
31)	T..T..Z	31)	D..W..X			31)	T..2..S			31)	L..D..N

CODES: P-PHYSICAL BIORHYTHM CURVE,E-EMOTIONAL BIORHYTHM CURVE,I-INTELLECTUAL BIORHYTHM CURVE

1959 1959

	...JANUARY..	..FEBRUARY..	....MARCH...	....APRIL...	.....MAY....	.....JUNE....
	P--E--I	P--E--I	P--E--I	P--E--I	P--E--I	P--E--I
1)	M..E..O	U..H..M	B..H..G	K..L..E	R..N..B	B..Q..8
2)	N..F..P	V..J..N	C..J..H	L..M..F	S..O..C	C..R..A
3)	O..G..Q	W..K..O	D..K..J	M..N..G	T..P..D	D..S..B
4)	P..H..R	X..L..P	E..L..K	N..O..H	U..Q..E	E..T..C
5)	Q..J..S	A..M..Q	F..M..L	O..P..J	V..R..F	F..U..D
6)	R..K..T	B..N..R	G..N..M	P..Q..K	W..S..G	G..V..E
7)	S..L..U	C..O..S	H..O..N	Q..R..L	X..T..H	H..W..F
8)	T..M..V	D..P..T	J..P..O	R..S..M	A..U..J	J..X..G
9)	U..N..W	E..Q..U	K..Q..P	S..T..N	B..V..K	K..Y..H
10)	V..O..X	F..R..V	L..R..Q	T..U..O	C..W..L	L..Z..J
11)	W..P..Y	G..S..W	M..S..R	U..V..P	D..X..M	M..1..K
12)	X..Q..Z	H..T..X	N..T..S	V..W..Q	E..Y..N	N..2..L
13)	A..R..1	J..U..Y	O..U..T	W..X..R	F..Z..O	O..3..M
14)	B..S..2	K..V..Z	P..V..U	X..Y..S	G..1..P	P..A..N
15)	C..T..3	L..W..1	Q..W..V	A..Z..T	H..2..Q	Q..B..O
16)	D..U..4	M..X..2	R..X..W	B..1..U	J..3..R	R..C..P
17)	E..V..5	N..Y..3	S..Y..X	C..2..V	K..A..S	S..D..Q
18)	F..W..6	O..Z..4	T..Z..Y	D..3..W	L..B..T	T..E..R
19)	G..X..7	P..1..5	U..1..Z	E..A..X	M..C..U	U..F..S
20)	H..Y..8	Q..2..6	V..2..1	F..B..Y	N..D..V	V..G..T
21)	J..Z..A	R..3..7	W..3..2	G..C..Z	O..E..W	W..H..U
22)	K..1..B	S..A..8	X..A..3	H..D..1	P..F..X	X..J..V
23)	L..2..C	T..B..A	A..B..4	J..E..2	Q..G..Y	A..K..W
24)	M..3..D	U..C..B	B..C..5	K..F..3	R..H..Z	B..L..X
25)	N..A..E	V..D..C	C..D..6	L..G..4	S..J..1	C..M..Y
26)	O..B..F	W..E..D	D..E..7	M..H..5	T..K..2	D..N..Z
27)	P..C..G	X..F..E	E..F..8	N..J..6	U..L..3	E..O..1
28)	Q..D..H	A..G..F	F..G..A	O..K..7	V..M..4	F..P..2
29)	R..E..J		G..H..B	P..L..8	W..N..5	G..Q..3
30)	S..F..K		H..J..C	Q..M..A	X..O..6	H..R..4
31)	T..G..L		J..K..D		A..P..7	

1959 | | | | | 1959

	....JULY....	...AUGUST...	..SEPTEMBER.	..OCTOBER...	..NOVEMBER..	..DECEMBER..
	P--E--I	P--E--I	P--E--I	P--E--I	P--E--I	P--E--I
1)	J..S..5	R..V..3	B..Y..1	J..1..X	R..A..V	A..C..S
2)	K..T..6	S..W..4	C..Z..2	K..2..Y	S..B..W	B..D..T
3)	L..U..7	T..X..5	D..1..3	L..3..Z	T..C..X	C..E..U
4)	M..V..8	U..Y..6	E..2..4	M..A..1	U..D..Y	D..F..V
5)	N..W..A	V..Z..7	F..3..5	N..B..2	V..E..Z	E..G..W
6)	O..X..B	W..1..8	G..A..6	O..C..3	W..F..1	F..H..X
7)	P..Y..C	X..2..A	H..B..7	P..D..4	X..G..2	G..J..Y
8)	Q..Z..D	A..3..B	J..C..8	Q..E..5	A..H..3	H..K..Z
9)	R..1..E	B..A..C	K..D..A	R..F..6	B..J..4	J..L..1
10)	S..2..F	C..B..D	L..E..B	S..G..7	C..K..5	K..M..2
11)	T..3..G	D..C..E	M..F..C	T..H..8	D..L..6	L..N..3
12)	U..A..H	E..D..F	N..G..D	U..J..A	E..M..7	M..O..4
13)	V..B..J	F..E..G	O..H..E	V..K..B	F..N..8	N..P..5
14)	W..C..K	G..F..H	P..J..F	W..L..C	G..O..A	O..Q..6
15)	X..D..L	H..G..J	Q..K..G	X..M..D	H..P..B	P..R..7
16)	A..E..M	J..H..K	R..L..H	A..N..E	J..Q..C	Q..S..8
17)	B..F..N	K..J..L	S..M..J	B..O..F	K..R..D	R..T..A
18)	C..G..O	L..K..M	T..N..K	C..P..G	L..S..E	S..U..B
19)	D..H..P	M..L..N	U..O..L	D..Q..H	M..T..F	T..V..C
20)	E..J..Q	N..M..O	V..P..M	E..R..J	N..U..G	U..W..D
21)	F..K..R	O..N..P	W..Q..N	F..S..K	O..V..H	V..X..E
22)	G..L..S	P..O..Q	X..R..O	G..T..L	P..W..J	W..Y..F
23)	H..M..T	Q..P..R	A..S..P	H..U..M	Q..X..K	X..Z..G
24)	J..N..U	R..Q..S	B..T..Q	J..V..N	R..Y..L	A..1..H
25)	K..O..V	S..R..T	C..U..R	K..W..O	S..Z..M	B..2..J
26)	L..P..W	T..S..U	D..V..S	L..X..P	T..1..N	C..3..K
27)	M..Q..X	U..T..V	E..W..T	M..Y..Q	U..2..O	D..A..L
28)	N..R..Y	V..U..W	F..X..U	N..Z..R	V..3..P	E..B..M
29)	O..S..Z	W..V..X	G..Y..V	O..1..S	W..A..Q	F..C..N
30)	P..T..1	X..W..Y	H..Z..W	P..2..T	X..B..R	G..D..O
31)	Q..U..2	A..X..Z		Q..3..U		H..E..P

CODES: P-PHYSICAL BIORHYTHM CURVE,E-EMOTIONAL BIORHYTHM CURVE,I-INTELLECTUAL BIORHYTHM CURVE

1960											1960
...JANUARY..		..FEBRUARY..		....MARCH...		....APRIL...		.....MAY....		....JUNE....
	P--E--I		P--E--I		P--E--I		P--E--I		P--E--I		P--E--I
1)	J..F..Q	1)	R..J..O	1)	X..K..K	1)	H..N..H	1)	P..P..E	1)	X..S..C
2)	K..G..R	2)	S..K..P	2)	A..L..L	2)	J..O..J	2)	Q..Q..F	2)	A..T..D
3)	L..H..S	3)	T..L..Q	3)	B..M..M	3)	K..P..K	3)	R..R..G	3)	B..U..E
4)	M..J..T	4)	U..M..R	4)	C..N..N	4)	L..Q..L	4)	S..S..H	4)	C..V..F
5)	N..K..U	5)	V..N..S	5)	D..O..O	5)	M..R..M	5)	T..T..J	5)	D..W..G
6)	O..L..V	6)	W..O..T	6)	E..P..P	6)	N..S..N	6)	U..U..K	6)	E..X..H
7)	P..M..W	7)	X..P..U	7)	F..Q..Q	7)	O..T..O	7)	V..V..L	7)	F..Y..J
8)	Q..N..X	8)	A..Q..V	8)	G..R..R	8)	P..U..P	8)	W..W..M	8)	G..Z..K
9)	R..O..Y	9)	B..R..W	9)	H..S..S	9)	Q..V..Q	9)	X..X..N	9)	H..1..L
10)	S..P..Z	10)	C..S..X	10)	J..T..T	10)	R..W..R	10)	A..Y..O	10)	J..2..M
11)	T..Q..1	11)	D..T..Y	11)	K..U..U	11)	S..X..S	11)	B..Z..P	11)	K..3..N
12)	U..R..2	12)	E..U..Z	12)	L..V..V	12)	T..Y..T	12)	C..1..Q	12)	L..A..O
13)	V..S..3	13)	F..V..1	13)	M..W..W	13)	U..Z..U	13)	D..2..R	13)	M..B..P
14)	W..T..4	14)	G..W..2	14)	N..X..X	14)	V..1..V	14)	E..3..S	14)	N..C..Q
15)	X..U..5	15)	H..X..3	15)	O..Y..Y	15)	W..2..W	15)	F..A..T	15)	O..D..R
16)	A..V..6	16)	J..Y..4	16)	P..Z..Z	16)	X..3..X	16)	G..B..U	16)	P..E..S
17)	B..W..7	17)	K..Z..5	17)	Q..1..1	17)	A..A..Y	17)	H..C..V	17)	Q..F..T
18)	C..X..8	18)	L..1..6	18)	R..2..2	18)	B..B..Z	18)	J..D..W	18)	R..G..U
19)	D..Y..A	19)	M..2..7	19)	S..3..3	19)	C..C..1	19)	K..E..X	19)	S..H..V
20)	E..Z..B	20)	N..3..8	20)	T..A..4	20)	D..D..2	20)	L..F..Y	20)	T..J..W
21)	F..1..C	21)	O..A..A	21)	U..B..5	21)	E..E..3	21)	M..G..Z	21)	U..K..X
22)	G..2..D	22)	P..B..B	22)	V..C..6	22)	F..F..4	22)	N..H..1	22)	V..L..Y
23)	H..3..E	23)	Q..C..C	23)	W..D..7	23)	G..G..5	23)	O..J..2	23)	W..M..Z
24)	J..A..F	24)	R..D..D	24)	X..E..8	24)	H..H..6	24)	P..K..3	24)	X..N..1
25)	K..B..G	25)	S..E..E	25)	A..F..A	25)	J..J..7	25)	Q..L..4	25)	A..O..2
26)	L..C..H	26)	T..F..F	26)	B..G..B	26)	K..K..8	26)	R..M..5	26)	B..P..3
27)	M..D..J	27)	U..G..G	27)	C..H..C	27)	L..L..A	27)	S..N..6	27)	C..Q..4
28)	N..E..K	28)	V..H..H	28)	D..J..D	28)	M..M..B	28)	T..O..7	28)	D..R..5
29)	O..F..L	29)	W..J..J	29)	E..K..E	29)	N..N..C	29)	U..P..8	29)	E..S..6
30)	P..G..M			30)	F..L..F	30)	O..O..D	30)	V..Q..A	30)	F..T..7
31)	Q..H..N			31)	G..M..G			31)	W..R..B		

1960 | 1960

Day	....JULY.... P--E--I	...AUGUST... P--E--I	..SEPTEMBER. P--E--I	..OCTOBER... P--E--I	..NOVEMBER.. P--E--I	..DECEMBER.. P--E--I
1)	G..U..8	P..X..6	X..1..4	G..3..1	P..C..Y	W..E..V
2)	H..V..A	Q..Y..7	A..2..5	H..A..2	Q..D..Z	X..F..W
3)	J..W..B	R..Z..8	B..3..6	J..B..3	R..E..1	A..G..X
4)	K..X..C	S..1..A	C..A..7	K..C..4	S..F..2	B..H..Y
5)	L..Y..D	T..2..B	D..B..8	L..D..5	T..G..3	C..J..Z
6)	M..Z..E	U..3..C	E..C..A	M..E..6	U..H..4	D..K..1
7)	N..1..F	V..A..D	F..D..B	N..F..7	V..J..5	E..L..2
8)	O..2..G	W..B..E	G..E..C	O..G..8	W..K..6	F..M..3
9)	P..3..H	X..C..F	H..F..D	P..H..A	X..L..7	G..N..4
10)	Q..A..J	A..D..G	J..G..E	Q..J..B	A..M..8	H..O..5
11)	R..B..K	B..E..H	K..H..F	R..K..C	B..N..A	J..P..6
12)	S..C..L	C..F..J	L..J..G	S..L..D	C..O..B	K..Q..7
13)	T..D..M	D..G..K	M..K..H	T..M..E	D..P..C	L..R..8
14)	U..E..N	E..H..L	N..L..J	U..N..F	E..Q..D	M..S..A
15)	V..F..O	F..J..M	O..M..K	V..O..G	F..R..E	N..T..B
16)	W..G..P	G..K..N	P..N..L	W..P..H	G..S..F	O..U..C
17)	X..H..Q	H..L..O	Q..O..M	X..Q..J	H..T..G	P..V..D
18)	A..J..R	J..M..P	R..P..N	A..R..K	J..U..H	Q..W..E
19)	B..K..S	K..N..Q	S..Q..O	B..S..L	K..V..J	R..X..F
20)	C..L..T	L..O..R	T..R..P	C..T..M	L..W..K	S..Y..G
21)	D..M..U	M..P..S	U..S..Q	D..U..N	M..X..L	T..Z..H
22)	E..N..V	N..Q..T	V..T..R	E..V..O	N..Y..M	U..1..J
23)	F..O..W	O..R..U	W..U..S	F..W..P	O..Z..N	V..2..K
24)	G..P..X	P..S..V	X..V..T	G..X..Q	P..1..O	W..3..L
25)	H..Q..Y	Q..T..W	A..W..U	H..Y..R	Q..2..P	X..A..M
26)	J..R..Z	R..U..X	B..X..V	J..Z..S	R..3..Q	A..B..N
27)	K..S..1	S..V..Y	C..Y..W	K..1..T	S..A..R	B..C..O
28)	L..T..2	T..W..Z	D..Z..X	L..2..U	T..B..S	C..D..P
29)	M..U..3	U..X..1	E..1..Y	M..3..V	U..C..T	D..E..Q
30)	N..V..4	V..Y..2	F..2..Z	N..A..W	V..D..U	E..F..R
31)	O..W..5	W..Z..3		O..B..X		F..G..S

CODES: P-PHYSICAL BIORHYTHM CURVE, E-EMOTIONAL BIORHYTHM CURVE, I-INTELLECTUAL BIORHYTHM CURVE

1961						1961

	...JANUARY..	..FEBRUARY..	....MARCH...	....APRIL...	.....MAY....	....JUNE....
	P--E--I	P--E--I	P--E--I	P--E--I	P--E--I	P--E--I
1)	G..H..T	P..L..R	U..L..M	E..O..K	M..Q..G	U..T..E
2)	H..J..U	Q..M..S	V..M..N	F..P..L	N..R..H	V..U..F
3)	J..K..V	R..N..T	W..N..O	G..Q..M	O..S..J	W..V..G
4)	K..L..W	S..O..U	X..O..P	H..R..N	P..T..K	X..W..H
5)	L..M..X	T..P..V	A..P..Q	J..S..O	Q..U..L	A..X..J
6)	M..N..Y	U..Q..W	B..Q..R	K..T..P	R..V..M	B..Y..K
7)	N..O..Z	V..R..X	C..R..S	L..U..Q	S..W..N	C..Z..L
8)	O..P..1	W..S..Y	D..S..T	M..V..R	T..X..O	D..1..M
9)	P..Q..2	X..T..Z	E..T..U	N..W..S	U..Y..P	E..2..N
10)	Q..R..3	A..U..1	F..U..V	O..X..T	V..Z..Q	F..3..O
11)	R..S..4	B..V..2	G..V..W	P..Y..U	W..1..R	G..A..P
12)	S..T..5	C..W..3	H..W..X	Q..Z..V	X..2..S	H..B..Q
13)	T..U..6	D..X..4	J..X..Y	R..1..W	A..3..T	J..C..R
14)	U..V..7	E..Y..5	K..Y..Z	S..2..X	B..A..U	K..D..S
15)	V..W..8	F..Z..6	L..Z..1	T..3..Y	C..B..V	L..E..T
16)	W..X..A	G..1..7	M..1..2	U..A..Z	D..C..W	M..F..U
17)	X..Y..B	H..2..8	N..2..3	V..B..1	E..D..X	N..G..V
18)	A..Z..C	J..3..A	O..3..4	W..C..2	F..E..Y	O..H..W
19)	B..1..D	K..A..B	P..A..5	X..D..3	G..F..Z	P..J..X
20)	C..2..E	L..B..C	Q..B..6	A..E..4	H..G..1	Q..K..Y
21)	D..3..F	M..C..D	R..C..7	B..F..5	J..H..2	R..L..Z
22)	E..A..G	N..D..E	S..D..8	C..G..6	K..J..3	S..M..1
23)	F..B..H	O..E..F	T..E..A	D..H..7	L..K..4	T..N..2
24)	G..C..J	P..F..G	U..F..B	E..J..8	M..L..5	U..O..3
25)	H..D..K	Q..G..H	V..G..C	F..K..A	N..M..6	V..P..4
26)	J..E..L	R..H..J	W..H..D	G..L..B	O..N..7	W..Q..5
27)	K..F..M	S..J..K	X..J..E	H..M..C	P..O..8	X..R..6
28)	L..G..N	T..K..L	A..K..F	J..N..D	Q..P..A	A..S..7
29)	M..H..O		B..L..G	K..O..E	R..Q..B	B..T..8
30)	N..J..P		C..M..H	L..P..F	S..R..C	C..U..A
31)	O..K..Q		D..N..J		T..S..D	

1961 1961

	....JULY....	...AUGUST...	..SEPTEMBER.	..OCTOBER...	..NOVEMBER..	..DECEMBER..
	P--E--I	P--E--I	P--E--I	P--E--I	P--E--I	P--E--I
1)	D..V..B	M..Y..8	U..2..6	D..A..3	M..D..1	T..F..X
2)	E..W..C	N..Z..A	V..3..7	E..B..4	N..E..2	U..G..Y
3)	F..X..D	O..1..B	W..A..8	F..C..5	O..F..3	V..H..Z
4)	G..Y..E	P..2..C	X..B..A	G..D..6	P..G..4	W..J..1
5)	H..Z..F	Q..3..D	A..C..B	H..E..7	Q..H..5	X..K..2
6)	J..1..G	R..A..E	B..D..C	J..F..8	R..J..6	A..L..3
7)	K..2..H	S..B..F	C..E..D	K..G..A	S..K..7	B..M..4
8)	L..3..J	T..C..G	D..F..E	L..H..B	T..L..8	C..N..5
9)	M..A..K	U..D..H	E..G..F	M..J..C	U..M..A	D..O..6
10)	N..B..L	V..E..J	F..H..G	N..K..D	V..N..B	E..P..7
11)	O..C..M	W..F..K	G..J..H	O..L..E	W..O..C	F..Q..8
12)	P..D..N	X..G..L	H..K..J	P..M..F	X..P..D	G..R..A
13)	Q..E..O	A..H..M	J..L..K	Q..N..G	A..Q..E	H..S..B
14)	R..F..P	B..J..N	K..M..L	R..O..H	B..R..F	J..T..C
15)	S..G..Q	C..K..O	L..N..M	S..P..J	C..S..G	K..U..D
16)	T..H..R	D..L..P	M..O..N	T..Q..K	D..T..H	L..V..E
17)	U..J..S	E..M..Q	M..P..O	U..R..L	E..U..J	M..W..F
18)	V..K..T	F..N..R	O..Q..P	V..S..M	F..V..K	N..X..G
19)	W..L..U	G..O..S	P..R..Q	W..T..N	G..W..L	O..Y..H
20)	X..M..V	H..P..T	Q..S..R	X..U..O	H..X..M	P..Z..J
21)	A..N..W	J..Q..U	R..T..S	A..V..P	J..Y..N	Q..1..K
22)	B..O..X	K..R..V	S..U..T	B..W..Q	K..Z..O	R..2..L
23)	C..P..Y	L..S..W	T..V..U	C..X..R	L..1..P	S..3..M
24)	D..Q..Z	M..T..X	U..W..V	D..Y..S	M..2..Q	T..A..N
25)	E..R..1	N..U..Y	V..X..W	E..Z..T	N..3..R	U..B..O
26)	F..S..2	O..V..Z	W..Y..X	F..1..U	O..A..S	V..C..P
27)	G..T..3	P..W..1	X..Z..Y	G..2..V	P..B..T	W..D..Q
28)	H..U..4	Q..X..2	A..1..Z	H..3..W	Q..C..U	X..E..R
29)	J..V..5	R..Y..3	B..2..1	J..A..X	R..D..V	A..F..S
30)	K..W..6	S..Z..4	C..3..2	K..B..Y	S..E..W	B..G..T
31)	L..X..7	T..1..5		L..C..Z		C..H..U

CODES: P-PHYSICAL BIORHYTHM CURVE, E-EMCTIONAL BIORHYTHM CURVE, I-INTELLECTUAL BIORHYTHM CURVE

1962 1962

	...JANUARY..	..FEBRUARY..	....MARCH...	....APRIL...	.....MAY....	....JUNE....
	P--E--I	P--E--I	P--E--I	P--E--I	P--E--I	P--E--I
1)	D..J..V	M..M..T	R..M..O	B..P..M	J..R..J	R..U..G
2)	E..K..W	N..N..U	S..N..P	C..Q..N	K..S..K	S..V..H
3)	F..L..X	O..O..V	T..O..Q	D..R..O	L..T..L	T..W..J
4)	G..M..Y	P..P..W	U..P..R	E..S..P	M..U..M	U..X..K
5)	H..N..Z	Q..Q..X	V..Q..S	F..T..Q	N..V..N	V..Y..L
6)	J..O..1	R..R..Y	W..R..T	G..U..R	O..W..O	W..Z..M
7)	K..P..2	S..S..Z	X..S..U	H..V..S	P..X..P	X..1..N
8)	L..Q..3	T..T..1	A..T..V	J..W..T	Q..Y..Q	A..2..O
9)	M..R..4	U..U..2	B..U..W	K..X..U	R..Z..R	B..3..P
10)	N..S..5	V..V..3	C..V..X	L..Y..V	S..1..S	C..A..Q
11)	O..T..6	W..W..4	D..W..Y	M..Z..W	T..2..T	D..B..R
12)	P..U..7	X..X..5	E..X..Z	N..1..X	U..3..U	E..C..S
13)	Q..V..8	A..Y..6	F..Y..1	O..2..Y	V..A..V	F..D..T
14)	R..W..A	B..Z..7	G..Z..2	P..3..Z	W..B..W	G..E..U
15)	S..X..B	C..1..8	H..1..3	Q..A..1	X..C..X	H..F..V
16)	T..Y..C	D..2..A	J..2..4	R..B..2	A..D..Y	J..G..W
17)	U..Z..D	E..3..B	K..3..5	S..C..3	B..E..Z	K..H..X
18)	V..1..E	F..A..C	L..A..6	T..D..4	C..F..1	L..J..Y
19)	W..2..F	G..B..D	M..B..7	U..E..5	D..G..2	M..K..Z
20)	X..3..G	H..C..E	N..C..8	V..F..6	E..H..3	N..L..1
21)	A..A..H	J..D..F	O..D..A	W..G..7	F..J..4	O..M..2
22)	B..B..J	K..E..G	P..E..B	X..H..8	G..K..5	P..N..3
23)	C..C..K	L..F..H	Q..F..C	A..J..A	H..L..6	Q..O..4
24)	D..D..L	M..G..J	R..G..D	B..K..B	J..M..7	R..P..5
25)	E..E..M	N..H..K	S..H..E	C..L..C	K..N..8	S..Q..6
26)	F..F..N	O..J..L	T..J..F	D..M..D	L..O..A	T..R..7
27)	G..G..O	P..K..M	U..K..G	E..N..E	M..P..B	U..S..8
28)	H..H..P	Q..L..N	V..L..H	F..O..F	N..Q..C	V..T..A
29)	J..J..Q		W..M..J	G..P..G	O..R..D	W..U..B
30)	K..K..R		X..N..K	H..Q..H	P..S..E	X..V..C
31)	L..L..S		A..O..L		Q..T..F	

1962 | 1962

Day	....JULY.... P--E--I	...AUGUST... P--E--I	..SEPTEMBER. P--E--I	..OCTOBER... P--E--I	..NOVEMBER.. P--E--I	..DECEMBER.. P--E--I
1)	A..W..D	J..Z..B	R..3..8	A..B..5	J..E..3	Q..G..Z
2)	B..X..E	K..1..C	S..A..A	B..C..6	K..F..4	R..H..1
3)	C..Y..F	L..2..D	T..B..B	C..D..7	L..G..5	S..J..2
4)	D..Z..G	M..3..E	U..C..C	D..E..8	M..H..6	T..K..3
5)	E..1..H	N..A..F	V..D..D	E..F..A	N..J..7	U..L..4
6)	F..2..J	O..B..G	W..E..E	F..G..B	O..K..8	V..M..5
7)	G..3..K	P..C..H	X..F..F	G..H..C	P..L..A	W..N..6
8)	H..A..L	Q..D..J	A..G..G	H..J..D	Q..M..B	X..O..7
9)	J..B..M	R..E..K	B..H..H	J..K..E	R..N..C	A..P..8
10)	K..C..N	S..F..L	C..J..J	K..L..F	S..O..D	B..Q..A
11)	L..D..O	T..G..M	D..K..K	L..M..G	T..P..E	C..R..B
12)	M..E..P	U..H..N	E..L..L	M..N..H	U..Q..F	D..S..C
13)	N..F..Q	V..J..O	F..M..M	N..O..J	V..R..G	E..T..D
14)	O..G..R	W..K..P	G..N..N	O..P..K	W..S..H	F..U..E
15)	P..H..S	X..L..Q	H..O..O	P..Q..L	X..T..J	G..V..F
16)	Q..J..T	A..M..R	J..P..P	Q..R..M	A..U..K	H..W..G
17)	R..K..U	B..N..S	K..Q..Q	R..S..N	B..V..L	J..X..H
18)	S..L..V	C..O..T	L..R..R	S..T..O	C..W..M	K..Y..J
19)	T..M..W	D..P..U	M..S..S	T..U..P	D..X..N	L..Z..K
20)	U..N..X	E..Q..V	N..T..T	U..V..Q	E..Y..O	M..1..L
21)	V..O..Y	F..R..W	O..U..U	V..W..R	F..Z..P	N..2..M
22)	W..P..Z	G..S..X	P..V..V	W..X..S	G..1..Q	O..3..N
23)	X..Q..1	H..T..Y	Q..W..W	X..Y..T	H..2..R	P..A..O
24)	A..R..2	J..U..Z	R..X..X	A..Z..U	J..3..S	Q..B..P
25)	B..S..3	K..V..1	S..Y..Y	B..1..V	K..A..T	R..C..Q
26)	C..T..4	L..W..2	T..Z..Z	C..2..W	L..B..U	S..D..R
27)	D..U..5	M..X..3	U..1..1	D..3..X	M..C..V	T..E..S
28)	E..V..6	N..Y..4	V..2..2	E..A..Y	N..D..W	U..F..T
29)	F..W..7	O..Z..5	W..3..3	F..B..Z	O..E..X	V..G..U
30)	G..X..8	P..1..6	X..A..4	G..C..1	P..F..Y	W..H..V
31)	H..Y..A	Q..2..7		H..D..2		X..J..W

CODES: P-PHYSICAL BIORHYTHM CURVE,E-EMOTIONAL BIORHYTHM CURVE,I-INTELLECTUAL BIORHYTHM CURVE

1963 | 1963

	...JANUARY..	..FEBRUARY..	....MARCH...	....APRIL...	.....MAY....	....JUNE....
	P--E--I	P--E--I	P--E--I	P--E--I	P--E--I	P--E--I
1)	A..K..X	J..N..V	O..N..Q	W..Q..O	F..S..L	O..V..J
2)	B..L..Y	K..O..W	P..O..R	X..R..P	G..T..M	P..W..K
3)	C..M..Z	L..P..X	Q..P..S	A..S..Q	H..U..N	Q..X..L
4)	D..N..1	M..Q..Y	R..Q..T	B..T..R	J..V..O	R..Y..M
5)	E..O..2	N..R..Z	S..R..U	C..U..S	K..W..P	S..Z..N
6)	F..P..3	O..S..1	T..S..V	D..V..T	L..X..Q	T..1..O
7)	G..Q..4	P..T..2	U..T..W	E..W..U	M..Y..R	U..2..P
8)	H..R..5	Q..U..3	V..U..X	F..X..V	N..Z..S	V..3..Q
9)	J..S..6	R..V..4	W..V..Y	G..Y..W	O..1..T	W..A..R
10)	K..T..7	S..W..5	X..W..Z	H..Z..X	P..2..U	X..B..S
11)	L..U..8	T..X..6	A..X..1	J..1..Y	Q..3..V	A..C..T
12)	M..V..A	U..Y..7	B..Y..2	K..2..Z	R..A..W	B..D..U
13)	N..W..B	V..Z..8	C..Z..3	L..3..1	S..B..X	C..E..V
14)	O..X..C	W..1..A	D..1..4	M..A..2	T..C..Y	D..F..W
15)	P..Y..D	X..2..B	E..2..5	N..B..3	U..D..Z	E..G..X
16)	Q..Z..E	A..3..C	F..3..6	O..C..4	V..E..1	F..H..Y
17)	R..1..F	B..A..D	G..A..7	P..D..5	W..F..2	G..J..Z
18)	S..2..G	C..B..E	H..B..8	Q..E..6	X..G..3	H..K..1
19)	T..3..H	D..C..F	J..C..A	R..F..7	A..H..4	J..L..2
20)	U..A..J	E..D..G	K..D..B	S..G..8	B..J..5	K..M..3
21)	V..B..K	F..E..H	L..E..C	T..H..A	C..K..6	L..N..4
22)	W..C..L	G..F..J	M..F..D	U..J..B	D..L..7	M..O..5
23)	X..D..M	H..G..K	N..G..E	V..K..C	E..M..8	N..P..6
24)	A..E..N	J..H..L	O..H..F	W..L..D	F..N..A	O..Q..7
25)	B..F..O	K..J..M	P..J..G	X..M..E	G..O..B	P..R..8
26)	C..G..P	L..K..N	Q..K..H	A..N..F	H..P..C	Q..S..A
27)	D..H..Q	M..L..O	R..L..J	B..O..G	J..Q..D	R..T..B
28)	E..J..R	N..M..P	S..M..K	C..P..H	K..R..E	S..U..C
29)	F..K..S		T..N..L	D..Q..J	L..S..F	T..V..D
30)	G..L..T		U..O..M	E..R..K	M..T..G	U..W..E
31)	H..M..U		V..P..N		N..U..H	

1963 | 1963

Day	....JULY.... P--E--I	...AUGUST... P--E--I	..SEPTEMBER. P--E--I	..OCTOBER... P--E--I	..NOVEMBER.. P--E--I	..DECEMBER.. P--E--I
1)	V..X..F	F..1..D	O..A..B	V..C..7	F..F..5	N..H..2
2)	W..Y..G	G..2..E	P..B..C	W..D..8	G..G..6	O..J..3
3)	X..Z..H	H..3..F	Q..C..D	X..E..A	H..H..7	P..K..4
4)	A..1..J	J..A..G	R..D..E	A..F..B	J..J..8	Q..L..5
5)	B..2..K	K..B..H	S..E..F	B..G..C	K..K..A	R..M..6
6)	C..3..L	L..C..J	T..F..G	C..H..D	L..L..B	S..N..7
7)	D..A..M	M..D..K	U..G..H	D..J..E	M..M..C	T..O..8
8)	E..B..N	N..E..L	V..H..J	E..K..F	N..N..D	U..P..A
9)	F..C..O	O..F..M	W..J..K	F..L..G	O..O..E	V..Q..B
10)	G..D..P	P..G..N	X..K..L	G..M..H	P..P..F	W..R..C
11)	H..E..Q	Q..H..O	A..L..M	H..N..J	Q..Q..G	X..S..D
12)	J..F..R	R..J..P	B..M..N	J..O..K	R..R..H	A..T..E
13)	K..G..S	S..K..Q	C..N..O	K..P..L	S..S..J	B..U..F
14)	L..H..T	T..L..R	D..O..P	L..Q..M	T..T..K	C..V..G
15)	M..J..U	U..M..S	E..P..Q	M..R..N	U..U..L	D..W..H
16)	N..K..V	V..N..T	F..Q..R	N..S..O	V..V..M	E..X..J
17)	O..L..W	W..O..U	G..R..S	O..T..P	W..W..N	F..Y..K
18)	P..M..X	X..P..V	H..S..T	P..U..Q	X..X..O	G..Z..L
19)	Q..N..Y	A..Q..W	J..T..U	Q..V..R	A..Y..P	H..1..M
20)	R..O..Z	B..R..X	K..U..V	R..W..S	B..Z..Q	J..2..N
21)	S..P..1	C..S..Y	L..V..W	S..X..T	C..1..R	K..3..O
22)	T..Q..2	D..T..Z	M..W..X	T..Y..U	D..2..S	L..A..P
23)	U..R..3	E..U..1	N..X..Y	U..Z..V	E..3..T	M..B..Q
24)	V..S..4	F..V..2	O..Y..Z	V..1..W	F..A..U	N..C..R
25)	W..T..5	G..W..3	P..Z..1	W..2..X	G..B..V	O..D..S
26)	X..U..6	H..X..4	Q..1..2	X..3..Y	H..C..W	P..E..T
27)	A..V..7	J..Y..5	R..2..3	A..A..Z	J..D..X	Q..F..U
28)	B..W..8	K..Z..6	S..3..4	B..B..1	K..E..Y	R..G..V
29)	C..X..A	L..1..7	T..A..5	C..C..2	L..F..Z	S..H..W
30)	D..Y..B	M..2..8	U..B..6	D..D..3	M..G..1	T..J..X
31)	E..Z..C	N..3..A		E..E..4		U..K..Y

CODES: P-PHYSICAL BIORHYTHM CURVE, E-EMOTIONAL BIORHYTHM CURVE, I-INTELLECTUAL BIORHYTHM CURVE

	...JANUARY..		..FEBRUARY..		....MARCH...		....APRIL...		.....MAY....		....JUNE....
	P--E--I		P--E--I		P--E--I		P--E--I		P--E--I		P--E--I
1)	V..L..Z	1)	F..O..X	1)	M..P..T	1)	U..S..R	1)	D..U..O	1)	M..X..M
2)	W..M..1	2)	G..P..Y	2)	N..Q..U	2)	V..T..S	2)	E..V..P	2)	N..Y..N
3)	X..N..2	3)	H..Q..Z	3)	O..R..V	3)	W..U..T	3)	F..W..Q	3)	O..Z..O
4)	A..O..3	4)	J..R..1	4)	P..S..W	4)	X..V..U	4)	G..X..R	4)	P..1..P
5)	B..P..4	5)	K..S..2	5)	Q..T..X	5)	A..W..V	5)	H..Y..S	5)	Q..2..Q
6)	C..Q..5	6)	L..T..3	6)	R..U..Y	6)	B..X..W	6)	J..Z..T	6)	R..3..R
7)	D..R..6	7)	M..U..4	7)	S..V..Z	7)	C..Y..X	7)	K..1..U	7)	S..A..S
8)	E..S..7	8)	N..V..5	8)	T..W..1	8)	D..Z..Y	8)	L..2..V	8)	T..B..T
9)	F..T..8	9)	O..W..6	9)	U..X..2	9)	E..1..Z	9)	M..3..W	9)	U..C..U
10)	G..U..A	10)	P..X..7	10)	V..Y..3	10)	F..2..1	10)	N..A..X	10)	V..D..V
11)	H..V..B	11)	Q..Y..8	11)	W..Z..4	11)	G..3..2	11)	O..B..Y	11)	W..E..W
12)	J..W..C	12)	R..Z..A	12)	X..1..5	12)	H..A..3	12)	P..C..Z	12)	X..F..X
13)	K..X..D	13)	S..1..B	13)	A..2..6	13)	J..B..4	13)	Q..D..1	13)	A..G..Y
14)	L..Y..E	14)	T..2..C	14)	B..3..7	14)	K..C..5	14)	R..E..2	14)	B..H..Z
15)	M..Z..F	15)	U..3..D	15)	C..A..8	15)	L..D..6	15)	S..F..3	15)	C..J..1
16)	N..1..G	16)	V..A..E	16)	D..B..A	16)	M..E..7	16)	T..G..4	16)	D..K..2
17)	O..2..H	17)	W..B..F	17)	E..C..B	17)	N..F..8	17)	U..H..5	17)	E..L..3
18)	P..3..J	18)	X..C..G	18)	F..D..C	18)	O..G..A	18)	V..J..6	18)	F..M..4
19)	Q..A..K	19)	A..D..H	19)	G..E..D	19)	P..H..B	19)	W..K..7	19)	G..N..5
20)	R..B..L	20)	B..E..J	20)	H..F..E	20)	Q..J..C	20)	X..L..8	20)	H..O..6
21)	S..C..M	21)	C..F..K	21)	J..G..F	21)	R..K..D	21)	A..M..A	21)	J..P..7
22)	T..D..N	22)	D..G..L	22)	K..H..G	22)	S..L..E	22)	B..N..B	22)	K..Q..8
23)	U..E..O	23)	E..H..M	23)	L..J..H	23)	T..M..F	23)	C..O..C	23)	L..R..A
24)	V..F..P	24)	F..J..N	24)	M..K..J	24)	U..N..G	24)	D..P..D	24)	M..S..B
25)	W..G..Q	25)	G..K..O	25)	N..L..K	25)	V..O..H	25)	E..Q..E	25)	N..T..C
26)	X..H..R	26)	H..L..P	26)	O..M..L	26)	W..P..J	26)	F..R..F	26)	O..U..D
27)	A..J..S	27)	J..M..Q	27)	P..N..M	27)	X..Q..K	27)	G..S..G	27)	P..V..E
28)	B..K..T	28)	K..N..R	28)	Q..O..N	28)	A..R..L	28)	H..T..H	28)	Q..W..F
29)	C..L..U	29)	L..O..S	29)	R..P..O	29)	B..S..M	29)	J..U..J	29)	R..X..G
30)	D..M..V			30)	S..Q..P	30)	C..T..N	30)	K..V..K	30)	S..Y..H
31)	E..N..W			31)	T..R..Q			31)	L..W..L		

1964 | 1964

	....JULY....		...AUGUST...		..SEPTEMBER.		..OCTOBER...		..NOVEMBER..		..DECEMBER..
	P--E--I		P--E--I		P--E--I		P--E--I		P--E--I		P--E--I
1)	T..Z..J	1)	D..3..G	1)	M..C..E	1)	T..E..B	1)	D..H..8	1)	L..K..5
2)	U..1..K	2)	E..A..H	2)	N..D..F	2)	U..F..C	2)	E..J..A	2)	M..L..6
3)	V..2..L	3)	F..B..J	3)	O..E..G	3)	V..G..D	3)	F..K..B	3)	N..M..7
4)	W..3..M	4)	G..C..K	4)	P..F..H	4)	W..H..E	4)	G..L..C	4)	O..N..8
5)	X..A..N	5)	H..D..L	5)	Q..G..J	5)	X..J..F	5)	H..M..D	5)	P..O..A
6)	A..B..O	6)	J..E..M	6)	R..H..K	6)	A..K..G	6)	J..N..E	6)	Q..P..B
7)	B..C..P	7)	K..F..N	7)	S..J..L	7)	B..L..H	7)	K..O..F	7)	R..Q..C
8)	C..D..Q	8)	L..G..O	8)	T..K..M	8)	C..M..J	8)	L..P..G	8)	S..R..D
9)	D..E..R	9)	M..H..P	9)	U..L..N	9)	D..N..K	9)	M..Q..H	9)	T..S..E
10)	E..F..S	10)	N..J..Q	10)	V..M..O	10)	E..O..L	10)	N..R..J	10)	U..T..F
11)	F..G..T	11)	O..K..R	11)	W..N..P	11)	F..P..M	11)	O..S..K	11)	V..U..G
12)	G..H..U	12)	P..L..S	12)	X..O..Q	12)	G..Q..N	12)	P..T..L	12)	W..V..H
13)	H..J..V	13)	Q..M..T	13)	A..P..R	13)	H..R..O	13)	Q..U..M	13)	X..W..J
14)	J..K..W	14)	R..N..U	14)	B..Q..S	14)	J..S..P	14)	R..V..N	14)	A..X..K
15)	K..L..X	15)	S..O..V	15)	C..R..T	15)	K..T..Q	15)	S..W..O	15)	B..Y..L
16)	L..M..Y	16)	T..P..W	16)	D..S..U	16)	L..U..R	16)	T..X..P	16)	C..Z..M
17)	M..N..Z	17)	U..Q..X	17)	E..T..V	17)	M..V..S	17)	U..Y..Q	17)	D..1..N
18)	N..O..1	18)	V..R..Y	18)	F..U..W	18)	N..W..T	18)	V..Z..R	18)	E..2..O
19)	O..P..2	19)	W..S..Z	19)	G..V..X	19)	O..X..U	19)	W..1..S	19)	F..3..P
20)	P..Q..3	20)	X..T..1	20)	H..W..Y	20)	P..Y..V	20)	X..2..T	20)	G..A..Q
21)	Q..R..4	21)	A..U..2	21)	J..X..Z	21)	Q..Z..W	21)	A..3..U	21)	H..B..R
22)	R..S..5	22)	B..V..3	22)	K..Y..1	22)	R..1..X	22)	B..A..V	22)	J..C..S
23)	S..T..6	23)	C..W..4	23)	L..Z..2	23)	S..2..Y	23)	C..B..W	23)	K..D..T
24)	T..U..7	24)	D..X..5	24)	M..1..3	24)	T..3..Z	24)	D..C..X	24)	L..E..U
25)	U..V..8	25)	E..Y..6	25)	N..2..4	25)	U..A..1	25)	E..D..Y	25)	M..F..V
26)	V..W..A	26)	F..Z..7	26)	O..3..5	26)	V..B..2	26)	F..E..Z	26)	N..G..W
27)	W..X..B	27)	G..1..8	27)	P..A..6	27)	W..C..3	27)	G..F..1	27)	O..H..X
28)	X..Y..C	28)	H..2..A	28)	Q..B..7	28)	X..D..4	28)	H..G..2	28)	P..J..Y
29)	A..Z..D	29)	J..3..B	29)	R..C..8	29)	A..E..5	29)	J..H..3	29)	Q..K..Z
30)	B..1..E	30)	K..A..C	30)	S..D..A	30)	B..F..6	30)	K..J..4	30)	R..L..1
31)	C..2..F	31)	L..B..D			31)	C..G..7			31)	S..M..2

CODES: P-PHYSICAL BIORHYTHM CURVE E-EMOTIONAL BIORHYTHM CURVE,I-INTELLECTUAL BIORHYTHM CURVE

1965 1965

	...JANUARY..	..FEBRUARY..	....MARCH...	....APRIL...	.....MAY....	....JUNE....
	P--E--I	P--E--I	P--E--I	P--E--I	P--E--I	P--E--I
1)	T..N..3	D..Q..1	J..Q..V	R..T..T	A..V..Q	J..Y..D
2)	U..O..4	E..R..2	K..R..W	S..U..U	B..W..R	K..Z..P
3)	V..P..5	F..S..3	L..S..X	T..V..V	C..X..S	L..1..Q
4)	W..Q..6	G..T..4	M..T..Y	U..W..W	D..Y..T	M..2..R
5)	X..R..7	H..U..5	N..U..Z	V..X..X	E..Z..U	N..3..S
6)	A..S..8	J..V..6	O..V..1	W..Y..Y	F..1..V	O..A..T
7)	B..T..A	K..W..7	P..W..2	X..Z..Z	G..2..W	P..B..U
8)	C..U..B	L..X..8	Q..X..3	A..1..1	H..3..X	Q..C..V
9)	D..V..C	M..Y..A	R..Y..4	B..2..2	J..A..Y	R..D..W
10)	E..W..D	N..Z..B	S..Z..5	C..3..3	K..B..Z	S..E..X
11)	F..X..E	O..1..C	T..1..6	D..A..4	L..C..1	T..F..Y
12)	G..Y..F	P..2..D	U..2..7	E..B..5	M..D..2	U..G..Z
13)	H..Z..G	Q..3..E	V..3..8	F..C..6	N..E..3	V..H..1
14)	J..1..H	R..A..F	W..A..A	G..D..7	O..F..4	W..J..2
15)	K..2..J	S..B..G	X..B..B	H..E..8	P..G..5	X..K..3
16)	L..3..K	T..C..H	A..C..C	J..F..A	Q..H..6	A..L..4
17)	M..A..L	U..D..J	B..D..D	K..G..B	R..J..7	B..M..5
18)	N..B..M	V..E..K	C..E..E	L..H..C	S..K..8	C..N..6
19)	O..C..N	W..F..L	D..F..F	M..J..D	T..L..A	D..O..7
20)	P..D..O	X..G..M	E..G..G	N..K..E	U..M..B	E..P..8
21)	Q..E..P	A..H..N	F..H..H	O..L..F	V..N..C	F..Q..A
22)	R..F..Q	B..J..O	G..J..J	P..M..G	W..O..D	G..R..B
23)	S..G..R	C..K..P	H..K..K	Q..N..H	X..P..E	H..S..C
24)	T..H..S	D..L..Q	J..L..L	R..O..J	A..Q..F	J..T..D
25)	U..J..T	E..M..R	K..M..M	S..P..K	B..R..G	K..U..E
26)	V..K..U	F..N..S	L..N..N	T..Q..L	C..S..H	L..V..F
27)	W..L..V	G..O..T	M..O..O	U..R..M	D..T..J	M..W..G
28)	X..M..W	H..P..U	N..P..P	V..S..N	E..U..K	N..X..H
29)	A..N..X		O..Q..Q	W..T..O	F..V..L	O..Y..J
30)	B..O..Y		P..R..R	X..U..P	G..W..M	P..Z..K
31)	C..P..Z		Q..S..S		H..X..N	

1965 1965

	....JULY....	...AUGUST...	..SEPTEMBER.	..OCTOBER...	..NOVEMBER..	..DECEMBER..
	P--E--I	P--E--I	P--E--I	P--E--I	P--E--I	P--E--I
1)	Q..1..L	A..A..J	J..D..G	Q..F..D	A..J..B	H..L..7
2)	R..2..M	B..B..K	K..E..H	R..G..E	B..K..C	J..M..8
3)	S..3..N	C..C..L	L..F..J	S..H..F	C..L..D	K..N..A
4)	T..A..O	D..D..M	M..G..K	T..J..G	D..M..E	L..O..B
5)	U..B..P	E..E..N	N..H..L	U..K..H	E..N..F	M..P..C
6)	V..C..Q	F..F..O	O..J..M	V..L..J	F..O..G	N..Q..D
7)	W..D..R	G..G..P	P..K..N	W..M..K	G..P..H	O..R..E
8)	X..E..S	H..H..Q	Q..L..O	X..N..L	H..Q..J	P..S..F
9)	A..F..T	J..J..R	R..M..P	A..O..M	J..R..K	Q..T..G
10)	B..G..U	K..K..S	S..N..Q	B..P..N	K..S..L	R..U..H
11)	C..H..V	L..L..T	T..O..R	C..Q..O	L..T..M	S..V..J
12)	D..J..W	M..M..U	U..P..S	D..R..P	M..U..N	T..W..K
13)	E..K..X	N..N..V	V..Q..T	E..S..Q	N..V..O	U..X..L
14)	F..L..Y	O..O..W	W..R..U	F..T..R	O..W..P	V..Y..M
15)	G..M..Z	P..P..X	X..S..V	G..U..S	P..X..Q	W..Z..N
16)	H..N..1	Q..Q..Y	A..T..W	H..V..T	Q..Y..R	X..1..O
17)	J..O..2	R..R..Z	B..U..X	J..W..U	R..Z..S	A..2..P
18)	K..P..3	S..S..1	C..V..Y	K..X..V	S..1..T	B..3..Q
19)	L..Q..4	T..T..2	D..W..Z	L..Y..W	T..2..U	C..A..R
20)	M..R..5	U..U..3	E..X..1	M..Z..X	U..3..V	D..B..S
21)	N..S..6	V..V..4	F..Y..2	N..1..Y	V..A..W	E..C..T
22)	O..T..7	W..W..5	G..Z..3	O..2..Z	W..B..X	F..D..U
23)	P..U..8	X..X..6	H..1..4	P..3..1	X..C..Y	G..E..V
24)	Q..V..A	A..Y..7	J..2..5	Q..A..2	A..D..Z	H..F..W
25)	R..W..B	B..Z..8	K..3..6	R..B..3	B..E..1	J..G..X
26)	S..X..C	C..1..A	L..A..7	S..C..4	C..F..2	K..H..Y
27)	T..Y..D	D..2..B	M..B..8	T..D..5	D..G..3	L..J..Z
28)	U..Z..E	E..3..C	N..C..A	U..E..6	E..H..4	M..K..1
29)	V..1..F	F..A..D	O..D..B	V..F..7	F..J..5	N..L..2
30)	W..2..G	G..B..E	P..E..C	W..G..8	G..K..6	O..M..3
31)	X..3..H	H..C..F		X..H..A		P..N..4

CODES: P-PHYSICAL BIORHYTHM CURVE,E-EMOTIONAL BIORHYTHM CURVE,I-INTELLECTUAL BIORHYTHM CURVE

1966 | 1966

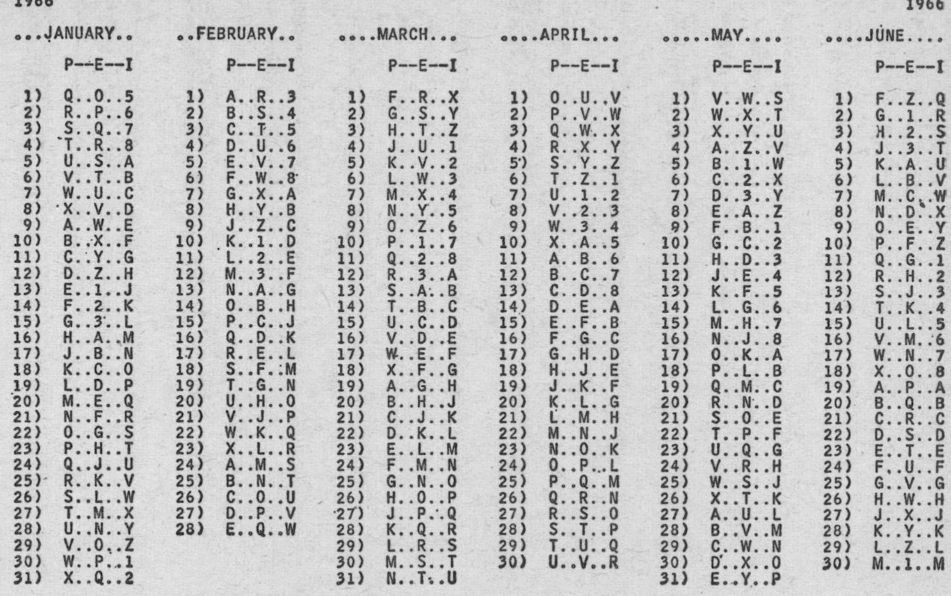

	...JANUARY..	..FEBRUARY..	....MARCH...	....APRIL...	.....MAY....	....JUNE....
	P--E--I	P--E--I	P--E--I	P--E--I	P--E--I	P--E--I
1)	Q..O..5	A..R..3	F..R..X	O..U..V	V..W..S	F..Z..Q
2)	R..P..6	B..S..4	G..S..Y	P..V..W	W..X..T	G..1..R
3)	S..Q..7	C..T..5	H..T..Z	Q..W..X	X..Y..U	H..2..S
4)	T..R..8	D..U..6	J..U..1	R..X..Y	A..Z..V	J..3..T
5)	U..S..A	E..V..7	K..V..2	S..Y..Z	B..1..W	K..A..U
6)	V..T..B	F..W..8	L..W..3	T..Z..1	C..2..X	L..B..V
7)	W..U..C	G..X..A	M..X..4	U..1..2	D..3..Y	M..C..W
8)	X..V..D	H..Y..B	N..Y..5	V..2..3	E..A..Z	N..D..X
9)	A..W..E	J..Z..C	O..Z..6	W..3..4	F..B..1	O..E..Y
10)	B..X..F	K..1..D	P..1..7	X..A..5	G..C..2	P..F..Z
11)	C..Y..G	L..2..E	Q..2..8	A..B..6	H..D..3	Q..G..1
12)	D..Z..H	M..3..F	R..3..A	B..C..7	J..E..4	R..H..2
13)	E..1..J	N..A..G	S..A..B	C..D..8	K..F..5	S..J..3
14)	F..2..K	O..B..H	T..B..C	D..E..A	L..G..6	T..K..4
15)	G..3..L	P..C..J	U..C..D	E..F..B	M..H..7	U..L..5
16)	H..A..M	Q..D..K	V..D..E	F..G..C	N..J..8	V..M..6
17)	J..B..N	R..E..L	W..E..F	G..H..D	O..K..A	W..N..7
18)	K..C..O	S..F..M	X..F..G	H..J..E	P..L..B	X..O..8
19)	L..D..P	T..G..N	A..G..H	J..K..F	Q..M..C	A..P..A
20)	M..E..Q	U..H..O	B..H..J	K..L..G	R..N..D	B..Q..B
21)	N..F..R	V..J..P	C..J..K	L..M..H	S..O..E	C..R..C
22)	O..G..S	W..K..Q	D..K..L	M..N..J	T..P..F	D..S..D
23)	P..H..T	X..L..R	E..L..M	N..O..K	U..Q..G	E..T..E
24)	Q..J..U	A..M..S	F..M..N	O..P..L	V..R..H	F..U..F
25)	R..K..V	B..N..T	G..N..O	P..Q..M	W..S..J	G..V..G
26)	S..L..W	C..O..U	H..O..P	Q..R..N	X..T..K	H..W..H
27)	T..M..X	D..P..V	J..P..Q	R..S..O	A..U..L	J..X..J
28)	U..N..Y	E..Q..W	K..Q..R	S..T..P	B..V..M	K..Y..K
29)	V..O..Z		L..R..S	T..U..Q	C..W..N	L..Z..L
30)	W..P..1		M..S..T	U..V..R	D..X..O	M..1..M
31)	X..Q..2		N..T..U		E..Y..P	

1966					1966
....JULY....	...AUGUST...	..SEPTEMBER.	..OCTOBER...	..NOVEMBER..	..DECEMBER..
P--E--I	P--E--I	P--E--I	P--E--I	P--E--I	P--E--I
1) N..2..N	1) V..B..L	1) F..E..J	1) N..G..F	1) V..K..D	1) E..M..A
2) O..3..O	2) W..C..M	2) G..F..K	2) O..H..G	2) W..L..E	2) F..N..B
3) P..A..P	3) X..D..N	3) H..G..L	3) P..J..H	3) X..M..F	3) G..O..C
4) Q..B..Q	4) A..E..O	4) J..H..M	4) Q..K..J	4) A..N..G	4) H..P..D
5) R..C..R	5) B..F..P	5) K..J..N	5) R..L..K	5) B..O..H	5) J..Q..E
6) S..D..S	6) C..G..Q	6) L..K..O	6) S..M..L	6) C..P..J	6) K..R..F
7) T..E..T	7) D..H..R	7) M..L..P	7) T..N..M	7) D..Q..K	7) L..S..G
8) U..F..U	8) E..J..S	8) N..M..Q	8) U..O..N	8) E..R..L	8) M..T..H
9) V..G..V	9) F..K..T	9) O..N..R	9) V..P..O	9) F..S..M	9) N..U..J
10) W..H..W	10) G..L..U	10) P..O..S	10) W..Q..P	10) G..T..N	10) O..V..K
11) X..J..X	11) H..M..V	11) Q..P..T	11) X..R..Q	11) H..U..O	11) P..W..L
12) A..K..Y	12) J..N..W	12) R..Q..U	12) A..S..R	12) J..V..P	12) Q..X..M
13) B..L..Z	13) K..O..X	13) S..R..V	13) B..T..S	13) K..W..Q	13) R..Y..N
14) C..M..1	14) L..P..Y	14) T..S..W	14) C..U..T	14) L..X..R	14) S..Z..O
15) D..N..2	15) M..Q..Z	15) U..T..X	15) D..V..U	15) M..Y..S	15) T..1..P
16) E..O..3	16) N..R..1	16) V..U..Y	16) E..W..V	16) N..Z..T	16) U..2..Q
17) F..P..4	17) O..S..2	17) W..V..Z	17) F..X..W	17) O..1..U	17) V..3..R
18) G..Q..5	18) P..T..3	18) X..W..1	18) G..Y..X	18) P..2..V	18) W..A..S
19) H..R..6	19) Q..U..4	19) A..X..2	19) H..Z..Y	19) Q..3..W	19) X..B..T
20) J..S..7	20) R..V..5	20) B..Y..3	20) J..1..Z	20) R..A..X	20) A..C..U
21) K..T..8	21) S..W..6	21) C..Z..4	21) K..2..1	21) S..B..Y	21) B..D..V
22) L..U..A	22) T..X..7	22) D..1..5	22) L..3..2	22) T..C..Z	22) C..E..W
23) M..V..B	23) U..Y..8	23) E..2..6	23) M..A..3	23) U..D..1	23) D..F..X
24) N..W..C	24) V..Z..A	24) F..3..7	24) N..B..4	24) V..E..2	24) E..G..Y
25) O..X..D	25) W..1..B	25) G..A..8	25) O..C..5	25) W..F..3	25) F..H..Z
26) P..Y..E	26) X..2..C	26) H..B..A	26) P..D..6	26) X..G..4	26) G..J..1
27) Q..Z..F	27) A..3..D	27) J..C..B	27) Q..E..7	27) A..H..5	27) H..K..2
28) R..1..G	28) B..A..E	28) K..D..C	28) R..F..8	28) B..J..6	28) J..L..3
29) S..2..H	29) C..B..F	29) L..E..D	29) S..G..A	29) C..K..7	29) K..M..4
30) T..3..J	30) D..C..G	30) M..F..E	30) T..H..B	30) D..L..8	30) L..N..5
31) U..A..K	31) E..D..H		31) U..J..C		31) M..O..6

CODES: P-PHYSICAL BIORHYTHM CURVE,E-EMOTIONAL BIORHYTHM CURVE,I-INTELLECTUAL BIORHYTHM CURVE

1967 | 1967

	...JANUARY.. P--E--I	..FEBRUARY.. P--E--I	....MARCH... P--E--I	....APRIL... P--E--I	.....MAY... P--E--I	....JUNE.... P--E--I
1)	N..P..7	V..S..5	C..S..Z	L..V..X	S..X..U	C..1..S
2)	O..Q..8	W..T..6	D..T..1	M..W..Y	T..Y..V	D..2..T
3)	P..R..A	X..U..7	E..U..2	N..X..Z	U..Z..W	E..3..U
4)	Q..S..B	A..V..8	F..V..3	O..Y..1	V..1..X	F..A..V
5)	R..T..C	B..W..A	G..W..4	P..Z..2	W..2..Y	G..B..W
6)	S..U..D	C..X..B	H..X..5	Q..1..3	X..3..Z	H..C..X
7)	T..V..E	D..Y..C	J..Y..6	R..2..4	A..A..1	J..D..Y
8)	U..W..F	E..Z..D	K..Z..7	S..3..5	B..B..2	K..E..Z
9)	V..X..G	F..1..E	L..1..8	T..A..6	C..C..3	L..F..1
10)	W..Y..H	G..2..F	M..2..A	U..B..7	D..D..4	M..G..2
11)	X..Z..J	H..3..G	N..3..B	V..C..8	E..E..5	N..H..3
12)	A..1..K	J..A..H	O..A..C	W..D..A	F..F..6	O..J..4
13)	B..2..L	K..B..J	P..B..D	X..E..B	G..G..7	P..K..5
14)	C..3..M	L..C..K	Q..C..E	A..F..C	H..H..8	Q..L..6
15)	D..A..N	M..D..L	R..D..F	B..G..D	J..J..A	R..M..7
16)	E..B..O	N..E..M	S..E..G	C..H..E	K..K..B	S..N..8
17)	F..C..P	O..F..N	T..F..H	D..J..F	L..L..C	T..O..A
18)	G..D..Q	P..G..O	U..G..J	E..K..G	M..M..D	U..P..B
19)	H..E..R	Q..H..P	V..H..K	F..L..H	N..N..E	V..Q..C
20)	J..F..S	R..J..Q	W..J..L	G..M..J	O..O..F	W..R..D
21)	K..G..T	S..K..R	X..K..M	H..N..K	P..P..G	X..S..E
22)	L..H..U	T..L..S	A..L..N	J..O..L	Q..Q..H	A..T..F
23)	M..J..V	U..M..T	B..M..O	K..P..M	R..R..J	B..U..G
24)	N..K..W	V..N..U	C..N..P	L..Q..N	S..S..K	C..V..H
25)	O..L..X	W..O..V	D..O..Q	M..R..O	T..T..L	D..W..J
26)	P..M..Y	X..P..W	E..P..R	N..S..P	U..U..M	E..X..K
27)	Q..N..Z	A..Q..X	F..Q..S	O..T..Q	V..V..N	F..Y..L
28)	R..O..1	B..R..Y	G..R..T	P..U..R	W..W..O	G..Z..M
29)	S..P..2		H..S..U	Q..V..S	X..X..P	H..1..N
30)	T..Q..3		J..T..V	R..W..T	A..Y..Q	J..2..O
31)	U..R..4		K..U..W		B..Z..R	

1967

	....JULY....	...AUGUST...	..SEPTEMBER.	..OCTOBER...	..NOVEMBER..	..DECEMBER..
	P--E--I	P--E--I	P--E--I	P--E--I	P--E--I	P--E--I
1)	K..3..P	S..C..N	C..F..L	K..H..H	S..L..F	B..N..C
2)	L..A..Q	T..D..O	D..G..M	L..J..J	T..M..G	C..O..D
3)	M..B..R	U..E..P	E..H..N	M..K..K	U..N..H	D..P..E
4)	N..C..S	V..F..Q	F..J..O	N..L..L	V..O..J	E..Q..F
5)	O..D..T	W..G..R	G..K..P	O..M..M	W..P..K	F..R..G
6)	P..E..U	X..H..S	H..L..Q	P..N..N	X..Q..L	G..S..H
7)	Q..F..V	A..J..T	J..M..R	Q..O..O	A..R..M	H..T..J
8)	R..G..W	B..K..U	K..N..S	R..P..P	B..S..N	J..U..K
9)	S..H..X	C..L..V	L..O..T	S..Q..Q	C..T..O	K..V..L
10)	T..J..Y	D..M..W	M..P..U	T..R..R	D..U..P	L..W..M
11)	U..K..Z	E..N..X	N..Q..V	U..S..S	E..V..Q	M..X..N
12)	V..L..1	F..O..Y	O..R..W	V..T..T	F..W..R	N..Y..O
13)	W..M..2	G..P..Z	P..S..X	W..U..U	G..X..S	O..Z..P
14)	X..N..3	H..Q..1	Q..T..Y	X..V..V	H..Y..T	P..1..Q
15)	A..O..4	J..R..2	R..U..Z	A..W..W	J..Z..U	Q..2..R
16)	B..P..5	K..S..3	S..V..1	B..X..X	K..1..V	R..3..S
17)	C..Q..6	L..T..4	T..W..2	C..Y..Y	L..2..W	S..A..T
18)	D..R..7	M..U..5	U..X..3	D..Z..Z	M..3..X	T..B..U
19)	E..S..8	N..V..6	V..Y..4	E..1..1	N..A..Y	U..C..V
20)	F..T..A	O..W..7	W..Z..5	F..2..2	O..B..Z	V..D..W
21)	G..U..B	P..X..8	X..1..6	G..3..3	P..C..1	W..E..X
22)	H..V..C	Q..Y..A	A..2..7	H..A..4	Q..D..2	X..F..Y
23)	J..W..D	R..Z..B	B..3..8	J..B..5	R..E..3	A..G..Z
24)	K..X..E	S..1..C	C..A..A	K..C..6	S..F..4	B..H..1
25)	L..Y..F	T..2..D	D..B..B	L..D..7	T..G..5	C..J..2
26)	M..Z..G	U..3..E	E..C..C	M..E..8	U..H..6	D..K..3
27)	N..1..H	V..A..F	F..D..D	N..F..A	V..J..7	E..L..4
28)	O..2..J	W..B..G	G..E..E	O..G..B	W..K..8	F..M..5
29)	P..3..K	X..C..H	H..F..F	P..H..C	X..L..A	G..N..6
30)	Q..A..L	A..D..J	J..G..G	Q..J..D	A..M..B	H..O..7
31)	R..B..M	B..E..K		R..K..E		J..P..8

CODES: P-PHYSICAL BIORHYTHM CURVE, E-EMOTIONAL BIORHYTHM CURVE, I-INTELLECTUAL BIORHYTHM CURVE

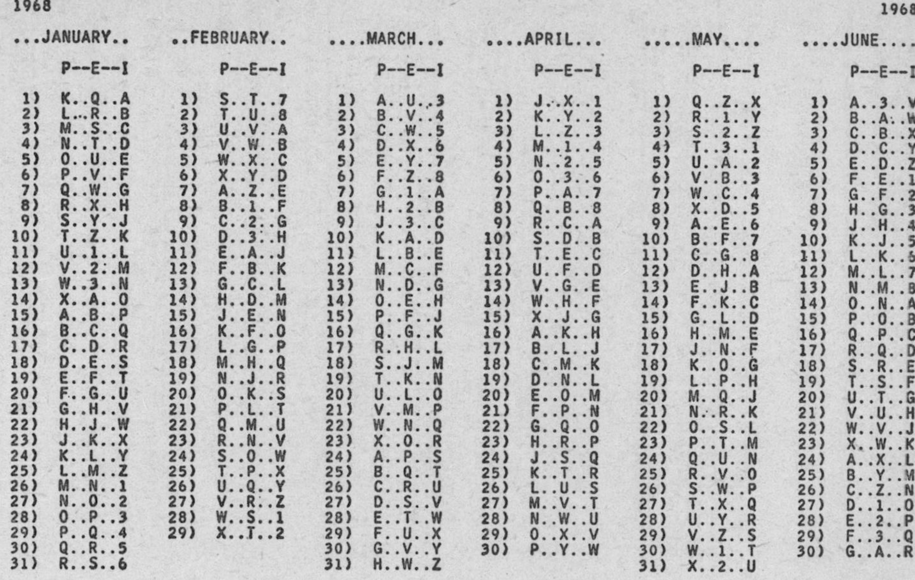

1968						1968
	...JANUARY..	..FEBRUARY..	....MARCH...	....APRIL...	.....MAY....	....JUNE....
	P--E--I	P--E--I	P--E--I	P--E--I	P--E--I	P--E--I
1)	K..Q..A	S..T..7	A..U..3	J..X..1	Q..Z..X	A..3..V
2)	L..R..B	T..U..8	B..V..4	K..Y..2	R..1..Y	B..A..W
3)	M..S..C	U..V..A	C..W..5	L..Z..3	S..2..Z	C..B..X
4)	N..T..D	V..W..B	D..X..6	M..1..4	T..3..1	D..C..Y
5)	O..U..E	W..X..C	E..Y..7	N..2..5	U..A..2	E..D..Z
6)	P..V..F	X..Y..D	F..Z..8	O..3..6	V..B..3	F..E..1
7)	Q..W..G	A..Z..E	G..1..A	P..A..7	W..C..4	G..F..2
8)	R..X..H	B..1..F	H..2..B	Q..B..8	X..D..5	H..G..3
9)	S..Y..J	C..2..G	J..3..C	R..C..A	A..E..6	J..H..4
10)	T..Z..K	D..3..H	K..A..D	S..D..B	B..F..7	K..J..5
11)	U..1..L	E..A..J	L..B..E	T..E..C	C..G..8	L..K..6
12)	V..2..M	F..B..K	M..C..F	U..F..D	D..H..A	M..L..7
13)	W..3..N	G..C..L	N..D..G	V..G..E	E..J..B	N..M..8
14)	X..A..O	H..D..M	O..E..H	W..H..F	F..K..C	O..N..A
15)	A..B..P	J..E..N	P..F..J	X..J..G	G..L..D	P..O..B
16)	B..C..Q	K..F..O	Q..G..K	A..K..H	H..M..E	Q..P..C
17)	C..D..R	L..G..P	R..H..L	B..L..J	J..N..F	R..Q..D
18)	D..E..S	M..H..Q	S..J..M	C..M..K	K..O..G	S..R..E
19)	E..F..T	N..J..R	T..K..N	D..N..L	L..P..H	T..S..F
20)	F..G..U	O..K..S	U..L..O	E..O..M	M..Q..J	U..T..G
21)	G..H..V	P..L..T	V..M..P	F..P..N	N..R..K	V..U..H
22)	H..J..W	Q..M..U	W..N..Q	G..Q..O	O..S..L	W..V..J
23)	J..K..X	R..N..V	X..O..R	H..R..P	P..T..M	X..W..K
24)	K..L..Y	S..O..W	A..P..S	J..S..Q	Q..U..N	A..X..L
25)	L..M..Z	T..P..X	B..Q..T	K..T..R	R..V..O	B..Y..M
26)	M..N..1	U..Q..Y	C..R..U	L..U..S	S..W..P	C..Z..N
27)	N..O..2	V..R..Z	D..S..V	M..V..T	T..X..Q	D..1..O
28)	O..P..3	W..S..1	E..T..W	N..W..U	U..Y..R	E..2..P
29)	P..Q..4	X..T..2	F..U..X	O..X..V	V..Z..S	F..3..Q
30)	Q..R..5		G..V..Y	P..Y..W	W..1..T	G..A..R
31)	R..S..6		H..W..Z		X..2..U	

1968 1968

	....JULY....	...AUGUST...	..SEPTEMBER.	..OCTOBER...	..NOVEMBER..	..DECEMBER..
	P--E--I	P--E--I	P--E--I	P--E--I	P--E--I	P--E--I
1)	H..B..S	Q..E..Q	A..H..O	H..K..L	Q..N..J	X..P..F
2)	J..C..T	R..F..R	B..J..P	J..L..M	R..O..K	A..Q..G
3)	K..D..U	S..G..S	C..K..Q	K..M..N	S..P..L	B..R..H
4)	L..E..V	T..H..T	D..L..R	L..N..O	T..Q..M	C..S..J
5)	M..F..W	U..J..U	E..M..S	M..O..P	U..R..N	D..T..K
6)	N..G..X	V..K..V	F..N..T	N..P..Q	V..S..O	E..U..L
7)	O..H..Y	W..L..W	G..O..U	O..Q..R	W..T..P	F..V..M
8)	P..J..Z	X..M..X	H..P..V	P..R..S	X..U..Q	G..W..N
9)	Q..K..1	A..N..Y	J..Q..W	Q..S..T	A..V..R	H..X..O
10)	R..L..2	B..O..Z	K..R..X	R..T..U	B..W..S	J..Y..P
11)	S..M..3	C..P..1	L..S..Y	S..U..V	C..X..T	K..Z..Q
12)	T..N..4	D..Q..2	M..T..Z	T..V..W	D..Y..U	L..1..R
13)	U..O..5	E..R..3	N..U..1	U..W..X	E..Z..V	M..2..S
14)	V..P..6	F..S..4	O..V..2	V..X..Y	F..1..W	N..3..T
15)	W..Q..7	G..T..5	P..W..3	W..Y..Z	G..2..X	O..A..U
16)	X..R..8	H..U..6	Q..X..4	X..Z..1	H..3..Y	P..B..V
17)	A..S..A	J..V..7	R..Y..5	A..1..2	J..A..Z	Q..C..W
18)	B..T..B	K..W..8	S..Z..6	B..2..3	K..B..1	R..D..X
19)	C..U..C	L..X..A	T..1..7	C..3..4	L..C..2	S..E..Y
20)	D..V..D	M..Y..B	U..2..8	D..A..5	M..D..3	T..F..Z
21)	E..W..E	N..Z..C	V..3..A	E..B..6	N..E..4	U..G..1
22)	F..X..F	O..1..D	W..A..B	F..C..7	O..F..5	V..H..2
23)	G..Y..G	P..2..E	X..B..C	G..D..8	P..G..6	W..J..3
24)	H..Z..H	Q..3..F	A..C..D	H..E..A	Q..H..7	X..K..4
25)	J..1..J	R..A..G	B..D..E	J..F..B	R..J..8	A..L..5
26)	K..2..K	S..B..H	C..E..F	K..G..C	S..K..A	B..M..6
27)	L..3..L	T..C..J	D..F..G	L..H..D	T..L..B	C..N..7
28)	M..A..M	U..D..K	E..G..H	M..J..E	U..M..C	D..O..8
29)	N..B..N	V..E..L	F..H..J	N..K..F	V..N..D	E..P..A
30)	O..C..O	W..F..M	G..J..K	O..L..G	W..O..E	F..Q..B
31)	P..D..P	X..G..N		P..M..H		G..R..C

CODES: P-PHYSICAL BIORHYTHM CURVE, E-EMOTIONAL BIORHYTHM CURVE, I-INTELLECTUAL BIORHYTHM CURVE

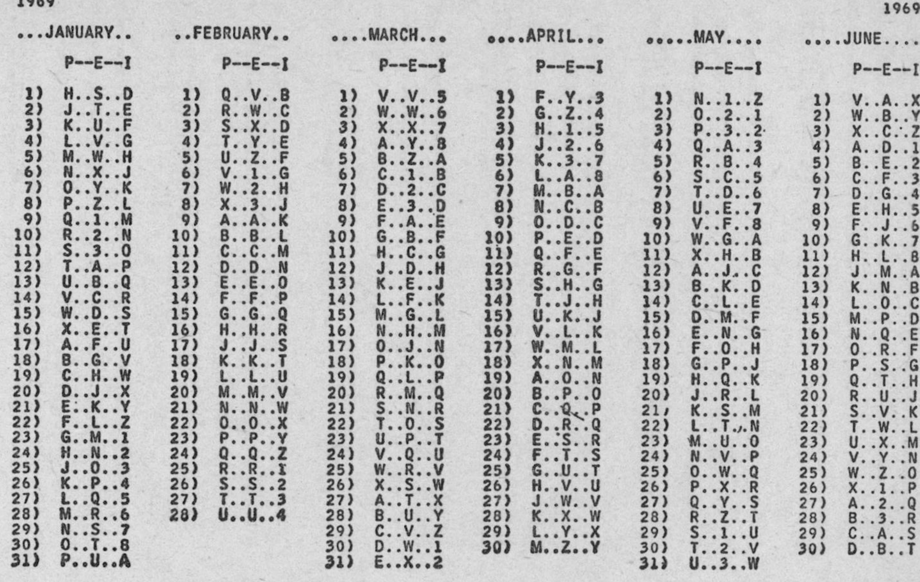

...JANUARY..	P--E--I	..FEBRUARY..	P--E--I	....MARCH...	P--E--I	....APRIL...	P--E--I	.....MAY....	P--E--I	....JUNE....	P--E--I
1)	H..S..D	1)	Q..V..B	1)	V..V..5	1)	F..Y..3	1)	N..1..Z	1)	V..A..X
2)	J..T..E	2)	R..W..C	2)	W..W..6	2)	G..Z..4	2)	O..2..1	2)	W..B..Y
3)	K..U..F	3)	S..X..D	3)	X..X..7	3)	H..1..5	3)	P..3..2	3)	X..C..Z
4)	L..V..G	4)	T..Y..E	4)	A..Y..8	4)	J..2..6	4)	Q..A..3	4)	A..D..1
5)	M..W..H	5)	U..Z..F	5)	B..Z..A	5)	K..3..7	5)	R..B..4	5)	B..E..2
6)	N..X..J	6)	V..1..G	6)	C..1..B	6)	L..A..8	6)	S..C..5	6)	C..F..3
7)	O..Y..K	7)	W..2..H	7)	D..2..C	7)	M..B..A	7)	T..D..6	7)	D..G..4
8)	P..Z..L	8)	X..3..J	8)	E..3..D	8)	N..C..B	8)	U..E..7	8)	E..H..5
9)	Q..1..M	9)	A..A..K	9)	F..A..E	9)	O..D..C	9)	V..F..8	9)	F..J..6
10)	R..2..N	10)	B..B..L	10)	G..B..F	10)	P..E..D	10)	W..G..A	10)	G..K..7
11)	S..3..O	11)	C..C..M	11)	H..C..G	11)	Q..F..E	11)	X..H..B	11)	H..L..8
12)	T..A..P	12)	D..D..N	12)	J..D..H	12)	R..G..F	12)	A..J..C	12)	J..M..A
13)	U..B..Q	13)	E..E..O	13)	K..E..J	13)	S..H..G	13)	B..K..D	13)	K..N..B
14)	V..C..R	14)	F..F..P	14)	L..F..K	14)	T..J..H	14)	C..L..E	14)	L..O..C
15)	W..D..S	15)	G..G..Q	15)	M..G..L	15)	U..K..J	15)	D..M..F	15)	M..P..D
16)	X..E..T	16)	H..H..R	16)	N..H..M	16)	V..L..K	16)	E..N..G	16)	N..Q..E
17)	A..F..U	17)	J..J..S	17)	O..J..N	17)	W..M..L	17)	F..O..H	17)	O..R..F
18)	B..G..V	18)	K..K..T	18)	P..K..O	18)	X..N..M	18)	G..P..J	18)	P..S..G
19)	C..H..W	19)	L..L..U	19)	Q..L..P	19)	A..O..N	19)	H..Q..K	19)	Q..T..H
20)	D..J..X	20)	M..M..V	20)	R..M..Q	20)	B..P..O	20)	J..R..L	20)	R..U..J
21)	E..K..Y	21)	N..N..W	21)	S..N..R	21)	C..Q..P	21)	K..S..M	21)	S..V..K
22)	F..L..Z	22)	O..O..X	22)	T..O..S	22)	D..R..Q	22)	L..T..N	22)	T..W..L
23)	G..M..1	23)	P..P..Y	23)	U..P..T	23)	E..S..R	23)	M..U..O	23)	U..X..M
24)	H..N..2	24)	Q..Q..Z	24)	V..Q..U	24)	F..T..S	24)	N..V..P	24)	V..Y..N
25)	J..O..3	25)	R..R..1	25)	W..R..V	25)	G..U..T	25)	O..W..Q	25)	W..Z..O
26)	K..P..4	26)	S..S..2	26)	X..S..W	26)	H..V..U	26)	P..X..R	26)	X..1..P
27)	L..Q..5	27)	T..T..3	27)	A..T..X	27)	J..W..V	27)	Q..Y..S	27)	A..2..Q
28)	M..R..6	28)	U..U..4	28)	B..U..Y	28)	K..X..W	28)	R..Z..T	28)	B..3..R
29)	N..S..7			29)	C..V..Z	29)	L..Y..X	29)	S..1..U	29)	C..A..S
30)	O..T..8			30)	D..W..1	30)	M..Z..Y	30)	T..2..V	30)	D..B..T
31)	P..U..A			31)	E..X..2			31)	U..3..W		

1969 | 1969

	....JULY....		...AUGUST...		..SEPTEMBER.		..OCTOBER...		..NOVEMBER..		..DECEMBER..
	P--E--I		P--E--I		P--E--I		P--E--I		P--E--I		P--E--I
1)	E..C..U	1)	N..F..S	1)	V..J..Q	1)	E..L..N	1)	N..O..L	1)	U..Q..H
2)	F..D..V	2)	O..G..T	2)	W..K..R	2)	F..M..O	2)	O..P..M	2)	V..R..J
3)	G..E..W	3)	P..H..U	3)	X..L..S	3)	G..N..P	3)	P..Q..N	3)	W..S..K
4)	H..F..X	4)	Q..J..V	4)	A..M..T	4)	H..O..Q	4)	Q..R..O	4)	X..T..L
5)	J..G..Y	5)	R..K..W	5)	B..N..U	5)	J..P..R	5)	R..S..P	5)	A..U..M
6)	K..H..Z	6)	S..L..X	6)	C..O..V	6)	K..Q..S	6)	S..T..Q	6)	B..V..N
7)	L..J..1	7)	T..M..Y	7)	D..P..W	7)	L..R..T	7)	T..U..R	7)	C..W..O
8)	M..K..2	8)	U..N..Z	8)	E..Q..X	8)	M..S..U	8)	U..V..S	8)	D..X..P
9)	N..L..3	9)	V..O..1	9)	F..R..Y	9)	N..T..V	9)	V..W..T	9)	E..Y..Q
10)	O..M..4	10)	W..P..2	10)	G..S..Z	10)	O..U..W	10)	W..X..U	10)	F..Z..R
11)	P..N..5	11)	X..Q..3	11)	H..T..1	11)	P..V..X	11)	X..Y..V	11)	G..1..S
12)	Q..O..6	12)	A..R..4	12)	J..U..2	12)	Q..W..Y	12)	A..Z..W	12)	H..2..T
13)	R..P..7	13)	B..S..5	13)	K..V..3	13)	R..X..Z	13)	B..1..X	13)	J..3..U
14)	S..Q..8	14)	C..T..6	14)	L..W..4	14)	S..Y..1	14)	C..2..Y	14)	K..A..V
15)	T..R..A	15)	D..U..7	15)	M..X..5	15)	T..Z..2	15)	D..3..Z	15)	L..B..W
16)	U..S..B	16)	E..V..8	16)	N..Y..6	16)	U..1..3	16)	E..A..1	16)	M..C..X
17)	V..T..C	17)	F..W..A	17)	O..Z..7	17)	V..2..4	17)	F..B..2	17)	N..D..Y
18)	W..U..D	18)	G..X..B	18)	P..1..8	18)	W..3..5	18)	G..C..3	18)	O..E..Z
19)	X..V..E	19)	H..Y..C	19)	Q..2..A	19)	X..A..6	19)	H..D..4	19)	P..F..1
20)	A..W..F	20)	J..Z..D	20)	R..3..B	20)	A..B..7	20)	J..E..5	20)	Q..G..2
21)	B..X..G	21)	K..1..E	21)	S..A..C	21)	B..C..8	21)	K..F..6	21)	R..H..3
22)	C..Y..H	22)	L..2..F	22)	T..B..D	22)	C..D..A	22)	L..G..7	22)	S..J..4
23)	D..Z..J	23)	M..3..G	23)	U..C..E	23)	D..E..B	23)	M..H..8	23)	T..K..5
24)	E..1..K	24)	N..A..H	24)	V..D..F	24)	E..F..C	24)	N..J..A	24)	U..L..6
25)	F..2..L	25)	O..B..J	25)	W..E..G	25)	F..G..D	25)	O..K..B	25)	V..M..7
26)	G..3..M	26)	P..C..K	26)	X..F..H	26)	G..H..E	26)	P..L..C	26)	W..N..8
27)	H..A..N	27)	Q..D..L	27)	A..G..J	27)	H..J..F	27)	Q..M..D	27)	X..O..A
28)	J..B..O	28)	R..E..M	28)	B..H..K	28)	J..K..G	28)	R..N..E	28)	A..P..B
29)	K..C..P	29)	S..F..N	29)	C..J..L	29)	K..L..H	29)	S..O..F	29)	B..Q..C
30)	L..D..Q	30)	T..G..O	30)	D..K..M	30)	L..M..J	30)	T..P..G	30)	C..R..D
31)	M..E..R	31)	U..H..P			31)	M..N..K			31)	D..S..E

CODES: P-PHYSICAL BIORHYTHM CURVE,E-EMOTIONAL BIORHYTHM CURVE,I-INTELLECTUAL BIORHYTHM CURVE

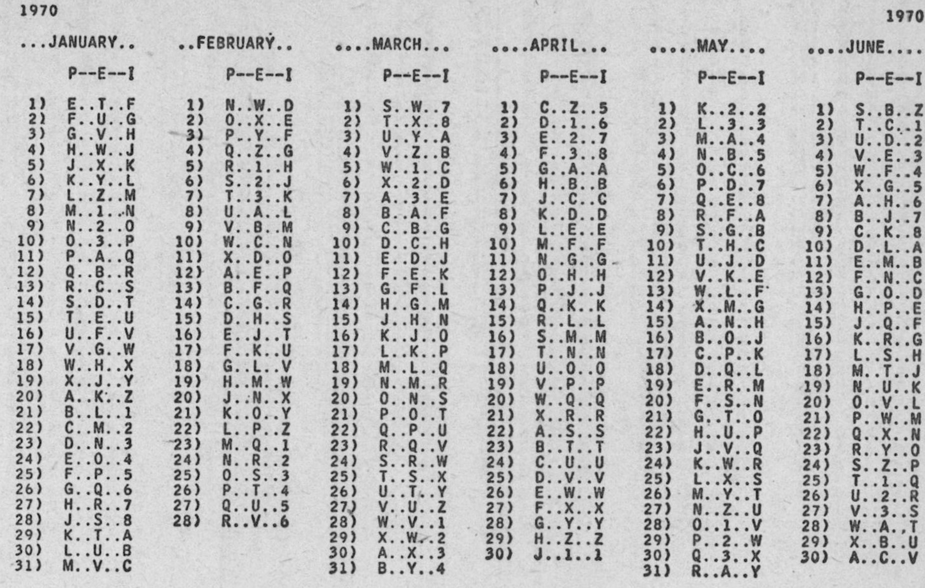

1970 1970

	...JANUARY..	..FEBRUARY..	....MARCH...	....APRIL...	.....MAY....	....JUNE....
	P--E--I	P--E--I	P--E--I	P--E--I	P--E--I	P--E--I
1)	E..T..F	N..W..D	S..W..7	C..Z..5	K..2..2	S..B..Z
2)	F..U..G	O..X..E	T..X..8	D..1..6	L..3..3	T..C..1
3)	G..V..H	P..Y..F	U..Y..A	E..2..7	M..A..4	U..D..2
4)	H..W..J	Q..Z..G	V..Z..B	F..3..8	N..B..5	V..E..3
5)	J..X..K	R..1..H	W..1..C	G..A..A	O..C..6	W..F..4
6)	K..Y..L	S..2..J	X..2..D	H..B..B	P..D..7	X..G..5
7)	L..Z..M	T..3..K	A..3..E	J..C..C	Q..E..8	A..H..6
8)	M..1..N	U..A..L	B..A..F	K..D..D	R..F..A	B..J..7
9)	N..2..O	V..B..M	C..B..G	L..E..E	S..G..B	C..K..8
10)	O..3..P	W..C..N	D..C..H	M..F..F	T..H..C	D..L..A
11)	P..A..Q	X..D..O	E..D..J	N..G..G	U..J..D	E..M..B
12)	Q..B..R	A..E..P	F..E..K	O..H..H	V..K..E	F..N..C
13)	R..C..S	B..F..Q	G..F..L	P..J..J	W..L..F	G..O..D
14)	S..D..T	C..G..R	H..G..M	Q..K..K	X..M..G	H..P..E
15)	T..E..U	D..H..S	J..H..N	R..L..L	A..N..H	J..Q..F
16)	U..F..V	E..J..T	K..J..O	S..M..M	B..O..J	K..R..G
17)	V..G..W	F..K..U	L..K..P	T..N..N	C..P..K	L..S..H
18)	W..H..X	G..L..V	M..L..Q	U..O..O	D..Q..L	M..T..J
19)	X..J..Y	H..M..W	N..M..R	V..P..P	E..R..M	N..U..K
20)	A..K..Z	J..N..X	O..N..S	W..Q..Q	F..S..N	O..V..L
21)	B..L..1	K..O..Y	P..O..T	X..R..R	G..T..O	P..W..M
22)	C..M..2	L..P..Z	Q..P..U	A..S..S	H..U..P	Q..X..N
23)	D..N..3	M..Q..1	R..Q..V	B..T..T	J..V..Q	R..Y..O
24)	E..O..4	N..R..2	S..R..W	C..U..U	K..W..R	S..Z..P
25)	F..P..5	O..S..3	T..S..X	D..V..V	L..X..S	T..1..Q
26)	G..Q..6	P..T..4	U..T..Y	E..W..W	M..Y..T	U..2..R
27)	H..R..7	Q..U..5	V..U..Z	F..X..X	N..Z..U	V..3..S
28)	J..S..8	R..V..6	W..V..1	G..Y..Y	O..1..V	W..A..T
29)	K..T..A		X..W..2	H..Z..Z	P..2..W	X..B..U
30)	L..U..B		A..X..3	J..1..1	Q..3..X	A..C..V
31)	M..V..C		B..Y..4		R..A..Y	

1970 | 1970

	....JULY....	...AUGUST...	..SEPTEMBER.	..OCTOBER...	..NOVEMBER..	..DECEMBER..
	P--E--I	P--E--I	P--E--I	P--E--I	P--E--I	P--E--I
1)	B..D..W	K..G..U	S..K..S	B..M..P	K..P..N	R..R..K
2)	C..E..X	L..H..V	T..L..T	C..N..Q	L..Q..O	S..S..L
3)	D..F..Y	M..J..W	U..M..U	D..O..R	M..R..P	T..T..M
4)	E..G..Z	N..K..X	V..N..V	E..P..S	N..S..Q	U..U..N
5)	F..H..1	O..L..Y	W..O..W	F..Q..T	O..T..R	V..V..O
6)	G..J..2	P..M..Z	X..P..X	G..R..U	P..U..S	W..W..P
7)	H..K..3	Q..N..1	A..Q..Y	H..S..V	Q..V..T	X..X..Q
8)	J..L..4	R..O..2	B..R..Z	J..T..W	R..W..U	A..Y..R
9)	K..M..5	S..P..3	C..S..1	K..U..X	S..X..V	B..Z..S
10)	L..N..6	T..Q..4	D..T..2	L..V..Y	T..Y..W	C..1..T
11)	M..O..7	U..R..5	E..U..3	M..W..Z	U..Z..X	D..2..U
12)	N..P..8	V..S..6	F..V..4	N..X..1	V..1..Y	E..3..V
13)	O..Q..A	W..T..7	G..W..5	O..Y..2	W..2..Z	F..A..W
14)	P..R..B	X..U..8	H..X..6	P..Z..3	X..3..1	G..B..X
15)	Q..S..C	A..V..A	J..Y..7	Q..1..4	A..A..2	H..C..Y
16)	R..T..D	B..W..B	K..Z..8	R..2..5	B..B..3	J..D..Z
17)	S..U..E	C..X..C	L..1..A	S..3..6	C..C..4	K..E..1
18)	T..V..F	D..Y..D	M..2..B	T..A..7	D..D..5	L..F..2
19)	U..W..G	E..Z..E	N..3..C	U..B..8	E..E..6	M..G..3
20)	V..X..H	F..1..F	O..A..D	V..C..A	F..F..7	N..H..4
21)	W..Y..J	G..2..G	P..B..E	W..D..B	G..G..8	O..J..5
22)	X..Z..K	H..3..H	Q..C..F	X..E..C	H..H..A	P..K..6
23)	A..1..L	J..A..J	R..D..G	A..F..D	J..J..B	Q..L..7
24)	B..2..M	K..B..K	S..E..H	B..G..E	K..K..C	R..M..8
25)	C..3..N	L..C..L	T..F..J	C..H..F	L..L..D	S..N..A
26)	D..A..O	M..D..M	U..G..K	D..J..G	M..M..E	T..O..B
27)	E..B..P	N..E..N	V..H..L	E..K..H	N..N..F	U..P..C
28)	F..C..Q	O..F..O	W..J..M	F..L..J	O..O..G	V..Q..D
29)	G..D..R	P..G..P	X..K..N	G..M..K	P..P..H	W..R..E
30)	H..E..S	Q..H..Q	A..L..O	H..N..L	Q..Q..J	X..S..F
31)	J..F..T	R..J..R		J..O..M		A..T..G

CODES: P-PHYSICAL BIORHYTHM CURVE,E-EMOTIONAL BIORHYTHM CURVE,I-INTELLECTUAL BIORHYTHM CURVE

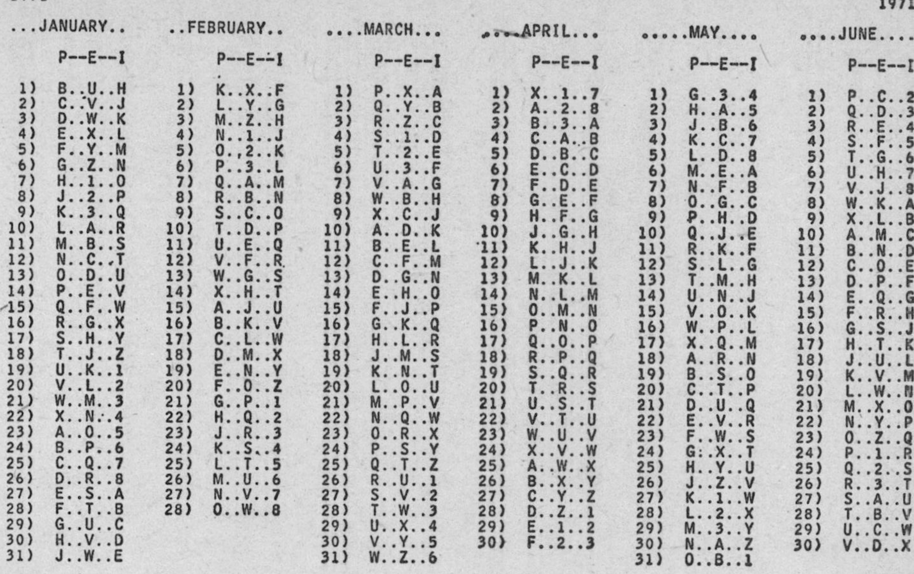

1971						1971
	...JANUARY..	..FEBRUARY..	....MARCH...	....APRIL...	.....MAY....	....JUNE....
	P--E--I	P--E--I	P--E--I	P--E--I	P--E--I	P--E--I
1)	B..U..H	K..X..F	P..X..A	X..1..7	G..3..4	P..C..2
2)	C..V..J	L..Y..G	Q..Y..B	A..2..8	H..A..5	Q..D..3
3)	D..W..K	M..Z..H	R..Z..C	B..3..A	J..B..6	R..E..4
4)	E..X..L	N..1..J	S..1..D	C..A..B	K..C..7	S..F..5
5)	F..Y..M	O..2..K	T..2..E	D..B..C	L..D..8	T..G..6
6)	G..Z..N	P..3..L	U..3..F	E..C..D	M..E..A	U..H..7
7)	H..1..O	Q..A..M	V..A..G	F..D..E	N..F..B	V..J..8
8)	J..2..P	R..B..N	W..B..H	G..E..F	O..G..C	W..K..A
9)	K..3..Q	S..C..O	X..C..J	H..F..G	P..H..D	X..L..B
10)	L..A..R	T..D..P	A..D..K	J..G..H	Q..J..E	A..M..C
11)	M..B..S	U..E..Q	B..E..L	K..H..J	R..K..F	B..N..D
12)	N..C..T	V..F..R	C..F..M	L..J..K	S..L..G	C..O..E
13)	O..D..U	W..G..S	D..G..N	M..K..L	T..M..H	D..P..F
14)	P..E..V	X..H..T	E..H..O	N..L..M	U..N..J	E..Q..G
15)	Q..F..W	A..J..U	F..J..P	O..M..N	V..O..K	F..R..H
16)	R..G..X	B..K..V	G..K..Q	P..N..O	W..P..L	G..S..J
17)	S..H..Y	C..L..W	H..L..R	Q..O..P	X..Q..M	H..T..K
18)	T..J..Z	D..M..X	J..M..S	R..P..Q	A..R..N	J..U..L
19)	U..K..1	E..N..Y	K..N..T	S..Q..R	B..S..O	K..V..M
20)	V..L..2	F..O..Z	L..O..U	T..R..S	C..T..P	L..W..N
21)	W..M..3	G..P..1	M..P..V	U..S..T	D..U..Q	M..X..O
22)	X..N..4	H..Q..2	N..Q..W	V..T..U	E..V..R	N..Y..P
23)	A..O..5	J..R..3	O..R..X	W..U..V	F..W..S	O..Z..Q
24)	B..P..6	K..S..4	P..S..Y	X..V..W	G..X..T	P..1..R
25)	C..Q..7	L..T..5	Q..T..Z	A..W..X	H..Y..U	Q..2..S
26)	D..R..8	M..U..6	R..U..1	B..X..Y	J..Z..V	R..3..T
27)	E..S..A	N..V..7	S..V..2	C..Y..Z	K..1..W	S..A..U
28)	F..T..B	O..W..8	T..W..3	D..Z..1	L..2..X	T..B..V
29)	G..U..C		U..X..4	E..1..2	M..3..Y	U..C..W
30)	H..V..D		V..Y..5	F..2..3	N..A..Z	V..D..X
31)	J..W..E		W..Z..6		O..B..1	

1971					1971
....JULY....	...AUGUST...	..SEPTEMBER.	..OCTOBER...	..NOVEMBER..	..DECEMBER..
P--E--I	P--E--I	P--E--I	P--E--I	P--E--I	P--E--I
1) W..E..Y	1) G..H..W	1) P..L..U	1) W..N..R	1) G..Q..P	1) O..S..M
2) X..F..Z	2) H..J..X	2) Q..M..V	2) X..O..S	2) H..R..Q	2) P..T..N
3) A..G..1	3) J..K..Y	3) R..N..W	3) A..P..T	3) J..S..R	3) Q..U..O
4) B..H..2	4) K..L..Z	4) S..O..X	4) B..Q..U	4) K..T..S	4) R..V..P
5) C..J..3	5) L..M..1	5) T..P..Y	5) C..R..V	5) L..U..T	5) S..W..Q
6) D..K..4	6) M..N..2	6) U..Q..Z	6) D..S..W	6) M..V..U	6) T..X..R
7) E..L..5	7) N..O..3	7) V..R..1	7) E..T..X	7) N..W..V	7) U..Y..S
8) F..M..6	8) O..P..4	8) W..S..2	8) F..U..Y	8) O..X..W	8) V..Z..T
9) G..N..7	9) P..Q..5	9) X..T..3	9) G..V..Z	9) P..Y..X	9) W..1..U
10) H..O..8	10) Q..R..6	10) A..U..4	10) H..W..1	10) Q..Z..Y	10) X..2..V
11) J..P..A	11) R..S..7	11) B..V..5	11) J..X..2	11) R..1..Z	11) A..3..W
12) K..Q..B	12) S..T..8	12) C..W..6	12) K..Y..3	12) S..2..1	12) B..A..X
13) L..R..C	13) T..U..A	13) D..X..7	13) L..Z..4	13) T..3..2	13) C..B..Y
14) M..S..D	14) U..V..B	14) E..Y..8	14) M..1..5	14) U..A..3	14) D..C..Z
15) N..T..E	15) V..W..C	15) F..Z..A	15) N..2..6	15) V..B..4	15) E..D..1
16) O..U..F	16) W..X..D	16) G..1..B	16) O..3..7	16) W..C..5	16) F..E..2
17) P..V..G	17) X..Y..E	17) H..2..C	17) P..A..8	17) X..D..6	17) G..F..3
18) Q..W..H	18) A..Z..F	18) J..3..D	18) Q..B..A	18) A..E..7	18) H..G..4
19) R..X..J	19) B..1..G	19) K..A..E	19) R..C..B	19) B..F..8	19) J..H..5
20) S..Y..K	20) C..2..H	20) L..B..F	20) S..D..C	20) C..G..A	20) K..J..6
21) T..Z..L	21) D..3..J	21) M..C..G	21) T..E..D	21) D..H..B	21) L..K..7
22) U..1..M	22) E..A..K	22) N..D..H	22) U..F..E	22) E..J..C	22) M..L..8
23) V..2..N	23) F..B..L	23) O..E..J	23) V..G..F	23) F..K..D	23) N..M..A
24) W..3..O	24) G..C..M	24) P..F..K	24) W..H..G	24) G..L..E	24) O..N..B
25) X..A..P	25) H..D..N	25) Q..G..L	25) X..J..H	25) H..M..F	25) P..O..C
26) A..B..Q	26) J..E..O	26) R..H..M	26) A..K..J	26) J..N..G	26) Q..P..D
27) B..C..R	27) K..F..P	27) S..J..N	27) B..L..K	27) K..O..H	27) R..Q..E
28) C..D..S	28) L..G..Q	28) T..K..O	28) C..M..L	28) L..P..J	28) S..R..F
29) D..E..T	29) M..H..R	29) U..L..P	29) D..N..M	29) M..Q..K	29) T..S..G
30) E..F..U	30) N..J..S	30) V..M..Q	30) E..O..N	30) N..R..L	30) U..T..H
31) F..G..V	31) O..K..T		31) F..P..O		31) V..U..J

CODES: P-PHYSICAL BIORHYTHM CURVE,E-EMOTIONAL BIORHYTHM CURVE,I-INTELLECTUAL BIORHYTHM CURVE

1972 | 1972

	...JANUARY..		..FEBRUARY..		....MARCH...		....APRIL...		.....MAY....		....JUNE....
	P--E--I		P--E--I		P--E--I		P--E--I		P--E--I		P--E--I
1)	W..V..K	1)	G..Y..H	1)	N..Z..D	1)	V..3..B	1)	E..B..7	1)	N..E..5
2)	X..W..L	2)	H..Z..J	2)	O..1..E	2)	W..A..C	2)	F..C..8	2)	O..F..6
3)	A..X..M	3)	J..1..K	3)	P..2..F	3)	X..B..D	3)	G..D..A	3)	P..G..7
4)	B..Y..N	4)	K..2..L	4)	Q..3..G	4)	A..C..E	4)	H..E..B	4)	Q..H..8
5)	C..Z..O	5)	L..3..M	5)	R..A..H	5)	B..D..F	5)	J..F..C	5)	R..J..A
6)	D..1..P	6)	M..A..N	6)	S..B..J	6)	C..E..G	6)	K..G..D	6)	S..K..B
7)	E..2..Q	7)	N..B..O	7)	T..C..K	7)	D..F..H	7)	L..H..E	7)	T..L..C
8)	F..3..R	8)	O..C..P	8)	U..D..L	8)	E..G..J	8)	M..J..F	8)	U..M..D
9)	G..A..S	9)	P..D..Q	9)	V..E..M	9)	F..H..K	9)	N..K..G	9)	V..N..E
10)	H..B..T	10)	Q..E..R	10)	W..F..N	10)	G..J..L	10)	O..L..H	10)	W..O..F
11)	J..C..U	11)	R..F..S	11)	X..G..O	11)	H..K..M	11)	P..M..J	11)	X..P..G
12)	K..D..V	12)	S..G..T	12)	A..H..P	12)	J..L..N	12)	Q..N..K	12)	A..Q..H
13)	L..E..W	13)	T..H..U	13)	B..J..Q	13)	K..M..O	13)	R..O..L	13)	B..R..J
14)	M..F..X	14)	U..J..V	14)	C..K..R	14)	L..N..P	14)	S..P..M	14)	C..S..K
15)	N..G..Y	15)	V..K..W	15)	D..L..S	15)	M..O..Q	15)	T..Q..N	15)	D..T..L
16)	O..H..Z	16)	W..L..X	16)	E..M..T	16)	N..P..R	16)	U..R..O	16)	E..U..M
17)	P..J..1	17)	X..M..Y	17)	F..N..U	17)	O..Q..S	17)	V..S..P	17)	F..V..N
18)	Q..K..2	18)	A..N..Z	18)	G..O..V	18)	P..R..T	18)	W..T..Q	18)	G..W..O
19)	R..L..3	19)	B..O..1	19)	H..P..W	19)	Q..S..U	19)	X..U..R	19)	H..X..P
20)	S..M..4	20)	C..P..2	20)	J..Q..X	20)	R..T..V	20)	A..V..S	20)	J..Y..Q
21)	T..N..5	21)	D..Q..3	21)	K..R..Y	21)	S..U..W	21)	B..W..T	21)	K..Z..R
22)	U..O..6	22)	E..R..4	22)	L..S..Z	22)	T..V..X	22)	C..X..U	22)	L..1..S
23)	V..P..7	23)	F..S..5	23)	M..T..1	23)	U..W..Y	23)	D..Y..V	23)	M..2..T
24)	W..Q..8	24)	G..T..6	24)	N..U..2	24)	V..X..Z	24)	E..Z..W	24)	N..3..U
25)	X..R..A	25)	H..U..7	25)	O..V..3	25)	W..Y..1	25)	F..1..X	25)	O..A..V
26)	A..S..B	26)	J..V..8	26)	P..W..4	26)	X..Z..2	26)	G..2..Y	26)	P..B..W
27)	B..T..C	27)	K..W..A	27)	Q..X..5	27)	A..1..3	27)	H..3..Z	27)	Q..C..X
28)	C..U..D	28)	L..X..B	28)	R..Y..6	28)	B..2..4	28)	J..A..1	28)	R..D..Y
29)	D..V..E	29)	M..Y..C	29)	S..Z..7	29)	C..3..5	29)	K..B..2	29)	S..E..Z
30)	E..W..F			30)	T..1..8	30)	D..A..6	30)	L..C..3	30)	T..F..1
31)	F..X..G			31)	U..2..A			31)	M..D..4		

1972											1972
....JULY....		...AUGUST...		..SEPTEMBER.		..OCTOBER...		..NOVEMBER..		..DECEMBER..	
	P--E--I		P--E--I		P--E--I		P--E--I		P--E--I		P--E--I
1)	U..G..2	1)	E..K..Z	1)	N..N..X	1)	U..P..U	1)	E..S..S	1)	M..U..P
2)	V..H..3	2)	F..L..1	2)	O..O..Y	2)	V..Q..V	2)	F..T..T	2)	N..V..Q
3)	W..J..4	3)	G..M..2	3)	P..P..Z	3)	W..R..W	3)	G..U..U	3)	O..W..R
4)	X..K..5	4)	H..N..3	4)	Q..Q..1	4)	X..S..X	4)	H..V..V	4)	P..X..S
5)	A..L..6	5)	J..O..4	5)	R..R..2	5)	A..T..Y	5)	J..W..W	5)	Q..Y..T
6)	B..M..7	6)	K..P..5	6)	S..S..3	6)	B..U..Z	6)	K..X..X	6)	R..Z..U
7)	C..N..8	7)	L..Q..6	7)	T..T..4	7)	C..V..1	7)	L..Y..Y	7)	S..1..V
8)	D..O..A	8)	M..R..7	8)	U..U..5	8)	D..W..2	8)	M..Z..Z	8)	T..2..W
9)	E..P..B	9)	N..S..8	9)	V..V..6	9)	E..X..3	9)	N..1..1	9)	U..3..X
10)	F..Q..C	10)	O..T..A	10)	W..W..7	10)	F..Y..4	10)	O..2..2	10)	V..A..Y
11)	G..R..D	11)	P..U..B	11)	X..X..8	11)	G..Z..5	11)	P..3..3	11)	W..B..Z
12)	H..S..E	12)	Q..V..C	12)	A..Y..A	12)	H..1..6	12)	Q..A..4	12)	X..C..1
13)	J..T..F	13)	R..W..D	13)	B..Z..B	13)	J..2..7	13)	R..B..5	13)	A..D..2
14)	K..U..G	14)	S..X..E	14)	C..1..C	14)	K..3..8	14)	S..C..6	14)	B..E..3
15)	L..V..H	15)	T..Y..F	15)	D..2..D	15)	L..A..A	15)	T..D..7	15)	C..F..4
16)	M..W..J	16)	U..Z..G	16)	E..3..E	16)	M..B..B	16)	U..E..8	16)	D..G..5
17)	N..X..K	17)	V..1..H	17)	F..A..F	17)	N..C..C	17)	V..F..A	17)	E..H..6
18)	O..Y..L	18)	W..2..J	18)	G..B..G	18)	O..D..D	18)	W..G..B	18)	F..J..7
19)	P..Z..M	19)	X..3..K	19)	H..C..H	19)	P..E..E	19)	X..H..C	19)	G..K..8
20)	Q..1..N	20)	A..A..L	20)	J..D..J	20)	Q..F..F	20)	A..J..D	20)	H..L..A
21)	R..2..O	21)	B..B..M	21)	K..E..K	21)	R..G..G	21)	B..K..E	21)	J..M..B
22)	S..3..P	22)	C..C..N	22)	L..F..L	22)	S..H..H	22)	C..L..F	22)	K..N..C
23)	T..A..Q	23)	D..D..O	23)	M..G..M	23)	T..J..J	23)	D..M..G	23)	L..O..D
24)	U..B..R	24)	E..E..P	24)	N..H..N	24)	U..K..K	24)	E..N..H	24)	M..P..E
25)	V..C..S	25)	F..F..Q	25)	O..J..O	25)	V..L..L	25)	F..O..J	25)	N..Q..F
26)	W..D..T	26)	G..G..R	26)	P..K..P	26)	W..M..M	26)	G..P..K	26)	O..R..G
27)	X..E..U	27)	H..H..S	27)	Q..L..Q	27)	X..N..N	27)	H..Q..L	27)	P..S..H
28)	A..F..V	28)	J..J..T	28)	R..M..R	28)	A..O..O	28)	J..R..M	28)	Q..T..J
29)	B..G..W	29)	K..K..U	29)	S..N..S	29)	B..P..P	29)	K..S..N	29)	R..U..K
30)	C..H..X	30)	L..L..V	30)	T..O..T	30)	C..Q..Q	30)	L..T..O	30)	S..V..L
31)	D..J..Y	31)	M..M..W			31)	D..R..R			31)	T..W..M

CODES: P-PHYSICAL BIORHYTHM CUEVE,E-EMOTIONAL BIORHYTHM CURVE,I-INTELLECTUAL BIORHYTHM CURVE

1973 | | | | | | 1973

	...JANUARY..	..FEBRUARY..	....MARCH...	....APRIL...	.....MAY....	....JUNE....
	P--E--I	P--E--I	P--E--I	P--E--I	P--E--I	P--E--I
1)	U..X..N	E..1..L	K..1..F	S..A..D	B..C..A	K..F..7
2)	V..Y..O	F..2..M	L..2..G	T..B..E	C..D..B	L..G..8
3)	W..Z..P	G..3..N	M..3..H	U..C..F	D..E..C	M..H..A
4)	X..1..Q	H..A..O	N..A..J	V..D..G	E..F..D	N..J..B
5)	A..2..R	J..B..P	O..B..K	W..E..H	F..G..E	O..K..C
6)	B..3..S	K..C..Q	P..C..L	X..F..J	G..H..F	P..L..D
7)	C..A..T	L..D..R	Q..D..M	A..G..K	H..J..G	Q..M..E
8)	D..B..U	M..E..S	R..E..N	B..H..L	J..K..H	R..N..F
9)	E..C..V	N..F..T	S..F..O	C..J..M	K..L..J	S..O..G
10)	F..D..W	O..G..U	T..G..P	D..K..N	L..M..K	T..P..H
11)	G..E..X	P..H..V	U..H..Q	E..L..O	M..N..L	U..Q..J
12)	H..F..Y	Q..J..W	V..J..R	F..M..P	N..O..M	V..R..K
13)	J..G..Z	R..K..X	W..K..S	G..N..Q	O..P..N	W..S..L
14)	K..H..1	S..L..Y	X..L..T	H..O..R	P..Q..O	X..T..M
15)	L..J..2	T..M..Z	A..M..U	J..P..S	Q..R..P	A..U..N
16)	M..K..3	U..N..1	B..N..V	K..Q..T	R..S..Q	B..V..O
17)	N..L..4	V..O..2	C..O..W	L..R..U	S..T..R	C..W..P
18)	O..M..5	W..P..3	D..P..X	M..S..V	T..U..S	D..X..Q
19)	P..N..6	X..Q..4	E..Q..Y	N..T..W	U..V..T	E..Y..R
20)	Q..O..7	A..R..5	F..R..Z	O..U..X	V..W..U	F..Z..S
21)	R..P..8	B..S..6	G..S..1	P..V..Y	W..X..V	G..1..T
22)	S..Q..A	C..T..7	H..T..2	Q..W..Z	X..Y..W	H..2..U
23)	T..R..B	D..U..8	J..U..3	R..X..1	A..Z..X	J..3..V
24)	U..S..C	E..V..A	K..V..4	S..Y..2	B..1..Y	K..A..W
25)	V..T..D	F..W..B	L..W..5	T..Z..3	C..2..Z	L..B..X
26)	W..U..E	G..X..C	M..X..6	U..1..4	D..3..1	M..C..Y
27)	X..V..F	H..Y..D	N..Y..7	V..2..5	E..A..2	N..D..Z
28)	A..W..G	J..Z..E	O..Z..8	W..3..6	F..B..3	O..E..1
29)	B..X..H		P..1..A	X..A..7	G..C..4	P..F..2
30)	C..Y..J		Q..2..B	A..B..8	H..D..5	Q..G..3
31)	D..Z..K		R..3..C		J..E..6	

1973 1973

	....JULY....	...AUGUST...	..SEPTEMBER.	..OCTOBER...	..NOVEMBER..	..DECEMBER..
	P--E--I	P--E--I	P--E--I	P--E--I	P--E--I	P--E--I
1)	R..H..4	B..L..2	K..O..Z	R..Q..W	B..T..U	J..V..R
2)	S..J..5	C..M..3	L..P..1	S..R..X	C..U..V	K..W..S
3)	T..K..6	D..N..4	M..Q..2	T..S..Y	D..V..W	L..X..T
4)	U..L..7	E..O..5	N..R..3	U..T..Z	E..W..X	M..Y..U
5)	V..M..8	F..P..6	O..S..4	V..U..1	F..X..Y	N..Z..V
6)	W..N..A	G..Q..7	P..T..5	W..V..2	G..Y..Z	O..1..W
7)	X..O..B	H..R..8	Q..U..6	X..W..3	H..Z..1	P..2..X
8)	A..P..C	J..S..A	R..V..7	A..X..4	J..1..2	Q..3..Y
9)	B..Q..D	K..T..B	S..W..8	B..Y..5	K..2..3	R..A..Z
10)	C..R..E	L..U..C	T..X..A	C..Z..6	L..3..4	S..B..1
11)	D..S..F	M..V..D	U..Y..B	D..1..7	M..A..5	T..C..2
12)	E..T..G	N..W..E	V..Z..C	E..2..8	N..B..6	U..D..3
13)	F..U..H	O..X..F	W..1..D	F..3..A	O..C..7	V..E..4
14)	G..V..J	P..Y..G	X..2..E	G..A..B	P..D..8	W..F..5
15)	H..W..K	Q..Z..H	A..3..F	H..B..C	Q..E..A	X..G..6
16)	J..X..L	R..1..J	B..A..G	J..C..D	R..F..B	A..H..7
17)	K..Y..M	S..2..K	C..B..H	K..D..E	S..G..C	B..J..8
18)	L..Z..N	T..3..L	D..C..J	L..E..F	T..H..D	C..K..A
19)	M..1..O	U..A..M	E..D..K	M..F..G	U..J..E	D..L..B
20)	N..2..P	V..B..N	F..E..L	N..G..H	V..K..F	E..M..C
21)	O..3..Q	W..C..O	G..F..M	O..H..J	W..L..G	F..N..D
22)	P..A..R	X..D..P	H..G..N	P..J..K	X..M..H	G..O..E
23)	Q..B..S	A..E..Q	J..H..O	Q..K..L	A..N..J	H..P..F
24)	R..C..T	B..F..R	K..J..P	R..L..M	B..O..K	J..Q..G
25)	S..D..U	C..G..S	L..K..Q	S..M..N	C..P..L	K..R..H
26)	T..E..V	D..H..T	M..L..R	T..N..O	D..Q..M	L..S..J
27)	U..F..W	E..J..U	N..M..S	U..O..P	E..R..N	M..T..K
28)	V..G..X	F..K..V	O..N..T	V..P..Q	F..S..O	N..U..L
29)	W..H..Y	G..L..W	P..O..U	W..Q..R	G..T..P	O..V..M
30)	X..J..Z	H..M..X	Q..P..V	X..R..S	H..U..Q	P..W..N
31)	A..K..1	J..N..Y		A..S..T		Q..X..O

CODES: P-PHYSICAL BIORHYTHM CURVE,E-EMOTIONAL BIORHYTHM CURVE,I-INTELLECTUAL BIORHYTHM CURVE

1974 | | | | | | | | | | | 1974

	...JANUARY..		..FEBRUARY..		....MARCH...		....APRIL...		.....MAY....		....JUNE....
	P--E--I		P--E--I		P--E--I		P--E--I		P--E--I		P--E--I
1)	R..Y..P	1)	B..2..N	1)	G..2..H	1)	P..B..F	1)	W..D..C	1)	G..G..A
2)	S..Z..Q	2)	C..3..O	2)	H..3..J	2)	Q..C..G	2)	X..E..D	2)	H..H..B
3)	T..1..R	3)	D..A..P	3)	J..A..K	3)	R..D..H	3)	A..F..E	3)	J..J..C
4)	U..2..S	4)	E..B..Q	4)	K..B..L	4)	S..E..J	4)	B..G..F	4)	K..K..D
5)	V..3..T	5)	F..C..R	5)	L..C..M	5)	T..F..K	5)	C..H..G	5)	L..L..E
6)	W..A..U	6)	G..D..S	6)	M..D..N	6)	U..G..L	6)	D..J..H	6)	M..M..F
7)	X..B..V	7)	H..E..T	7)	N..E..O	7)	V..H..M	7)	E..K..J	7)	N..N..G
8)	A..C..W	8)	J..F..U	8)	O..F..P	8)	W..J..N	8)	F..L..K	8)	O..O..H
9)	B..D..X	9)	K..G..V	9)	P..G..Q	9)	X..K..O	9)	G..M..L	9)	P..P..J
10)	C..E..Y	10)	L..H..W	10)	Q..H..R	10)	A..L..P	10)	H..N..M	10)	Q..Q..K
11)	D..F..Z	11)	M..J..X	11)	R..J..S	11)	B..M..Q	11)	J..O..N	11)	R..R..L
12)	E..G..1	12)	N..K..Y	12)	S..K..T	12)	C..N..R	12)	K..P..O	12)	S..S..M
13)	F..H..2	13)	O..L..Z	13)	T..L..U	13)	D..O..S	13)	L..Q..P	13)	T..T..N
14)	G..J..3	14)	P..M..1	14)	U..M..V	14)	E..P..T	14)	M..R..Q	14)	U..U..O
15)	H..K..4	15)	Q..N..2	15)	V..N..W	15)	F..Q..U	15)	N..S..R	15)	V..V..P
16)	J..L..5	16)	R..O..3	16)	W..O..X	16)	G..R..V	16)	O..T..S	16)	W..W..Q
17)	K..M..6	17)	S..P..4	17)	X..P..Y	17)	H..S..W	17)	P..U..T	17)	X..X..R
18)	L..N..7	18)	T..Q..5	18)	A..Q..Z	18)	J..T..X	18)	Q..V..U	18)	A..Y..S
19)	M..O..8	19)	U..R..6	19)	B..R..1	19)	K..U..Y	19)	R..W..V	19)	B..Z..T
20)	N..P..A	20)	V..S..7	20)	C..S..2	20)	L..V..Z	20)	S..X..W	20)	C..1..U
21)	O..Q..B	21)	W..T..8	21)	D..T..3	21)	M..W..1	21)	T..Y..X	21)	D..2..V
22)	P..R..C	22)	X..U..A	22)	E..U..4	22)	N..X..2	22)	U..Z..Y	22)	E..3..W
23)	Q..S..D	23)	A..V..B	23)	F..V..5	23)	O..Y..3	23)	V..1..Z	23)	F..A..X
24)	R..T..E	24)	B..W..C	24)	G..W..6	24)	P..Z..4	24)	W..2..1	24)	G..B..Y
25)	S..U..F	25)	C..X..D	25)	H..X..7	25)	Q..1..5	25)	X..3..2	25)	H..C..Z
26)	T..V..G	26)	D..Y..E	26)	J..Y..8	26)	R..2..6	26)	A..A..3	26)	J..D..1
27)	U..W..H	27)	E..Z..F	27)	K..Z..A	27)	S..3..7	27)	B..B..4	27)	K..E..2
28)	V..X..J	28)	F..1..G	28)	L..1..B	28)	T..A..8	28)	C..C..5	28)	L..F..3
29)	W..Y..K			29)	M..2..C	29)	U..B..A	29)	D..D..6	29)	M..G..4
30)	X..Z..L			30)	N..3..D	30)	V..C..B	30)	E..E..7	30)	N..H..5
31)	A..1..M			31)	O..A..E			31)	F..F..8		

1974 | 1974

Day	....JULY.... P--E--I	...AUGUST... P--E--I	..SEPTEMBER. P--E--I	..OCTOBER... P--E--I	..NOVEMBER.. P--E--I	..DECEMBER.. P--E--I
1)	O..J..6	W..M..4	G..P..2	O..R..Y	W..U..W	F..W..T
2)	P..K..7	X..N..5	H..Q..3	P..S..Z	X..V..X	G..X..U
3)	Q..L..8	A..O..6	J..R..4	Q..T..1	A..W..Y	H..Y..V
4)	R..M..A	B..P..7	K..S..5	R..U..2	B..X..Z	J..Z..W
5)	S..N..B	C..Q..8	L..T..6	S..V..3	C..Y..1	K..1..X
6)	T..O..C	D..R..A	M..U..7	T..W..4	D..Z..2	L..2..Y
7)	U..P..D	E..S..B	N..V..8	U..X..5	E..1..3	M..3..Z
8)	V..Q..E	F..T..C	O..W..A	V..Y..6	F..2..4	N..A..1
9)	W..R..F	G..U..D	P..X..B	W..Z..7	G..3..5	O..B..2
10)	X..S..G	H..V..E	Q..Y..C	X..1..8	H..A..6	P..C..3
11)	A..T..H	J..W..F	R..Z..D	A..2..A	J..B..7	Q..D..4
12)	B..U..J	K..X..G	S..1..E	B..3..B	K..C..8	R..E..5
13)	C..V..K	L..Y..H	T..2..F	C..A..C	L..D..A	S..F..6
14)	D..W..L	M..Z..J	U..3..G	D..B..D	M..E..B	T..G..7
15)	E..X..M	N..1..K	V..A..H	E..C..E	N..F..C	U..H..8
16)	F..Y..N	O..2..L	W..B..J	F..D..F	O..G..D	V..J..A
17)	G..Z..O	P..3..M	X..C..K	G..E..G	P..H..E	W..K..B
18)	H..1..P	Q..A..N	A..D..L	H..F..H	Q..J..F	X..L..C
19)	J..2..Q	R..B..O	B..E..M	J..G..J	R..K..G	A..M..D
20)	K..3..R	S..C..P	C..F..N	K..H..K	S..L..H	B..N..E
21)	L..A..S	T..D..Q	D..G..O	L..J..L	T..M..J	C..O..F
22)	M..B..T	U..E..R	E..H..P	M..K..M	U..N..K	D..P..G
23)	N..C..U	V..F..S	F..J..Q	N..L..N	V..O..L	E..Q..H
24)	O..D..V	W..G..T	G..K..R	O..M..O	W..P..M	F..R..J
25)	P..E..W	X..H..U	H..L..S	P..N..P	X..Q..N	G..S..K
26)	Q..F..X	A..J..V	J..M..T	Q..O..Q	A..R..O	H..T..L
27)	R..G..Y	B..K..W	K..N..U	R..P..R	B..S..P	J..U..M
28)	S..H..Z	C..L..X	L..O..V	S..Q..S	C..T..Q	K..V..N
29)	T..J..1	D..M..Y	M..P..W	T..R..T	D..U..R	L..W..O
30)	U..K..2	E..N..Z	N..Q..X	U..S..U	E..V..S	M..X..P
31)	V..L..3	F..O..1		V..T..V		N..Y..Q

CODES: P-PHYSICAL BIORHYTHM CURVE,E-EMOTIONAL BIORHYTHM CURVE,I-INTELLECTUAL BIORHYTHM CURVE

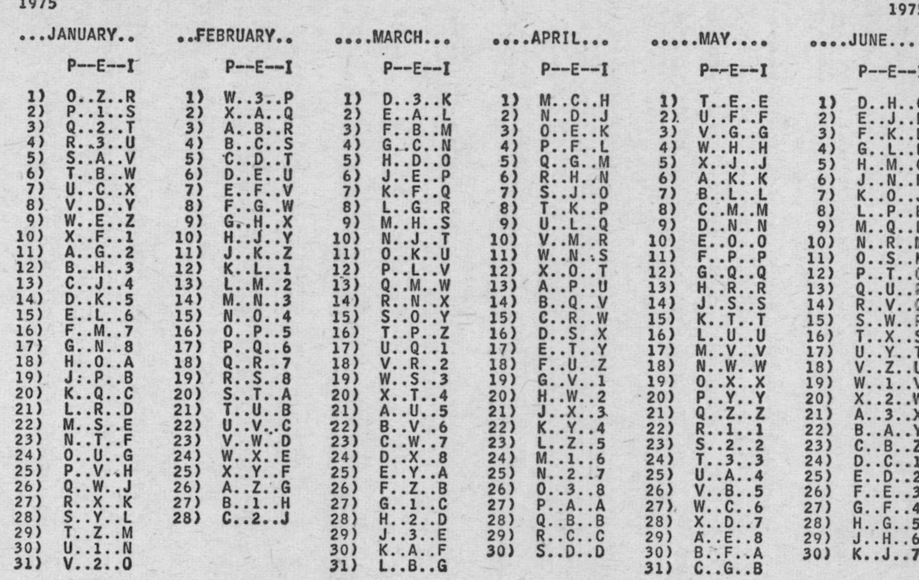

1975						1975
	...JANUARY..	..FEBRUARY..	....MARCH...	....APRIL...	.....MAY....	....JUNE....
	P--E--I	P--E--I	P--E--I	P--E--I	P--E--I	P--E--I
1)	O..Z..R	W..3..P	D..3..K	M..C..H	T..E..E	D..H..C
2)	P..1..S	X..A..Q	E..A..L	N..D..J	U..F..F	E..J..D
3)	Q..2..T	A..B..R	F..B..M	O..E..K	V..G..G	F..K..E
4)	R..3..U	B..C..S	G..C..N	P..F..L	W..H..H	G..L..F
5)	S..A..V	C..D..T	H..D..O	Q..G..M	X..J..J	H..M..G
6)	T..B..W	D..E..U	J..E..P	R..H..N	A..K..K	J..N..H
7)	U..C..X	E..F..V	K..F..Q	S..J..O	B..L..L	K..O..J
8)	V..D..Y	F..G..W	L..G..R	T..K..P	C..M..M	L..P..K
9)	W..E..Z	G..H..X	M..H..S	U..L..Q	D..N..N	M..Q..L
10)	X..F..1	H..J..Y	N..J..T	V..M..R	E..O..O	N..R..M
11)	A..G..2	J..K..Z	O..K..U	W..N..S	F..P..P	O..S..N
12)	B..H..3	K..L..1	P..L..V	X..O..T	G..Q..Q	P..T..O
13)	C..J..4	L..M..2	Q..M..W	A..P..U	H..R..R	Q..U..P
14)	D..K..5	M..N..3	R..N..X	B..Q..V	J..S..S	R..V..Q
15)	E..L..6	N..O..4	S..O..Y	C..R..W	K..T..T	S..W..R
16)	F..M..7	O..P..5	T..P..Z	D..S..X	L..U..U	T..X..S
17)	G..N..8	P..Q..6	U..Q..1	E..T..Y	M..V..V	U..Y..T
18)	H..O..A	Q..R..7	V..R..2	F..U..Z	N..W..W	V..Z..U
19)	J..P..B	R..S..8	W..S..3	G..V..1	O..X..X	W..1..V
20)	K..Q..C	S..T..A	X..T..4	H..W..2	P..Y..Y	X..2..W
21)	L..R..D	T..U..B	A..U..5	J..X..3	Q..Z..Z	A..3..X
22)	M..S..E	U..V..C	B..V..6	K..Y..4	R..1..1	B..A..Y
23)	N..T..F	V..W..D	C..W..7	L..Z..5	S..2..2	C..B..Z
24)	O..U..G	W..X..E	D..X..8	M..1..6	T..3..3	D..C..1
25)	P..V..H	X..Y..F	E..Y..A	N..2..7	U..A..4	E..D..2
26)	Q..W..J	A..Z..G	F..Z..B	O..3..8	V..B..5	F..E..3
27)	R..X..K	B..1..H	G..1..C	P..A..A	W..C..6	G..F..4
28)	S..Y..L	C..2..J	H..2..D	Q..B..B	X..D..7	H..G..5
29)	T..Z..M		J..3..E	R..C..C	A..E..8	J..H..6
30)	U..1..N		K..A..F	S..D..D	B..F..A	K..J..7
31)	V..2..O		L..B..G		C..G..B	

Day	....JULY.... P--E--I	...AUGUST... P--E--I	..SEPTEMBER. P--E--I	..OCTOBER... P--E--I	..NOVEMBER.. P--E--I	..DECEMBER.. P--E--I
1)	L..K..8	T..N..6	D..Q..4	L..S..1	T..V..Y	C..X..V
2)	M..L..A	U..O..7	E..R..5	M..T..2	U..W..Z	D..Y..W
3)	N..M..B	V..P..8	F..S..6	N..U..3	V..X..1	E..Z..X
4)	O..N..C	W..Q..A	G..T..7	O..V..4	W..Y..2	F..1..Y
5)	P..O..D	X..R..B	H..U..8	P..W..5	X..Z..3	G..2..Z
6)	Q..P..E	A..S..C	J..V..A	Q..X..6	A..1..4	H..3..1
7)	R..Q..F	B..T..D	K..W..B	R..Y..7	B..2..5	J..A..2
8)	S..R..G	C..U..E	L..X..C	S..Z..8	C..3..6	K..B..3
9)	T..S..H	D..V..F	M..Y..D	T..1..A	D..A..7	L..C..4
10)	U..T..J	E..W..G	N..Z..E	U..2..B	E..B..8	M..D..5
11)	V..U..K	F..X..H	O..1..F	V..3..C	F..C..A	N..E..6
12)	W..V..L	G..Y..J	P..2..G	W..A..D	G..D..B	O..F..7
13)	X..W..M	H..Z..K	Q..3..H	X..B..E	H..E..C	P..G..8
14)	A..X..N	J..1..L	R..A..J	A..C..F	J..F..D	Q..H..A
15)	B..Y..O	K..2..M	S..B..K	B..D..G	K..G..E	R..J..B
16)	C..Z..P	L..3..N	T..C..L	C..E..H	L..H..F	S..K..C
17)	D..1..Q	M..A..O	U..D..M	D..F..J	M..J..G	T..L..D
18)	E..2..R	N..B..P	V..E..N	E..G..K	N..K..H	U..M..E
19)	F..3..S	O..C..Q	W..F..O	F..H..L	O..L..J	V..N..F
20)	G..A..T	P..D..R	X..G..P	G..J..M	P..M..K	W..O..G
21)	H..B..U	Q..E..S	A..H..Q	H..K..N	Q..N..L	X..P..H
22)	J..C..V	R..F..T	B..J..R	J..L..O	R..O..M	A..Q..J
23)	K..D..W	S..G..U	C..K..S	K..M..P	S..P..N	B..R..K
24)	L..E..X	T..H..V	D..L..T	L..N..Q	T..Q..O	C..S..L
25)	M..F..Y	U..J..W	E..M..U	M..O..R	U..R..P	D..T..M
26)	N..G..Z	V..K..X	F..N..V	N..P..S	V..S..Q	E..U..N
27)	O..H..1	W..L..Y	G..O..W	O..Q..T	W..T..R	F..V..O
28)	P..J..2	X..M..Z	H..P..X	P..R..U	X..U..S	G..W..P
29)	Q..K..3	A..N..1	J..Q..Y	Q..S..V	A..V..T	H..X..Q
30)	R..L..4	B..O..2	K..R..Z	R..T..W	B..W..U	J..Y..R
31)	S..M..5	C..P..3		S..U..X		K..Z..S

CODES: P-PHYSICAL BIORHYTHM CURVE,E-EMCTIONAL BIORHYTHM CURVE,I-INTELLECTUAL BIORHYTHM CURVE

Appendix B
Biorhythm Charts

JANUARY 1978 — PHYSICAL

1 SU	2 M	3 TU	4 W	5 TH	6 F	7 SA	8 SU	9 M	10 TU	11 W	12 TH	13 F	14 SA	15 SU
AX	BA	CB	DC	ED	FE	GF	HG	JH	KJ	LK	ML	NM	ON	PO
BW	CX	DA	EB	FC	GD	HE	JF	KG	LH	MJ	NK	OL	PM	QN
CV	DW	EX	FA	GB	HC	JD	KE	LF	MG	NH	OJ	PK	QL	RM
DU	EV	FW	GX	HA	JB	KC	LD	ME	NF	OG	PH	QJ	RK	SL
ET	FU	GV	HW	JX	KA	LB	MC	ND	OE	PF	QG	RH	SJ	TK
FS	GT	HU	JV	KW	LX	MA	NB	OC	PD	QE	RF	SG	TH	UJ
GR	HS	JT	KU	LV	MW	NX	OA	PB	QC	RD	SE	TF	UG	VH
HQ	JR	KS	LT	MU	NV	OW	PX	QA	RB	SC	TD	UE	VF	WG
JP	KQ	LR	MS	NT	OU	PV	QW	RX	SA	TB	UC	VD	WE	XF
KO	LP	MQ	NR	OS	PT	QU	RV	SW	TX	UA	VB	WC	XD	AE
LN	MO	NP	OQ	PR	QS	RT	SU	TV	UW	VX	WA	XB	AC	BD
M	N	O	P	Q	R	S	T	U	V	W	X	A	B	C

JANUARY 1978 — EMOTIONAL

1 SU	2 M	3 TU	4 W	5 TH	6 F	7 SA	8 SU	9 M	10 TU	11 W	12 TH	13 F	14 SA	15 SU	
W	X	Y	Z	1	2	3	A	B	C	D	E	F	G	H	
XV	YW	ZX	1Y	2Z	31	A2	B3	CA	DB	EC	FD	GE	HF	JG	
YU	ZV	1W	2X	3Y	AZ	B1	C2	D3	EA	FB	GC	HD	JE	KF	
ZT	1U	2V	3W	AX	BY	CZ	D1	E2	F3	GA	HB	JC	KD	LE	
1S	2T	3U	AV	BW	CX	DY	EZ	F1	G2	H3	JA	KB	LC	MD	+
2R	3S	AT	BU	CV	DW	EX	FY	GZ	H1	J2	K3	LA	MB	NC	
3Q	AR	BS	CT	DU	EV	FW	GX	HY	JZ	K1	L2	M3	NA	OB	
AP	BQ	CR	DS	ET	FU	GV	HW	JX	KY	LZ	M1	N2	O3	PA	o
BO	CP	DQ	ER	FS	GT	HU	JV	KW	LX	MY	NZ	O1	P2	Q3	
CN	DO	EP	FQ	GR	HS	JT	KU	LV	MW	NX	OY	PZ	Q1	R2	
DM	EN	FO	GP	HQ	JR	KS	LT	MU	NV	OW	PX	QY	RZ	S1	−
EL	FM	GN	HO	JP	KQ	LR	MS	NT	OU	PV	QW	RX	SY	TZ	
FK	GL	HM	JN	KO	LP	MQ	NR	OS	PT	QU	RV	SW	TX	UY	
GJ	HK	JL	KM	LN	MO	NP	OQ	PR	QS	RT	SU	TV	UW	VX	
H	J	K	L	M	N	O	P	Q	R	S	T	U	V	W	

JANUARY 1978 — INTELLECTUAL

1 SU	2 M	3 TU	4 W	5 TH	6 F	7 SA	8 SU	9 M	10 TU	11 W	12 TH	13 F	14 SA	15 SU
Q	R	S	T	U	V	W	X	Y	Z	1	2	3	4	5
RP	SQ	TR	US	VT	WU	XV	YW	ZX	1Y	2Z	31	42	53	64
SO	TP	UQ	VR	WS	XT	YU	ZV	1W	2X	3Y	4Z	51	62	73
TN	UO	VP	WQ	XR	YS	ZT	1U	2V	3W	4X	5Y	6Z	71	82
UM	VN	WO	XP	YQ	ZR	1S	2T	3U	4V	5W	6X	7Y	8Z	A1
VL	WM	XN	YO	ZP	1Q	2R	3S	4T	5U	6V	7W	8X	AY	BZ
WK	XL	YM	ZN	1O	2P	3Q	4R	5S	6T	7U	8V	AW	BX	CY
XJ	YK	ZL	1M	2N	3O	4P	5Q	6R	7S	8T	AU	BV	CW	DX
YH	ZJ	1K	2L	3M	4N	5O	6P	7Q	8R	AS	BT	CU	DV	EW
ZG	1H	2J	3K	4L	5M	6N	7O	8P	AQ	BR	CS	DT	EU	FV
1F	2G	3H	4J	5K	6L	7M	8N	AO	BP	CQ	DR	ES	FT	GU
2E	3F	4G	5H	6J	7K	8L	AM	BN	CO	DP	EQ	FR	GS	HT
3D	4E	5F	6G	7H	8J	AK	BL	CM	DN	EO	FP	GQ	HR	JS
4C	5D	6E	7F	8G	AH	BJ	CK	DL	EM	FN	GO	HP	JQ	KR
5B	6C	7D	8E	AF	BG	CH	DJ	EK	FL	GM	HN	JO	KP	LQ
6A	7B	8C	AD	BE	CF	DG	EH	FJ	GK	HL	JM	KN	LO	MP
78	8A	AB	BC	CD	DE	EF	FG	GH	HJ	JK	KL	LM	MN	NO

JANUARY 1978 — PHYSICAL

16 M	17 TU	18 W	19 TH	20 F	21 SA	22 SU	23 M	24 TU	25 W	26 TH	27 F	28 SA	29 SU	30 M	31 TU
QP	RQ	SR	TS	UT	VU	WV	XW	AX	BA	CB	DC	ED	FE	GF	HG
RO	SP	TQ	UR	VS	WT	XU	AV	BW	CX	DA	EB	FC	GD	HE	JF
SN	TO	UP	VQ	WR	XS	AT	BU	CV	DW	EX	FA	GB	HC	JD	KE
TM	UN	VO	WP	XQ	AR	BS	CT	DU	EV	FW	GX	HA	JB	KC	LD
UL	VM	WN	XO	AP	BQ	CR	DS	ET	FU	GV	HW	JX	KA	LB	MC
VK	WL	XM	AN	BO	CP	DQ	ER	FS	GT	HU	JV	KW	LX	MA	NB
WJ	XK	AL	BM	CN	DO	EP	FQ	GR	HS	JT	KU	LV	MW	NX	OA
XH	AJ	BK	CL	DM	EN	FO	GP	HQ	JR	KS	LT	MU	NV	OW	PX
AG	BH	CJ	DK	EL	FM	GN	HO	JP	KQ	LR	MS	NT	OU	PV	QW
BF	CG	DH	EJ	FK	GL	HM	JN	KO	LP	MQ	NR	OS	PT	QU	RV
CE	DF	EG	FH	GJ	HK	JL	KM	LN	MO	NP	OQ	PR	QS	RT	SU
D	E	F	G	H	J	K	L	M	N	O	P	Q	R	S	T

JANUARY 1978 — EMOTIONAL

16 M	17 TU	18 W	19 TH	20 F	21 SA	22 SU	23 M	24 TU	25 W	26 TH	27 F	28 SA	29 SU	30 M	31 TU	
J	K	L	M	N	O	P	Q	R	S	T	U	V	W	X	Y	
KH	LJ	MK	NL	OM	PN	QO	RP	SQ	TR	US	VT	WU	XV	YW	ZX	
LG	MH	NJ	OK	PL	QM	RN	SO	TP	UQ	VR	WS	XT	YU	ZV	1W	
MF	NG	OH	PJ	QK	RL	SM	TN	UO	VP	WQ	XR	YS	ZT	1U	2V	
NE	OF	PG	QH	RJ	SK	TL	UM	VN	WO	XP	YQ	ZR	1S	2T	3U	+
OD	PE	QF	RG	SH	TJ	UK	VL	WM	XN	YO	ZP	1Q	2R	3S	AT	
PC	QD	RE	SF	TG	UH	VJ	WK	XL	YM	ZN	1O	2P	3Q	AR	BS	
QB	RC	SD	TE	UF	VG	WH	XJ	YK	ZL	1M	2N	3O	AP	BQ	CR	o
RA	SB	TC	UD	VE	WF	XG	YH	ZJ	1K	2L	3M	AN	BO	CP	DQ	
S3	TA	UB	VC	WD	XE	YF	ZG	1H	2J	3K	AL	BM	CN	DO	EP	
T2	U3	VA	WB	XC	YD	ZE	1F	2G	3H	AJ	BK	CL	DM	EN	FO	–
U1	V2	W3	XA	YB	ZC	1D	2E	3F	AG	BH	CJ	DK	EL	FM	GN	
VZ	W1	X2	Y3	ZA	1B	2C	3D	AE	BF	CG	DH	EJ	FK	GL	HM	
WY	XZ	Y1	Z2	13	2A	3B	AC	BD	CE	DF	EG	FH	GJ	HK	JL	
X	Y	Z	1	2	3	A	B	C	D	E	F	G	H	J	K	

JANUARY 1978 — INTELLECTUAL

16 M	17 TU	18 W	19 TH	20 F	21 SA	22 SU	23 M	24 TU	25 W	26 TH	27 F	28 SA	29 SU	30 M	31 TU
6	7	8	A	B	C	D	E	F	G	H	J	K	L	M	N
75	86	A7	B8	CA	DB	EC	FD	GE	HF	JG	KH	LJ	MK	NL	OM
84	A5	B6	C7	D8	EA	FB	GC	HD	JE	KF	LG	MH	NJ	OK	PL
A3	B4	C5	D6	E7	F8	GA	HB	JC	KD	LE	MF	NG	OH	PJ	QK
B2	C3	D4	E5	F6	G7	H8	JA	KB	LC	MD	NE	OF	PG	QH	RJ
C1	D2	E3	F4	G5	H6	J7	K8	LA	MB	NC	OD	PE	QF	RG	SH
DZ	E1	F2	G3	H4	J5	K6	L7	M8	NA	OB	PC	QD	RE	SF	TG
EY	FZ	G1	H2	J3	K4	L5	M6	N7	O8	PA	QB	RC	SD	TE	UF
FX	GY	HZ	J1	K2	L3	M4	N5	O6	P7	Q8	RA	SB	TC	UD	VE
GW	HX	JY	KZ	L1	M2	N3	O4	P5	Q6	R7	S8	TA	UB	VC	WD
HV	JW	KX	LY	MZ	N1	O2	P3	Q4	R5	S6	T7	U8	VA	WB	XC
JU	KV	LW	MX	NY	OZ	P1	Q2	R3	S4	T5	U6	V7	W8	XA	YB
KT	LU	MV	NW	OX	PY	QZ	R1	S2	T3	U4	V5	W6	X7	Y8	ZA
LS	MT	NU	OV	PW	QX	RY	SZ	T1	U2	V3	W4	X5	Y6	Z7	18
MR	NS	OT	PU	QV	RW	SX	TY	UZ	V1	W2	X3	Y4	Z5	16	27
NQ	OR	PS	QT	RU	SV	TW	UX	VY	WZ	X1	Y2	Z3	14	25	36
OP	PQ	QR	RS	ST	TU	UV	VW	WX	XY	YZ	Z1	12	23	34	45

FEBRUARY 1978 — PHYSICAL

1 W	2 TH	3 F	4 SA	5 SU	6 M	7 TU	8 W	9 TH	10 F	11 SA	12 SU	13 M	14 TU	15 W
JH	KJ	LK	ML	NM	ON	PO	QP	RQ	SR	TS	UT	VU	WV	XW
KG	LH	MJ	NK	OL	PM	QN	RO	SP	TQ	UR	VS	WT	XU	AV
LF	MG	NH	OJ	PK	QL	RM	SN	TO	UP	VQ	WR	XS	AT	BU
ME	NF	OG	PH	QJ	RK	SL	TM	UN	VO	WP	XQ	AR	BS	CT
ND	OE	PF	QG	RH	SJ	TK	UL	VM	WN	XO	AP	BQ	CR	DS
OC	PD	QE	RF	SG	TH	UJ	VK	WL	XM	AN	BO	CP	DQ	ER
PB	QC	RD	SE	TF	UG	VH	WJ	XK	AL	BM	CN	DO	EP	FQ
QA	RB	SC	TD	UE	VF	WG	XH	AJ	BK	CL	DM	EN	FO	GP
RX	SA	TB	UC	VD	WE	XF	AG	BH	CJ	DK	EL	FM	GN	HO
SW	TX	UA	VB	WC	XD	AE	BF	CG	DH	EJ	FK	GL	HM	JN
TV	UW	VX	WA	XB	AC	BD	CE	DF	EG	FH	GJ	HK	JL	KM
U	V	W	X	A	B	C	D	E	F	G	H	J	K	L

FEBRUARY 1978 — EMOTIONAL

1 W	2 TH	3 F	4 SA	5 SU	6 M	7 TU	8 W	9 TH	10 F	11 SA	12 SU	13 M	14 TU	15 W	
Z	1	2	3	A	B	C	D	E	F	G	H	J	K	L	
1Y	2Z	31	A2	B3	CA	DB	EC	FD	GE	HF	JG	KH	LJ	MK	
2X	3Y	AZ	B1	C2	D3	EA	FB	GC	HD	JE	KF	LG	MH	NJ	
3W	AX	BY	CZ	D1	E2	F3	GA	HB	JC	KD	LE	MF	NG	OH	
AV	BW	CX	DY	EZ	F1	G2	H3	JA	KB	LC	MD	NE	OF	PG	+
BU	CV	DW	EX	FY	GZ	H1	J2	K3	LA	MB	NC	OD	PE	QF	
CT	DU	EV	FW	GX	HY	JZ	K1	L2	M3	NA	OB	PC	QD	RE	
DS	ET	FU	GV	HW	JX	KY	LZ	M1	N2	O3	PA	QB	RC	SD	0
ER	FS	GT	HU	JV	KW	LX	MY	NZ	O1	P2	Q3	RA	SB	TC	
FQ	GR	HS	JT	KU	LV	MW	NX	OY	PZ	Q1	R2	S3	TA	UB	
GP	HQ	JR	KS	LT	MU	NV	OW	PX	QY	RZ	S1	T2	U3	VA	−
HO	JP	KQ	LR	MS	NT	OU	PV	QW	RX	SY	TZ	U1	V2	W3	
JN	KO	LP	MQ	NR	OS	PT	QU	RV	SW	TX	UY	VZ	W1	X2	
KM	LN	MO	NP	OQ	PR	QS	RT	SU	TV	UW	VX	WY	XZ	Y1	
L	M	N	O	P	Q	R	S	T	U	V	W	X	Y	Z	

FEBRUARY 1978 — INTELLECTUAL

1 W	2 TH	3 F	4 SA	5 SU	6 M	7 TU	8 W	9 TH	10 F	11 SA	12 SU	13 M	14 TU	15 W
O	P	Q	R	S	T	U	V	W	X	Y	Z	1	2	3
PN	QO	RP	SQ	TR	US	VT	WU	XV	YW	ZX	1Y	2Z	31	42
QM	RN	SO	TP	UQ	VR	WS	XT	YU	ZV	1W	2X	3Y	4Z	51
RL	SM	TN	UO	VP	WQ	XR	YS	ZT	1U	2V	3W	4X	5Y	6Z
SK	TL	UM	VN	WO	XP	YQ	ZR	1S	2T	3U	4V	5W	6X	7Y
TJ	UK	VL	WM	XN	YO	ZP	1Q	2R	3S	4T	5U	6V	7W	8X
UH	VJ	WK	XL	YM	ZN	1O	2P	3Q	4R	5S	6T	7U	8V	AW
VG	WH	XJ	YK	ZL	1M	2N	3O	4P	5Q	6R	7S	8T	AU	BV
WF	XG	YH	ZJ	1K	2L	3M	4N	5O	6P	7Q	8R	AS	BT	CU
XE	YF	ZG	1H	2J	3K	4L	5M	6N	7O	8P	AQ	BR	CS	DT
YD	ZE	1F	2G	3H	4J	5K	6L	7M	8N	AO	BP	CQ	DR	ES
ZC	1D	2E	3F	4G	5H	6J	7K	8L	AM	BN	CO	DP	EQ	FR
1B	2C	3D	4E	5F	6G	7H	8J	AK	BL	CM	DN	EO	FP	GQ
2A	3B	4C	5D	6E	7F	8G	AH	BJ	CK	DL	EM	FN	GO	HP
38	4A	5B	6C	7D	8E	AF	BG	CH	DJ	EK	FL	GM	HN	JO
47	58	6A	7B	8C	AD	BE	CF	DG	EH	FJ	GK	HL	JM	KN
56	67	78	8A	AB	BC	CD	DE	EF	FG	GH	HJ	JK	KL	LM

FEBRUARY 1978 — PHYSICAL

16 TH	17 F	18 SA	19 SU	20 M	21 TU	22 W	23 TH	24 F	25 SA	26 SU	27 M	28 TU
AX	BA	CB	DC	ED	FE	GF	HG	JH	KJ	LK	ML	NM
BW	CX	DA	EB	FC	GD	HE	JF	KG	LH	MJ	NK	OL
CV	DW	EX	FA	GB	HC	JD	KE	LF	MG	NH	OJ	PK
DU	EV	FW	GX	HA	JB	KC	LD	ME	NF	OG	PH	QJ
ET	FU	GV	HW	JX	KA	LB	MC	ND	OE	PF	QG	RH
FS	GT	HU	JV	KW	LX	MA	NB	OC	PD	QE	RF	SG
GR	HS	JT	KU	LV	MW	NX	OA	PB	QC	RD	SE	TF
HQ	JR	KS	LT	MU	NV	OW	PX	QA	RB	SC	TD	UE
JP	KQ	LR	MS	NT	OU	PV	QW	RX	SA	TB	UC	VD
KO	LP	MQ	NR	OS	PT	QU	RV	SW	TX	UA	VB	WC
LN	MO	NP	OQ	PR	QS	RT	SU	TV	UW	VX	WA	XB
M	N	O	P	Q	R	S	T	U	V	W	X	A

FEBRUARY 1978 — EMOTIONAL

16 TH	17 F	18 SA	19 SU	20 M	21 TU	22 W	23 TH	24 F	25 SA	26 SU	27 M	28 TU	
M	N	O	P	Q	R	S	T	U	V	W	X	Y	
NL	OM	PN	QO	RP	SQ	TR	US	VT	WU	XV	YW	ZX	
OK	PL	QM	RN	SO	TP	UQ	VR	WS	XT	YU	ZV	1W	
PJ	QK	RL	SM	TN	UO	VP	WQ	XR	YS	ZT	1U	2V	
QH	RJ	SK	TL	UM	VN	WO	XP	YQ	ZR	1S	2T	3U	+
RG	SH	TJ	UK	VL	WM	XN	YO	ZP	1Q	2R	3S	AT	
SF	TG	UH	VJ	WK	XL	YM	ZN	1O	2P	3Q	AR	BS	
TE	UF	VG	WH	XJ	YK	ZL	1M	2N	3O	AP	BQ	CR	0
UD	VE	WF	XG	YH	ZJ	1K	2L	3M	AN	BO	CP	DQ	
VC	WD	XE	YF	ZG	1H	2J	3K	AL	BM	CN	DO	EP	
WB	XC	YD	ZE	1F	2G	3H	AJ	BK	CL	DM	EN	FO	−
XA	YB	ZC	1D	2E	3F	AG	BH	CJ	DK	EL	FM	GN	
Y3	ZA	1B	2C	3D	AE	BF	CG	DH	EJ	FK	GL	HM	
Z2	13	2A	3B	AC	DD	CE	DF	EG	FH	GJ	HK	JL	
1	2	3	A	B	C	D	E	F	G	H	J	K	

FEBRUARY 1978 — INTELLECTUAL

16 TH	17 F	18 SA	19 SU	20 M	21 TU	22 W	23 TH	24 F	25 SA	26 SU	27 M	28 TU
4	5	6	7	8	A	B	C	D	E	F	G	H
53	64	75	86	A7	B8	CA	DB	EC	FD	GE	HF	JG
62	73	84	A5	B6	C7	D8	EA	FB	GC	HD	JE	KF
71	82	A3	B4	C5	D6	E7	F8	GA	HB	JC	KD	LE
8Z	A1	B2	C3	D4	E5	F6	G7	H8	JA	KB	LC	MD
AY	BZ	C1	D2	E3	F4	G5	H6	J7	K8	LA	MB	NC
BX	CY	DZ	E1	F2	G3	H4	J5	K6	L7	M8	NA	OB
CW	DX	EY	FZ	G1	H2	J3	K4	L5	M6	N7	O8	PA
DV	EW	FX	GY	HZ	J1	K2	L3	M4	N5	O6	P7	Q8
EU	FV	GW	HX	JY	KZ	L1	M2	N3	O4	P5	Q6	R7
FT	GU	HV	JW	KX	LY	MZ	N1	O2	P3	Q4	R5	S6
GS	HT	JU	KV	LW	MX	NY	OZ	P1	Q2	R3	S4	T5
HR	JS	KT	LU	MV	NW	OX	PY	QZ	R1	S2	T3	U4
JQ	KR	LS	MT	NU	OV	PW	QX	RY	SZ	T1	U2	V3
KP	LQ	MR	NS	OT	PU	QV	RW	SX	TY	UZ	V1	W2
LO	MP	NQ	OR	PS	QT	RU	SV	TW	UX	VY	WZ	X1
MN	NO	OP	PQ	QR	RS	ST	TU	UV	VW	WX	XY	YZ

MARCH 1978 PHYSICAL

1 W	2 TH	3 F	4 SA	5 SU	6 M	7 TU	8 W	9 TH	10 F	11 SA	12 SU	13 M	14 TU	15 W
ON	PO	QP	RQ	SR	TS	UT	VU	WV	XW	AX	BA	CB	DC	ED
PM	QN	RO	SP	TQ	UR	VS	WT	XU	AV	BW	CX	DA	EB	FC
QL	RM	SN	TO	UP	VQ	WR	XS	AT	BU	CV	DW	EX	FA	GB
RK	SL	TM	UN	VO	WP	XQ	AR	BS	CT	DU	EV	FW	GX	HA
SJ	TK	UL	VM	WN	XO	AP	BQ	CR	DS	ET	FU	GV	HW	JX
TH	UJ	VK	WL	XM	AN	BO	CP	DQ	ER	FS	GT	HU	JV	KW
UG	VH	WJ	XK	AL	BM	CN	DO	EP	FQ	GR	HS	JT	KU	LV
VF	WG	XH	AJ	BK	CL	DM	EN	FO	GP	HQ	JR	KS	LT	MU
WE	XF	AG	BH	CJ	DK	EL	FM	GN	HO	JP	KQ	LR	MS	NT
XD	AE	BF	CG	DH	EJ	FK	GL	HM	JN	KO	LP	MQ	NR	OS
AC	BD	CE	DF	EG	FH	GJ	HK	JL	KM	LN	MO	NP	OQ	PR
B	C	D	E	F	G	H	J	K	L	M	N	O	P	Q

MARCH 1978 EMOTIONAL

1 W	2 TH	3 F	4 SA	5 SU	6 M	7 TU	8 W	9 TH	10 F	11 SA	12 SU	13 M	14 TU	15 W	
Z	1	2	3	A	B	C	D	E	F	G	H	J	K	L	
1Y	2Z	31	A2	B3	CA	DB	EC	FD	GE	HF	JG	KH	LJ	MK	
2X	3Y	AZ	B1	C2	D3	EA	FB	GC	HD	JE	KF	LG	MH	NJ	
3W	AX	BY	CZ	D1	E2	F3	GA	HB	JC	KD	LE	MF	NG	OH	
AV	BW	CX	DY	EZ	F1	G2	H3	JA	KB	LC	MD	NE	OF	PG	+
BU	CV	DW	EX	FY	GZ	H1	J2	K3	LA	MB	NC	OD	PE	QF	
CT	DU	EV	FW	GX	HY	JZ	K1	L2	M3	NA	OB	PC	QD	RE	
DS	ET	FU	GV	HW	JX	KY	LZ	M1	N2	O3	PA	QB	RC	SD	o
ER	FS	GT	HU	JV	KW	LX	MY	NZ	O1	P2	Q3	RA	SB	TC	
FQ	GR	HS	JT	KU	LV	MW	NX	OY	PZ	Q1	R2	S3	TA	UB	
GP	HQ	JR	KS	LT	MU	NV	OW	PX	QY	RZ	S1	T2	U3	VA	–
HO	JP	KQ	LR	MS	NT	OU	PV	QW	RX	SY	TZ	U1	V2	W3	
JN	KO	LP	MQ	NR	OS	PT	QU	RV	SW	TX	UY	VZ	W1	X2	
KM	LN	MO	NP	OQ	PR	QS	RT	SU	TV	UW	VX	WY	XZ	Y1	
L	M	N	O	P	Q	R	S	T	U	V	W	X.	Y	Z	

MARCH 1978 INTELLECTUAL

1 W	2 TH	3 F	4 SA	5 SU	6 M	7 TU	8 W	9 TH	10 F	11 SA	12 SU	13 M	14 TU	15 W
J	K	L	M	N	O	P	Q	R	S	T	U	V	W	X
KH	LJ	MK	NL	OM	PN	QO	RP	SQ	TR	US	VT	WU	XV	YW
LG	MH	NJ	OK	PL	QM	RN	SO	TP	UQ	VR	WS	XT	YU	ZV
MF	NG	OH	PJ	QK	RL	SM	TN	UO	VP	WQ	XR	YS	ZT	1U
NE	OF	PG	QH	RJ	SK	TL	UM	VN	WO	XP	YQ	ZR	1S	2T
OD	PE	QF	RG	SH	TJ	UK	VL	WM	XN	YO	ZP	1Q	2R	3S
PC	QD	RE	SF	TG	UH	VJ	WK	XL	YM	ZN	1O	2P	3Q	4R
QB	RC	SD	TE	UF	VG	WH	XJ	YK	ZL	1M	2N	3O	4P	5Q
RA	SB	TC	UD	VE	WF	XG	YH	ZJ	1K	2L	3M	4N	5O	6P
S8	TA	UB	VC	WD	XE	YF	ZG	1H	2J	3K	4L	5M	6N	7O
T7	U8	VA	WB	XC	YD	ZE	1F	2G	3H	4J	5K	6L	7M	8N
U6	V7	W8	XA	YB	ZC	1D	2E	3F	4G	5H	6J	7K	8L	AM
V5	W6	X7	Y8	ZA	1B	2C	3D	4E	5F	6G	7H	8J	AK	BL
W4	X5	Y6	Z7	18	2A	3B	4C	5D	6E	7F	8G	AH	BJ	CK
X3	Y4	Z5	16	27	38	4A	5B	6C	7D	8E	AF	BG	CH	DJ
Y2	Z3	14	25	36	47	58	6A	7B	8C	AD	BE	CF	DG	EH
Z1	12	23	34	45	56	67	78	8A	AB	BC	CD	DE	EF	FG

MARCH 1978 **PHYSICAL**

16 TH	17 F	18 SA	19 SU	20 M	21 TU	22 W	23 TH	24 F	25 SA	26 SU	27 M	28 TU	29 W	30 TH	31 F
FE	GF	HG	JH	KJ	LK	ML	NM	ON	PO	QP	RQ	SR	TS	UT	VU
GD	HE	JF	KG	LH	MJ	NK	OL	PM	QN	RO	SP	TQ	UR	VS	WT
HC	JD	KE	LF	MG	NH	OJ	PK	QL	RM	SN	TO	UP	VQ	WR	XS
JB	KC	LD	ME	NF	OG	PH	QJ	RK	SL	TM	UN	VO	WP	XQ	AR
KA	LB	MC	ND	OE	PF	QG	RH	SJ	TK	UL	VM	WN	XO	AP	BQ
LX	MA	NB	OC	PD	QE	RF	SG	TH	UJ	VK	WL	XM	AN	BO	CP
MW	NX	OA	PB	QC	RD	SE	TF	UG	VH	WJ	XK	AL	BM	CN	DO
NV	OW	PX	QA	RB	SC	TD	UE	VF	WG	XH	AJ	BK	CL	DM	EN
OU	PV	QW	RX	SA	TB	UC	VD	WE	XF	AG	BH	CJ	DK	EL	FM
PT	QU	RV	SW	TX	UA	VB	WC	XD	AE	BF	CG	DH	EJ	FK	GL
QS	RT	SU	TV	UW	VX	WA	XB	AC	BD	CE	DF	EG	FH	GJ	HK
R	S	T	U	V	W	X	A	B	C	D	E	F	G	H	J

MARCH 1978 **EMOTIONAL**

16 TH	17 F	18 SA	19 SU	20 M	21 TU	22 W	23 TH	24 F	25 SA	26 SU	27 M	28 TU	29 W	30 TH	31 F	
M	N	O	P	Q	R	S	T	U	V	W	X	Y	Z	1	2	
NL	OM	PN	QO	RP	SQ	TR	US	VT	WU	XV	YW	ZX	1Y	2Z	31	
OK	PL	QM	RN	SO	TP	UQ	VR	WS	XT	YU	ZV	1W	2X	3Y	AZ	
PJ	QK	RL	SM	TN	UO	VP	WQ	XR	YS	ZT	1U	2V	3W	AX	BY	
QH	RJ	SK	TL	UM	VN	WO	XP	YQ	ZR	1S	2T	3U	AV	BW	CX	+
RG	SH	TJ	UK	VL	WM	XN	YO	ZP	1Q	2R	3S	AT	BU	CV	DW	
SF	TG	UH	VJ	WK	XL	YM	ZN	1O	2P	3Q	AR	BS	CT	DU	EV	
TE	UF	VG	WH	XJ	YK	ZL	1M	2N	3O	AP	BQ	CR	DS	ET	FU	o
UD	VE	WF	XG	YH	ZJ	1K	2L	3M	AN	BO	CP	DQ	ER	FS	GT	
VC	WD	XE	YF	ZG	1H	2J	3K	AL	BM	CN	DO	EP	FQ	GR	HS	
WB	XC	YD	ZE	1F	2G	3H	AJ	BK	CL	DM	EN	FO	GP	HQ	JR	–
XA	YB	ZC	1D	2E	3F	AG	BH	CJ	DK	EL	FM	GN	HO	JP	KQ	
Y3	ZA	1B	2C	3D	AE	BF	CG	DH	EJ	FK	GL	HM	JN	KO	LP	
Z2	13	2A	3B	AC	BD	CE	DF	EG	FH	GJ	HK	JL	KM	LN	MO	
1	2	3	A	B	C	D	E	F	G	H	J	K	L	M	N	

MARCH 1978 **INTELLECTUAL**

16 TH	17 F	18 SA	19 SU	20 M	21 TU	22 W	23 TH	24 F	25 SA	26 SU	27 M	28 TU	29 W	30 TH	31 F
Y	Z	1	2	3	4	5	6	7	8	A	B	C	D	E	F
ZX	1Y	2Z	31	42	53	64	75	86	A7	B8	CA	DB	EC	FD	GE
1W	2X	3Y	4Z	51	62	73	84	A5	B6	C7	D8	EA	FB	GC	HD
2V	3W	4X	5Y	6Z	71	82	A3	B4	C5	D6	E7	F8	GA	HB	JC
3U	4V	5W	6X	7Y	8Z	A1	B2	C3	D4	E5	F6	G7	H8	JA	KB
4T	5U	6V	7W	8X	AY	BZ	C1	D2	E3	F4	G5	H6	J7	K8	LA
5S	6T	7U	8V	AW	BX	CY	DZ	E1	F2	G3	H4	J5	K6	L7	M8
6R	7S	8T	AU	BV	CW	DX	EY	FZ	G1	H2	J3	K4	L5	M6	N7
7Q	8R	AS	BT	CU	DV	EW	FX	GY	HZ	J1	K2	L3	M4	N5	O6
8P	AQ	BR	CS	DT	EU	FV	GW	HX	JY	KZ	L1	M2	N3	O4	P5
AO	BP	CQ	DR	ES	FT	GU	HV	JW	KX	LY	MZ	N1	O2	P3	Q4
BN	CO	DP	EQ	FR	GS	HT	JU	KV	LW	MX	NY	OZ	P1	Q2	R3
CM	DN	EO	FP	GQ	HR	JS	KT	LU	MV	NW	OX	PY	QZ	R1	S2
DL	EM	FN	GO	HP	JQ	KR	LS	MT	NU	OV	PW	QX	RY	SZ	T1
EK	FL	GM	HN	JO	KP	LQ	MR	NS	OT	PU	QV	RW	SX	TY	UZ
FJ	GK	HL	JM	KN	LO	MP	NQ	OR	PS	QT	RU	SV	TW	UX	VY
GH	HJ	JK	KL	LM	MN	NO	OP	PQ	QR	RS	ST	TU	UV	VW	WX

APRIL 1978 — PHYSICAL

1 SA	2 SU	3 M	4 TU	5 W	6 TH	7 F	8 SA	9 SU	10 M	11 TU	12 W	13 TH	14 F	15 SA
WV	XW	AX	BA	CB	DC	ED	FE	GF	HG	JH	KJ	LK	ML	NM
XU	AV	BW	CX	DΛ	EB	FC	GD	HE	JF	KG	LH	MJ	NK	OL
AT	BU	CV	DW	EX	FA	GB	HC	JD	KE	LF	MG	NH	OJ	PK
BS	CT	DU	EV	FW	GX	HA	JB	KC	LD	ME	NF	OG	PH	QJ
CR	DS	ET	FU	GV	HW	JX	KA	LB	MC	ND	OE	PF	QG	RH
DQ	ER	FS	GT	HU	JV	KW	LX	MA	NB	OC	PD	QE	RF	SG
EP	FQ	GR	HS	JT	KU	LV	MW	NX	OA	PB	QC	RD	SE	TF
FO	GP	HQ	JR	KS	LT	MU	NV	OW	PX	QA	RB	SC	TD	UE
GN	HO	JP	KQ	LR	MS	NT	OU	PV	QW	RX	SA	TB	UC	VD
HM	JN	KO	LP	MQ	NR	OS	PT	QU	RV	SW	TX	UA	VB	WC
JL	KM	LN	MO	NP	OQ	PR	QS	RT	SU	TV	UW	VX	WA	XB
K	L	M	N	O	P	Q	R	S	T	U	V	W	X	A

APRIL 1978 — EMOTIONAL

1 SA	2 SU	3 M	4 TU	5 W	6 TH	7 F	8 SA	9 SU	10 M	11 TU	12 W	13 TH	14 F	15 SA	
3	A	B	C	D	E	F	G	H	J	K	L	M	N	O	
A2	B3	CA	DB	EC	FD	GE	HF	JG	KH	LJ	MK	NL	OM	PN	
B1	C2	D3	EA	FB	GC	HD	JE	KF	LG	MH	NJ	OK	PL	QM	
CZ	D1	E2	F3	GA	HB	JC	KD	LE	MF	NG	OH	PJ	QK	RL	
DY	EZ	F1	G2	H3	JA	KB	LC	MD	NE	OF	PG	QH	RJ	SK	+
EX	FY	GZ	H1	J2	K3	LA	MB	NC	OD	PE	QF	RG	SH	TJ	
FW	GX	HY	JZ	K1	L2	M3	NA	OB	PC	QD	RE	SF	TG	UH	
GV	HW	JX	KY	LZ	M1	N2	O3	PA	QB	RC	SD	TE	UF	VG	o
HU	JV	KW	LX	MY	NZ	O1	P2	Q3	RA	SB	TC	UD	VE	WF	
JT	KU	LV	MW	NX	OY	PZ	Q1	R2	S3	TA	UB	VC	WD	XE	
KS	LT	MU	NV	OW	PX	QY	RZ	S1	T2	U3	VA	WB	XC	YD	−
LR	MS	NT	OU	PV	QW	RX	SY	TZ	U1	V2	W3	XA	YB	ZC	
MQ	NR	OS	PT	QU	RV	SW	TX	UY	VZ	W1	X2	Y3	ZA	1B	
NP	OQ	PR	QS	RT	SU	TV	UW	VX	WY	XZ	Y1	Z2	13	2A	
O	P	Q	R	S	T	U	V	W	X	Y	Z	1	2	3	

APRIL 1978 — INTELLECTUAL

1 SA	2 SU	3 M	4 TU	5 W	6 TH	7 F	8 SA	9 SU	10 M	11 TU	12 W	13 TH	14 F	15 SA
G	H	J	K	L	M	N	O	P	Q	R	S	T	U	V
HF	JG	KH	LJ	MK	NL	OM	PN	QO	RP	SQ	TR	US	VT	WU
JE	KF	LG	MH	NJ	OK	PL	QM	RN	SO	TP	UQ	VR	WS	XT
KD	LE	MF	NG	OH	PJ	QK	RL	SM	TN	UO	VP	WQ	XR	YS
LC	MD	NE	OF	PG	QH	RJ	SK	TL	UM	VN	WO	XP	YQ	ZR
MB	NC	OD	PE	QF	RG	SH	TJ	UK	VL	WM	XN	YO	ZP	1Q
NA	OB	PC	QD	RE	SF	TG	UH	VJ	WK	XL	YM	ZN	1O	2P
O8	PA	QB	RC	SD	TE	UF	VG	WH	XJ	YK	ZL	1M	2N	3O
P7	Q8	RA	SB	TC	UD	VE	WF	XG	YH	ZJ	1K	2L	3M	4N
Q6	R7	S8	TA	UB	VC	WD	XE	YF	ZG	1H	2J	3K	4L	5M
R5	S6	T7	U8	VA	WB	XC	YD	ZE	1F	2G	3H	4J	5K	6L
S4	T5	U6	V7	W8	XA	YB	ZC	1D	2E	3F	4G	5H	6J	7K
T3	U4	V5	W6	X7	Y8	ZA	1B	2C	3D	4E	5F	6G	7H	8J
U2	V3	W4	X5	Y6	Z7	18	2A	3B	4C	5D	6E	7F	8G	AH
V1	W2	X3	Y4	Z5	16	27	38	4A	5B	6C	7D	8E	AF	BG
WZ	X1	Y2	Z3	14	25	36	47	58	6A	7B	8C	AD	BE	CF
XY	YZ	Z1	12	23	34	45	56	67	78	8A	AB	BC	CD	DE

APRIL 1978 — PHYSICAL

16 SU	17 M	18 TU	19 W	20 TH	21 F	22 SA	23 SU	24 M	25 TU	26 W	27 TH	28 F	29 SA	30 SU
ON	PO	QP	RQ	SR	TS	UT	VU	WV	XW	AX	BA	CB	DC	ED
PM	QN	RO	SP	TQ	UR	VS	WT	XU	AV	BW	CX	DA	EB	FC
QL	RM	SN	TO	UP	VQ	WR	XS	AT	BU	CV	DW	EX	FA	GB
RK	SL	TM	UN	VO	WP	XQ	AR	BS	CT	DU	EV	FW	GX	HA
SJ	TK	UL	VM	WN	XO	AP	BQ	CR	DS	ET	FU	GV	HW	JX
TH	UJ	VK	WL	XM	AN	BO	CP	DQ	ER	FS	GT	HU	JV	KW
UG	VH	WJ	XK	AL	BM	CN	DO	EP	FQ	GR	HS	JT	KU	LV
VF	WG	XH	AJ	BK	CL	DM	EN	FO	GP	HQ	JR	KS	LT	MU
WE	XF	AG	BH	CJ	DK	EL	FM	GN	HO	JP	KQ	LR	MS	NT
XD	AE	BF	CG	DH	EJ	FK	GL	HM	JN	KO	LP	MQ	NR	OS
AC	BD	CE	DF	EG	FH	GJ	HK	JL	KM	LN	MO	NP	OQ	PR
B	C	D	E	F	G	H	J	K	L	M	N	O	P	Q

APRIL 1978 — EMOTIONAL

16 SU	17 M	18 TU	19 W	20 TH	21 F	22 SA	23 SU	24 M	25 TU	26 W	27 TH	28 F	29 SA	30 SU	
P	Q	R	S	T	U	V	W	X	Y	Z	1	2	3	A	
QO	RP	SQ	TR	US	VT	WU	XV	YW	ZX	1Y	2Z	31	A2	B3	
RN	SO	TP	UQ	VR	WS	XT	YU	ZV	1W	2X	3Y	AZ	B1	C2	
SM	TN	UO	VP	WQ	XR	YS	ZT	1U	2V	3W	AX	BY	CZ	D1	
TL	UM	VN	WO	XP	YQ	ZR	1S	2T	3U	AV	BW	CX	DY	EZ	+
UK	VL	WM	XN	YO	ZP	1Q	2R	3S	AT	BU	CV	DW	EX	FY	
VJ	WK	XL	YM	ZN	1O	2P	3Q	AR	BS	CT	DU	EV	FW	GX	
WH	XJ	YK	ZL	1M	2N	3O	AP	BQ	CR	DS	ET	FU	GV	HW	o
XG	YH	ZJ	1K	2L	3M	AN	BO	CP	DQ	ER	FS	GT	HU	JV	
YF	ZG	1H	2J	3K	AL	BM	CN	DO	EP	FQ	GR	HS	JT	KU	
ZE	1F	2G	3H	AJ	BK	CL	DM	EN	FO	GP	HQ	JR	KS	LT	−
1D	2E	3F	AG	BH	CJ	DK	EL	FM	GN	HO	JP	KQ	LR	MS	
2C	3D	AE	BF	CG	DH	EJ	FK	GL	HM	JN	KO	LP	MQ	NR	
3B	AC	BD	CE	DF	EG	FH	GJ	HK	JL	KM	LN	MO	NP	OQ	
A	B	C	D	E	F	G	H	J	K	L	M	N	O	P	

APRIL 1978 — INTELLECTUAL

16 SU	17 M	18 TU	19 W	20 TH	21 F	22 SA	23 SU	24 M	25 TU	26 W	27 TH	28 F	29 SA	30 SU
W	X	Y	Z	1	2	3	4	5	6	7	8	A	B	C
XV	YW	ZX	1Y	2Z	31	42	53	64	75	86	A7	B8	CA	DB
YU	ZV	1W	2X	3Y	4Z	51	62	73	84	A5	B6	C7	D8	EA
ZT	1U	2V	3W	4X	5Y	6Z	71	82	A3	B4	C5	D6	E7	F8
1S	2T	3U	4V	5W	6X	7Y	8Z	A1	B2	C3	D4	E5	F6	G7
2R	3S	4T	5U	6V	7W	8X	AY	BZ	C1	D2	E3	F4	G5	H6
3Q	4R	5S	6T	7U	8V	AW	BX	CY	DZ	E1	F2	G3	H4	J5
4P	5Q	6R	7S	8T	AU	BV	CW	DX	EY	FZ	G1	H2	J3	K4
5O	6P	7Q	8R	AS	BT	CU	DV	EW	FX	GY	HZ	J1	K2	L3
6N	7O	8P	AQ	BR	CS	DT	EU	FV	GW	HX	JY	KZ	L1	M2
7M	8N	AO	BP	CQ	DR	ES	FT	GU	HV	JW	KX	LY	MZ	N1
8L	AM	BN	CO	DP	EQ	FR	GS	HT	JU	KV	LW	MX	NY	OZ
AK	BL	CM	DN	EO	FP	GQ	HR	JS	KT	LU	MV	NW	OX	PY
BJ	CK	DL	EM	FN	GO	HP	JQ	KR	LS	MT	NU	OV	PW	QX
CH	DJ	EK	FL	GM	HN	JO	KP	LQ	MR	NS	OT	PU	QV	RW
DG	EH	FJ	GK	HL	JM	KN	LO	MP	NQ	OR	PS	QT	RU	SV
EF	FG	GH	HJ	JK	KL	LM	MN	NO	OP	PQ	QR	RS	ST	TU

MAY 1978 PHYSICAL

1 M	2 TU	3 W	4 TH	5 F	6 SA	7 SU	8 M	9 TU	10 W	11 TH	12 F	13 SA	14 SU	15 M
FE	GF	HG	JH	KJ	LK	ML	NM	ON	PO	QP	RQ	SR	TS	UT
GD	HE	JF	KG	LH	MJ	NK	OL	PM	QN	RO	SP	TQ	UR	VS
HC	JD	KE	LF	MG	NH	OJ	PK	QL	RM	SN	TO	UP	VQ	WR
JB	KC	LD	ME	NF	OG	PH	QJ	RK	SL	TM	UN	VO	WP	XQ
KA	LB	MC	ND	OE	PF	QG	RH	SJ	TK	UL	VM	WN	XO	AP
LX	MA	NB	OC	PD	QE	RF	SG	TH	UJ	VK	WL	XM	AN	BO
MW	NX	OA	PB	QC	RD	SE	TF	UG	VH	WJ	XK	AL	BM	CN
NV	OW	PX	QA	RB	SC	TD	UE	VF	WG	XH	AJ	BK	CL	DM
OU	PV	QW	RX	SA	TB	UC	VD	WE	XF	AG	BH	CJ	DK	EL
PT	QU	RV	SW	TX	UA	VB	WC	XD	AE	BF	CG	DH	EJ	FK
QS	RT	SU	TV	UW	VX	WA	XB	AC	BD	CE	DF	EG	FH	GJ
R	S	T	U	V	W	X	A	B	C	D	E	F	G	H

MAY 1978 EMOTIONAL

1 M	2 TU	3 W	4 TH	5 F	6 SA	7 SU	8 M	9 TU	10 W	11 TH	12 F	13 SA	14 SU	15 M	
B	C	D	E	F	G	H	J	K	L	M	N	O	P	Q	
CA	DB	EC	FD	GE	HF	JG	KH	LJ	MK	NL	OM	PN	QO	RP	
D3	EA	FB	GC	HD	JE	KF	LG	MH	NJ	OK	PL	QM	RN	SO	
E2	F3	GA	HB	JC	KD	LE	MF	NG	OH	PJ	QK	RL	SM	TN	
F1	G2	H3	JA	KB	LC	MD	NE	OF	PG	QH	RJ	SK	TL	UM	+
GZ	H1	J2	K3	LA	MB	NC	OD	PE	QF	RG	SH	TJ	UK	VL	
HY	JZ	K1	L2	M3	NA	OB	PC	QD	RE	SF	TG	UH	VJ	WK	
JX	KY	LZ	M1	N2	O3	PA	QB	RC	SD	TE	UF	VG	WH	XJ	o
KW	LX	MY	NZ	O1	P2	Q3	RA	SB	TC	UD	VE	WF	XG	YH	
LV	MW	NX	OY	PZ	Q1	R2	S3	TA	UB	VC	WD	XE	YF	ZG	
MU	NV	OW	PX	QY	RZ	S1	T2	U3	VA	WB	XC	YD	ZE	1F	−
NT	OU	PV	QW	RX	SY	TZ	U1	V2	W3	XA	YB	ZC	1D	2E	
OS	PT	QU	RV	SW	TX	UY	VZ	W1	X2	Y3	ZA	1B	2C	3D	
PR	QS	RT	SU	TV	UW	VX	WY	XZ	Y1	Z2	13	2A	3B	AC	
Q	R	S	T	U	V	W	X	Y	Z	1	2	3	A	B	

MAY 1978 INTELLECTUAL

1 M	2 TU	3 W	4 TH	5 F	6 SA	7 SU	8 M	9 TU	10 W	11 TH	12 F	13 SA	14 SU	15 M
D	E	F	G	H	J	K	L	M	N	O	P	Q	R	S
EC	FD	GE	HF	JG	KH	LJ	MK	NL	OM	PN	QO	RP	SQ	TR
FB	GC	HD	JE	KF	LG	MH	NJ	OK	PL	QM	RN	SO	TP	UQ
GA	HB	JC	KD	LE	MF	NG	OH	PJ	QK	RL	SM	TN	UO	VP
H8	JA	KB	LC	MD	NE	OF	PG	QH	RJ	SK	TL	UM	VN	WO
J7	K8	LA	MB	NC	OD	PE	QF	RG	SH	TJ	UK	VL	WM	XN
K6	L7	M8	NA	OB	PC	QD	RE	SF	TG	UH	VJ	WK	XL	YM
L5	M6	N7	O8	PA	QB	RC	SD	TE	UF	VG	WH	XJ	YK	ZL
M4	N5	O6	P7	Q8	RA	SB	TC	UD	VE	WF	XG	YH	ZJ	1K
N3	O4	P5	Q6	R7	S8	TA	UB	VC	WD	XE	YF	ZG	1H	2J
O2	P3	Q4	R5	S6	T7	U8	VA	WB	XC	YD	ZE	1F	2G	3H
P1	Q2	R3	S4	T5	U6	V7	W8	XA	YB	ZC	1D	2E	3F	4G
QZ	R1	S2	T3	U4	V5	W6	X7	Y8	ZA	1B	2C	3D	4E	5F
RY	SZ	T1	U2	V3	W4	X5	Y6	Z7	18	2A	3B	4C	5D	6E
SX	TY	UZ	V1	W2	X3	Y4	Z5	16	27	38	4A	5B	6C	7D
TW	UX	VY	WZ	X1	Y2	Z3	14	25	36	47	58	6A	7B	8C
UV	VW	WX	XY	YZ	Z1	12	23	34	45	56	67	78	8A	AB

MAY 1978 — PHYSICAL

16 TU	17 W	18 TH	19 F	20 SA	21 SU	22 M	23 TU	24 W	25 TH	26 F	27 SA	28 SU	29 M	30 TU	31 W
VU	WV	XW	AX	BA	CB	DC	ED	FE	GF	HG	JH	KJ	LK	ML	NM
WT	XU	AV	BW	CX	DA	EB	FC	GD	HE	JF	KG	LH	MJ	NK	OL
XS	AT	BU	CV	DW	EX	FA	GB	HC	JD	KE	LF	MG	NH	OJ	PK
AR	BS	CT	DU	EV	FW	GX	HA	JB	KC	LD	ME	NF	OG	PH	QJ
BQ	CR	DS	ET	FU	GV	HW	JX	KA	LB	MC	ND	OE	PF	QG	RH
CP	DQ	ER	FS	GT	HU	JV	KW	LX	MA	NB	OC	PD	QE	RF	SG
DO	EP	FQ	GR	HS	JT	KU	LV	MW	NX	OA	PB	QC	RD	SE	TF
EN	FO	GP	HQ	JR	KS	LT	MU	NV	OW	PX	QA	RB	SC	TD	UE
FM	GN	HO	JP	KQ	LR	MS	NT	OU	PV	QW	RX	SA	TB	UC	VD
GL	HM	JN	KO	LP	MQ	NR	OS	PT	QU	RV	SW	TX	UA	VB	WC
HK	JL	KM	LN	MO	NP	OQ	PR	QS	RT	SU	TV	UW	VX	WA	XB
J	K	L	M	N	O	P	Q	R	S	T	U	V	W	X	A

MAY 1978 — EMOTIONAL

16 TU	17 W	18 TH	19 F	20 SA	21 SU	22 M	23 TU	24 W	25 TH	26 F	27 SA	28 SU	29 M	30 TU	31 W	
R	S	T	U	V	W	X	Y	Z	1	2	3	A	B	C	D	
SQ	TR	US	VT	WU	XV	YW	ZX	1Y	2Z	31	A2	B3	CA	DB	EC	
TP	UQ	VR	WS	XT	YU	ZV	1W	2X	3Y	AZ	B1	C2	D3	EA	FB	
UO	VP	WQ	XR	YS	ZT	1U	2V	3W	AX	BY	CZ	D1	E2	F3	GA	
VN	WO	XP	YQ	ZR	1S	2T	3U	AV	BW	CX	DY	EZ	F1	G2	H3	+
WM	XN	YO	ZP	1Q	2R	3S	AT	BU	CV	DW	EX	FY	GZ	H1	J2	
XL	YM	ZN	1O	2P	3Q	AR	BS	CT	DU	EV	FW	GX	HY	JZ	K1	
YK	ZL	1M	2N	3O	AP	BQ	CR	DS	ET	FU	GV	HW	JX	KY	LZ	o
ZJ	1K	2L	3M	AN	BO	CP	DQ	ER	FS	GT	HU	JV	KW	LX	MY	
1H	2J	3K	AL	BM	CN	DO	EP	FQ	GR	HS	JT	KU	LV	MW	NX	
2G	3H	AJ	BK	CL	DM	EN	FO	GP	HQ	JR	KS	LT	MU	NV	OW	−
3F	AG	BH	CJ	DK	EL	FM	GN	HO	JP	KQ	LR	MS	NT	OU	PV	
AE	BF	CG	DH	EJ	FK	GL	HM	JN	KO	LP	MQ	NR	OS	PT	QU	
BD	CE	DF	EG	FH	GJ	HK	JL	KM	LN	MO	NP	OQ	PR	QS	RT	
C	D	E	F	G	H	J	K	L	M	N	O	P	Q	R	S	

MAY 1978 — INTELLECTUAL

16 TU	17 W	18 TH	19 F	20 SA	21 SU	22 M	23 TU	24 W	25 TH	26 F	27 SA	28 SU	29 M	30 TU	31 W
T	U	V	W	X	Y	Z	1	2	3	4	5	6	7	8	A
US	VT	WU	XV	YW	ZX	1Y	2Z	31	42	53	64	75	86	A7	B8
VR	WS	XT	YU	ZV	1W	2X	3Y	4Z	51	62	73	84	A5	B6	C7
WQ	XR	YS	ZT	1U	2V	3W	4X	5Y	6Z	71	82	A3	B4	C5	D6
XP	YQ	ZR	1S	2T	3U	4V	5W	6X	7Y	8Z	A1	B2	C3	D4	E5
YO	ZP	1Q	2R	3S	4T	5U	6V	7W	8X	AY	BZ	C1	D2	E3	F4
ZN	1O	2P	3Q	4R	5S	6T	7U	8V	AW	BX	CY	DZ	E1	F2	G3
1M	2N	3O	4P	5Q	6R	7S	8T	AU	BV	CW	DX	EY	FZ	G1	H2
2L	3M	4N	5O	6P	7Q	8R	AS	BT	CU	DV	EW	FX	GY	HZ	J1
3K	4L	5M	6N	7O	8P	AQ	BR	CS	DT	EU	FV	GW	HX	JY	KZ
4J	5K	6L	7M	8N	AO	BP	CQ	DR	ES	FT	GU	HV	JW	KX	LY
5H	6J	7K	8L	AM	BN	CO	DP	EQ	FR	GS	HT	JU	KV	LW	MX
6G	7H	8J	AK	BL	CM	DN	EO	FP	GQ	HR	JS	KT	LU	MV	NW
7F	8G	AH	BJ	CK	DL	EM	FN	GO	HP	JQ	KR	LS	MT	NU	OV
8E	AF	BG	CH	DJ	EK	FL	GM	HN	JO	KP	LQ	MR	NS	OT	PU
AD	BE	CF	DG	EH	FJ	GK	HL	JM	KN	LO	MP	NQ	OR	PS	QT
BC	CD	DE	EF	FG	GH	HJ	JK	KL	LM	MN	NO	OP	PQ	QR	RS

JUNE 1978 **PHYSICAL**

1 TH	2 F	3 SA	4 SU	5 M	6 TU	7 W	8 TH	9 F	10 SA	11 SU	12 M	13 TU	14 W	15 TH
ON	PO	QP	RQ	SR	TS	UT	VU	WV	XW	AX	BA	CB	DC	ED
PM	QN	RO	SP	TQ	UR	VS	WT	XU	AV	BW	CX	DA	EB	FC
QL	RM	SN	TO	UP	VQ	WR	XS	AT	BU	CV	DW	EX	FA	GB
RK	SL	TM	UN	VO	WP	XQ	AR	BS	CT	DU	EV	FW	GX	HA
SJ	TK	UL	VM	WN	XO	AP	BQ	CR	DS	ET	FU	GV	HW	JX
TH	UJ	VK	WL	XM	AN	BO	CP	DQ	ER	FS	GT	HU	JV	KW
UG	VH	WJ	XK	AL	BM	CN	DO	EP	FQ	GR	HS	JT	KU	LV
VF	WG	XH	AJ	BK	CL	DM	EN	FO	GP	HQ	JR	KS	LT	MU
WE	XF	AG	BH	CJ	DK	EL	FM	GN	HO	JP	KQ	LR	MS	NT
XD	AE	BF	CG	DH	EJ	FK	GL	HM	JN	KO	LP	MQ	NR	OS
AC	BD	CE	DF	EG	FH	GJ	HK	JL	KM	LN	MO	NP	OQ	PR
B	C	D	E	F	G	H	J	K	L	M	N	O	P	Q

JUNE 1978 **EMOTIONAL**

1 TH	2 F	3 SA	4 SU	5 M	6 TU	7 W	8 TH	9 F	10 SA	11 SU	12 M	13 TU	14 W	15 TH	
E	F	G	H	J	K	L	M	N	O	P	Q	R	S	T	
FD	GE	HF	JG	KH	LJ	MK	NL	OM	PN	QO	RP	SQ	TR	US	
GC	HD	JE	KF	LG	MH	NJ	OK	PL	QM	RN	SO	TP	UQ	VR	
HB	JC	KD	LE	MF	NG	OH	PJ	QK	RL	SM	TN	UO	VP	WQ	
JA	KB	LC	MD	NE	OF	PG	QH	RJ	SK	TL	UM	VN	WO	XP	+
K3	LA	MB	NC	OD	PE	QF	RG	SH	TJ	UK	VL	WM	XN	YO	
L2	M3	NA	OB	PC	QD	RE	SF	TG	UH	VJ	WK	XL	YM	ZN	
M1	N2	O3	PA	QB	RC	SD	TE	UF	VG	WH	XJ	YK	ZL	1M	o
NZ	O1	P2	Q3	RA	SB	TC	UD	VE	WF	XG	YH	ZJ	1K	2L	
OY	PZ	Q1	R2	S3	TA	UB	VC	WD	XE	YF	ZG	1H	2J	3K	
PX	QY	RZ	S1	T2	U3	VA	WB	XC	YD	ZE	1F	2G	3H	AJ	−
QW	RX	SY	TZ	U1	V2	W3	XA	YB	ZC	1D	2E	3F	AG	BH	
RV	SW	TX	UY	VZ	W1	X2	Y3	ZA	1B	2C	3D	AE	BF	CG	
SU	TV	UW	VX	WY	XZ	Y1	Z2	13	2A	3B	AC	BD	CE	DF	
T	U	V	W	X	Y	Z	1	2	3	A	B	C	D	E	

JUNE 1978 **INTELLECTUAL**

1 TH	2 F	3 SA	4 SU	5 M	6 TU	7 W	8 TH	9 F	10 SA	11 SU	12 M	13 TU	14 W	15 TH
B	C	D	E	F	G	H	J	K	L	M	N	O	P	Q
CA	DB	EC	FD	GE	HF	JG	KH	LJ	MK	NL	OM	PN	QO	RP
D8	EA	FB	GC	HD	JE	KF	LG	MH	NJ	OK	PL	QM	RN	SO
E7	F8	GA	HB	JC	KD	LE	MF	NG	OH	PJ	QK	RL	SM	TN
F6	G7	H8	JA	KB	LC	MD	NE	OF	PG	QH	RJ	SK	TL	UM
G5	H6	J7	K8	LA	MB	NC	OD	PE	QF	RG	SH	TJ	UK	VL
H4	J5	K6	L7	M8	NA	OB	PC	QD	RE	SF	TG	UH	VJ	WK
J3	K4	L5	M6	N7	O8	PA	QB	RC	SD	TE	UF	VG	WH	XJ
K2	L3	M4	N5	O6	P7	Q8	RA	SB	TC	UD	VE	WF	XG	YH
L1	M2	N3	O4	P5	Q6	R7	S8	TA	UB	VC	WD	XE	YF	ZG
MZ	N1	O2	P3	Q4	R5	S6	T7	U8	VA	WB	XC	YD	ZE	1F
NY	OZ	P1	Q2	R3	S4	T5	U6	V7	W8	XA	YB	ZC	1D	2E
OX	PY	QZ	R1	S2	T3	U4	V5	W6	X7	Y8	ZA	1B	2C	3D
PW	QX	RY	SZ	T1	U2	V3	W4	X5	Y6	Z7	18	2A	3B	4C
QV	RW	SX	TY	UZ	V1	W2	X3	Y4	Z5	16	27	38	4A	5B
RU	SV	TW	UX	VY	WZ	X1	Y2	Z3	14	25	36	47	58	6A
ST	TU	UV	VW	WX	XY	YZ	Z1	12	23	34	45	56	67	78

JUNE 1978 PHYSICAL

16 F	17 SA	18 SU	19 M	20 TU	21 W	22 TH	23 F	24 SA	25 SU	26 M	27 TU	28 W	29 TH	30 F
FE	GF	HG	JH	KJ	LK	ML	NM	ON	PO	QP	RQ	SR	TS	UT
GD	HE	JF	KG	LH	MJ	NK	OL	PM	QN	RO	SP	TQ	UR	VS
HC	JD	KE	LF	MG	NH	OJ	PK	QL	RM	SN	TO	UP	VQ	WR
JB	KC	LD	ME	NF	OG	PH	QJ	RK	SL	TM	UN	VO	WP	XQ
KA	LB	MC	ND	OE	PF	QG	RH	SJ	TK	UL	VM	WN	XO	AP
LX	MA	NB	OC	PD	QE	RF	SG	TH	UJ	VK	WL	XM	AN	BO
MW	NX	OA	PB	QC	RD	SE	TF	UG	VH	WJ	XK	AL	BM	CN
NV	OW	PX	QA	RB	SC	TD	UE	VF	WG	XH	AJ	BK	CL	DM
OU	PV	QW	RX	SA	TB	UC	VD	WE	XF	AG	BH	CJ	DK	EL
PT	QU	RV	SW	TX	UA	VB	WC	XD	AE	BF	CG	DH	EJ	FK
QS	RT	SU	TV	UW	VX	WA	XB	AC	BD	CE	DF	EG	FH	GJ
R	S	T	U	V	W	X	A	B	C	D	E	F	G	H

JUNE 1978 EMOTIONAL

16 F	17 SA	18 SU	19 M	20 TU	21 W	22 TH	23 F	24 SA	25 SU	26 M	27 TU	28 W	29 TH	30 F	
U	V	W	X	Y	Z	1	2	3	A	B	C	D	E	F	
VT	WU	XV	YW	ZX	1Y	2Z	31	A2	B3	CA	DB	EC	FD	GE	
WS	XT	YU	ZV	1W	2X	3Y	AZ	B1	C2	D3	EA	FB	GC	HD	
XR	YS	ZT	1U	2V	3W	AX	BY	CZ	D1	E2	F3	GA	HB	JC	
YQ	ZR	1S	2T	3U	AV	BW	CX	DY	EZ	F1	G2	H3	JA	KB	+
ZP	1Q	2R	3S	AT	BU	CV	DW	EX	FY	GZ	H1	J2	K3	LA	
1O	2P	3Q	AR	BS	CT	DU	EV	FW	GX	HY	JZ	K1	L2	M3	
2N	3O	AP	BQ	CR	DS	ET	FU	GV	HW	JX	KY	LZ	M1	N2	o
3M	AN	BO	CP	DQ	ER	FS	GT	HU	JV	KW	LX	MY	NZ	O1	
AL	BM	CN	DO	EP	FQ	GR	HS	JT	KU	LV	MW	NX	OY	PZ	
BK	CL	DM	EN	FO	GP	HQ	JR	KS	LT	MU	NV	OW	PX	QY	−
CJ	DK	EL	FM	GN	HO	JP	KQ	LR	MS	NT	OU	PV	QW	RX	
DH	EJ	FK	GL	HM	JN	KO	LP	MQ	NR	OS	PT	QU	RV	SW	
EG	FH	GJ	HK	JL	KM	LN	MO	NP	OQ	PR	QS	RT	SU	TV	
F	G	H	J	K	L	M	N	O	P	Q	R	S	T	U	

JUNE 1978 INTELLECTUAL

16 F	17 SA	18 SU	19 M	20 TU	21 W	22 TH	23 F	24 SA	25 SU	26 M	27 TU	28 W	29 TH	30 F
R	S	T	U	V	W	X	Y	Z	1	2	3	4	5	6
SQ	TR	US	VT	WU	XV	YW	ZX	1Y	2Z	31	42	53	64	75
TP	UQ	VR	WS	XT	YU	ZV	1W	2X	3Y	4Z	51	62	73	84
UO	VP	WQ	XR	YS	ZT	1U	2V	3W	4X	5Y	6Z	71	82	A3
VN	WO	XP	YQ	ZR	1S	2T	3U	4V	5W	6X	7Y	8Z	A1	B2
WM	XN	YO	ZP	1Q	2R	3S	4T	5U	6V	7W	8X	AY	BZ	C1
XL	YM	ZN	1O	2P	3Q	4R	5S	6T	7U	8V	AW	BX	CY	DZ
YK	ZL	1M	2N	3O	4P	5Q	6R	7S	8T	AU	BV	CW	DX	EY
ZJ	1K	2L	3M	4N	5O	6P	7Q	8R	AS	BT	CU	DV	EW	FX
1H	2J	3K	4L	5M	6N	7O	8P	AQ	BR	CS	DT	EU	FV	GW
2G	3H	4J	5K	6L	7M	8N	AO	BP	CQ	DR	ES	FT	GU	HV
3F	4G	5H	6J	7K	8L	AM	BN	CO	DP	EQ	FR	GS	HT	JU
4E	5F	6G	7H	8J	AK	BL	CM	DN	EO	FP	GQ	HR	JS	KT
5D	6E	7F	8G	AH	BJ	CK	DL	EM	FN	GO	HP	JQ	KR	LS
6C	7D	8E	AF	BG	CH	DJ	EK	FL	GM	HN	JO	KP	LQ	MR
7B	8C	AD	BE	CF	DG	EH	FJ	GK	HL	JM	KN	LO	MP	NQ
8A	AB	BC	CD	DE	EF	FG	GH	HJ	JK	KL	LM	MN	NO	OP

JULY 1978 **PHYSICAL**

1 SA	2 SU	3 M	4 TU	5 W	6 TH	7 F	8 SA	9 SU	10 M	11 TU	12 W	13 TH	14 F	15 SA
VU	WV	XW	AX	BA	CB	DC	ED	FE	GF	HG	JH	KJ	LK	ML
WT	XU	AV	BW	CX	DA	EB	FC	GD	HE	JF	KG	LH	MJ	NK
XS	AT	BU	CV	DW	EX	FA	GB	HC	JD	KE	LF	MG	NH	OJ
AR	BS	CT	DU	EV	FW	GX	HA	JB	KC	LD	ME	NF	OG	PH
BQ	CR	DS	ET	FU	GV	HW	JX	KA	LB	MC	ND	OE	PF	QG
CP	DQ	ER	FS	GT	HU	JV	KW	LX	MA	NB	OC	PD	QE	RF
DO	EP	FQ	GR	HS	JT	KU	LV	MW	NX	OA	PB	QC	RD	SE
EN	FO	GP	HQ	JR	KS	LT	MU	NV	OW	PX	QA	RB	SC	TD
FM	GN	HO	JP	KQ	LR	MS	NT	OU	PV	QW	RX	SA	TB	UC
GL	HM	JN	KO	LP	MQ	NR	OS	PT	QU	RV	SW	TX	UA	VB
HK	JL	KM	LN	MO	NP	OQ	PR	QS	RT	SU	TV	UW	VX	WA
J	K	L	M	N	O	P	Q	R	S	T	U	V	W	X

JULY 1978 **EMOTIONAL**

1 SA	2 SU	3 M	4 TU	5 W	6 TH	7 F	8 SA	9 SU	10 M	11 TU	12 W	13 TH	14 F	15 SA	
G	H	J	K	L	M	N	O	P	Q	R	S	T	U	V	
HF	JG	KH	LJ	MK	NL	OM	PN	QO	RP	SQ	TR	US	VT	WU	
JE	KF	LG	MH	NJ	OK	PL	QM	RN	SO	TP	UQ	VR	WS	XT	
KD	LE	MF	NG	OH	PJ	QK	RL	SM	TN	UO	VP	WQ	XR	YS	
LC	MD	NE	OF	PG	QH	RJ	SK	TL	UM	VN	WO	XP	YQ	ZR	+
MB	NC	OD	PE	QF	RG	SH	TJ	UK	VL	WM	XN	YO	ZP	1Q	
NA	OB	PC	QD	RE	SF	TG	UH	VJ	WK	XL	YM	ZN	1O	2P	
O3	PA	QB	RC	SD	TE	UF	VG	WH	XJ	YK	ZL	1M	2N	3O	0
P2	Q3	RA	SB	TC	UD	VE	WF	XG	YH	ZJ	1K	2L	3M	AN	
Q1	R2	S3	TA	UB	VC	WD	XE	YF	ZG	1H	2J	3K	AL	BM	
RZ	S1	T2	U3	VA	WB	XC	YD	ZE	1F	2G	3H	AJ	BK	CL	−
SY	TZ	U1	V2	W3	XA	YB	ZC	1D	2E	3F	AG	BH	CJ	DK	
TX	UY	VZ	W1	X2	Y3	ZA	1B	2C	3D	AE	BF	CG	DH	EJ	
UW	VX	WY	XZ	Y1	Z2	13	2A	3B	AC	BD	CE	DF	EG	FH	
V	W	X	Y	Z	1	2	3	A	B	C	D	E	F	G	

JULY 1978 **INTELLECTUAL**

1 SA	2 SU	3 M	4 TU	5 W	6 TH	7 F	8 SA	9 SU	10 M	11 TU	12 W	13 TH	14 F	15 SA
7	8	A	B	C	D	E	F	G	H	J	K	L	M	N
86	A7	B8	CA	DB	EC	FD	GE	HF	JG	KH	LJ	MK	NL	OM
A5	B6	C7	D8	EA	FB	GC	HD	JE	KF	LG	MH	NJ	OK	PL
B4	C5	D6	E7	F8	GA	HB	JC	KD	LE	MF	NG	OH	PJ	QK
C3	D4	E5	F6	G7	H8	JA	KB	LC	MD	NE	OF	PG	QH	RJ
D2	E3	F4	G5	H6	J7	K8	LA	MB	NC	OD	PE	QF	RG	SH
E1	F2	G3	H4	J5	K6	L7	M8	NA	OB	PC	QD	RE	SF	TG
FZ	G1	H2	J3	K4	L5	M6	N7	O8	PA	QB	RC	SD	TE	UF
GY	HZ	J1	K2	L3	M4	N5	O6	P7	Q8	RA	SB	TC	UD	VE
HX	JY	KZ	L1	M2	N3	O4	P5	Q6	R7	S8	TA	UB	VC	WD
JW	KX	LY	MZ	N1	O2	P3	Q4	R5	S6	T7	U8	VA	WB	XC
KV	LW	MX	NY	OZ	P1	Q2	R3	S4	T5	U6	V7	W8	XA	YB
LU	MV	NW	OX	PY	QZ	R1	S2	T3	U4	V5	W6	X7	Y8	ZA
MT	NU	OV	PW	QX	RY	SZ	T1	U2	V3	W4	X5	Y6	Z7	18
NS	OT	PU	QV	RW	SX	TY	UZ	V1	W2	X3	Y4	Z5	16	27
OR	PS	QT	RU	SV	TW	UX	VY	WZ	X1	Y2	Z3	14	25	36
PQ	QR	RS	ST	TU	UV	VW	WX	XY	YZ	Z1	12	23	34	45

JULY 1978 — PHYSICAL

16 SU	17 M	18 TU	19 W	20 TH	21 F	22 SA	23 SU	24 M	25 TU	26 W	27 TH	28 F	29 SA	30 SU	31 M
NM	ON	PO	QP	RQ	SR	TS	UT	VU	WV	XW	AX	BA	CB	DC	ED
OL	PM	QN	RO	SP	TQ	UR	VS	WT	XU	AV	BW	CX	DA	EB	FC
PK	QL	RM	SN	TO	UP	VQ	WR	XS	AT	BU	CV	DW	EX	FA	GB
QJ	RK	SL	TM	UN	VO	WP	XQ	AR	BS	CT	DU	EV	FW	GX	HA
RH	SJ	TK	UL	VM	WN	XO	AP	BQ	CR	DS	ET	FU	GV	HW	JX
SG	TH	UJ	VK	WL	XM	AN	BO	CP	DQ	ER	FS	GT	HU	JV	KW
TF	UG	VH	WJ	XK	AL	BM	CN	DO	EP	FQ	GR	HS	JT	KU	LV
UE	VF	WG	XH	AJ	BK	CL	DM	EN	FO	GP	HQ	JR	KS	LT	MU
VD	WE	XF	AG	BH	CJ	DK	EL	FM	GN	HO	JP	KQ	LR	MS	NT
WC	XD	AE	BF	CG	DH	EJ	FK	GL	HM	JN	KO	LP	MQ	NR	OS
XB	AC	BD	CE	DF	EG	FH	GJ	HK	JL	KM	LN	MO	NP	OQ	PR
A	B	C	D	E	F	G	H	J	K	L	M	N	O	P	Q

JULY 1978 — EMOTIONAL

16 SU	17 M	18 TU	19 W	20 TH	21 F	22 SA	23 SU	24 M	25 TU	26 W	27 TH	28 F	29 SA	30 SU	31 M	
W	X	Y	Z	1	2	3	A	B	C	D	E	F	G	H	J	
XV	YW	ZX	1Y	2Z	31	A2	B3	CA	DB	EC	FD	GE	HF	JG	KH	
YU	ZV	1W	2X	3Y	AZ	B1	C2	D3	EA	FB	GC	HD	JE	KF	LG	
ZT	1U	2V	3W	AX	BY	CZ	D1	E2	F3	GA	HB	JC	KD	LE	MF	
1S	2T	3U	AV	BW	CX	DY	EZ	F1	G2	H3	JA	KB	LC	MD	NE	+
2R	3S	AT	BU	CV	DW	EX	FY	GZ	H1	J2	K3	LA	MB	NC	OD	
3Q	AR	BS	CT	DU	EV	FW	GX	HY	JZ	K1	L2	M3	NA	OB	PC	
AP	BQ	CR	DS	ET	FU	GV	HW	JX	KY	LZ	M1	N2	O3	PA	QB	o
BO	CP	DQ	ER	FS	GT	HU	JV	KW	LX	MY	NZ	O1	P2	Q3	RA	
CN	DO	EP	FQ	GR	HS	JT	KU	LV	MW	NX	OY	PZ	Q1	R2	S3	
DM	EN	FO	GP	HQ	JR	KS	LT	MU	NV	OW	PX	QY	RZ	S1	T2	−
EL	FM	GN	HO	JP	KQ	LR	MS	NT	OU	PV	QW	RX	SY	TZ	U1	
FK	GL	HM	JN	KO	LP	MQ	NR	OS	PT	QU	RV	SW	TX	UY	VZ	
GJ	HK	JL	KM	LN	MO	NP	OQ	PR	OS	RT	SU	TV	UW	VX	WY	
H	J	K	L	M	N	O	P	Q	R	S	T	U	V	W	X	

JULY 1978 — INTELLECTUAL

16 SU	17 M	18 TU	19 W	20 TH	21 F	22 SA	23 SU	24 M	25 TU	26 W	27 TH	28 F	29 SA	30 SU	31 M
O	P	Q	R	S	T	U	V	W	X	Y	Z	1	2	3	4
PN	QO	RP	SQ	TR	US	VT	WU	XV	YW	ZX	1Y	2Z	31	42	53
QM	RN	SO	TP	UQ	VR	WS	XT	YU	ZV	1W	2X	3Y	4Z	51	62
RL	SM	TN	UO	VP	WQ	XR	YS	ZT	1U	2V	3W	4X	5Y	6Z	71
SK	TL	UM	VN	WO	XP	YQ	ZR	1S	2T	3U	4V	5W	6X	7Y	8Z
TJ	UK	VL	WM	XN	YO	ZP	1Q	2R	3S	4T	5U	6V	7W	8X	AY
UH	VJ	WK	XL	YM	ZN	1O	2P	3Q	4R	5S	6T	7U	8V	AW	BX
VG	WH	XJ	YK	ZL	1M	2N	3O	4P	5Q	6R	7S	8T	AU	BV	CW
WF	XG	YH	ZJ	1K	2L	3M	4N	5O	6P	7Q	8R	AS	BT	CU	DV
XE	YF	ZG	1H	2J	3K	4L	5M	6N	7O	8P	AQ	BR	CS	DT	EU
YD	ZE	1F	2G	3H	4J	5K	6L	7M	8N	AO	BP	CQ	DR	ES	FT
ZC	1D	2E	3F	4G	5H	6J	7K	8L	AM	BN	CO	DP	EQ	FR	GS
1B	2C	3D	4E	5F	6G	7H	8J	AK	BL	CM	DN	EO	FP	GQ	HR
2A	3B	4C	5D	6E	7F	8G	AH	BJ	CK	DL	EM	FN	GO	HP	JQ
38	4A	5B	6C	7D	8E	AF	BG	CH	DJ	EK	FL	GM	HN	JO	KP
47	58	6A	7B	8C	AD	BE	CF	DG	EH	FJ	GK	HL	JM	KN	LO
56	67	78	8A	AB	BC	CD	DE	EF	FG	GH	HJ	JK	KL	LM	MN

AUGUST 1978 **PHYSICAL**

1 TU	2 W	3 TH	4 F	5 SA	6 SU	7 M	8 TU	9 W	10 TH	11 F	12 SA	13 SU	14 M	15 TU
FE	GF	HG	JH	KJ	LK	ML	NM	ON	PO	QP	RQ	SR	TS	UT
GD	HE	JF	KG	LH	MJ	NK	OL	PM	QN	RO	SP	TQ	UR	VS
HC	JD	KE	LF	MG	NH	OJ	PK	QL	RM	SN	TO	UP	VQ	WR
JB	KC	LD	ME	NF	OG	PH	QJ	RK	SL	TM	UN	VO	WP	XQ
KA	LB	MC	ND	OE	PF	QG	RH	SJ	TK	UL	VM	WN	XO	AP
LX	MA	NB	OC	PD	QE	RF	SG	TH	UJ	VK	WL	XM	AN	BO
MW	NX	OA	PB	QC	RD	SE	TF	UG	VH	WJ	XK	AL	BM	CN
NV	OW	PX	QA	RB	SC	TD	UE	VF	WG	XH	AJ	BK	CL	DM
OU	PV	QW	RX	SA	TB	UC	VD	WE	XF	AG	BH	CJ	DK	EL
PT	QU	RV	SW	TX	UA	VB	WC	XD	AE	BF	CG	DH	EJ	FK
QS	RT	SU	TV	UW	VX	WA	XB	AC	BD	CE	DF	EG	FH	GJ
R	S	T	U	V	W	X	A	B	C	D	E	F	G	H

AUGUST 1978 **EMOTIONAL**

1 TU	2 W	3 TH	4 F	5 SA	6 SU	7 M	8 TU	9 W	10 TH	11 F	12 SA	13 SU	14 M	15 TU	
K	L	M	N	O	P	Q	R	S	T	U	V	W	X	Y	
LJ	MK	NL	OM	PN	QO	RP	SQ	TR	US	VT	WU	XV	YW	ZX	
MH	NJ	OK	PL	QM	RN	SO	TP	UQ	VR	WS	XT	YU	ZV	1W	
NG	OH	PJ	QK	RL	SM	TN	UO	VP	WQ	XR	YS	ZT	1U	2V	
OF	PG	QH	RJ	SK	TL	UM	VN	WO	XP	YQ	ZR	1S	2T	3U	+
PE	QF	RG	SH	TJ	UK	VL	WM	XN	YO	ZP	1Q	2R	3S	AT	
QD	RE	SF	TG	UH	VJ	WK	XL	YM	ZN	1O	2P	3Q	AR	BS	
RC	SD	TE	UF	VG	WH	XJ	YK	ZL	1M	2N	3O	AP	BQ	CR	O
SB	TC	UD	VE	WF	XG	YH	ZJ	1K	2L	3M	AN	BO	CP	DQ	
TA	UB	VC	WD	XE	YF	ZG	1H	2J	3K	AL	BM	CN	DO	EP	
U3	VA	WB	XC	YD	ZE	1F	2G	3H	AJ	BK	CL	DM	EN	FO	−
V2	W3	XA	YB	ZC	1D	2E	3F	AG	BH	CJ	DK	EL	FM	GN	
W1	X2	Y3	ZA	1B	2C	3D	AE	BF	CG	DH	EJ	FK	GL	HM	
XZ	Y1	Z2	13	2A	3B	AC	BD	CE	DF	EG	FH	GJ	HK	JL	
Y	Z	1	2	3	A	B	C	D	E	F	G	H	J	K	

AUGUST 1978 **INTELLECTUAL**

1 TU	2 W	3 TH	4 F	5 SA	6 SU	7 M	8 TU	9 W	10 TH	11 F	12 SA	13 SU	14 M	15 TU
5	6	7	8	A	B	C	D	E	F	G	H	J	K	L
64	75	86	A7	B8	CA	DB	EC	FD	GE	HF	JG	KH	LJ	MK
73	84	A5	B6	C7	D8	EA	FB	GC	HD	JE	KF	LG	MH	NJ
82	A3	B4	C5	D6	E7	F8	GA	HB	JC	KD	LE	MF	NG	OH
A1	B2	C3	D4	E5	F6	G7	H8	JA	KB	LC	MD	NE	OF	PG
BZ	C1	D2	E3	F4	G5	H6	J7	K8	LA	MB	NC	OD	PE	QF
CY	DZ	E1	F2	G3	H4	J5	K6	L7	M8	NA	OB	PC	QD	RE
DX	EY	FZ	G1	H2	J3	K4	L5	M6	N7	O8	PA	QB	RC	SD
EW	FX	GY	HZ	J1	K2	L3	M4	N5	O6	P7	Q8	RA	SB	TC
FV	GW	HX	JY	KZ	L1	M2	N3	O4	P5	Q6	R7	S8	TA	UB
GU	HV	JW	KX	LY	MZ	N1	O2	P3	Q4	R5	S6	T7	U8	VA
HT	JU	KV	LW	MX	NY	OZ	P1	Q2	R3	S4	T5	U6	V7	W8
JS	KT	LU	MV	NW	OX	PY	QZ	R1	S2	T3	U4	V5	W6	X7
KR	LS	MT	NU	OV	PW	QX	RY	SZ	T1	U2	V3	W4	X5	Y6
LQ	MR	NS	OT	PU	QV	RW	SX	TY	UZ	V1	W2	X3	Y4	Z5
MP	NQ	OR	PS	QT	RU	SV	TW	UX	VY	WZ	X1	Y2	Z3	14
NO	OP	PQ	QR	RS	ST	TU	UV	VW	WX	XY	YZ	Z1	12	23

AUGUST 1978 **PHYSICAL**

16 W	17 TH	18 F	19 SA	20 SU	21 M	22 TU	23 W	24 TH	25 F	26 SA	27 SU	28 M	29 TU	30 W	31 TH
VU	WV	XW	AX	BA	CB	DC	ED	FE	GF	HG	JH	KJ	LK	ML	NM
WT	XU	AV	BW	CX	DA	EB	FC	GD	HE	JF	KG	LH	MJ	NK	OL
XS	AT	BU	CV	DW	EX	FA	GB	HC	JD	KE	LF	MG	NH	OJ	PK
AR	BS	CT	DU	EV	FW	GX	HA	JB	KC	LD	ME	NF	OG	PH	QJ
BQ	CR	DS	ET	FU	GV	HW	JX	KA	LB	MC	ND	OE	PF	QG	RH
CP	DQ	ER	FS	GT	HU	JV	KW	LX	MA	NB	OC	PD	QE	RF	SG
DO	EP	FQ	GR	HS	JT	KU	LV	MW	NX	OA	PB	QC	RD	SE	TF
EN	FO	GP	HQ	JR	KS	LT	MU	NV	OW	PX	QA	RB	SC	TD	UE
FM	GN	HO	JP	KQ	LR	MS	NT	OU	PV	QW	RX	SA	TB	UC	VD
GL	HM	JN	KO	LP	MQ	NR	OS	PT	QU	RV	SW	TX	UA	VB	WC
HK	JL	KM	LN	MO	NP	OQ	PR	QS	RT	SU	TV	UW	VX	WA	XB
J	K	L	M	N	O	P	Q	R	S	T	U	V	W	X	A

AUGUST 1978 **EMOTIONAL**

16 W	17 TH	18 F	19 SA	20 SU	21 M	22 TU	23 W	24 TH	25 F	26 SA	27 SU	28 M	29 TU	30 W	31 TH	
Z	1	2	3	A	B	C	D	E	F	G	H	J	K	L	M	
1Y	2Z	31	A2	B3	CA	DB	EC	FD	GE	HF	JG	KH	LJ	MK	NL	
2X	3Y	AZ	B1	C2	D3	EA	FB	GC	HD	JE	KF	LG	MH	NJ	OK	
3W	AX	BY	CZ	D1	E2	F3	GA	HB	JC	KD	LE	MF	NG	OH	PJ	
AV	BW	CX	DY	EZ	F1	G2	H3	JA	KB	LC	MD	NE	OF	PG	QH	+
BU	CV	DW	EX	FY	GZ	H1	J2	K3	LA	MB	NC	OD	PE	QF	RG	
CT	DU	EV	FW	GX	HY	JZ	K1	L2	M3	NA	OB	PC	QD	RE	SF	
DS	ET	FU	GV	HW	JX	KY	LZ	M1	N2	O3	PA	QB	RC	SD	TE	0
ER	FS	GT	HU	JV	KW	LX	MY	NZ	O1	P2	Q3	RA	SB	TC	UD	
FQ	GR	HS	JT	KU	LV	MW	NX	OY	PZ	Q1	R2	S3	TA	UB	VC	
GP	HQ	JR	KS	LT	MU	NV	OW	PX	QY	RZ	S1	T2	U3	VA	WB	−
HO	JP	KQ	LR	MS	NT	OU	PV	QW	RX	SY	TZ	U1	V2	W3	XA	
JN	KO	LP	MQ	NR	OS	PT	QU	RV	SW	TX	UY	VZ	W1	X2	Y3	
KM	LN	MO	NP	OQ	PR	QS	RT	SU	TV	UW	VX	WY	XZ	Y1	Z2	
L	M	N	O	P	Q	R	S	T	U	V	W	X	Y	Z	1	

AUGUST 1978 **INTELLECTUAL**

16 W	17 TH	18 F	19 SA	20 SU	21 M	22 TU	23 W	24 TH	25 F	26 SA	27 SU	28 M	29 TU	30 W	31 TH
M	N	O	P	Q	R	S	T	U	V	W	X	Y	Z	1	2
NL	OM	PN	QO	RP	SQ	TR	US	VT	WU	XV	YW	ZX	1Y	2Z	31
OK	PL	QM	RN	SO	TP	UQ	VR	WS	XT	YU	ZV	1W	2X	3Y	4Z
PJ	QK	RL	SM	TN	UO	VP	WQ	XR	YS	ZT	1U	2V	3W	4X	5Y
QH	RJ	SK	TL	UM	VN	WO	XP	YQ	ZR	1S	2T	3U	4V	5W	6X
RG	SH	TJ	UK	VL	WM	XN	YO	ZP	1Q	2R	3S	4T	5U	6V	7W
SF	TG	UH	VJ	WK	XL	YM	ZN	1O	2P	3Q	4R	5S	6T	7U	8V
TE	UF	VG	WH	XJ	YK	ZL	1M	2N	3O	4P	5Q	6R	7S	8T	AU
UD	VE	WF	XG	YH	ZJ	1K	2L	3M	4N	5O	6P	7Q	8R	AS	BT
VC	WD	XE	YF	ZG	1H	2J	3K	4L	5M	6N	7O	8P	AQ	BR	CS
WB	XC	YD	ZE	1F	2G	3H	4J	5K	6L	7M	8N	AO	BP	CQ	DR
XA	YB	ZC	1D	2E	3F	4G	5H	6J	7K	8L	AM	BN	CO	DP	EQ
Y8	ZA	1B	2C	3D	4E	5F	6G	7H	8J	AK	BL	CM	DN	EO	FP
Z7	18	2A	3B	4C	5D	6E	7F	8G	AH	BJ	CK	DL	EM	FN	GO
16	27	38	4A	5B	6C	7D	8E	AF	BG	CH	DJ	EK	FL	GM	HN
25	36	47	58	6A	7B	8C	AD	BE	CF	DG	EH	FJ	GK	HL	JM
34	45	56	67	78	8A	AB	BC	CD	DE	EF	FG	GH	HJ	JK	KL

SEPTEMBER 1978 — PHYSICAL

1 F	2 SA	3 SU	4 M	5 TU	6 W	7 TH	8 F	9 SA	10 SU	11 M	12 TU	13 W	14 TH	15 F
ON	PO	QP	RQ	SR	TS	UT	VU	WV	XW	AX	BA	CB	DC	ED
PM	QN	RO	SP	TQ	UR	VS	WT	XU	AV	BW	CX	DA	EB	FC
QL	RM	SN	TO	UP	VQ	WR	XS	AT	BU	CV	DW	EX	FA	GB
RK	SL	TM	UN	VO	WP	XQ	AR	BS	CT	DU	EV	FW	GX	HA
SJ	TK	UL	VM	WN	XO	AP	BQ	CR	DS	ET	FU	GV	HW	JX
TH	UJ	VK	WL	XM	AN	BO	CP	DQ	ER	FS	GT	HU	JV	KW
UG	VH	WJ	XK	AL	BM	CN	DO	EP	FQ	GR	HS	JT	KU	LV
VF	WG	XH	AJ	BK	CL	DM	EN	FO	GP	HQ	JR	KS	LT	MU
WE	XF	AG	BH	CJ	DK	EL	FM	GN	HO	JP	KQ	LR	MS	NT
XD	AE	BF	CG	DH	EJ	FK	GL	HM	JN	KO	LP	MQ	NR	OS
AC	BD	CE	DF	EG	FH	GJ	HK	JL	KM	LN	MO	NP	OQ	PR
B	C	D	E	F	G	H	J	K	L	M	N	O	P	Q

SEPTEMBER 1978 — EMOTIONAL

1 F	2 SA	3 SU	4 M	5 TU	6 W	7 TH	8 F	9 SA	10 SU	11 M	12 TU	13 W	14 TH	15 F	
N	O	P	Q	R	S	T	U	V	W	X	Y	Z	1	2	
OM	PN	QO	RP	SQ	TR	US	VT	WU	XV	YW	ZX	1Y	2Z	31	
PL	QM	RN	SO	TP	UQ	VR	WS	XT	YU	ZV	1W	2X	3Y	AZ	
QK	RL	SM	TN	UO	VP	WQ	XR	YS	ZT	1U	2V	3W	AX	BY	
RJ	SK	TL	UM	VN	WO	XP	YQ	ZR	1S	2T	3U	AV	BW	CX	+
SH	TJ	UK	VL	WM	XN	YO	ZP	1Q	2R	3S	AT	BU	CV	DW	
TG	UH	VJ	WK	XL	YM	ZN	1O	2P	3Q	AR	BS	CT	DU	EV	
UF	VG	WH	XJ	YK	ZL	1M	2N	3O	AP	BQ	CR	DS	ET	FU	o
VE	WF	XG	YH	ZJ	1K	2L	3M	AN	BO	CP	DQ	ER	FS	GT	
WD	XE	YF	ZG	1H	2J	3K	AL	BM	CN	DO	EP	FQ	GR	HS	
XC	YD	ZE	1F	2G	3H	AJ	BK	CL	DM	EN	FO	GP	HQ	JR	−
YB	ZC	1D	2E	3F	AG	BH	CJ	DK	EL	FM	GN	HO	JP	KQ	
ZA	1B	2C	3D	AE	BF	CG	DH	EJ	FK	GL	HM	JN	KO	LP	
13	2A	3B	AC	BD	CE	DF	EG	FH	GJ	HK	JL	KM	LN	MO	
2	3	A	B	C	D	E	F	G	H	J	K	L	M	N	

SEPTEMBER 1978 — INTELLECTUAL

1 F	2 SA	3 SU	4 M	5 TU	6 W	7 TH	8 F	9 SA	10 SU	11 M	12 TU	13 W	14 TH	15 F
3	4	5	6	7	8	A	B	C	D	E	F	G	H	J
42	53	64	75	86	A7	B8	CA	DB	EC	FD	GE	HF	JG	KH
51	62	73	84	A5	B6	C7	D8	EA	FB	GC	HD	JE	KF	LG
6Z	71	82	A3	B4	C5	D6	E7	F8	GA	HB	JC	KD	LE	MF
7Y	8Z	A1	B2	C3	D4	E5	F6	G7	H8	JA	KB	LC	MD	NE
8X	AY	BZ	C1	D2	E3	F4	G5	H6	J7	K8	LA	MB	NC	OD
AW	BX	CY	DZ	E1	F2	G3	H4	J5	K6	L7	M8	NA	OB	PC
BV	CW	DX	EY	FZ	G1	H2	J3	K4	L5	M6	N7	O8	PA	QB
CU	DV	EW	FX	GY	HZ	J1	K2	L3	M4	N5	O6	P7	Q8	RA
DT	EU	FV	GW	HX	JY	KZ	L1	M2	N3	O4	P5	Q6	R7	S8
ES	FT	GU	HV	JW	KX	LY	MZ	N1	O2	P3	Q4	R5	S6	T7
FR	GS	HT	JU	KV	LW	MX	NY	OZ	P1	Q2	R3	S4	T5	U6
GQ	HR	JS	KT	LU	MV	NW	OX	PY	QZ	R1	S2	T3	U4	V5
HP	JQ	KR	LS	MT	NU	OV	PW	QX	RY	SZ	T1	U2	V3	W4
JO	KP	LQ	MR	NS	OT	PU	QV	RW	SX	TY	UZ	V1	W2	X3
KN	LO	MP	NQ	OR	PS	QT	RU	SV	TW	UX	VY	WZ	X1	Y2
LM	MN	NO	OP	PQ	QR	RS	ST	TU	UV	VW	WX	XY	YZ	Z1

SEPTEMBER 1978 — PHYSICAL

16 SA	17 SU	18 M	19 TU	20 W	21 TH	22 F	23 SA	24 SU	25 M	26 TU	27 W	28 TH	29 F	30 SA
FE	GF	HG	JH	KJ	LK	ML	NM	ON	PO	QP	RQ	SR	TS	UT
GD	HE	JF	KG	LH	MJ	NK	OL	PM	QN	RO	SP	TQ	UR	VS
HC	JD	KE	LF	MG	NH	OJ	PK	QL	RM	SN	TO	UP	VQ	WR
JB	KC	LD	ME	NF	OG	PH	QJ	RK	SL	TM	UN	VO	WP	XQ
KA	LB	MC	ND	OE	PF	QG	RH	SJ	TK	UL	VM	WN	XO	AP
LX	MA	NB	OC	PD	QE	RF	SG	TH	UJ	VK	WL	XM	AN	BO
MW	NX	OA	PB	QC	RD	SE	TF	UG	VH	WJ	XK	AL	BM	CN
NV	OW	PX	QA	RB	SC	TD	UE	VF	WG	XH	AJ	BK	CL	DM
OU	PV	QW	RX	SA	TB	UC	VD	WE	XF	AG	BH	CJ	DK	EL
PT	QU	RV	SW	TX	UA	VB	WC	XD	AE	BF	CG	DH	EJ	FK
QS	RT	SU	TV	UW	VX	WA	XB	AC	BD	CE	DF	EG	FH	GJ
R	S	T	U	V	W	X	A	B	C	D	E	F	G	H

SEPTEMBER 1978 — EMOTIONAL

16 SA	17 SU	18 M	19 TU	20 W	21 TH	22 F	23 SA	24 SU	25 M	26 TU	27 W	28 TH	29 F	30 SA	
3	A	B	C	D	E	F	G	H	J	K	L	M	N	O	
A2	B3	CA	DB	EC	FD	GE	HF	JG	KH	LJ	MK	NL	OM	PN	
B1	C2	D3	EA	FB	GC	HD	JE	KF	LG	MH	NJ	OK	PL	QM	
CZ	D1	E2	F3	GA	HB	JC	KD	LE	MF	NG	OH	PJ	QK	RL	
DY	EZ	F1	G2	H3	JA	KB	LC	MD	NE	OF	PG	QH	RJ	SK	+
EX	FY	GZ	H1	J2	K3	LA	MB	NC	OD	PE	QF	RG	SH	TJ	
FW	GX	HY	JZ	K1	L2	M3	NA	OB	PC	QD	RE	SF	TG	UH	
GV	HW	JX	KY	LZ	M1	N2	O3	PA	QB	RC	SD	TE	UF	VG	0
HU	JV	KW	LX	MY	NZ	O1	P2	Q3	RA	SB	TC	UD	VE	WF	
JT	KU	LV	MW	NX	OY	PZ	Q1	R2	S3	TA	UB	VC	WD	XE	
KS	LT	MU	NV	OW	PX	QY	RZ	S1	T2	U3	VA	WB	XC	YD	–
LR	MS	NT	OU	PV	QW	RX	SY	TZ	U1	V2	W3	XA	YB	ZC	
MQ	NR	OS	PT	QU	RV	SW	TX	UY	VZ	W1	X2	Y3	ZA	1B	
NP	OQ	PR	QS	RT	SU	TV	UW	VX	WY	XZ	Y1	Z2	13	2A	
O	P	Q	R	S	T	U	V	W	X	Y	Z	1	2	3	

SEPTEMBER 1978 — INTELLECTUAL

16 SA	17 SU	18 M	19 TU	20 W	21 TH	22 F	23 SA	24 SU	25 M	26 TU	27 W	28 TH	29 F	30 SA
K	L	M	N	O	P	Q	R	S	T	U	V	W	X	Y
LJ	MK	NL	OM	PN	QO	RP	SQ	TR	US	VT	WU	XV	YW	ZX
MH	NJ	OK	PL	QM	RN	SO	TP	UQ	VR	WS	XT	YU	ZV	1W
NG	OH	PJ	QK	RL	SM	TN	UO	VP	WQ	XR	YS	ZT	1U	2V
OF	PG	QH	RJ	SK	TL	UM	VN	WO	XP	YQ	ZR	1S	2T	3U
PE	QF	RG	SH	TJ	UK	VL	WM	XN	YO	ZP	1Q	2R	3S	4T
QD	RE	SF	TG	UH	VJ	WK	XL	YM	ZN	1O	2P	3Q	4R	5S
RC	SD	TE	UF	VG	WH	XJ	YK	ZL	1M	2N	3O	4P	5Q	6R
SB	TC	UD	VE	WF	XG	YH	ZJ	1K	2L	3M	4N	5O	6P	7Q
TA	UB	VC	WD	XE	YF	ZG	1H	2J	3K	4L	5M	6N	7O	8P
U8	VA	WB	XC	YD	ZE	1F	2G	3H	4J	5K	6L	7M	8N	AO
V7	W8	XA	YB	ZC	1D	2E	3F	4G	5H	6J	7K	8L	AM	BN
W6	X7	Y8	ZA	1B	2C	3D	4E	5F	6G	7H	8J	AK	BL	CM
X5	Y6	Z7	18	2A	3B	4C	5D	6E	7F	8G	AH	BJ	CK	DL
Y4	Z5	16	27	38	4A	5B	6C	7D	8E	AF	BG	CH	DJ	EK
Z3	14	25	36	47	58	6A	7B	8C	AD	BE	CF	DG	EH	FJ
12	23	34	45	56	67	78	8A	AB	BC	CD	DE	EF	FG	GH

OCTOBER 1978 — PHYSICAL

1 SU	2 M	3 TU	4 W	5 TH	6 F	7 SA	8 SU	9 M	10 TU	11 W	12 TH	13 F	14 SA	15 SU
VU	WV	XW	AX	BA	CB	DC	ED	FE	GF	HG	JH	KJ	LK	ML
WT	XU	AV	BW	CX	DA	EB	FC	GD	HE	JF	KG	LH	MJ	NK
XS	AT	BU	CV	DW	EX	FA	GB	HC	JD	KE	LF	MG	NH	OJ
AR	BS	CT	DU	EV	FW	GX	HA	JB	KC	LD	ME	NF	OG	PH
BQ	CR	DS	ET	FU	GV	HW	JX	KA	LB	MC	ND	OE	PF	QG
CP	DQ	ER	FS	GT	HU	JV	KW	LX	MA	NB	OC	PD	QE	RF
DO	EP	FQ	GR	HS	JT	KU	LV	MW	NX	OA	PB	QC	RD	SE
EN	FO	GP	HQ	JR	KS	LT	MU	NV	OW	PX	QA	RB	SC	TD
FM	GN	HO	JP	KQ	LR	MS	NT	OU	PV	QW	RX	SA	TB	UC
GL	HM	JN	KO	LP	MQ	NR	OS	PT	QU	RV	SW	TX	UA	VB
HK	JL	KM	LN	MO	NP	OQ	PR	QS	RT	SU	TV	UW	VX	WA
J	K	L	M	N	O	P	Q	R	S	T	U	V	W	X

OCTOBER 1978 — EMOTIONAL

1 SU	2 M	3 TU	4 W	5 TH	6 F	7 SA	8 SU	9 M	10 TU	11 W	12 TH	13 F	14 SA	15 SU	
P	Q	R	S	T	U	V	W	X	Y	Z	1	2	3	A	
QO	RP	SQ	TR	US	VT	WU	XV	YW	ZX	1Y	2Z	31	A2	B3	
RN	SO	TP	UQ	VR	WS	XT	YU	ZV	1W	2X	3Y	AZ	B1	C2	
SM	TN	UO	VP	WQ	XR	YS	ZT	1U	2V	3W	AX	BY	CZ	D1	
TL	UM	VN	WO	XP	YQ	ZR	1S	2T	3U	AV	BW	CX	DY	EZ	+
UK	VL	WM	XN	YO	ZP	1Q	2R	3S	AT	BU	CV	DW	EX	FY	
VJ	WK	XL	YM	ZN	1O	2P	3Q	AR	BS	CT	DU	EV	FW	GX	
WH	XJ	YK	ZL	1M	2N	3O	AP	BQ	CR	DS	ET	FU	GV	HW	o
XG	YH	ZJ	1K	2L	3M	AN	BO	CP	DQ	ER	FS	GT	HU	JV	
YF	ZG	1H	2J	3K	AL	BM	CN	DO	EP	FQ	GR	HS	JT	KU	
ZE	1F	2G	3H	AJ	BK	CL	DM	EN	FO	GP	HQ	JR	KS	LT	−
1D	2E	3F	AG	BH	CJ	DK	EL	FM	GN	HO	JP	KQ	LR	MS	
2C	3D	AE	BF	CG	DH	EJ	FK	GL	HM	JN	KO	LP	MQ	NR	
3B	AC	BD	CE	DF	EG	FH	GJ	HK	JL	KM	LN	MO	NP	OQ	
A	B	C	D	E	F	G	H	J	K	L	M	N	O	P	

OCTOBER 1978 — INTELLECTUAL

1 SU	2 M	3 TU	4 W	5 TH	6 F	7 SA	8 SU	9 M	10 TU	11 W	12 TH	13 F	14 SA	15 SU
Z	1	2	3	4	5	6	7	8	A	B	C	D	E	F
1Y	2Z	31	42	53	64	75	86	A7	B8	CA	DB	EC	FD	GE
2X	3Y	4Z	51	62	73	84	A5	B6	C7	D8	EA	FB	GC	HD
3W	4X	5Y	6Z	71	82	A3	B4	C5	D6	E7	F8	GA	HB	JC
4V	5W	6X	7Y	8Z	A1	B2	C3	D4	E5	F6	G7	H8	JA	KB
5U	6V	7W	8X	AY	BZ	C1	D2	E3	F4	G5	H6	J7	K8	LA
6T	7U	8V	AW	BX	CY	DZ	E1	F2	G3	H4	J5	K6	L7	M8
7S	8T	AU	BV	CW	DX	EY	FZ	G1	H2	J3	K4	L5	M6	N7
8R	AS	BT	CU	DV	EW	FX	GY	HZ	J1	K2	L3	M4	N5	O6
AQ	BR	CS	DT	EU	FV	GW	HX	JY	KZ	L1	M2	N3	O4	P5
BP	CQ	DR	ES	FT	GU	HV	JW	KX	LY	MZ	N1	O2	P3	Q4
CO	DP	EQ	FR	GS	HT	JU	KV	LW	MX	NY	OZ	P1	Q2	R3
DN	EO	FP	GQ	HR	JS	KT	LU	MV	NW	OX	PY	QZ	R1	S2
EM	FN	GO	HP	JQ	KR	LS	MT	NU	OV	PW	QX	RY	SZ	T1
FL	GM	HN	JO	KP	LQ	MR	NS	OT	PU	QV	RW	SX	TY	UZ
GK	HL	JM	KN	LO	MP	NQ	OR	PS	QT	RU	SV	TW	UX	VY
HJ	JK	KL	LM	MN	NO	OP	PQ	QR	RS	ST	TU	UV	VW	WX

OCTOBER 1978 — PHYSICAL

16 M	17 TU	18 W	19 TH	20 F	21 SA	22 SU	23 M	24 TU	25 W	26 TH	27 F	28 SA	29 SU	30 M	31 TU
NM	ON	PO	QP	RQ	SR	TS	UT	VU	WV	XW	AX	BA	CB	DC	ED
OL	PM	QN	RO	SP	TQ	UR	VS	WT	XU	AV	BW	CX	DA	EB	FC
PK	QL	RM	SN	TO	UP	VQ	WR	XS	AT	BU	CV	DW	EX	FA	GB
QJ	RK	SL	TM	UN	VO	WP	XQ	AR	BS	CT	DU	EV	FW	GX	HA
RH	SJ	TK	UL	VM	WN	XO	AP	BQ	CR	DS	ET	FU	GV	HW	JX
SG	TH	UJ	VK	WL	XM	AN	BO	CP	DQ	ER	FS	GT	HU	JV	KW
TF	UG	VH	WJ	XK	AL	BM	CN	DO	EP	FQ	GR	HS	JT	KU	LV
UE	VF	WG	XH	AJ	BK	CL	DM	EN	FO	GP	HQ	JR	KS	LT	MU
VD	WE	XF	AG	BH	CJ	DK	EL	FM	GN	HO	JP	KQ	LR	MS	NT
WC	XD	AE	BF	CG	DH	EJ	FK	GL	HM	JN	KO	LP	MQ	NR	OS
XB	AC	BD	CE	DF	EG	FH	GJ	HK	JL	KM	LN	MO	NP	OQ	PR
A	B	C	D	E	F	G	H	J	K	L	M	N	O	P	Q

OCTOBER 1978 — EMOTIONAL

16 M	17 TU	18 W	19 TH	20 F	21 SA	22 SU	23 M	24 TU	25 W	26 TH	27 F	28 SA	29 SU	30 M	31 TU	
B	C	D	E	F	G	H	J	K	L	M	N	O	P	Q	R	
CA	DB	EC	FD	GE	HF	JG	KH	LJ	MK	NL	OM	PN	QO	RP	SQ	
D3	EA	FB	GC	HD	JE	KF	LG	MH	NJ	OK	PL	QM	RN	SO	TP	
E2	F3	GA	HB	JC	KD	LE	MF	NG	OH	PJ	QK	RL	SM	TN	UO	
F1	G2	H3	JA	KB	LC	MD	NE	OF	PG	QH	RJ	SK	TL	UM	VN	+
GZ	H1	J2	K3	LA	MB	NC	OD	PE	QF	RG	SH	TJ	UK	VL	WM	
HY	JZ	K1	L2	M3	NA	OB	PC	QD	RE	SF	TG	UH	VJ	WK	XL	
JX	KY	LZ	M1	N2	O3	PA	QB	RC	SD	TE	UF	VG	WH	XJ	YK	o
KW	LX	MY	NZ	O1	P2	Q3	RA	SB	TC	UD	VE	WF	XG	YH	ZJ	
LV	MW	NX	OY	PZ	Q1	R2	S3	TA	UB	VC	WD	XE	YF	ZG	1H	
MU	NV	OW	PX	QY	RZ	S1	T2	U3	VA	WB	XC	YD	ZE	1F	2G	−
NT	OU	PV	QW	RX	SY	TZ	U1	V2	W3	XA	YB	ZC	1D	2E	3F	
OS	PT	QU	RV	SW	TX	UY	VZ	W1	X2	Y3	ZA	1B	2C	3D	AE	
PR	QS	RT	SU	TV	UW	VX	WY	XZ	Y1	Z2	13	2A	3B	AC	BD	
Q	R	S	T	U	V	W	X	Y	Z	1	2	3	A	B	C	

OCTOBER 1978 — INTELLECTUAL

16 M	17 TU	18 W	19 TH	20 F	21 SA	22 SU	23 M	24 TU	25 W	26 TH	27 F	28 SA	29 SU	30 M	31 TU
G	H	J	K	L	M	N	O	P	Q	R	S	T	U	V	W
HF	JG	KH	LJ	MK	NL	OM	PN	QO	RP	SQ	TR	US	VT	WU	XV
JE	KF	LG	MH	NJ	OK	PL	QM	RN	SO	TP	UQ	VR	WS	XT	YU
KD	LE	MF	NG	OH	PJ	QK	RL	SM	TN	UO	VP	WQ	XR	YS	ZT
LC	MD	NE	OF	PG	QH	RJ	SK	TL	UM	VN	WO	XP	YQ	ZR	1S
MB	NC	OD	PE	QF	RG	SH	TJ	UK	VL	WM	XN	YO	ZP	1Q	2R
NA	OB	PC	QD	RE	SF	TG	UH	VJ	WK	XL	YM	ZN	1O	2P	3Q
O8	PA	QB	RC	SD	TE	UF	VG	WH	XJ	YK	ZL	1M	2N	3O	4P
P7	Q8	RA	SB	TC	UD	VE	WF	XG	YH	ZJ	1K	2L	3M	4N	5O
Q6	R7	S8	TA	UB	VC	WD	XE	YF	ZG	1H	2J	3K	4L	5M	6N
R5	S6	T7	U8	VA	WB	XC	YD	ZE	1F	2G	3H	4J	5K	6L	7M
S4	T5	U6	V7	W8	XA	YB	ZC	1D	2E	3F	4G	5H	6J	7K	8L
T3	U4	V5	W6	X7	Y8	ZA	1B	2C	3D	4E	5F	6G	7H	8J	AK
U2	V3	W4	X5	Y6	Z7	18	2A	3B	4C	5D	6E	7F	8G	AH	BJ
V1	W2	X3	Y4	Z5	16	27	38	4A	5B	6C	7D	8E	AF	BG	CH
WZ	X1	Y2	Z3	14	25	36	47	58	6A	7B	8C	AD	BE	CF	DG
XY	YZ	Z1	12	23	34	45	56	67	78	8A	AB	BC	CD	DE	EF

NOVEMBER 1978 PHYSICAL

1 W	2 TH	3 F	4 SA	5 SU	6 M	7 TU	8 W	9 TH	10 F	11 SA	12 SU	13 M	14 TU	15 W
FE	GF	HG	JH	KJ	LK	ML	NM	ON	PO	QP	RQ	SR	TS	UT
GD	HE	JF	KG	LH	MJ	NK	OL	PM	QN	RO	SP	TQ	UR	VS
HC	JD	KE	LF	MG	NH	OJ	PK	QL	RM	SN	TO	UP	VQ	WR
JB	KC	LD	ME	NF	OG	PH	QJ	RK	SL	TM	UN	VO	WP	XQ
KA	LB	MC	ND	OE	PF	QG	RH	SJ	TK	UL	VM	WN	XO	AP
LX	MA	NB	OC	PD	QE	RF	SG	TH	UJ	VK	WL	XM	AN	BO
MW	NX	OA	PB	QC	RD	SE	TF	UG	VH	WJ	XK	AL	BM	CN
NV	OW	PX	QA	RB	SC	TD	UE	VF	WG	XH	AJ	BK	CL	DM
OU	PV	QW	RX	SA	TB	UC	VD	WE	XF	AG	BH	CJ	DK	EL
PT	QU	RV	SW	TX	UA	VB	WC	XD	AE	BF	CG	DH	EJ	FK
QS	RT	SU	TV	UW	VX	WA	XB	AC	BD	CE	DF	EG	FH	GJ
R	S	T	U	V	W	X	A	B	C	D	E	F	G	H

NOVEMBER 1978 EMOTIONAL

1 W	2 TH	3 F	4 SA	5 SU	6 M	7 TU	8 W	9 TH	10 F	11 SA	12 SU	13 M	14 TU	15 W	
S	T	U	V	W	X	Y	Z	1	2	3	A	B	C	D	
TR	US	VT	WU	XV	YW	ZX	1Y	2Z	31	A2	B3	CA	DB	EC	
UQ	VR	WS	XT	YU	ZV	1W	2X	3Y	AZ	B1	C2	D3	EA	FB	
VP	WQ	XR	YS	ZT	1U	2V	3W	AX	BY	CZ	D1	E2	F3	GA	
WO	XP	YQ	ZR	1S	2T	3U	AV	BW	CX	DY	EZ	F1	G2	H3	+
XN	YO	ZP	1Q	2R	3S	AT	BU	CV	DW	EX	FY	GZ	H1	J2	
YM	ZN	1O	2P	3Q	AR	BS	CT	DU	EV	FW	GX	HY	JZ	K1	
ZL	1M	2N	3O	AP	BQ	CR	DS	ET	FU	GV	HW	JX	KY	LZ	o
1K	2L	3M	AN	BO	CP	DQ	ER	FS	GT	HU	JV	KW	LX	MY	
2J	3K	AL	BM	CN	DO	EP	FQ	GR	HS	JT	KU	LV	MW	NX	
3H	AJ	BK	CL	DM	EN	FO	GP	HQ	JR	KS	LT	MU	NV	OW	–
AG	BH	CJ	DK	EL	FM	GN	HO	JP	KQ	LR	MS	NT	OU	PV	
BF	CG	DH	EJ	FK	GL	HM	JN	KO	LP	MQ	NR	OS	PT	QU	
CE	DF	EG	FH	GJ	HK	JL	KM	LN	MO	NP	OQ	PR	QS	RT	
D	E	F	G	H	J	K	L	M	N	O	P	Q	R	S	

NOVEMBER 1978 INTELLECTUAL

1 W	2 TH	3 F	4 SA	5 SU	6 M	7 TU	8 W	9 TH	10 F	11 SA	12 SU	13 M	14 TU	15 W
X	Y	Z	1	2	3	4	5	6	7	8	A	B	C	D
YW	ZX	1Y	2Z	31	42	53	64	75	86	A7	B8	CA	DB	EC
ZV	1W	2X	3Y	4Z	51	62	73	84	A5	B6	C7	D8	EA	FB
1U	2V	3W	4X	5Y	6Z	71	82	A3	B4	C5	D6	E7	F8	GA
2T	3U	4V	5W	6X	7Y	8Z	A1	B2	C3	D4	E5	F6	G7	H8
3S	4T	5U	6V	7W	8X	AY	BZ	C1	D2	E3	F4	G5	H6	J7
4R	5S	6T	7U	8V	AW	BX	CY	DZ	E1	F2	G3	H4	J5	K6
5Q	6R	7S	8T	AU	BV	CW	DX	EY	FZ	G1	H2	J3	K4	L5
6P	7Q	8R	AS	BT	CU	DV	EW	FX	GY	HZ	J1	K2	L3	M4
7O	8P	AQ	BR	CS	DT	EU	FV	GW	HX	JY	KZ	L1	M2	N3
8N	AO	BP	CQ	DR	ES	FT	GU	HV	JW	KX	LY	MZ	N1	O2
AM	BN	CO	DP	EQ	FR	GS	HT	JU	KV	LW	MX	NY	OZ	P1
BL	CM	DN	EO	FP	GQ	HR	JS	KT	LU	MV	NW	OX	PY	QZ
CK	DL	EM	FN	GO	HP	JQ	KR	LS	MT	NU	OV	PW	QX	RY
DJ	EK	FL	GM	HN	JO	KP	LQ	MR	NS	OT	PU	QV	RW	SX
EH	FJ	GK	HL	JM	KN	LO	MP	NQ	OR	PS	QT	RU	SV	TW
FG	GH	HJ	JK	KL	LM	MN	NO	OP	PQ	QR	RS	ST	TU	UV

NOVEMBER 1978 PHYSICAL

16 TH	17 F	18 SA	19 SU	20 M	21 TU	22 W	23 TH	24 F	25 SA	26 SU	27 M	28 TU	29 W	30 TH
VU	WV	XW	AX	BA	CB	DC	ED	FE	GF	HG	JH	KJ	LK	ML
WT	XU	AV	BW	CX	DA	EB	FC	GD	HE	JF	KG	LH	MJ	NK
XS	AT	BU	CV	DW	EX	FA	GB	HC	JD	KE	LF	MG	NH	OJ
AR	BS	CT	DU	EV	FW	GX	HA	JB	KC	LD	ME	NF	OG	PH
BQ	CR	DS	ET	FU	GV	HW	JX	KA	LB	MC	ND	OE	PF	QG
CP	DQ	ER	FS	GT	HU	JV	KW	LX	MA	NB	OC	PD	QE	RF
DO	EP	FQ	GR	HS	JT	KU	LV	MW	NX	OA	PB	QC	RD	SE
EN	FO	GP	HQ	JR	KS	LT	MU	NV	OW	PX	QA	RB	SC	TD
FM	GN	HO	JP	KQ	LR	MS	NT	OU	PV	QW	RX	SA	TB	UC
GL	HM	JN	KO	LP	MQ	NR	OS	PT	QU	RV	SW	TX	UA	VB
HK	JL	KM	LN	MO	NP	OQ	PR	QS	RT	SU	TV	UW	VX	WA
J	K	L	M	N	O	P	Q	R	S	T	U	V	W	X

NOVEMBER 1978 EMOTIONAL

16 TH	17 F	18 SA	19 SU	20 M	21 TU	22 W	23 TH	24 F	25 SA	26 SU	27 M	28 TU	29 W	30 TH	
E	F	G	H	J	K	L	M	N	O	P	Q	R	S	T	
FD	GE	HF	JG	KH	LJ	MK	NL	OM	PN	QO	RP	SQ	TR	US	
GC	HD	JE	KF	LG	MH	NJ	OK	PL	QM	RN	SO	TP	UQ	VR	
HB	JC	KD	LE	MF	NG	OH	PJ	QK	RL	SM	TN	UO	VP	WQ	
JA	KB	LC	MD	NE	OF	PG	QH	RJ	SK	TL	UM	VN	WO	XP	+
K3	LA	MB	NC	OD	PE	QF	RG	SH	TJ	UK	VL	WM	XN	YO	
L2	M3	NA	OB	PC	QD	RE	SF	TG	UH	VJ	WK	XL	YM	ZN	
M1	N2	O3	PA	QB	RC	SD	TE	UF	VG	WH	XJ	YK	ZL	1M	o
NZ	O1	P2	Q3	RA	SB	TC	UD	VE	WF	XG	YH	ZJ	1K	2L	
OY	PZ	Q1	R2	S3	TA	UB	VC	WD	XE	YF	ZG	1H	2J	3K	
PX	QY	RZ	S1	T2	U3	VA	WB	XC	YD	ZE	1F	2G	3H	AJ	−
QW	RX	SY	TZ	U1	V2	W3	XA	YB	ZC	1D	2E	3F	AG	BH	
RV	SW	TX	UY	VZ	W1	X2	Y3	ZA	1B	2C	3D	AE	BF	CG	
SU	TV	UW	VX	WY	XZ	Y1	Z2	13	2A	3B	AC	BD	CE	DF	
T	U	V	W	X	Y	Z	1	2	3	A	B	C	D	E	

NOVEMBER 1978 INTELLECTUAL

16 TH	17 F	18 SA	19 SU	20 M	21 TU	22 W	23 TH	24 F	25 SA	26 SU	27 M	28 TU	29 W	30 TH
E	F	G	H	J	K	L	M	N	O	P	Q	R	S	T
FD	GE	HF	JG	KH	LJ	MK	NL	OM	PN	QO	RP	SQ	TR	US
GC	HD	JE	KF	LG	MH	NJ	OK	PL	QM	RN	SO	TP	UQ	VR
HB	JC	KD	LE	MF	NG	OH	PJ	QK	RL	SM	TN	UO	VP	WQ
JA	KB	LC	MD	NE	OF	PG	QH	RJ	SK	TL	UM	VN	WO	XP
K8	LA	MB	NC	OD	PE	QF	RG	SH	TJ	UK	VL	WM	XN	YO
L7	M8	NA	OB	PC	QD	RE	SF	TG	UH	VJ	WK	XL	YM	ZN
M6	N7	O8	PA	QB	RC	SD	TE	UF	VG	WH	XJ	YK	ZL	1M
N5	O6	P7	Q8	RA	SB	TC	UD	VE	WF	XG	YH	ZJ	1K	2L
O4	P5	Q6	R7	S8	TA	UB	VC	WD	XE	YF	ZG	1H	2J	3K
P3	Q4	R5	S6	T7	U8	VA	WB	XC	YD	ZE	1F	2G	3H	4J
Q2	R3	S4	T5	U6	V7	W8	XA	YB	ZC	1D	2E	3F	4G	5H
R1	S2	T3	U4	V5	W6	X7	Y8	ZA	1B	2C	3D	4E	5F	6G
SZ	T1	U2	V3	W4	X5	Y6	Z7	18	2A	3B	4C	5D	6E	7F
TY	UZ	V1	W2	X3	Y4	Z5	16	27	38	4A	5B	6C	7D	8E
UX	VY	WZ	X1	Y2	Z3	14	25	36	47	58	6A	7B	8C	AD
VW	WX	XY	YZ	Z1	12	23	34	45	56	67	78	8A	AB	BC

DECEMBER 1978 — PHYSICAL

1 F	2 SA	3 SU	4 M	5 TU	6 W	7 TH	8 F	9 SA	10 SU	11 M	12 TU	13 W	14 TH	15 F
NM	ON	PO	QP	RQ	SR	TS	UT	VU	WV	XW	AX	BA	CB	DC
OL	PM	QN	RO	SP	TQ	UR	VS	WT	XU	AV	BW	CX	DA	EB
PK	QL	RM	SN	TO	UP	VQ	WR	XS	AT	BU	CV	DW	EX	FA
QJ	RK	SL	TM	UN	VO	WP	XQ	AR	BS	CT	DU	EV	FW	GX
RH	SJ	TK	UL	VM	WN	XO	AP	BQ	CR	DS	ET	FU	GV	HW
SG	TH	UJ	VK	WL	XM	AN	BO	CP	DQ	ER	FS	GT	HU	JV
TF	UG	VH	WJ	XK	AL	BM	CN	DO	EP	FQ	GR	HS	JT	KU
UE	VF	WG	XH	AJ	BK	CL	DM	EN	FO	GP	HQ	JR	KS	LT
VD	WE	XF	AG	BH	CJ	DK	EL	FM	GN	HO	JP	KQ	LR	MS
WC	XD	AE	BF	CG	DH	EJ	FK	GL	HM	JN	KO	LP	MQ	NR
XB	AC	BD	CE	DF	EG	FH	GJ	HK	JL	KM	LN	MO	NP	OQ
A	B	C	D	E	F	G	H	J	K	L	M	N	O	P

DECEMBER 1978 — EMOTIONAL

1 F	2 SA	3 SU	4 M	5 TU	6 W	7 TH	8 F	9 SA	10 SU	11 M	12 TU	13 W	14 TH	15 F	
U	V	W	X	Y	Z	1	2	3	A	B	C	D	E	F	
VT	WU	XV	YW	ZX	1Y	2Z	31	A2	B3	CA	DB	EC	FD	GE	
WS	XT	YU	ZV	1W	2X	3Y	AZ	B1	C2	D3	EA	FB	GC	HD	
XR	YS	ZT	1U	2V	3W	AX	BY	CZ	D1	E2	F3	GA	HB	JC	
YQ	ZR	1S	2T	3U	AV	BW	CX	DY	EZ	F1	G2	H3	JA	KB	+
ZP	1Q	2R	3S	AT	BU	CV	DW	EX	FY	GZ	H1	J2	K3	LA	
1O	2P	3Q	AR	BS	CT	DU	EV	FW	GX	HY	JZ	K1	L2	M3	
2N	3O	AP	BQ	CR	DS	ET	FU	GV	HW	JX	KY	LZ	M1	N2	o
3M	AN	BO	CP	DQ	ER	FS	GT	HU	JV	KW	LX	MY	NZ	O1	
AL	BM	CN	DO	EP	FQ	GR	HS	JT	KU	LV	MW	NX	OY	PZ	
BK	CL	DM	EN	FO	GP	HQ	JR	KS	LT	MU	NV	OW	PX	QY	−
CJ	DK	EL	FM	GN	HO	JP	KQ	LR	MS	NT	OU	PV	QW	RX	
DH	EJ	FK	GL	HM	JN	KO	LP	MQ	NR	OS	PT	QU	RV	SW	
EG	FH	GJ	HK	JL	KM	LN	MO	NP	OQ	PR	QS	RT	SU	TV	
F	G	H	J	K	L	M	N	O	P	Q	R	S	T	U	

DECEMBER 1978 — INTELLECTUAL

1 F	2 SA	3 SU	4 M	5 TU	6 W	7 TH	8 F	9 SA	10 SU	11 M	12 TU	13 W	14 TH	15 F
U	V	W	X	Y	Z	1	2	3	4	5	6	7	8	A
VT	WU	XV	YW	ZX	1Y	2Z	31	42	53	64	75	86	A7	B8
WS	XT	YU	ZV	1W	2X	3Y	4Z	51	62	73	84	A5	B6	C7
XR	YS	ZT	1U	2V	3W	4X	5Y	6Z	71	82	A3	B4	C5	D6
YQ	ZR	1S	2T	3U	4V	5W	6X	7Y	8Z	A1	B2	C3	D4	E5
ZP	1Q	2R	3S	4T	5U	6V	7W	8X	AY	BZ	C1	D2	E3	F4
1O	2P	3Q	4R	5S	6T	7U	8V	AW	BX	CY	DZ	E1	F2	G3
2N	3O	4P	5Q	6R	7S	8T	AU	BV	CW	DX	EY	FZ	G1	H2
3M	4N	5O	6P	7Q	8R	AS	BT	CU	DV	EW	FX	GY	HZ	J1
4L	5M	6N	7O	8P	AQ	BR	CS	DT	EU	FV	GW	HX	JY	KZ
5K	6L	7M	8N	AO	BP	CQ	DR	ES	FT	GU	HV	JW	KX	LY
6J	7K	8L	AM	BN	CO	DP	EQ	FR	GS	HT	JU	KV	LW	MX
7H	8J	AK	BL	CM	DN	EO	FP	GQ	HR	JS	KT	LU	MV	NW
8G	AH	BJ	CK	DL	EM	FN	GO	HP	JQ	KR	LS	MT	NU	OV
AF	BG	CH	DJ	EK	FL	GM	HN	JO	KP	LQ	MR	NS	OT	PU
BE	CF	DG	EH	FJ	GK	HL	JM	KN	LO	MP	NQ	OR	PS	QT
CD	DE	EF	FG	GH	HJ	JK	KL	LM	MN	NO	OP	PQ	QR	RS

DECEMBER 1978 — PHYSICAL

16 SA	17 SU	18 M	19 TU	20 W	21 TH	22 F	23 SA	24 SU	25 M	26 TU	27 W	28 TH	29 F	30 SA	31 SU
ED	FE	GF	HG	JH	KJ	LK	ML	NM	ON	PO	QP	RQ	SR	TS	UT
FC	GD	HE	JF	KG	LH	MJ	NK	OL	PM	QN	RO	SP	TQ	UR	VS
GB	HC	JD	KE	LF	MG	NH	OJ	PK	QL	RM	SN	TO	UP	VQ	WR
HA	JB	KC	LD	ME	NF	OG	PH	QJ	RK	SL	TM	UN	VO	WP	XQ
JX	KA	LB	MC	ND	OE	PF	QG	RH	SJ	TK	UL	VM	WN	XO	AP
KW	LX	MA	NB	OC	PD	QE	RF	SG	TH	UJ	VK	WL	XM	AN	BO
LV	MW	NX	OA	PB	QC	RD	SE	TF	UG	VH	WJ	XK	AL	BM	CN
MU	NV	OW	PX	QA	RB	SC	TD	UE	VF	WG	XH	AJ	BK	CL	DM
NT	OU	PV	QW	RX	SA	TB	UC	VD	WE	XF	AG	BH	CJ	DK	EL
OS	PT	QU	RV	SW	TX	UA	VB	WC	XD	AE	BF	CG	DH	EJ	FK
PR	QS	RT	SU	TV	UW	VX	WA	XB	AC	BD	CE	DF	EG	FH	GJ
Q	R	S	T	U	V	W	X	A	B	C	D	E	F	G	H

DECEMBER 1978 — EMOTIONAL

16 SA	17 SU	18 M	19 TU	20 W	21 TH	22 F	23 SA	24 SU	25 M	26 TU	27 W	28 TH	29 F	30 SA	31 SU	
G	H	J	K	L	M	N	O	P	Q	R	S	T	U	V	W	
HF	JG	KH	LJ	MK	NL	OM	PN	QO	RP	SQ	TR	US	VT	WU	XV	
JE	KF	LG	MH	NJ	OK	PL	QM	RN	SO	TP	UQ	VR	WS	XT	YU	
KD	LE	MF	NG	OH	PJ	QK	RL	SM	TN	UO	VP	WQ	XR	YS	ZT	
LC	MD	NE	OF	PG	QH	RJ	SK	TL	UM	VN	WO	XP	YQ	ZR	1S	+
MB	NC	OD	PE	QF	RG	SH	TJ	UK	VL	WM	XN	YO	ZP	1Q	2R	
NA	OB	PC	QD	RE	SF	TG	UH	VJ	WK	XL	YM	ZN	1O	2P	3Q	
O3	PA	QB	RC	SD	TE	UF	VG	WH	XJ	YK	ZL	1M	2N	3O	AP	o
P2	Q3	RA	SB	TC	UD	VE	WF	XG	YH	ZJ	1K	2L	3M	AN	BO	
Q1	R2	S3	TA	UB	VC	WD	XE	YF	ZG	1H	2J	3K	AL	BM	CN	−
RZ	S1	T2	U3	VA	WB	XC	YD	ZE	1F	2G	3H	AJ	BK	CL	DM	
SY	TZ	U1	V2	W3	XA	YB	ZC	1D	2E	3F	AG	BH	CJ	DK	EL	
TX	UY	VZ	W1	X2	Y3	ZA	1B	2C	3D	AE	BF	CG	DH	EJ	FK	
UW	VX	WY	XZ	Y1	Z2	13	2A	3B	AC	BD	CE	DF	EG	FH	GJ	
V	W	X	Y	Z	1	2	3	A	B	C	D	E	F	G	H	

DECEMBER 1978 — INTELLECTUAL

16 SA	17 SU	18 M	19 TU	20 W	21 TH	22 F	23 SA	24 SU	25 M	26 TU	27 W	28 TH	29 F	30 SA	31 SU
B	C	D	E	F	G	H	J	K	L	M	N	O	P	Q	R
CA	DB	EC	FD	GE	HF	JG	KH	LJ	MK	NL	OM	PN	QO	RP	SQ
D8	EA	FB	GC	HD	JE	KF	LG	MH	NJ	OK	PL	QM	RN	SO	TP
E7	F8	GA	HB	JC	KD	LE	MF	NG	OH	PJ	QK	RL	SM	TN	UO
F6	G7	H8	JA	KB	LC	MD	NE	OF	PG	QH	RJ	SK	TL	UM	VN
G5	H6	J7	K8	LA	MB	NC	OD	PE	QF	RG	SH	TJ	UK	VL	WM
H4	J5	K6	L7	M8	NA	OB	PC	QD	RE	SF	TG	UH	VJ	WK	XL
J3	K4	L5	M6	N7	O8	PA	QB	RC	SD	TE	UF	VG	WH	XJ	YK
K2	L3	M4	N5	O6	P7	Q8	RA	SB	TC	UD	VE	WF	XG	YH	ZJ
L1	M2	N3	O4	P5	Q6	R7	S8	TA	UB	VC	WD	XE	YF	ZG	1H
MZ	N1	O2	P3	Q4	R5	S6	T7	U8	VA	WB	XC	YD	ZE	1F	2G
NY	OZ	P1	Q2	R3	S4	T5	U6	V7	W8	XA	YB	ZC	1D	2E	3F
OX	PY	QZ	R1	S2	T3	U4	V5	W6	X7	Y8	ZA	1B	2C	3D	4E
PW	QX	RY	SZ	T1	U2	V3	W4	X5	Y6	Z7	18	2A	3B	4C	5D
QV	RW	SX	TY	UZ	V1	W2	X3	Y4	Z5	16	27	38	4A	5B	6C
RU	SV	TW	UX	VY	WZ	X1	Y2	Z3	14	25	36	47	58	6A	7B
ST	TU	UV	VW	WX	XY	YZ	Z1	12	23	34	45	56	67	78	8A

JANUARY 1979 **PHYSICAL**

1 M	2 TU	3 W	4 TH	5 F	6 SA	7 SU	8 M	9 TU	10 W	11 TH	12 F	13 SA	14 SU	15 M
VU	WV	XW	AX	BA	CB	DC	ED	FE	GF	HG	JH	KJ	LK	ML
WT	XU	AV	BW	CX	DA	EB	FC	GD	HE	JF	KG	LH	MJ	NK
XS	AT	BU	CV	DW	EX	FA	GB	HC	JD	KE	LF	MG	NH	OJ
AR	BS	CT	DU	EV	FW	GX	HA	JB	KC	LD	ME	NF	OG	PH
BQ	CR	DS	ET	FU	GV	HW	JX	KA	LB	MC	ND	OE	PF	QG
CP	DQ	ER	FS	GT	HU	JV	KW	LX	MA	NB	OC	PD	QE	RF
DO	EP	FQ	GR	HS	JT	KU	LV	MW	NX	OA	PB	QC	RD	SE
EN	FO	GP	HQ	JR	KS	LT	MU	NV	OW	PX	QA	RB	SC	TD
FM	GN	HO	JP	KQ	LR	MS	NT	OU	PV	QW	RX	SA	TB	UC
GL	HM	JN	KO	LP	MQ	NR	OS	PT	QU	RV	SW	TX	UA	VB
HK	JL	KM	LN	MO	NP	OQ	PR	QS	RT	SU	TV	UW	VX	WA
J	K	L	M	N	O	P	Q	R	S	T	U	V	W	X

JANUARY 1979 **EMOTIONAL**

1 M	2 TU	3 W	4 TH	5 F	6 SA	7 SU	8 M	9 TU	10 W	11 TH	12 F	13 SA	14 SU	15 M	
X	Y	Z	1	2	3	A	B	C	D	E	F	G	H	J	
YW	ZX	1Y	2Z	31	A2	B3	CA	DB	EC	FD	GE	HF	JG	KH	
ZV	1W	2X	3Y	AZ	B1	C2	D3	EA	FB	GC	HD	JE	KF	LG	
1U	2V	3W	AX	BY	CZ	D1	E2	F3	GA	HB	JC	KD	LE	MF	
2T	3U	AV	BW	CX	DY	EZ	F1	G2	H3	JA	KB	LC	MD	NE	+
3S	AT	BU	CV	DW	EX	FY	GZ	H1	J2	K3	LA	MB	NC	OD	
AR	BS	CT	DU	EV	FW	GX	HY	JZ	K1	L2	M3	NA	OB	PC	
BQ	CR	DS	ET	FU	GV	HW	JX	KY	LZ	M1	N2	O3	PA	QB	o
CP	DQ	ER	FS	GT	HU	JV	KW	LX	MY	NZ	O1	P2	Q3	RA	
DO	EP	FQ	GR	HS	JT	KU	LV	MW	NX	OY	PZ	Q1	R2	S3	
EN	FO	GP	HQ	JR	KS	LT	MU	NV	OW	PX	QY	RZ	S1	T2	–
FM	GN	HO	JP	KQ	LR	MS	NT	OU	PV	QW	RX	SY	TZ	U1	
GL	HM	JN	KO	LP	MQ	NR	OS	PT	QU	RV	SW	TX	UY	VZ	
HK	JL	KM	LN	MO	NP	OQ	PR	QS	RT	SU	TV	UW	VX	WY	
J	K	L	M	N	O	P	Q	R	S	T	U	V	W	X	

JANUARY 1979 **INTELLECTUAL**

1 M	2 TU	3 W	4 TH	5 F	6 SA	7 SU	8 M	9 TU	10 W	11 TH	12 F	13 SA	14 SU	15 M
S	T	U	V	W	X	Y	Z	1	2	3	4	5	6	7
TR	US	VT	WU	XV	YW	ZX	1Y	2Z	31	42	53	64	75	86
UQ	VR	WS	XT	YU	ZV	1W	2X	3Y	4Z	51	62	73	84	A5
VP	WQ	XR	YS	ZT	1U	2V	3W	4X	5Y	6Z	71	82	A3	B4
WO	XP	YQ	ZR	1S	2T	3U	4V	5W	6X	7Y	8Z	A1	B2	C3
XN	YO	ZP	1Q	2R	3S	4T	5U	6V	7W	8X	AY	BZ	C1	D2
YM	ZN	1O	2P	3Q	4R	5S	6T	7U	8V	AW	BX	CY	DZ	E1
ZL	1M	2N	3O	4P	5Q	6R	7S	8T	AU	BV	CW	DX	EY	FZ
1K	2L	3M	4N	5O	6P	7Q	8R	AS	BT	CU	DV	EW	FX	GY
2J	3K	4L	5M	6N	7O	8P	AQ	BR	CS	DT	EU	FV	GW	HX
3H	4J	5K	6L	7M	8N	AO	BP	CQ	DR	ES	FT	GU	HV	JW
4G	5H	6J	7K	8L	AM	BN	CO	DP	EQ	FR	GS	HT	JU	KV
5F	6G	7H	8J	AK	BL	CM	DN	EO	FP	GQ	HR	JS	KT	LU
6E	7F	8G	AH	BJ	CK	DL	EM	FN	GO	HP	JQ	KR	LS	MT
7D	8E	AF	BG	CH	DJ	EK	FL	GM	HN	JO	KP	LQ	MR	NS
8C	AD	BE	CF	DG	EH	FJ	GK	HL	JM	KN	LO	MP	NQ	OR
AB	BC	CD	DE	EF	FG	GH	HJ	JK	KL	LM	MN	NO	OP	PQ

JANUARY 1979 — PHYSICAL

16 TU	17 W	18 TH	19 F	20 SA	21 SU	22 M	23 TU	24 W	25 TH	26 F	27 SA	28 SU	29 M	30 TU	31 W
NM	ON	PO	QP	RQ	SR	TS	UT	VU	WV	XW	AX	BA	CB	DC	ED
OL	PM	QN	RO	SP	TQ	UR	VS	WT	XU	AV	BW	CX	DA	EB	FC
PK	QL	RM	SN	TO	UP	VQ	WR	XS	AT	BU	CV	DW	EX	FA	GB
QJ	RK	SL	TM	UN	VO	WP	XQ	AR	BS	CT	DU	EV	FW	GX	HA
RH	SJ	TK	UL	VM	WN	XO	AP	BQ	CR	DS	ET	FU	GV	HW	JX
SG	TH	UJ	VK	WL	XM	AN	BO	CP	DQ	ER	FS	GT	HU	JV	KW
TF	UG	VH	WJ	XK	AL	BM	CN	DO	EP	FQ	GR	HS	JT	KU	LV
UE	VF	WG	XH	AJ	BK	CL	DM	EN	FO	GP	HQ	JR	KS	LT	MU
VD	WE	XF	AG	BH	CJ	DK	EL	FM	GN	HO	JP	KQ	LR	MS	NT
WC	XD	AE	BF	CG	DH	EJ	FK	GL	HM	JN	KO	LP	MQ	NR	OS
XB	AC	BD	CE	DF	EG	FH	GJ	HK	JL	KM	LN	MO	NP	OQ	PR
A	B	C	D	E	F	G	H	J	K	L	M	N	O	P	Q

JANUARY 1979 — EMOTIONAL

16 TU	17 W	18 TH	19 F	20 SA	21 SU	22 M	23 TU	24 W	25 TH	26 F	27 SA	28 SU	29 M	30 TU	31 W	
K	L	M	N	O	P	Q	R	S	T	U	V	W	X	Y	Z	
LJ	MK	NL	OM	PN	QO	RP	SQ	TR	US	VT	WU	XV	YW	ZX	1Y	
MH	NJ	OK	PL	QM	RN	SO	TP	UQ	VR	WS	XT	YU	ZV	1W	2X	
NG	OH	PJ	QK	RL	SM	TN	UO	VP	WQ	XR	YS	ZT	1U	2V	3W	
OF	PG	QH	RJ	SK	TL	UM	VN	WO	XP	YQ	ZR	1S	2T	3U	AV	+
PE	QF	RG	SH	TJ	UK	VL	WM	XN	YO	ZP	1Q	2R	3S	AT	BU	
QD	RE	SF	TG	UH	VJ	WK	XL	YM	ZN	1O	2P	3Q	AR	BS	CT	
RC	SD	TE	UF	VG	WH	XJ	YK	ZL	1M	2N	3O	AP	BQ	CR	DS	o
SB	TC	UD	VE	WF	XG	YH	ZJ	1K	2L	3M	AN	BO	CP	DQ	ER	
TA	UB	VC	WD	XE	YF	ZG	1H	2J	3K	AL	BM	CN	DO	EP	FQ	
U3	VA	WB	XC	YD	ZE	1F	2G	3H	AJ	BK	CL	DM	EN	FO	GP	−
V2	W3	XA	YB	ZC	1D	2F	3F	AG	BH	CJ	DK	EL	FM	GN	HO	
W1	X2	Y3	ZA	1B	2C	3D	AE	BF	CG	DH	EJ	FK	GL	HM	JN	
XZ	Y1	Z2	13	2A	3B	AC	BD	CE	DF	EG	FH	GJ	HK	JL	KM	
Y	Z	1	2	3	A	B	C	D	E	F	G	H	J	K	L	

JANUARY 1979 — INTELLECTUAL

16 TU	17 W	18 TH	19 F	20 SA	21 SU	22 M	23 TU	24 W	25 TH	26 F	27 SA	28 SU	29 M	30 TU	31 W
8	A	B	C	D	E	F	G	H	J	K	L	M	N	O	P
A7	B8	CA	DB	EC	FD	GE	HF	JG	KH	LJ	MK	NL	OM	PN	QO
B6	C7	D8	EA	FB	GC	HD	JE	KF	LG	MH	NJ	OK	PL	QM	RN
C5	D6	E7	F8	GA	HB	JC	KD	LE	MF	NG	OH	PJ	QK	RL	SM
D4	E5	F6	G7	H8	JA	KB	LC	MD	NE	OF	PG	QH	RJ	SK	TL
E3	F4	G5	H6	J7	K8	LA	MB	NC	OD	PE	QF	RG	SH	TJ	UK
F2	G3	H4	J5	K6	L7	M8	NA	OB	PC	QD	RE	SF	TG	UH	VJ
G1	H2	J3	K4	L5	M6	N7	O8	PA	QB	RC	SD	TE	UF	VG	WH
HZ	J1	K2	L3	M4	N5	O6	P7	Q8	RA	SB	TC	UD	VE	WF	XG
JY	KZ	L1	M2	N3	O4	P5	Q6	R7	S8	TA	UB	VC	WD	XE	YF
KX	LY	MZ	N1	O2	P3	Q4	R5	S6	T7	U8	VA	WB	XC	YD	ZE
LW	MX	NY	OZ	P1	Q2	R3	S4	T5	U6	V7	W8	XA	YB	ZC	1D
MV	NW	OX	PY	QZ	R1	S2	T3	U4	V5	W6	X7	Y8	ZA	1B	2C
NU	OV	PW	QX	RY	SZ	T1	U2	V3	W4	X5	Y6	Z7	18	2A	3B
OT	PU	QV	RW	SX	TY	UZ	V1	W2	X3	Y4	Z5	16	27	38	4A
PS	QT	RU	SV	TW	UX	VY	WZ	X1	Y2	Z3	14	25	36	47	58
QR	RS	ST	TU	UV	VW	WX	XY	YZ	Z1	12	23	34	45	56	67

FEBRUARY 1979 — PHYSICAL

1 TH	2 F	3 SA	4 SU	5 M	6 TU	7 W	8 TH	9 F	10 SA	11 SU	12 M	13 TU	14 W	15 TH
FE	GF	HG	JH	KJ	LK	ML	NM	ON	PO	QP	RQ	SR	TS	UT
GD	HE	JF	KG	LH	MJ	NK	OL	PM	QN	RO	SP	TQ	UR	VS
HC	JD	KE	LF	MG	NH	OJ	PK	QL	RM	SN	TO	UP	VQ	WR
JB	KC	LD	ME	NF	OG	PH	QJ	RK	SL	TM	UN	VO	WP	XQ
KA	LB	MC	ND	OE	PF	QG	RH	SJ	TK	UL	VM	WN	XO	AP
LX	MA	NB	OC	PD	QE	RF	SG	TH	UJ	VK	WL	XM	AN	BO
MW	NX	OA	PB	QC	RD	SE	TF	UG	VH	WJ	XK	AL	BM	CN
NV	OW	PX	QA	RB	SC	TD	UE	VF	WG	XH	AJ	BK	CL	DM
OU	PV	QW	RX	SA	TB	UC	VD	WE	XF	AG	BH	CJ	DK	EL
PT	QU	RV	SW	TX	UA	VB	WC	XD	AE	BF	CG	DH	EJ	FK
QS	RT	SU	TV	UW	VX	WA	XB	AC	BD	CE	DF	EG	FH	GJ
R	S	T	U	V	W	X	A	B	C	D	E	F	G	H

FEBRUARY 1979 — EMOTIONAL

1 TH	2 F	3 SA	4 SU	5 M	6 TU	7 W	8 TH	9 F	10 SA	11 SU	12 M	13 TU	14 W	15 TH	
1	2	3	A	B	C	D	E	F	G	H	J	K	L	M	
2Z	31	A2	B3	CA	DB	EC	FD	GE	HF	JG	KH	LJ	MK	NL	
3Y	AZ	B1	C2	D3	EA	FB	GC	HD	JE	KF	LG	MH	NJ	OK	
AX	BY	CZ	D1	E2	F3	GA	HB	JC	KD	LE	MF	NG	OH	PJ	
BW	CX	DY	EZ	F1	G2	H3	JA	KB	LC	MD	NE	OF	PG	QH	+
CV	DW	EX	FY	GZ	H1	J2	K3	LA	MB	NC	OD	PE	QF	RG	
DU	EV	FW	GX	HY	JZ	K1	L2	M3	NA	OB	PC	QD	RE	SF	
ET	FU	GV	HW	JX	KY	LZ	M1	N2	O3	PA	QB	RC	SD	TE	o
FS	GT	HU	JV	KW	LX	MY	NZ	O1	P2	Q3	RA	SB	TC	UD	
GR	HS	JT	KU	LV	MW	NX	OY	PZ	Q1	R2	S3	TA	UB	VC	
HQ	JR	KS	LT	MU	NV	OW	PX	QY	RZ	S1	T2	U3	VA	WB	−
JP	KQ	LR	MS	NT	OU	PV	QW	RX	SY	TZ	U1	V2	W3	XA	
KO	LP	MQ	NR	OS	PT	QU	RV	SW	TX	UY	VZ	W1	X2	Y3	
LN	MO	NP	OQ	PR	QS	RT	SU	TV	UW	VX	WY	XZ	Y1	Z2	
M	N	O	P	Q	R	S	T	U	V	W	X	Y	Z	1	

FEBRUARY 1979 — INTELLECTUAL

1 TH	2 F	3 SA	4 SU	5 M	6 TU	7 W	8 TH	9 F	10 SA	11 SU	12 M	13 TU	14 W	15 TH
Q	R	S	T	U	V	W	X	Y	Z	1	2	3	4	5
RP	SQ	TR	US	VT	WU	XV	YW	ZX	1Y	2Z	31	42	53	64
SO	TP	UQ	VR	WS	XT	YU	ZV	1W	2X	3Y	4Z	51	62	73
TN	UO	VP	WQ	XR	YS	ZT	1U	2V	3W	4X	5Y	6Z	71	82
UM	VN	WO	XP	YQ	ZR	1S	2T	3U	4V	5W	6X	7Y	8Z	A1
VL	WM	XN	YO	ZP	1Q	2R	3S	4T	5U	6V	7W	8X	AY	BZ
WK	XL	YM	ZN	1O	2P	3Q	4R	5S	6T	7U	8V	AW	BX	CY
XJ	YK	ZL	1M	2N	3O	4P	5Q	6R	7S	8T	AU	BV	CW	DX
YH	ZJ	1K	2L	3M	4N	5O	6P	7Q	8R	AS	BT	CU	DV	EW
ZG	1H	2J	3K	4L	5M	6N	7O	8P	AQ	BR	CS	DT	EU	FV
1F	2G	3H	4J	5K	6L	7M	8N	AO	BP	CQ	DR	ES	FT	GU
2E	3F	4G	5H	6J	7K	8L	AM	BN	CO	DP	EQ	FR	GS	HT
3D	4E	5F	6G	7H	8J	AK	BL	CM	DN	EO	FP	GQ	HR	JS
4C	5D	6E	7F	8G	AH	BJ	CK	DL	EM	FN	GO	HP	JQ	KR
5B	6C	7D	8E	AF	BG	CH	DJ	EK	FL	GM	HN	JO	KP	LQ
6A	7B	8C	AD	BE	CF	DG	EH	FJ	GK	HL	JM	KN	LO	MP
78	8A	AB	BC	CD	DE	EF	FG	GH	HJ	JK	KL	LM	MN	NO

FEBRUARY 1979 — PHYSICAL

16 F	17 SA	18 SU	19 M	20 TU	21 W	22 TH	23 F	24 SA	25 SU	26 M	27 TU	28 W
VU	WV	XW	AX	BA	CB	DC	ED	FE	GF	HG	JH	KJ
WT	XU	AV	BW	CX	DA	EB	FC	GD	HE	JF	KG	LH
XS	AT	BU	CV	DW	EX	FA	GB	HC	JD	KE	LF	MG
AR	BS	CT	DU	EV	FW	GX	HA	JB	KC	LD	ME	NF
BQ	CR	DS	ET	FU	GV	HW	JX	KA	LB	MC	ND	OE
CP	DQ	ER	FS	GT	HU	JV	KW	LX	MA	NB	OC	PD
DO	EP	FQ	GR	HS	JT	KU	LV	MW	NX	OA	PB	QC
EN	FO	GP	HQ	JR	KS	LT	MU	NV	OW	PX	QA	RB
FM	GN	HO	JP	KQ	LR	MS	NT	OU	PV	QW	RX	SA
GL	HM	JN	KO	LP	MQ	NR	OS	PT	QU	RV	SW	TX
HK	JL	KM	LN	MO	NP	OQ	PR	QS	RT	SU	TV	UW
J	K	L	M	N	O	P	Q	R	S	T	U	V

FEBRUARY 1979 — EMOTIONAL

16 F	17 SA	18 SU	19 M	20 TU	21 W	22 TH	23 F	24 SA	25 SU	26 M	27 TU	28 W	
N	O	P	Q	R	S	T	U	V	W	X	Y	Z	
OM	PN	QO	RP	SQ	TR	US	VT	WU	XV	YW	ZX	1Y	
PL	QM	RN	SO	TP	UQ	VR	WS	XT	YU	ZV	1W	2X	
QK	RL	SM	TN	UO	VP	WQ	XR	YS	ZT	1U	2V	3W	
RJ	SK	TL	UM	VN	WO	XP	YQ	ZR	1S	2T	3U	AV	+
SH	TJ	UK	VL	WM	XN	YO	ZP	1Q	2R	3S	AT	BU	
TG	UH	VJ	WK	XL	YM	ZN	1O	2P	3Q	AR	BS	CT	
UF	VG	WH	XJ	YK	ZL	1M	2N	3O	AP	BQ	CR	DS	o
VE	WF	XG	YH	ZJ	1K	2L	3M	AN	BO	CP	DQ	ER	
WD	XE	YF	ZG	1H	2J	3K	AL	BM	CN	DO	EP	FQ	
XC	YD	ZE	1F	2G	3H	AJ	BK	CL	DM	EN	FO	GP	–
YB	ZC	1D	2E	3F	AG	BH	CJ	DK	EL	FM	GN	HO	
ZA	1B	2C	3D	AE	BF	CG	DH	EJ	FK	GL	HM	JN	
13	2A	3B	AC	BD	CE	DF	EG	FH	GJ	HK	JL	KM	
2	3	A	B	C	D	E	F	G	H	J	K	L	

FEBRUARY 1979 — INTELLECTUAL

16 F	17 SA	18 SU	19 M	20 TU	21 W	22 TH	23 F	24 SA	25 SU	26 M	27 TU	28 W
6	7	8	A	B	C	D	E	F	G	H	J	K
75	86	A7	B8	CA	DB	EC	FD	GE	HF	JG	KH	LJ
84	A5	B6	C7	D8	EA	FB	GC	HD	JE	KF	LG	MH
A3	B4	C5	D6	E7	F8	GA	HB	JC	KD	LE	MF	NG
B2	C3	D4	E5	F6	G7	H8	JA	KB	LC	MD	NE	OF
C1	D2	E3	F4	G5	H6	J7	K8	LA	MB	NC	OD	PE
DZ	E1	F2	G3	H4	J5	K6	L7	M8	NA	OB	PC	QD
EY	FZ	G1	H2	J3	K4	L5	M6	N7	O8	PA	QB	RC
FX	GY	HZ	J1	K2	L3	M4	N5	O6	P7	Q8	RA	SB
GW	HX	JY	KZ	L1	M2	N3	O4	P5	Q6	R7	S8	TA
HV	JW	KX	LY	MZ	N1	O2	P3	Q4	R5	S6	T7	U8
JU	KV	LW	MX	NY	OZ	P1	Q2	R3	S4	T5	U6	V7
KT	LU	MV	NW	OX	PY	QZ	R1	S2	T3	U4	V5	W6
LS	MT	NU	OV	PW	QX	RY	SZ	T1	U2	V3	W4	X5
MR	NS	OT	PU	QV	RW	SX	TY	UZ	V1	W2	X3	Y4
NQ	OR	PS	QT	RU	SV	TW	UX	VY	WZ	X1	Y2	Z3
OP	PQ	QR	RS	ST	TU	UV	VW	WX	XY	YZ	Z1	12

MARCH 1979 — PHYSICAL

1 TH	2 F	3 SA	4 SU	5 M	6 TU	7 W	8 TH	9 F	10 SA	11 SU	12 M	13 TU	14 W	15 TH
LK	ML	NM	ON	PO	QP	RQ	SR	TS	UT	VU	WV	XW	AX	BA
MJ	NK	OL	PM	QN	RO	SP	TQ	UR	VS	WT	XU	AV	BW	CX
NH	OJ	PK	QL	RM	SN	TO	UP	VQ	WR	XS	AT	BU	CV	DW
OG	PH	QJ	RK	SL	TM	UN	VO	WP	XQ	AR	BS	CT	DU	EV
PF	QG	RH	SJ	TK	UL	VM	WN	XO	AP	BQ	CR	DS	ET	FU
QE	RF	SG	TH	UJ	VK	WL	XM	AN	BO	CP	DQ	ER	FS	GT
RD	SE	TF	UG	VH	WJ	XK	AL	BM	CN	DO	EP	FQ	GR	HS
SC	TD	UE	VF	WG	XH	AJ	BK	CL	DM	EN	FO	GP	HQ	JR
TB	UC	VD	WE	XF	AG	BH	CJ	DK	EL	FM	GN	HO	JP	KQ
UA	VB	WC	XD	AE	BF	CG	DH	EJ	FK	GL	HM	JN	KO	LP
VX	WA	XB	AC	BD	CE	DF	EG	FH	GJ	HK	JL	KM	LN	MO
W	X	A	B	C	D	E	F	G	H	J	K	L	M	N

MARCH 1979 — EMOTIONAL

1 TH	2 F	3 SA	4 SU	5 M	6 TU	7 W	8 TH	9 F	10 SA	11 SU	12 M	13 TU	14 W	15 TH	
1	2	3	A	B	C	D	E	F	G	H	J	K	L	M	
2Z	31	A2	B3	CA	DB	EC	FD	GE	HF	JG	KH	LJ	MK	NL	
3Y	AZ	B1	C2	D3	EA	FB	GC	HD	JE	KF	LG	MH	NJ	OK	
AX	BY	CZ	D1	E2	F3	GA	HB	JC	KD	LE	MF	NG	OH	PJ	
BW	CX	DY	EZ	F1	G2	H3	JA	KB	LC	MD	NE	OF	PG	QH	+
CV	DW	EX	FY	GZ	H1	J2	K3	LA	MB	NC	OD	PE	QF	RG	
DU	EV	FW	GX	HY	JZ	K1	L2	M3	NA	OB	PC	QD	RE	SF	
ET	FU	GV	HW	JX	KY	LZ	M1	N2	O3	PA	QB	RC	SD	TE	0
FS	GT	HU	JV	KW	LX	MY	NZ	O1	P2	Q3	RA	SB	TC	UD	
GR	HS	JT	KU	LV	MW	NX	OY	PZ	Q1	R2	S3	TA	UB	VC	
HQ	JR	KS	LT	MU	NV	OW	PX	QY	RZ	S1	T2	U3	VA	WB	–
JP	KQ	LR	MS	NT	OU	PV	QW	RX	SY	TZ	U1	V2	W3	XA	
KO	LP	MQ	NR	OS	PT	QU	RV	SW	TX	UY	VZ	W1	X2	Y3	
LN	MO	NP	OQ	PR	QS	RT	SU	TV	UW	VX	WY	XZ	Y1	Z2	
M	N	O	P	Q	R	S	T	U	V	W	X	Y	Z	1	

MARCH 1979 — INTELLECTUAL

1 TH	2 F	3 SA	4 SU	5 M	6 TU	7 W	8 TH	9 F	10 SA	11 SU	12 M	13 TU	14 W	15 TH
L	M	N	O	P	Q	R	S	T	U	V	W	X	Y	Z
MK	NL	OM	PN	QO	RP	SQ	TR	US	VT	WU	XV	YW	ZX	1Y
NJ	OK	PL	QM	RN	SO	TP	UQ	VR	WS	XT	YU	ZV	1W	2X
OH	PJ	QK	RL	SM	TN	UO	VP	WQ	XR	YS	ZT	1U	2V	3W
PG	QH	RJ	SK	TL	UM	VN	WO	XP	YQ	ZR	1S	2T	3U	4V
QF	RG	SH	TJ	UK	VL	WM	XN	YO	ZP	1Q	2R	3S	4T	5U
RE	SF	TG	UH	VJ	WK	XL	YM	ZN	1O	2P	3Q	4R	5S	6T
SD	TE	UF	VG	WH	XJ	YK	ZL	1M	2N	3O	4P	5Q	6R	7S
TC	UD	VE	WF	XG	YH	ZJ	1K	2L	3M	4N	5O	6P	7Q	8R
UB	VC	WD	XE	YF	ZG	1H	2J	3K	4L	5M	6N	7O	8P	AQ
VA	WB	XC	YD	ZE	1F	2G	3H	4J	5K	6L	7M	8N	AO	BP
W8	XA	YB	ZC	1D	2E	3F	4G	5H	6J	7K	8L	AM	BN	CO
X7	Y8	ZA	1B	2C	3D	4E	5F	6G	7H	8J	AK	BL	CM	DN
Y6	Z7	18	2A	3B	4C	5D	6E	7F	8G	AH	BJ	CK	DL	EM
Z5	16	27	38	4A	5B	6C	7D	8E	AF	BG	CH	DJ	EK	FL
14	25	36	47	58	6A	7B	8C	AD	BE	CF	DG	EH	FJ	GK
23	34	45	56	67	78	8A	AB	BC	CD	DE	EF	FG	GH	HJ

MARCH 1979 — PHYSICAL

16 F	17 SA	18 SU	19 M	20 TU	21 W	22 TH	23 F	24 SA	25 SU	26 M	27 TU	28 W	29 TH	30 F	31 SA
CB	DC	ED	FE	GF	HG	JH	KJ	LK	ML	NM	ON	PO	QP	RQ	SR
DA	EB	FC	GD	HE	JF	KG	LH	MJ	NK	OL	PM	QN	RO	SP	TQ
EX	FA	GB	HC	JD	KE	LF	MG	NH	OJ	PK	QL	RM	SN	TO	UP
FW	GX	HA	JB	KC	LD	ME	NF	OG	PH	QJ	RK	SL	TM	UN	VO
GV	HW	JX	KA	LB	MC	ND	OE	PF	QG	RH	SJ	TK	UL	VM	WN
HU	JV	KW	LX	MA	NB	OC	PD	QE	RF	SG	TH	UJ	VK	WL	XM
JT	KU	LV	MW	NX	OA	PB	QC	RD	SE	TF	UG	VH	WJ	XK	AL
KS	LT	MU	NV	OW	PX	QA	RB	SC	TD	UE	VF	WG	XH	AJ	BK
LR	MS	NT	OU	PV	QW	RX	SA	TB	UC	VD	WE	XF	AG	BH	CJ
MQ	NR	OS	PT	QU	RV	SW	TX	UA	VB	WC	XD	AE	BF	CG	DH
NP	OQ	PR	QS	RT	SU	TV	UW	VX	WA	XB	AC	BD	CE	DF	EG
O	P	Q	R	S	T	U	V	W	X	A	B	C	D	E	F

MARCH 1979 — EMOTIONAL

16 F	17 SA	18 SU	19 M	20 TU	21 W	22 TH	23 F	24 SA	25 SU	26 M	27 TU	28 W	29 TH	30 F	31 SA	
N	O	P	Q	R	S	T	U	V	W	X	Y	Z	1	2	3	
OM	PN	QO	RP	SQ	TR	US	VT	WU	XV	YW	ZX	1Y	2Z	31	A2	
PL	QM	RN	SO	TP	UQ	VR	WS	XT	YU	ZV	1W	2X	3Y	AZ	B1	
QK	RL	SM	TN	UO	VP	WQ	XR	YS	ZT	1U	2V	3W	AX	BY	CZ	
RJ	SK	TL	UM	VN	WO	XP	YQ	ZR	1S	2T	3U	AV	BW	CX	DY	+
SH	TJ	UK	VL	WM	XN	YO	ZP	1Q	2R	3S	AT	BU	CV	DW	EX	
TG	UH	VJ	WK	XL	YM	ZN	1O	2P	3Q	AR	BS	CT	DU	EV	FW	
UF	VG	WH	XJ	YK	ZL	1M	2N	3O	AP	BQ	CR	DS	ET	FU	GV	o
VE	WF	XG	YH	ZJ	1K	2L	3M	AN	BO	CP	DQ	ER	FS	GT	HU	
WD	XE	YF	ZG	1H	2J	3K	AL	BM	CN	DO	EP	FQ	GR	HS	JT	
XC	YD	ZE	1F	2G	3H	AJ	BK	CL	DM	EN	FO	GP	HQ	JR	KS	−
YB	ZC	1D	2E	3F	AG	BH	CJ	DK	EL	FM	GN	HO	JP	KQ	LR	
ZA	1D	2C	3D	AE	BF	CG	DH	EJ	FK	GL	HM	JN	KO	LP	MQ	
13	2A	3B	AC	BD	CE	DF	EG	FH	GJ	HK	JL	KM	LN	MO	NP	
2	3	A	B	C	D	E	F	G	H	J	K	L	M	N	O	

MARCH 1979 — INTELLECTUAL

16 F	17 SA	18 SU	19 M	20 TU	21 W	22 TH	23 F	24 SA	25 SU	26 M	27 TU	28 W	29 TH	30 F	31 SA
1	2	3	4	5	6	7	8	A	B	C	D	E	F	G	H
2Z	31	42	53	64	75	86	A7	B8	CA	DB	EC	FD	GE	HF	JG
3Y	4Z	51	62	73	84	A5	B6	C7	D8	EA	FB	GC	HD	JE	KF
4X	5Y	6Z	71	82	A3	B4	C5	D6	E7	F8	GA	HB	JC	KD	LE
5W	6X	7Y	8Z	A1	B2	C3	D4	E5	F6	G7	H8	JA	KB	LC	MD
6V	7W	8X	AY	BZ	C1	D2	E3	F4	G5	H6	J7	K8	LA	MB	NC
7U	8V	AW	BX	CY	DZ	E1	F2	G3	H4	J5	K6	L7	M8	NA	OB
8T	AU	BV	CW	DX	EY	FZ	G1	H2	J3	K4	L5	M6	N7	O8	PA
AS	BT	CU	DV	EW	FX	GY	HZ	J1	K2	L3	M4	N5	O6	P7	Q8
BR	CS	DT	EU	FV	GW	HX	JY	KZ	L1	M2	N3	O4	P5	Q6	R7
CQ	DR	ES	FT	GU	HV	JW	KX	LY	MZ	N1	O2	P3	Q4	R5	S6
DP	EQ	FR	GS	HT	JU	KV	LW	MX	NY	OZ	P1	Q2	R3	S4	T5
EO	FP	GQ	HR	JS	KT	LU	MV	NW	OX	PY	QZ	R1	S2	T3	U4
FN	GO	HP	JQ	KR	LS	MT	NU	OV	PW	QX	RY	SZ	T1	U2	V3
GM	HN	JO	KP	LQ	MR	NS	OT	PU	QV	RW	SX	TY	UZ	V1	W2
HL	JM	KN	LO	MP	NQ	OR	PS	QT	RU	SV	TW	UX	VY	WZ	X1
JK	KL	LM	MN	NO	OP	PQ	QR	RS	ST	TU	UV	VW	WX	XY	YZ

APRIL 1979 **PHYSICAL**

1 SU	2 M	3 TU	4 W	5 TH	6 F	7 SA	8 SU	9 M	10 TU	11 W	12 TH	13 F	14 SA	15 SU
TS	UT	VU	WV	XW	AX	BA	CB	DC	ED	FE	GF	HG	JH	KJ
UR	VS	WT	XU	AV	BW	CX	DA	EB	FC	GD	HE	JF	KG	LH
VQ	WR	XS	AT	BU	CV	DW	EX	FA	GB	HC	JD	KE	LF	MG
WP	XQ	AR	BS	CT	DU	EV	FW	GX	HA	JB	KC	LD	ME	NF
XO	AP	BQ	CR	DS	ET	FU	GV	HW	JX	KA	LB	MC	ND	OE
AN	BO	CP	DQ	ER	FS	GT	HU	JV	KW	LX	MA	NB	OC	PD
BM	CN	DO	EP	FQ	GR	HS	JT	KU	LV	MW	NX	OA	PB	QC
CL	DM	EN	FO	GP	HQ	JR	KS	LT	MU	NV	OW	PX	QA	RB
DK	EL	FM	GN	HO	JP	KQ	LR	MS	NT	OU	PV	QW	RX	SA
EJ	FK	GL	HM	JN	KO	LP	MQ	NR	OS	PT	QU	RV	SW	TX
FH	GJ	HK	JL	KM	LN	MO	NP	OQ	PR	QS	RT	SU	TV	UW
G	H	J	K	L	M	N	O	P	Q	R	S	T	U	V

APRIL 1979 **EMOTIONAL**

1 SU	2 M	3 TU	4 W	5 TH	6 F	7 SA	8 SU	9 M	10 TU	11 W	12 TH	13 F	14 SA	15 SU	
A	B	C	D	E	F	G	H	J	K	L	M	N	O	P	
B3	CA	DB	EC	FD	GE	HF	JG	KH	LJ	MK	NL	OM	PN	QO	
C2	D3	EA	FB	GC	HD	JE	KF	LG	MH	NJ	OK	PL	QM	RN	
D1	E2	F3	GA	HB	JC	KD	LE	MF	NG	OH	PJ	QK	RL	SM	
EZ	F1	G2	H3	JA	KB	LC	MD	NE	OF	PG	QH	RJ	SK	TL	+
FY	GZ	H1	J2	K3	LA	MB	NC	OD	PE	QF	RG	SH	TJ	UK	
GX	HY	JZ	K1	L2	M3	NA	OB	PC	QD	RE	SF	TG	UH	VJ	
HW	JX	KY	LZ	M1	N2	O3	PA	QB	RC	SD	TE	UF	VG	WH	o
JV	KW	LX	MY	NZ	O1	P2	Q3	RA	SB	TC	UD	VE	WF	XG	
KU	LV	MW	NX	OY	PZ	Q1	R2	S3	TA	UB	VC	WD	XE	YF	
LT	MU	NV	OW	PX	QY	RZ	S1	T2	U3	VA	WB	XC	YD	ZE	−
MS	NT	OU	PV	QW	RX	SY	TZ	U1	V2	W3	XA	YB	ZC	1D	
NR	OS	PT	QU	RV	SW	TX	UY	VZ	W1	X2	Y3	ZA	1B	2C	
OQ	PR	QS	RT	SU	TV	UW	VX	WY	XZ	Y1	Z2	13	2A	3B	
P	Q	R	S	T	U	V	W	X	Y	Z	1	2	3	A	

APRIL 1979 **INTELLECTUAL**

1 SU	2 M	3 TU	4 W	5 TH	6 F	7 SA	8 SU	9 M	10 TU	11 W	12 TH	13 F	14 SA	15 SU
J	K	L	M	N	O	P	Q	R	S	T	U	V	W	X
KH	LJ	MK	NL	OM	PN	QO	RP	SQ	TR	US	VT	WU	XV	YW
LG	MH	NJ	OK	PL	QM	RN	SO	TP	UQ	VR	WS	XT	YU	ZV
MF	NG	OH	PJ	QK	RL	SM	TN	UO	VP	WQ	XR	YS	ZT	1U
NE	OF	PG	QH	RJ	SK	TL	UM	VN	WO	XP	YQ	ZR	1S	2T
OD	PE	QF	RG	SH	TJ	UK	VL	WM	XN	YO	ZP	1Q	2R	3S
PC	QD	RE	SF	TG	UH	VJ	WK	XL	YM	ZN	1O	2P	3Q	4R
QB	RC	SD	TE	UF	VG	WH	XJ	YK	ZL	1M	2N	3O	4P	5Q
RA	SB	TC	UD	VE	WF	XG	YH	ZJ	1K	2L	3M	4N	5O	6P
S8	TA	UB	VC	WD	XE	YF	ZG	1H	2J	3K	4L	5M	6N	7O
T7	U8	VA	WB	XC	YD	ZE	1F	2G	3H	4J	5K	6L	7M	8N
U6	V7	W8	XA	YB	ZC	1D	2E	3F	4G	5H	6J	7K	8L	AM
V5	W6	X7	Y8	ZA	1B	2C	3D	4E	5F	6G	7H	8J	AK	BL
W4	X5	Y6	Z7	18	2A	3B	4C	5D	6E	7F	8G	AH	BJ	CK
X3	Y4	Z5	16	27	38	4A	5B	6C	7D	8E	AF	BG	CH	DJ
Y2	Z3	14	25	36	47	58	6A	7B	8C	AD	BE	CF	DG	EH
Z1	12	23	34	45	56	67	78	8A	AB	BC	CD	DE	EF	FG

APRIL 1979 — PHYSICAL

16 M.	17 TU	18 W	19 TH	20 F	21 SA	22 SU	23 M	24 TU	25 W	26 TH	27 F	28 SA	29 SU	30 M
LK	ML	NM	ON	PO	QP	RQ	SR	TS	UT	VU	WV	XW	AX	BA
MJ	NK	OL	PM	QN	RO	SP	TQ	UR	VS	WT	XU	AV	BW	CX
NH	OJ	PK	QL	RM	SN	TO	UP	VQ	WR	XS	AT	BU	CV	DW
OG	PH	QJ	RK	SL	TM	UN	VO	WP	XQ	AR	BS	CT	DU	EV
PF	QG	RH	SJ	TK	UL	VM	WN	XO	AP	BQ	CR	DS	ET	FU
QE	RF	SG	TH	UJ	VK	WL	XM	AN	BO	CP	DQ	ER	FS	GT
RD	SE	TF	UG	VH	WJ	XK	AL	BM	CN	DO	EP	FQ	GR	HS
SC	TD	UE	VF	WG	XH	AJ	BK	CL	DM	EN	FO	GP	HQ	JR
TB	UC	VD	WE	XF	AG	BH	CJ	DK	EL	FM	GN	HO	JP	KQ
UA	VB	WC	XD	AE	BF	CG	DH	EJ	FK	GL	HM	JN	KO	LP
VX	WA	XB	AC	BD	CE	DF	EG	FH	GJ	HK	JL	KM	LN	MO
W	X	A	B	C	D	E	F	G	H	J	K	L	M	N

APRIL 1979 — EMOTIONAL

16 M	17 TU	18 W	19 TH	20 F	21 SA	22 SU	23 M	24 TU	25 W	26 TH	27 F	28 SA	29 SU	30 M	
Q	R	S	T	U	V	W	X	Y	Z	1	2	3	A	B	
RP	SQ	TR	US	VT	WU	XV	YW	ZX	1Y	2Z	31	A2	B3	CA	
SO	TP	UQ	VR	WS	XT	YU	ZV	1W	2X	3Y	AZ	B1	C2	D3	
TN	UO	VP	WQ	XR	YS	ZT	1U	2V	3W	AX	BY	CZ	D1	E2	
UM	VN	WO	XP	YQ	ZR	1S	2T	3U	AV	BW	CX	DY	EZ	F1	+
VL	WM	XN	YO	ZP	1Q	2R	3S	AT	BU	CV	DW	EX	FY	GZ	
WK	XL	YM	ZN	1O	2P	3Q	AR	BS	CT	DU	EV	FW	GX	HY	
XJ	YK	ZL	1M	2N	3O	AP	BQ	CR	DS	ET	FU	GV	HW	JX	0
YH	ZJ	1K	2L	3M	AN	BO	CP	DQ	ER	FS	GT	HU	JV	KW	
ZG	1H	2J	3K	AL	BM	CN	DO	EP	FQ	GR	HS	JT	KU	LV	
1F	2G	3H	AJ	BK	CL	DM	EN	FO	GP	HQ	JR	KS	LT	MU	–
2E	3F	AG	BH	CJ	DK	EL	FM	GN	HO	JP	KQ	LR	MS	NT	
3D	AE	BF	CG	DH	EJ	FK	GL	HM	JN	KO	LP	MQ	NR	OS	
AC	BD	CE	DF	EG	FH	GJ	HK	JL	KM	LN	MO	NP	OQ	PR	
B	C	D	E	F	G	H	J	K	L	M	N	O	P	Q	

APRIL 1979 — INTELLECTUAL

16 M	17 TU	18 W	19 TH	20 F	21 SA	22 SU	23 M	24 TU	25 W	26 TH	27 F	28 SA	29 SU	30 M
Y	Z	1	2	3	4	5	6	7	8	A	B	C	D	E
ZX	1Y	2Z	31	42	53	64	75	86	A7	B8	CA	DB	EC	FD
1W	2X	3Y	4Z	51	62	73	84	A5	B6	C7	D8	EA	FB	GC
2V	3W	4X	5Y	6Z	71	82	A3	B4	C5	D6	E7	F8	GA	HB
3U	4V	5W	6X	7Y	8Z	A1	B2	C3	D4	E5	F6	G7	H8	JA
4T	5U	6V	7W	8X	AY	BZ	C1	D2	E3	F4	G5	H6	J7	K8
5S	6T	7U	8V	AW	BX	CY	DZ	E1	F2	G3	H4	J5	K6	L7
6R	7S	8T	AU	BV	CW	DX	EY	FZ	G1	H2	J3	K4	L5	M6
7Q	8R	AS	BT	CU	DV	EW	FX	GY	HZ	J1	K2	L3	M4	N5
8P	AQ	BR	CS	DT	EU	FV	GW	HX	JY	KZ	L1	M2	N3	O4
AO	BP	CQ	DR	ES	FT	GU	HV	JW	KX	LY	MZ	N1	O2	P3
BN	CO	DP	EQ	FR	GS	HT	JU	KV	LW	MX	NY	OZ	P1	Q2
CM	DN	EO	FP	GQ	HR	JS	KT	LU	MV	NW	OX	PY	QZ	R1
DL	EM	FN	GO	HP	JQ	KR	LS	MT	NU	OV	PW	QX	RY	SZ
EK	FL	GM	HN	JO	KP	LQ	MR	NS	OT	PU	QV	RW	SX	TY
FJ	GK	HL	JM	KN	LO	MP	NQ	OR	PS	QT	RU	SV	TW	UX
GH	HJ	JK	KL	LM	MN	NO	OP	PQ	QR	RS	ST	TU	UV	VW

MAY 1979 — PHYSICAL

1 TU	2 W	3 TH	4 F	5 SA	6 SU	7 M	8 TU	9 W	10 TH	11 F	12 SA	13 SU	14 M	15 TU
CB	DC	ED	FE	GF	HG	JH	KJ	LK	ML	NM	ON	PO	QP	RQ
DA	EB	FC	GD	HE	JF	KG	LH	MJ	NK	OL	PM	QN	RO	SP
EX	FA	GB	HC	JD	KE	LF	MG	NH	OJ	PK	QL	RM	SN	TO
FW	GX	HA	JB	KC	LD	ME	NF	OG	PH	QJ	RK	SL	TM	UN
GV	HW	JX	KA	LB	MC	ND	OE	PF	QG	RH	SJ	TK	UL	VM
HU	JV	KW	LX	MA	NB	OC	PD	QE	RF	SG	TH	UJ	VK	WL
JT	KU	LV	MW	NX	OA	PB	QC	RD	SE	TF	UG	VH	WJ	XK
KS	LT	MU	NV	OW	PX	QA	RB	SC	TD	UE	VF	WG	XH	AJ
LR	MS	NT	OU	PV	QW	RX	SA	TB	UC	VD	WE	XF	AG	BH
MQ	NR	OS	PT	QU	RV	SW	TX	UA	VB	WC	XD	AE	BF	CG
NP	OQ	PR	QS	RT	SU	TV	UW	VX	WA	XB	AC	BD	CE	DF
O	P	Q	R	S	T	U	V	W	X	A	B	C	D	E

MAY 1979 — EMOTIONAL

1 TU	2 W	3 TH	4 F	5 SA	6 SU	7 M	8 TU	9 W	10 TH	11 F	12 SA	13 SU	14 M	15 TU	
C	D	E	F	G	H	J	K	L	M	N	O	P	Q	R	
DB	EC	FD	GE	HF	JG	KH	LJ	MK	NL	OM	PN	QO	RP	SQ	
EA	FB	GC	HD	JE	KF	LG	MH	NJ	OK	PL	QM	RN	SO	TP	
F3	GA	HB	JC	KD	LE	MF	NG	OH	PJ	QK	RL	SM	TN	UO	
G2	H3	JA	KB	LC	MD	NE	OF	PG	QH	RJ	SK	TL	UM	VN	+
H1	J2	K3	LA	MB	NC	OD	PE	QF	RG	SH	TJ	UK	VL	WM	
JZ	K1	L2	M3	NA	OB	PC	QD	RE	SF	TG	UH	VJ	WK	XL	
KY	LZ	M1	N2	O3	PA	QB	RC	SD	TE	UF	VG	WH	XJ	YK	o
LX	MY	NZ	O1	P2	Q3	RA	SB	TC	UD	VE	WF	XG	YH	ZJ	
MW	NX	OY	PZ	Q1	R2	S3	TA	UB	VC	WD	XE	YF	ZG	1H	
NV	OW	PX	QY	RZ	S1	T2	U3	VA	WB	XC	YD	ZE	1F	2G	−
OU	PV	QW	RX	SY	TZ	U1	V2	W3	XA	YB	ZC	1D	2E	3F	
PT	QU	RV	SW	TX	UY	VZ	W1	X2	Y3	ZA	1B	2C	3D	AE	
QS	RT	SU	TV	UW	VX	WY	XZ	Y1	Z2	13	2A	3B	AC	BD	
R	S	T	U	V	W	X	Y	Z	1	2	3	A	B	C	

MAY 1979 — INTELLECTUAL

1 TU	2 W	3 TH	4 F	5 SA	6 SU	7 M	8 TU	9 W	10 TH	11 F	12 SA	13 SU	14 M	15 TU
F	G	H	J	K	L	M	N	O	P	Q	R	S	T	U
GE	HF	JG	KH	LJ	MK	NL	OM	PN	QO	RP	SQ	TR	US	VT
HD	JE	KF	LG	MH	NJ	OK	PL	QM	RN	SO	TP	UQ	VR	WS
JC	KD	LE	MF	NG	OH	PJ	QK	RL	SM	TN	UO	VP	WQ	XR
KB	LC	MD	NE	OF	PG	QH	RJ	SK	TL	UM	VN	WO	XP	YQ
LA	MB	NC	OD	PE	QF	RG	SH	TJ	UK	VL	WM	XN	YO	ZP
M8	NA	OB	PC	QD	RE	SF	TG	UH	VJ	WK	XL	YM	ZN	1O
N7	O8	PA	QB	RC	SD	TE	UF	VG	WH	XJ	YK	ZL	1M	2N
O6	P7	Q8	RA	SB	TC	UD	VE	WF	XG	YH	ZJ	1K	2L	3M
P5	Q6	R7	S8	TA	UB	VC	WD	XE	YF	ZG	1H	2J	3K	4L
Q4	R5	S6	T7	U8	VA	WB	XC	YD	ZE	1F	2G	3H	4J	5K
R3	S4	T5	U6	V7	W8	XA	YB	ZC	1D	2E	3F	4G	5H	6J
S2	T3	U4	V5	W6	X7	Y8	ZA	1B	2C	3D	4E	5F	6G	7H
T1	U2	V3	W4	X5	Y6	Z7	18	2A	3B	4C	5D	6E	7F	8G
UZ	V1	W2	X3	Y4	Z5	16	27	38	4A	5B	6C	7D	8E	AF
VY	WZ	X1	Y2	Z3	14	25	36	47	58	6A	7B	8C	AD	BE
WX	XY	YZ	Z1	12	23	34	45	56	67	78	8A	AB	BC	CD

MAY 1979 — PHYSICAL

16 W	17 TH	18 F	19 SA	20 SU	21 M	22 TU	23 W	24 TH	25 F	26 SA	27 SU	28 M	29 TU	30 W	31 TH
SR	TS	UT	VU	WV	XW	AX	BA	CB	DC	ED	FE	GF	HG	JH	KJ
TQ	UR	VS	WT	XU	AV	BW	CX	DA	EB	FC	GD	HE	JF	KG	LH
UP	VQ	WR	XS	AT	BU	CV	DW	EX	FA	GB	HC	JD	KE	LF	MG
VO	WP	XQ	AR	BS	CT	DU	EV	FW	GX	HA	JB	KC	LD	ME	NF
WN	XO	AP	BQ	CR	DS	ET	FU	GV	HW	JX	KA	LB	MC	ND	OE
XM	AN	BO	CP	DQ	ER	FS	GT	HU	JV	KW	LX	MA	NB	OC	PD
AL	BM	CN	DO	EP	FQ	GR	HS	JT	KU	LV	MW	NX	OA	PB	QC
BK	CL	DM	EN	FO	GP	HQ	JR	KS	LT	MU	NV	OW	PX	QA	RB
CJ	DK	EL	FM	GN	HO	JP	KQ	LR	MS	NT	OU	PV	QW	RX	SA
DH	EJ	FK	GL	HM	JN	KO	LP	MQ	NR	OS	PT	QU	RV	SW	TX
EG	FH	GJ	HK	JL	KM	LN	MO	NP	OQ	PR	QS	RT	SU	TV	UW
F	G	H	J	K	L	M	N	O	P	Q	R	S	T	U	V

MAY 1979 — EMOTIONAL

16 W	17 TH	18 F	19 SA	20 SU	21 M	22 TU	23 W	24 TH	25 F	26 SA	27 SU	28 M	29 TU	30 W	31 TH	
S	T	U	V	W	X	Y	Z	1	2	3	A	B	C	D	E	
TR	US	VT	WU	XV	YW	ZX	1Y	2Z	31	A2	B3	CA	DB	EC	FD	
UQ	VR	WS	XT	YU	ZV	1W	2X	3Y	AZ	B1	C2	D3	EA	FB	GC	
VP	WQ	XR	YS	ZT	1U	2V	3W	AX	BY	CZ	D1	E2	F3	GA	HB	
WO	XP	YQ	ZR	1S	2T	3U	AV	BW	CX	DY	EZ	F1	G2	H3	JA	+
XN	YO	ZP	1Q	2R	3S	AT	BU	CV	DW	EX	FY	GZ	H1	J2	K3	
YM	ZN	1O	2P	3Q	AR	BS	CT	DU	EV	FW	GX	HY	JZ	K1	L2	
ZL	1M	2N	3O	AP	BQ	CR	DS	ET	FU	GV	HW	JX	KY	LZ	M1	o
1K	2L	3M	AN	BO	CP	DQ	ER	FS	GT	HU	JV	KW	LX	MY	NZ	
2J	3K	AL	BM	CN	DO	EP	FQ	GR	HS	JT	KU	LV	MW	NX	OY	
3H	AJ	BK	CL	DM	EN	FO	GP	HQ	JR	KS	LT	MU	NV	OW	PX	–
AG	BH	CJ	DK	EL	FM	GN	HO	JP	KQ	LR	MS	NT	OU	PV	QW	
BF	CG	DH	EJ	FK	GL	HM	JN	KO	LP	MQ	NR	OS	PT	QU	RV	
CE	DF	EG	FH	GJ	HK	JL	KM	LN	MO	NP	OQ	PR	QS	RT	SU	
D	E	F	G	H	J	K	L	M	N	O	P	Q	R	S	T	

MAY 1979 — INTELLECTUAL

16 W	17 TH	18 F	19 SA	20 SU	21 M	22 TU	23 W	24 TH	25 F	26 SA	27 SU	28 M	29 TU	30 W	31 TH
V	W	X	Y	Z	1	2	3	4	5	6	7	8	A	B	C
WU	XV	YW	ZX	1Y	2Z	31	42	53	64	75	86	A7	B8	CA	DB
XT	YU	ZV	1W	2X	3Y	4Z	51	62	73	84	A5	B6	C7	D8	EA
YS	ZT	1U	2V	3W	4X	5Y	6Z	71	82	A3	B4	C5	D6	E7	F8
ZR	1S	2T	3U	4V	5W	6X	7Y	8Z	A1	B2	C3	D4	E5	F6	G7
1Q	2R	3S	4T	5U	6V	7W	8X	AY	BZ	C1	D2	E3	F4	G5	H6
2P	3Q	4R	5S	6T	7U	8V	AW	BX	CY	DZ	E1	F2	G3	H4	J5
3O	4P	5Q	6R	7S	8T	AU	BV	CW	DX	EY	FZ	G1	H2	J3	K4
4N	5O	6P	7Q	8R	AS	BT	CU	DV	EW	FX	GY	HZ	J1	K2	L3
5M	6N	7O	8P	AQ	BR	CS	DT	EU	FV	GW	HX	JY	KZ	L1	M2
6L	7M	8N	AO	BP	CQ	DR	ES	FT	GU	HV	JW	KX	LY	MZ	N1
7K	8L	AM	BN	CO	DP	EQ	FR	GS	HT	JU	KV	LW	MX	NY	OZ
8J	AK	BL	CM	DN	EO	FP	GQ	HR	JS	KT	LU	MV	NW	OX	PY
AH	BJ	CK	DL	EM	FN	GO	HP	JQ	KR	LS	MT	NU	OV	PW	QX
BG	CH	DJ	EK	FL	GM	HN	JO	KP	LQ	MR	NS	OT	PU	QV	RW
CF	DG	EH	FJ	GK	HL	JM	KN	LO	MP	NQ	OR	PS	QT	RU	SV
DE	EF	FG	GH	HJ	JK	KL	LM	MN	NO	OP	PQ	QR	RS	ST	TU

JUNE 1979 PHYSICAL

1 F	2 SA	3 SU	4 M	5 TU	6 W	7 TH	8 F	9 SA	10 SU	11 M	12 TU	13 W	14 TH	15 F
LK	ML	NM	ON	PO	QP	RQ	SR	TS	UT	VU	WV	XW	AX	BA
MJ	NK	OL	PM	QN	RO	SP	TQ	UR	VS	WT	XU	AV	BW	CX
NH	OJ	PK	QL	RM	SN	TO	UP	VQ	WR	XS	AT	BU	CV	DW
OG	PH	QJ	RK	SL	TM	UN	VO	WP	XQ	AR	BS	CT	DU	EV
PF	QG	RH	SJ	TK	UL	VM	WN	XO	AP	BQ	CR	DS	ET	FU
QE	RF	SG	TH	UJ	VK	WL	XM	AN	BO	CP	DQ	ER	FS	GT
RD	SE	TF	UG	VH	WJ	XK	AL	BM	CN	DO	EP	FQ	GR	HS
SC	TD	UE	VF	WG	XH	AJ	BK	CL	DM	EN	FO	GP	HQ	JR
TB	UC	VD	WE	XF	AG	BH	CJ	DK	EL	FM	GN	HO	JP	KQ
UA	VB	WC	XD	AE	BF	CG	DH	EJ	FK	GL	HM	JN	KO	LP
VX	WA	XB	AC	BD	CE	DF	EG	FH	GJ	HK	JL	KM	LN	MO
W	X	A	B	C	D	E	F	G	H	J	K	L	M	N

JUNE 1979 EMOTIONAL

1 F	2 SA	3 SU	4 M	5 TU	6 W	7 TH	8 F	9 SA	10 SU	11 M	12 TU	13 W	14 TH	15 F	
F	G	H	J	K	L	M	N	O	P	Q	R	S	T	U	
GE	HF	JG	KH	LJ	MK	NL	OM	PN	QO	RP	SQ	TR	US	VT	
HD	JE	KF	LG	MH	NJ	OK	PL	QM	RN	SO	TP	UQ	VR	WS	
JC	KD	LE	MF	NG	OH	PJ	QK	RL	SM	TN	UO	VP	WQ	XR	
KB	LC	MD	NE	OF	PG	QH	RJ	SK	TL	UM	VN	WO	XP	YQ	+
LA	MB	NC	OD	PE	QF	RG	SH	TJ	UK	VL	WM	XN	YO	ZP	
M3	NA	OB	PC	QD	RE	SF	TG	UH	VJ	WK	XL	YM	ZN	1O	
N2	O3	PA	QB	RC	SD	TE	UF	VG	WH	XJ	YK	ZL	1M	2N	0
O1	P2	Q3	RA	SB	TC	UD	VE	WF	XG	YH	ZJ	1K	2L	3M	
PZ	Q1	R2	S3	TA	UB	VC	WD	XE	YF	ZG	1H	2J	3K	AL	
QY	RZ	S1	T2	U3	VA	WB	XC	YD	ZE	1F	2G	3H	AJ	BK	−
RX	SY	TZ	U1	V2	W3	XA	YB	ZC	1D	2E	3F	AG	BH	CJ	
SW	TX	UY	VZ	W1	X2	Y3	ZA	1B	2C	3D	AE	BF	CG	DH	
TV	UW	VX	WY	XZ	Y1	Z2	13	2A	3B	AC	BD	CE	DF	EG	
U	V	W	X	Y	Z	1	2	3	A	B	C	D	E	F	

JUNE 1979 INTELLECTUAL

1 F	2 SA	3 SU	4 M	5 TU	6 W	7 TH	8 F	9 SA	10 SU	11 M	12 TU	13 W	14 TH	15 F
D	E	F	G	H	J	K	L	M	N	O	P	Q	R	S
EC	FD	GE	HF	JG	KH	LJ	MK	NL	OM	PN	QO	RP	SQ	TR
FB	GC	HD	JE	KF	LG	MH	NJ	OK	PL	QM	RN	SO	TP	UQ
GA	HB	JC	KD	LE	MF	NG	OH	PJ	QK	RL	SM	TN	UO	VP
H8	JA	KB	LC	MD	NE	OF	PG	QH	RJ	SK	TL	UM	VN	WO
J7	K8	LA	MB	NC	OD	PE	QF	RG	SH	TJ	UK	VL	WM	XN
K6	L7	M8	NA	OB	PC	QD	RE	SF	TG	UH	VJ	WK	XL	YM
L5	M6	N7	O8	PA	QB	RC	SD	TE	UF	VG	WH	XJ	YK	ZL
M4	N5	O6	P7	Q8	RA	SB	TC	UD	VE	WF	XG	YH	ZJ	1K
N3	O4	P5	Q6	R7	S8	TA	UB	VC	WD	XE	YF	ZG	1H	2J
O2	P3	Q4	R5	S6	T7	U8	VA	WB	XC	YD	ZE	1F	2G	3H
P1	Q2	R3	S4	T5	U6	V7	W8	XA	YB	ZC	1D	2E	3F	4G
QZ	R1	S2	T3	U4	V5	W6	X7	Y8	ZA	1B	2C	3D	4E	5F
RY	SZ	T1	U2	V3	W4	X5	Y6	Z7	18	2A	3B	4C	5D	6E
SX	TY	UZ	V1	W2	X3	Y4	Z5	16	27	38	4A	5B	6C	7D
TW	UX	VY	WZ	X1	Y2	Z3	14	25	36	47	58	6A	7B	8C
UV	VW	WX	XY	YZ	Z1	12	23	34	45	56	67	78	8A	AB

JUNE 1979 — PHYSICAL

16 SA	17 SU	18 M	19 TU	20 W	21 TH	22 F	23 SA	24 SU	25 M	26 TU	27 W	28 TH	29 F	30 SA
CB	DC	ED	FE	GF	HG	JH	KJ	LK	ML	NM	ON	PO	QP	RQ
DA	EB	FC	GD	HE	JF	KG	LH	MJ	NK	OL	PM	QN	RO	SP
EX	FA	GB	HC	JD	KE	LF	MG	NH	OJ	PK	QL	RM	SN	TO
FW	GX	HA	JB	KC	LD	ME	NF	OG	PH	QJ	RK	SL	TM	UN
GV	HW	JX	KA	LB	MC	ND	OE	PF	QG	RH	SJ	TK	UL	VM
HU	JV	KW	LX	MA	NB	OC	PD	QE	RF	SG	TH	UJ	VK	WL
JT	KU	LV	MW	NX	OA	PB	QC	RD	SE	TF	UG	VH	WJ	XK
KS	LT	MU	NV	OW	PX	QA	RB	SC	TD	UE	VF	WG	XH	AJ
LR	MS	NT	OU	PV	QW	RX	SA	TB	UC	VD	WE	XF	AG	BH
MQ	NR	OS	PT	QU	RV	SW	TX	UA	VB	WC	XD	AE	BF	CG
NP	OQ	PR	QS	RT	SU	TV	UW	VX	WA	XB	AC	BD	CE	DF
O	P	Q	R	S	T	U	V	W	X	A	B	C	D	E

JUNE 1979 — EMOTIONAL

16 SA	17 SU	18 M	19 TU	20 W	21 TH	22 F	23 SA	24 SU	25 M	26 TU	27 W	28 TH	29 F	30 SA	
V	W	X	Y	Z	1	2	3	A	B	C	D	E	F	G	
WU	XV	YW	ZX	1Y	2Z	31	A2	B3	CA	DB	EC	FD	GE	HF	
XT	YU	ZV	1W	2X	3Y	AZ	B1	C2	D3	EA	FB	GC	HD	JE	
YS	ZT	1U	2V	3W	AX	BY	CZ	D1	E2	F3	GA	HB	JC	KD	
ZR	1S	2T	3U	AV	BW	CX	DY	EZ	F1	G2	H3	JA	KB	LC	+
1Q	2R	3S	AT	BU	CV	DW	EX	FY	GZ	H1	J2	K3	LA	MB	
2P	3Q	AR	BS	CT	DU	EV	FW	GX	HY	JZ	K1	L2	M3	NA	
3O	AP	BQ	CR	DS	ET	FU	GV	HW	JX	KY	LZ	M1	N2	O3	o
AN	BO	CP	DQ	ER	FS	GT	HU	JV	KW	LX	MY	NZ	O1	P2	
BM	CN	DO	EP	FQ	GR	HS	JT	KU	LV	MW	NX	OY	PZ	Q1	
CL	DM	EN	FO	GP	HQ	JR	KS	LT	MU	NV	OW	PX	QY	RZ	−
DK	EL	FM	GN	HO	JP	KQ	LR	MS	NT	OU	PV	QW	RX	SY	
EJ	FK	GL	HM	JN	KO	LP	MQ	NR	OS	PT	QU	RV	SW	TX	
FH	GJ	HK	JL	KM	LN	MO	NP	OQ	PR	QS	RT	SU	TV	UW	
G	H	J	K	L	M	N	O	P	Q	R	S	T	U	V	

JUNE 1979 — INTELLECTUAL

16 SA	17 SU	18 M	19 TU	20 W	21 TH	22 F	23 SA	24 SU	25 M	26 TU	27 W	28 TH	29 F	30 SA
T	U	V	W	X	Y	Z	1	2	3	4	5	6	7	8
US	VT	WU	XV	YW	ZX	1Y	2Z	31	42	53	64	75	86	A7
VR	WS	XT	YU	ZV	1W	2X	3Y	4Z	51	62	73	84	A5	B6
WQ	XR	YS	ZT	1U	2V	3W	4X	5Y	6Z	71	82	A3	B4	C5
XP	YQ	ZR	1S	2T	3U	4V	5W	6X	7Y	8Z	A1	B2	C3	D4
YO	ZP	1Q	2R	3S	4T	5U	6V	7W	8X	AY	BZ	C1	D2	E3
ZN	1O	2P	3Q	4R	5S	6T	7U	8V	AW	BX	CY	DZ	E1	F2
1M	2N	3O	4P	5Q	6R	7S	8T	AU	BV	CW	DX	EY	FZ	G1
2L	3M	4N	5O	6P	7Q	8R	AS	BT	CU	DV	EW	FX	GY	HZ
3K	4L	5M	6N	7O	8P	AQ	BR	CS	DT	EU	FV	GW	HX	JY
4J	5K	6L	7M	8N	AO	BP	CQ	DR	ES	FT	GU	HV	JW	KX
5H	6J	7K	8L	AM	BN	CO	DP	EQ	FR	GS	HT	JU	KV	LW
6G	7H	8J	AK	BL	CM	DN	EO	FP	GQ	HR	JS	KT	LU	MV
7F	8G	AH	BJ	CK	DL	EM	FN	GO	HP	JQ	KR	LS	MT	NU
8E	AF	BG	CH	DJ	EK	FL	GM	HN	JO	KP	LQ	MR	NS	OT
AD	BE	CF	DG	EH	FJ	GK	HL	JM	KN	LO	MP	NQ	OR	PS
BC	CD	DE	EF	FG	GH	HJ	JK	KL	LM	MN	NO	OP	PQ	QR

JULY 1979 PHYSICAL

1 SU	2 M	3 TU	4 W	5 TH	6 F	7 SA	8 SU	9 M	10 TU	11 W	12 TH	13 F	14 SA	15 SU
SR	TS	UT	VU	WV	XW	AX	BA	CB	DC	ED	FE	GF	HG	JH
TQ	UR	VS	WT	XU	AV	BW	CX	DA	EB	FC	GD	HE	JF	KG
UP	VQ	WR	XS	AT	BU	CV	DW	EX	FA	GB	HC	JD	KE	LF
VO	WP	XQ	AR	BS	CT	DU	EV	FW	GX	HA	JB	KC	LD	ME
WN	XO	AP	BQ	CR	DS	ET	FU	GV	HW	JX	KA	LB	MC	ND
XM	AN	BO	CP	DQ	ER	FS	GT	HU	JV	KW	LX	MA	NB	OC
AL	BM	CN	DO	EP	FQ	GR	HS	JT	KU	LV	MW	NX	OA	PB
BK	CL	DM	EN	FO	GP	HQ	JR	KS	LT	MU	NV	OW	PX	QA
CJ	DK	EL	FM	GN	HO	JP	KQ	LR	MS	NT	OU	PV	QW	RX
DH	EJ	FK	GL	HM	JN	KO	LP	MQ	NR	OS	PT	QU	RV	SW
EG	FH	GJ	HK	JL	KM	LN	MO	NP	OQ	PR	QS	RT	SU	TV
F	G	H	J	K	L	M	N	O	P	Q	R	S	T	U

JULY 1979 EMOTIONAL

1 SU	2 M	3 TU	4 W	5 TH	6 F	7 SA	8 SU	9 M	10 TU	11 W	12 TH	13 F	14 SA	15 SU	
H	J	K	L	M	N	O	P	Q	R	S	T	U	V	W	
JG	KH	LJ	MK	NL	OM	PN	QO	RP	SQ	TR	US	VT	WU	XV	
KF	LG	MH	NJ	OK	PL	QM	RN	SO	TP	UQ	VR	WS	XT	YU	
LE	MF	NG	OH	PJ	QK	RL	SM	TN	UO	VP	WQ	XR	YS	ZT	
MD	NE	OF	PG	QH	RJ	SK	TL	UM	VN	WO	XP	YQ	ZR	1S	+
NC	OD	PE	QF	RG	SH	TJ	UK	VL	WM	XN	YO	ZP	1Q	2R	
OB	PC	QD	RE	SF	TG	UH	VJ	WK	XL	YM	ZN	1O	2P	3Q	
PA	QB	RC	SD	TE	UF	VG	WH	XJ	YK	ZL	1M	2N	3O	AP	0
Q3	RA	SB	TC	UD	VE	WF	XG	YH	ZJ	1K	2L	3M	AN	BO	
R2	S3	TA	UB	VC	WD	XE	YF	ZG	1H	2J	3K	AL	BM	CN	
S1	T2	U3	VA	WB	XC	YD	ZE	1F	2G	3H	AJ	BK	CL	DM	−
TZ	U1	V2	W3	XA	YB	ZC	1D	2E	3F	AG	BH	CJ	DK	EL	
UY	VZ	W1	X2	Y3	ZA	1B	2C	3D	AE	BF	CG	DH	EJ	FK	
VX	WY	XZ	Y1	Z2	13	2A	3B	AC	BD	CE	DF	EG	FH	GJ	
W	X	Y	Z	1	2	3	A	B	C	D	E	F	G	H	

JULY 1979 INTELLECTUAL

1 SU	2 M	3 TU	4 W	5 TH	6 F	7 SA	8 SU	9 M	10 TU	11 W	12 TH	13 F	14 SA	15 SU
A	B	C	D	E	F	G	H	J	K	L	M	N	O	P
B8	CA	DB	EC	FD	GE	HF	JG	KH	LJ	MK	NL	OM	PN	QO
C7	D8	EA	FB	GC	HD	JE	KF	LG	MH	NJ	OK	PL	QM	RN
D6	E7	F8	GA	HB	JC	KD	LE	MF	NG	OH	PJ	QK	RL	SM
E5	F6	G7	H8	JA	KB	LC	MD	NE	OF	PG	QH	RJ	SK	TL
F4	G5	H6	J7	K8	LA	MB	NC	OD	PE	QF	RG	SH	TJ	UK
G3	H4	J5	K6	L7	M8	NA	OB	PC	QD	RE	SF	TG	UH	VJ
H2	J3	K4	L5	M6	N7	O8	PA	QB	RC	SD	TE	UF	VG	WH
J1	K2	L3	M4	N5	O6	P7	Q8	RA	SB	TC	UD	VE	WF	XG
KZ	L1	M2	N3	O4	P5	Q6	R7	S8	TA	UB	VC	WD	XE	YF
LY	MZ	N1	O2	P3	Q4	R5	S6	T7	U8	VA	WB	XC	YD	ZE
MX	NY	OZ	P1	Q2	R3	S4	T5	U6	V7	W8	XA	YB	ZC	1D
NW	OX	PY	QZ	R1	S2	T3	U4	V5	W6	X7	Y8	ZA	1B	2C
OV	PW	QX	RY	SZ	T1	U2	V3	W4	X5	Y6	Z7	18	2A	3B
PU	QV	RW	SX	TY	UZ	V1	W2	X3	Y4	Z5	16	27	38	4A
QT	RU	SV	TW	UX	VY	WZ	X1	Y2	Z3	14	25	36	47	58
RS	ST	TU	UV	VW	WX	XY	YZ	Z1	12	23	34	45	56	67

JULY 1979 **PHYSICAL**

16 M	17 TU	18 W	19 TH	20 F	21 SA	22 SU	23 M	24 TU	25 W	26 TH	27 F	28 SA	29 SU	30 M	31 TU
KJ	LK	ML	NM	ON	PO	QP	RQ	SR	TS	UT	VU	WV	XW	AX	BA
LH	MJ	NK	OL	PM	QN	RO	SP	TQ	UR	VS	WT	XU	AV	BW	CX
MG	NH	OJ	PK	QL	RM	SN	TO	UP	VQ	WR	XS	AT	BU	CV	DW
NF	OG	PH	QJ	RK	SL	TM	UN	VO	WP	XQ	AR	BS	CT	DU	EV
OE	PF	QG	RH	SJ	TK	UL	VM	WN	XO	AP	BQ	CR	DS	ET	FU
PD	QE	RF	SG	TH	UJ	VK	WL	XM	AN	BO	CP	DQ	ER	FS	GT
QC	RD	SE	TF	UG	VH	WJ	XK	AL	BM	CN	DO	EP	FQ	GR	HS
RB	SC	TD	UE	VF	WG	XH	AJ	BK	CL	DM	EN	FO	GP	HQ	JR
SA	TB	UC	VD	WE	XF	AG	BH	CJ	DK	EL	FM	GN	HO	JP	KQ
TX	UA	VB	WC	XD	AE	BF	CG	DH	EJ	FK	GL	HM	JN	KO	LP
UW	VX	WA	XB	AC	BD	CE	DF	EG	FH	GJ	HK	JL	KM	LN	MO
V	W	X	A	B	C	D	E	F	G	H	J	K	L	M	N

JULY 1979 **EMOTIONAL**

16 M	17 TU	18 W	19 TH	20 F	21 SA	22 SU	23 M	24 TU	25 W	26 TH	27 F	28 SA	29 SU	30 M	31 TU	
X	Y	Z	1	2	3	A	B	C	D	E	F	G	H	J	K	
YW	ZX	1Y	2Z	31	A2	B3	CA	DB	EC	FD	GE	HF	JG	KH	LJ	
ZV	1W	2X	3Y	AZ	B1	C2	D3	EA	FB	GC	HD	JE	KF	LG	MH	
1U	2V	3W	AX	BY	CZ	D1	E2	F3	GA	HB	JC	KD	LE	MF	NG	
2T	3U	AV	BW	CX	DY	EZ	F1	G2	H3	JA	KB	LC	MD	NE	OF	+
3S	AT	BU	CV	DW	EX	FY	GZ	H1	J2	K3	LA	MB	NC	OD	PE	
AR	BS	CT	DU	EV	FW	GX	HY	JZ	K1	L2	M3	NA	OB	PC	QD	
BQ	CR	DS	ET	FU	GV	HW	JX	KY	LZ	M1	N2	O3	PA	QB	RC	o
CP	DQ	ER	FS	GT	HU	JV	KW	LX	MY	NZ	O1	P2	Q3	RA	SB	
DO	EP	FQ	GR	HS	JT	KU	LV	MW	NX	OY	PZ	Q1	R2	S3	TA	
EN	FO	GP	HQ	JR	KS	LT	MU	NV	OW	PX	QY	RZ	S1	T2	U3	−
FM	GN	HO	JP	KQ	LR	MS	NT	OU	PV	QW	RX	SY	TZ	U1	V2	
GL	HM	JN	KO	LP	MQ	NR	OS	PT	QU	RV	SW	TX	UY	VZ	W1	
HK	JL	KM	LN	MO	NP	OQ	PR	QS	RT	SU	TV	UW	VX	WY	XZ	
J	K	L	M	N	O	P	Q	R	S	T	U	V	W	X	Y	

JULY 1979 **INTELLECTUAL**

16 M	17 TU	18 W	19 TH	20 F	21 SA	22 SU	23 M	24 TU	25 W	26 TH	27 F	28 SA	29 SU	30 M	31 TU
Q	R	S	T	U	V	W	X	Y	Z	1	2	3	4	5	6
RP	SQ	TR	US	VT	WU	XV	YW	ZX	1Y	2Z	31	42	53	64	75
SO	TP	UQ	VR	WS	XT	YU	ZV	1W	2X	3Y	4Z	51	62	73	84
TN	UO	VP	WQ	XR	YS	ZT	1U	2V	3W	4X	5Y	6Z	71	82	A3
UM	VN	WO	XP	YQ	ZR	1S	2T	3U	4V	5W	6X	7Y	8Z	A1	B2
VL	WM	XN	YO	ZP	1Q	2R	3S	4T	5U	6V	7W	8X	AY	BZ	C1
WK	XL	YM	ZN	1O	2P	3Q	4R	5S	6T	7U	8V	AW	BX	CY	DZ
XJ	YK	ZL	1M	2N	3O	4P	5Q	6R	7S	8T	AU	BV	CW	DX	EY
YH	ZJ	1K	2L	3M	4N	5O	6P	7Q	8R	AS	BT	CU	DV	EW	FX
ZG	1H	2J	3K	4L	5M	6N	7O	8P	AQ	BR	CS	DT	EU	FV	GW
1F	2G	3H	4J	5K	6L	7M	8N	AO	BP	CQ	DR	ES	FT	GU	HV
2E	3F	4G	5H	6J	7K	8L	AM	BN	CO	DP	EQ	FR	GS	HT	JU
3D	4E	5F	6G	7H	8J	AK	BL	CM	DN	EO	FP	GQ	HR	JS	KT
4C	5D	6E	7F	8G	AH	BJ	CK	DL	EM	FN	GO	HP	JQ	KR	LS
5B	6C	7D	8E	AF	BG	CH	DJ	EK	FL	GM	HN	JO	KP	LQ	MR
6A	7B	8C	AD	BE	CF	DG	EH	FJ	GK	HL	JM	KN	LO	MP	NQ
78	8A	AB	BC	CD	DE	EF	FG	GH	HJ	JK	KL	LM	MN	NO	OP

AUGUST 1979 — PHYSICAL

1 W	2 TH	3 F	4 SA	5 SU	6 M	7 TU	8 W	9 TH	10 F	11 SA	12 SU	13 M	14 TU	15 W
CB	DC	ED	FE	GF	HG	JH	KJ	LK	ML	NM	ON	PO	QP	RQ
DA	EB	FC	GD	HE	JF	KG	LH	MJ	NK	OL	PM	QN	RO	SP
EX	FA	GB	HC	JD	KE	LF	MG	NH	OJ	PK	QL	RM	SN	TO
FW	GX	HA	JB	KC	LD	ME	NF	OG	PH	QJ	RK	SL	TM	UN
GV	HW	JX	KA	LB	MC	ND	OE	PF	QG	RH	SJ	TK	UL	VM
HU	JV	KW	LX	MA	NB	OC	PD	QE	RF	SG	TH	UJ	VK	WL
JT	KU	LV	MW	NX	OA	PB	QC	RD	SE	TF	UG	VH	WJ	XK
KS	LT	MU	NV	OW	PX	QA	RB	SC	TD	UE	VF	WG	XH	AJ
LR	MS	NT	OU	PV	QW	RX	SA	TB	UC	VD	WE	XF	AG	BH
MQ	NR	OS	PT	QU	RV	SW	TX	UA	VB	WC	XD	AE	BF	CG
NP	OQ	PR	QS	RT	SU	TV	UW	VX	WA	XB	AC	BD	CE	DF
O	P	Q	R	S	T	U	V	W	X	A	B	C	D	E

AUGUST 1979 — EMOTIONAL

1 W	2 TH	3 F	4 SA	5 SU	6 M	7 TU	8 W	9 TH	10 F	11 SA	12 SU	13 M	14 TU	15 W	
L	M	N	O	P	Q	R	S	T	U	V	W	X	Y	Z	
MK	NL	OM	PN	QO	RP	SQ	TR	US	VT	WU	XV	YW	ZX	1Y	
NJ	OK	PL	QM	RN	SO	TP	UQ	VR	WS	XT	YU	ZV	1W	2X	
OH	PJ	QK	RL	SM	TN	UO	VP	WQ	XR	YS	ZT	1U	2V	3W	
PG	QH	RJ	SK	TL	UM	VN	WO	XP	YQ	ZR	1S	2T	3U	AV	+
QF	RG	SH	TJ	UK	VL	WM	XN	YO	ZP	1Q	2R	3S	AT	BU	
RE	SF	TG	UH	VJ	WK	XL	YM	ZN	1O	2P	3Q	AR	BS	CT	
SD	TE	UF	VG	WH	XJ	YK	ZL	1M	2N	3O	AP	BQ	CR	DS	o
TC	UD	VE	WF	XG	YH	ZJ	1K	2L	3M	AN	BO	CP	DQ	ER	
UB	VC	WD	XE	YF	ZG	1H	2J	3K	AL	BM	CN	DO	EP	FQ	
VA	WB	XC	YD	ZE	1F	2G	3H	AJ	BK	CL	DM	EN	FO	GP	−
W3	XA	YB	ZC	1D	2E	3F	AG	BH	CJ	DK	EL	FM	GN	HO	
X2	Y3	ZA	1B	2C	3D	AE	BF	CG	DH	EJ	FK	GL	HM	JN	
Y1	Z2	13	2A	3B	AC	BD	CE	DF	EG	FH	GJ	HK	JL	KM	
Z	1	2	3	A	B	C	D	E	F	G	H	J	K	L	

AUGUST 1979 — INTELLECTUAL

1 W	2 TH	3 F	4 SA	5 SU	6 M	7 TU	8 W	9 TH	10 F	11 SA	12 SU	13 M	14 TU	15 W
7	8	A	B	C	D	E	F	G	H	J	K	L	M	N
86	A7	B8	CA	DB	EC	FD	GE	HF	JG	KH	LJ	MK	NL	OM
A5	B6	C7	D8	EA	FB	GC	HD	JE	KF	LG	MH	NJ	OK	PL
B4	C5	D6	E7	F8	GA	HB	JC	KD	LE	MF	NG	OH	PJ	QK
C3	D4	E5	F6	G7	H8	JA	KB	LC	MD	NE	OF	PG	QH	RJ
D2	E3	F4	G5	H6	J7	K8	LA	MB	NC	OD	PE	QF	RG	SH
E1	F2	G3	H4	J5	K6	L7	M8	NA	OB	PC	QD	RE	SF	TG
FZ	G1	H2	J3	K4	L5	M6	N7	O8	PA	QB	RC	SD	TE	UF
GY	HZ	J1	K2	L3	M4	N5	O6	P7	Q8	RA	SB	TC	UD	VE
HX	JY	KZ	L1	M2	N3	O4	P5	Q6	R7	S8	TA	UB	VC	WD
JW	KX	LY	MZ	N1	O2	P3	Q4	R5	S6	T7	U8	VA	WB	XC
KV	LW	MX	NY	OZ	P1	Q2	R3	S4	T5	U6	V7	W8	XA	YB
LU	MV	NW	OX	PY	QZ	R1	S2	T3	U4	V5	W6	X7	Y8	ZA
MT	NU	OV	PW	QX	RY	SZ	T1	U2	V3	W4	X5	Y6	Z7	18
NS	OT	PU	QV	RW	SX	TY	UZ	V1	W2	X3	Y4	Z5	16	27
OR	PS	QT	RU	SV	TW	UX	VY	WZ	X1	Y2	Z3	14	25	36
PQ	QR	RS	ST	TU	UV	VW	WX	XY	YZ	Z1	12	23	34	45

AUGUST 1979 — PHYSICAL

16 TH	17 F	18 SA	19 SU	20 M	21 TU	22 W	23 TH	24 F	25 SA	26 SU	27 M	28 TU	29 W	30 TH	31 F
SR	TS	UT	VU	WV	XW	AX	BA	CB	DC	ED	FE	GF	HG	JH	KJ
TQ	UR	VS	WT	XU	AV	BW	CX	DA	EB	FC	GD	HE	JF	KG	LH
UP	VQ	WR	XS	AT	BU	CV	DW	EX	FA	GB	HC	JD	KE	LF	MG
VO	WP	XQ	AR	BS	CT	DU	EV	FW	GX	HA	JB	KC	LD	ME	NF
WN	XO	AP	BQ	CR	DS	ET	FU	GV	HW	JX	KA	LB	MC	ND	OE
XM	AN	BO	CP	DQ	ER	FS	GT	HU	JV	KW	LX	MA	NB	OC	PD
AL	BM	CN	DO	EP	FQ	GR	HS	JT	KU	LV	MW	NX	OA	PB	QC
BK	CL	DM	EN	FO	GP	HQ	JR	KS	LT	MU	NV	OW	PX	QA	RB
CJ	DK	EL	FM	GN	HO	JP	KQ	LR	MS	NT	OU	PV	QW	RX	SA
DH	EJ	FK	GL	HM	JN	KO	LP	MQ	NR	OS	PT	QU	RV	SW	TX
EG	FH	GJ	HK	JL	KM	LN	MO	NP	OQ	PR	QS	RT	SU	TV	UW
F	G	H	J	K	L	M	N	O	P	Q	R	S	T	U	V

AUGUST 1979 — EMOTIONAL

16 TH	17 F	18 SA	19 SU	20 M	21 TU	22 W	23 TH	24 F	25 SA	26 SU	27 M	28 TU	29 W	30 TH	31 F	
1	2	3	A	B	C	D	E	F	G	H	J	K	L	M	N	
2Z	31	A2	B3	CA	DB	EC	FD	GE	HF	JG	KH	LJ	MK	NL	OM	
3Y	AZ	B1	C2	D3	EA	FB	GC	HD	JE	KF	LG	MH	NJ	OK	PL	
AX	BY	CZ	D1	E2	F3	GA	HB	JC	KD	LE	MF	NG	OH	PJ	QK	
BW	CX	DY	EZ	F1	G2	H3	JA	KB	LC	MD	NE	OF	PG	QH	RJ	+
CV	DW	EX	FY	GZ	H1	J2	K3	LA	MB	NC	OD	PE	QF	RG	SH	
DU	EV	FW	GX	HY	JZ	K1	L2	M3	NA	OB	PC	QD	RE	SF	TG	
ET	FU	GV	HW	JX	KY	LZ	M1	N2	O3	PA	QB	RC	SD	TE	UF	0
FS	GT	HU	JV	KW	LX	MY	NZ	O1	P2	Q3	RA	SB	TC	UD	VE	
GR	HS	JT	KU	LV	MW	NX	OY	PZ	Q1	R2	S3	TA	UB	VC	WD	
HQ	JR	KS	LT	MU	NV	OW	PX	QY	RZ	S1	T2	U3	VA	WB	XC	−
JP	KQ	LR	MS	NT	OU	PV	QW	RX	SY	TZ	U1	V2	W3	XA	YB	
KO	LP	MQ	NR	OS	PT	QU	RV	SW	TX	UY	VZ	W1	X2	Y3	ZA	
LN	MO	NP	OQ	PR	QS	RT	SU	TV	UW	VX	WY	XZ	Y1	Z2	13	
M	N	O	P	Q	R	S	T	U	V	W	X	Y	Z	1	2	

AUGUST 1979 — INTELLECTUAL

16 TH	17 F	18 SA	19 SU	20 M	21 TU	22 W	23 TH	24 F	25 SA	26 SU	27 M	28 TU	29 W	30 TH	31 F
O	P	Q	R	S	T	U	V	W	X	Y	Z	1	2	3	4
PN	QO	RP	SQ	TR	US	VT	WU	XV	YW	ZX	1Y	2Z	31	42	53
QM	RN	SO	TP	UQ	VR	WS	XT	YU	ZV	1W	2X	3Y	4Z	51	62
RL	SM	TN	UO	VP	WQ	XR	YS	ZT	1U	2V	3W	4X	5Y	6Z	71
SK	TL	UM	VN	WO	XP	YQ	ZR	1S	2T	3U	4V	5W	6X	7Y	8Z
TJ	UK	VL	WM	XN	YO	ZP	1Q	2R	3S	4T	5U	6V	7W	8X	AY
UH	VJ	WK	XL	YM	ZN	1O	2P	3Q	4R	5S	6T	7U	8V	AW	BX
VG	WH	XJ	YK	ZL	1M	2N	3O	4P	5Q	6R	7S	8T	AU	BV	CW
WF	XG	YH	ZJ	1K	2L	3M	4N	5O	6P	7Q	8R	AS	BT	CU	DV
XE	YF	ZG	1H	2J	3K	4L	5M	6N	7O	8P	AQ	BR	CS	DT	EU
YD	ZE	1F	2G	3H	4J	5K	6L	7M	8N	AO	BP	CQ	DR	ES	FT
ZC	1D	2E	3F	4G	5H	6J	7K	8L	AM	BN	CO	DP	EQ	FR	GS
1B	2C	3D	4E	5F	6G	7H	8J	AK	BL	CM	DN	EO	FP	GQ	HR
2A	3B	4C	5D	6E	7F	8G	AH	BJ	CK	DL	EM	FN	GO	HP	JQ
38	4A	5B	6C	7D	8E	AF	BG	CH	DJ	EK	FL	GM	HN	JO	KP
47	58	6A	7B	8C	AD	BE	CF	DG	EH	FJ	GK	HL	JM	KN	LO
56	67	78	8A	AB	BC	CD	DE	EF	FG	GH	HJ	JK	KL	LM	MN

SEPTEMBER 1979 — PHYSICAL

1 SA	2 SU	3 M	4 TU	5 W	6 TH	7 F	8 SA	9 SU	10 M	11 TU	12 W	13 TH	14 F	15 SA
LK	ML	NM	ON	PO	QP	RQ	SR	TS	UT	VU	WV	XW	AX	BA
MJ	NK	OL	PM	QN	RO	SP	TQ	UR	VS	WT	XU	AV	BW	CX
NH	OJ	PK	QL	RM	SN	TO	UP	VQ	WR	XS	AT	BU	CV	DW
OG	PH	QJ	RK	SL	TM	UN	VO	WP	XQ	AR	BS	CT	DU	EV
PF	QG	RH	SJ	TK	UL	VM	WN	XO	AP	BQ	CR	DS	ET	FU
QE	RF	SG	TH	UJ	VK	WL	XM	AN	BO	CP	DQ	ER	FS	GT
RD	SE	TF	UG	VH	WJ	XK	AL	BM	CN	DO	EP	FQ	GR	HS
SC	TD	UE	VF	WG	XH	AJ	BK	CL	DM	EN	FO	GP	HQ	JR
TB	UC	VD	WE	XF	AG	BH	CJ	DK	EL	FM	GN	HO	JP	KQ
UA	VB	WC	XD	AE	BF	CG	DH	EJ	FK	GL	HM	JN	KO	LP
VX	WA	XB	AC	BD	CE	DF	EG	FH	GJ	HK	JL	KM	LN	MO
W	X	A	B	C	D	E	F	G	H	J	K	L	M	N

SEPTEMBER 1979 — EMOTIONAL

1 SA	2 SU	3 M	4 TU	5 W	6 TH	7 F	8 SA	9 SU	10 M	11 TU	12 W	13 TH	14 F	15 SA	
O	P	Q	R	S	T	U	V	W	X	Y	Z	1	2	3	
PN	QO	RP	SQ	TR	US	VT	WU	XV	YW	ZX	1Y	2Z	31	A2	
QM	RN	SO	TP	UQ	VR	WS	XT	YU	ZV	1W	2X	3Y	AZ	B1	
RL	SM	TN	UO	VP	WQ	XR	YS	ZT	1U	2V	3W	AX	BY	CZ	
SK	TL	UM	VN	WO	XP	YQ	ZR	1S	2T	3U	AV	BW	CX	DY	+
TJ	UK	VL	WM	XN	YO	ZP	1Q	2R	3S	AT	BU	CV	DW	EX	
UH	VJ	WK	XL	YM	ZN	1O	2P	3Q	AR	BS	CT	DU	EV	FW	
VG	WH	XJ	YK	ZL	1M	2N	3O	AP	BQ	CR	DS	ET	FU	GV	o
WF	XG	YH	ZJ	1K	2L	3M	AN	BO	CP	DQ	ER	FS	GT	HU	
XE	YF	ZG	1H	2J	3K	AL	BM	CN	DO	EP	FQ	GR	HS	JT	
YD	ZE	1F	2G	3H	AJ	BK	CL	DM	EN	FO	GP	HQ	JR	KS	–
ZC	1D	2E	3F	AG	BH	CJ	DK	EL	FM	GN	HO	JP	KQ	LR	
1B	2C	3D	AE	BF	CG	DH	EJ	FK	GL	HM	JN	KO	LP	MQ	
2A	3B	AC	BD	CE	DF	EG	FH	GJ	HK	JL	KM	LN	MO	NP	
3	A	B	C	D	E	F	G	H	J	K	L	M	N	O	

SEPTEMBER 1979 — INTELLECTUAL

1 SA	2 SU	3 M	4 TU	5 W	6 TH	7 F	8 SA	9 SU	10 M	11 TU	12 W	13 TH	14 F	15 SA
5	6	7	8	A	B	C	D	E	F	G	H	J	K	L
64	75	86	A7	B8	CA	DB	EC	FD	GE	HF	JG	KH	LJ	MK
73	84	A5	B6	C7	D8	EA	FB	GC	HD	JE	KF	LG	MH	NJ
82	A3	B4	C5	D6	E7	F8	GA	HB	JC	KD	LE	MF	NG	OH
A1	B2	C3	D4	E5	F6	G7	H8	JA	KB	LC	MD	NE	OF	PG
BZ	C1	D2	E3	F4	G5	H6	J7	K8	LA	MB	NC	OD	PE	QF
CY	DZ	E1	F2	G3	H4	J5	K6	L7	M8	NA	OB	PC	QD	RE
DX	EY	FZ	G1	H2	J3	K4	L5	M6	N7	O8	PA	QB	RC	SD
EW	FX	GY	HZ	J1	K2	L3	M4	N5	O6	P7	Q8	RA	SB	TC
FV	GW	HX	JY	KZ	L1	M2	N3	O4	P5	Q6	R7	S8	TA	UB
GU	HV	JW	KX	LY	MZ	N1	O2	P3	Q4	R5	S6	T7	U8	VA
HT	JU	KV	LW	MX	NY	OZ	P1	Q2	R3	S4	T5	U6	V7	W8
JS	KT	LU	MV	NW	OX	PY	QZ	R1	S2	T3	U4	V5	W6	X7
KR	LS	MT	NU	OV	PW	QX	RY	SZ	T1	U2	V3	W4	X5	Y6
LQ	MR	NS	OT	PU	QV	RW	SX	TY	UZ	V1	W2	X3	Y4	Z5
MP	NQ	OR	PS	QT	RU	SV	TW	UX	VY	WZ	X1	Y2	Z3	14
NO	OP	PQ	QR	RS	ST	TU	UV	VW	WX	XY	YZ	Z1	12	23

SEPTEMBER 1979 — PHYSICAL

16 SU	17 M	18 TU	19 W	20 TH	21 F	22 SA	23 SU	24 M	25 TU	26 W	27 TH	28 F	29 SA	30 SU
CB	DC	ED	FE	GF	HG	JH	KJ	LK	ML	NM	ON	PO	QP	RQ
DA	EB	FC	GD	HE	JF	KG	LH	MJ	NK	OL	PM	QN	RO	SP
EX	FA	GB	HC	JD	KE	LF	MG	NH	OJ	PK	QL	RM	SN	TO
FW	GX	HA	JB	KC	LD	ME	NF	OG	PH	QJ	RK	SL	TM	UN
GV	HW	JX	KA	LB	MC	ND	OE	PF	QG	RH	SJ	TK	UL	VM
HU	JV	KW	LX	MA	NB	OC	PD	QE	RF	SG	TH	UJ	VK	WL
JT	KU	LV	MW	NX	OA	PB	QC	RD	SE	TF	UG	VH	WJ	XK
KS	LT	MU	NV	OW	PX	QA	RB	SC	TD	UE	VF	WG	XH	AJ
LR	MS	NT	OU	PV	QW	RX	SA	TB	UC	VD	WE	XF	AG	BH
MQ	NR	OS	PT	QU	RV	SW	TX	UA	VB	WC	XD	AE	BF	CG
NP	OQ	PR	QS	RT	SU	TV	UW	VX	WA	XB	AC	BD	CE	DF
O	P	Q	R	S	T	U	V	W	X	A	B	C	D	E

SEPTEMBER 1979 — EMOTIONAL

16 SU	17 M	18 TU	19 W	20 TH	21 F	22 SA	23 SU	24 M	25 TU	26 W	27 TH	28 F	29 SA	30 SU	
A	B	C	D	E	F	G	H	J	K	L	M	N	O	P	
B3	CA	DB	EC	FD	GE	HF	JG	KH	LJ	MK	NL	OM	PN	QO	
C2	D3	EA	FB	GC	HD	JE	KF	LG	MH	NJ	OK	PL	QM	RN	
D1	E2	F3	GA	HB	JC	KD	LE	MF	NG	OH	PJ	QK	RL	SM	
EZ	F1	G2	H3	JA	KB	LC	MD	NE	OF	PG	QH	RJ	SK	TL	+
FY	GZ	H1	J2	K3	LA	MB	NC	OD	PE	QF	RG	SH	TJ	UK	
GX	HY	JZ	K1	L2	M3	NA	OB	PC	QD	RE	SF	TG	UH	VJ	
HW	JX	KY	LZ	M1	N2	O3	PA	QB	RC	SD	TE	UF	VG	WH	o
JV	KW	LX	MY	NZ	O1	P2	Q3	RA	SB	TC	UD	VE	WF	XG	
KU	LV	MW	NX	OY	PZ	Q1	R2	S3	TA	UB	VC	WD	XE	YF	
LT	MU	NV	OW	PX	QY	RZ	S1	T2	U3	VA	WB	XC	YD	ZE	−
MS	NT	OU	PV	QW	RX	SY	TZ	U1	V2	W3	XA	YB	ZC	1D	
NR	OS	PT	QU	RV	SW	TX	UY	VZ	W1	X2	Y3	ZA	1B	2C	
OQ	PR	QS	RT	SU	TV	UW	VX	WY	XZ	Y1	Z2	13	2A	3B	
P	Q	R	S	T	U	V	W	X	Y	Z	1	2	3	A	

SEPTEMBER 1979 — INTELLECTUAL

16 SU	17 M	18 TU	19 W	20 TH	21 F	22 SA	23 SU	24 M	25 TU	26 W	27 TH	28 F	29 SA	30 SU
M	N	O	P	Q	R	S	T	U	V	W	X	Y	Z	1
NL	OM	PN	QO	RP	SQ	TR	US	VT	WU	XV	YW	ZX	1Y	2Z
OK	PL	QM	RN	SO	TP	UQ	VR	WS	XT	YU	ZV	1W	2X	3Y
PJ	QK	RL	SM	TN	UO	VP	WQ	XR	YS	ZT	1U	2V	3W	4X
QH	RJ	SK	TL	UM	VN	WO	XP	YQ	ZR	1S	2T	3U	4V	5W
RG	SH	TJ	UK	VL	WM	XN	YO	ZP	1Q	2R	3S	4T	5U	6V
SF	TG	UH	VJ	WK	XL	YM	ZN	1O	2P	3Q	4R	5S	6T	7U
TE	UF	VG	WH	XJ	YK	ZL	1M	2N	3O	4P	5Q	6R	7S	8T
UD	VE	WF	XG	YH	ZJ	1K	2L	3M	4N	5O	6P	7Q	8R	AS
VC	WD	XE	YF	ZG	1H	2J	3K	4L	5M	6N	7O	8P	AQ	BR
WB	XC	YD	ZE	1F	2G	3H	4J	5K	6L	7M	8N	AO	BP	CQ
XA	YB	ZC	1D	2E	3F	4G	5H	6J	7K	8L	AM	BN	CO	DP
Y8	ZA	1B	2C	3D	4E	5F	6G	7H	8J	AK	BL	CM	DN	EO
Z7	18	2A	3B	4C	5D	6E	7F	8G	AH	BJ	CK	DL	EM	FN
16	27	38	4A	5B	6C	7D	8E	AF	BG	CH	DJ	EK	FL	GM
25	36	47	58	6A	7B	8C	AD	BE	CF	DG	EH	FJ	GK	HL
34	45	56	67	78	8A	AB	BC	CD	DE	EF	FG	GH	HJ	JK

OCTOBER 1979 — PHYSICAL

1 M	2 TU	3 W	4 TH	5 F	6 SA	7 SU	8 M	9 TU	10 W	11 TH	12 F	13 SA	14 SU	15 M
SR	TS	UT	VU	WV	XW	AX	BA	CB	DC	ED	FE	GF	HG	JH
TQ	UR	VS	WT	XU	AV	BW	CX	DA	EB	FC	GD	HE	JF	KG
UP	VQ	WR	XS	AT	BU	CV	DW	EX	FA	GB	HC	JD	KE	LF
VO	WP	XQ	AR	BS	CT	DU	EV	FW	GX	HA	JB	KC	LD	ME
WN	XO	AP	BQ	CR	DS	ET	FU	GV	HW	JX	KA	LB	MC	ND
XM	AN	BO	CP	DQ	ER	FS	GT	HU	JV	KW	LX	MA	NB	OC
AL	BM	CN	DO	EP	FQ	GR	HS	JT	KU	LV	MW	NX	OA	PB
BK	CL	DM	EN	FO	GP	HQ	JR	KS	LT	MU	NV	OW	PX	QA
CJ	DK	EL	FM	GN	HO	JP	KQ	LR	MS	NT	OU	PV	QW	RX
DH	EJ	FK	GL	HM	JN	KO	LP	MQ	NR	OS	PT	QU	RV	SW
EG	FH	GJ	HK	JL	KM	LN	MO	NP	OQ	PR	QS	RT	SU	TV
F	G	H	J	K	L	M	N	O	P	Q	R	S	T	U

OCTOBER 1979 — EMOTIONAL

1 M	2 TU	3 W	4 TH	5 F	6 SA	7 SU	8 M	9 TU	10 W	11 TH	12 F	13 SA	14 SU	15 M	
Q	R	S	T	U	V	W	X	Y	Z	1	2	3	A	B	
RP	SQ	TR	US	VT	WU	XV	YW	ZX	1Y	2Z	31	A2	B3	CA	
SO	TP	UQ	VR	WS	XT	YU	ZV	1W	2X	3Y	AZ	B1	C2	D3	
TN	UO	VP	WQ	XR	YS	ZT	1U	2V	3W	AX	BY	CZ	D1	E2	
UM	VN	WO	XP	YQ	ZR	1S	2T	3U	AV	BW	CX	DY	EZ	F1	+
VL	WM	XN	YO	ZP	1Q	2R	3S	AT	BU	CV	DW	EX	FY	GZ	
WK	XL	YM	ZN	1O	2P	3Q	AR	BS	CT	DU	EV	FW	GX	HY	
XJ	YK	ZL	1M	2N	3O	AP	BQ	CR	DS	ET	FU	GV	HW	JX	o
YH	ZJ	1K	2L	3M	AN	BO	CP	DQ	ER	FS	GT	HU	JV	KW	
ZG	1H	2J	3K	AL	BM	CN	DO	EP	FQ	GR	HS	JT	KU	LV	
1F	2G	3H	AJ	BK	CL	DM	EN	FO	GP	HQ	JR	KS	LT	MU	–
2E	3F	AG	BH	CJ	DK	EL	FM	GN	HO	JP	KQ	LR	MS	NT	
3D	AE	BF	CG	DH	EJ	FK	GL	HM	JN	KO	LP	MQ	NR	OS	
AC	BD	CE	DF	EG	FH	GJ	HK	JL	KM	LN	MO	NP	OQ	PR	
B	C	D	E	F	G	H	J	K	L	M	N	O	P	Q	

OCTOBER 1979 — INTELLECTUAL

1 M	2 TU	3 W	4 TH	5 F	6 SA	7 SU	8 M	9 TU	10 W	11 TH	12 F	13 SA	14 SU	15 M
2	3	4	5	6	7	8	A	B	C	D	E	F	G	H
31	42	53	64	75	86	A7	B8	CA	DB	EC	FD	GE	HF	JG
4Z	51	62	73	84	A5	B6	C7	D8	EA	FB	GC	HD	JE	KF
5Y	6Z	71	82	A3	B4	C5	D6	E7	F8	GA	HB	JC	KD	LE
6X	7Y	8Z	A1	B2	C3	D4	E5	F6	G7	H8	JA	KB	LC	MD
7W	8X	AY	BZ	C1	D2	E3	F4	G5	H6	J7	K8	LA	MB	NC
8V	AW	BX	CY	DZ	E1	F2	G3	H4	J5	K6	L7	M8	NA	OB
AU	BV	CW	DX	EY	FZ	G1	H2	J3	K4	L5	M6	N7	O8	PA
BT	CU	DV	EW	FX	GY	HZ	J1	K2	L3	M4	N5	O6	P7	Q8
CS	DT	EU	FV	GW	HX	JY	KZ	L1	M2	N3	O4	P5	Q6	R7
DR	ES	FT	GU	HV	JW	KX	LY	MZ	N1	O2	P3	Q4	R5	S6
EQ	FR	GS	HT	JU	KV	LW	MX	NY	OZ	P1	Q2	R3	S4	T5
FP	GQ	HR	JS	KT	LU	MV	NW	OX	PY	QZ	R1	S2	T3	U4
GO	HP	JQ	KR	LS	MT	NU	OV	PW	QX	RY	SZ	T1	U2	V3
HN	JO	KP	LQ	MR	NS	OT	PU	QV	RW	SX	TY	UZ	V1	W2
JM	KN	LO	MP	NQ	OR	PS	QT	RU	SV	TW	UX	VY	WZ	X1
KL	LM	MN	NO	OP	PQ	QR	RS	ST	TU	UV	VW	WX	XY	YZ

OCTOBER 1979 **PHYSICAL**

16 TU	17 W	18 TH	19 F	20 SA	21 SU	22 M	23 TU	24 W	25 TH	26 F	27 SA	28 SU	29 M	30 TU	31 W
KJ	LK	ML	NM	ON	PO	QP	RQ	SR	TS	UT	VU	WV	XW	AX	BA
LH	MJ	NK	OL	PM	QN	RO	SP	TQ	UR	VS	WT	XU	AV	BW	CX
MG	NH	OJ	PK	QL	RM	SN	TO	UP	VQ	WR	XS	AT	BU	CV	DW
NF	OG	PH	QJ	RK	SL	TM	UN	VO	WP	XQ	AR	BS	CT	DU	EV
OE	PF	QG	RH	SJ	TK	UL	VM	WN	XO	AP	BQ	CR	DS	ET	FU
PD	QE	RF	SG	TH	UJ	VK	WL	XM	AN	BO	CP	DQ	ER	FS	GT
QC	RD	SE	TF	UG	VH	WJ	XK	AL	BM	CN	DO	EP	FQ	GR	HS
RB	SC	TD	UE	VF	WG	XH	AJ	BK	CL	DM	EN	FO	GP	HQ	JR
SA	TB	UC	VD	WE	XF	AG	BH	CJ	DK	EL	FM	GN	HO	JP	KQ
TX	UA	VB	WC	XD	AE	BF	CG	DH	EJ	FK	GL	HM	JN	KO	LP
UW	VX	WA	XB	AC	BD	CE	DF	EG	FH	GJ	HK	JL	KM	LN	MO
V	W	X	A	B	C	D	E	F	G	H	J	K	L	M	N

OCTOBER 1979 **EMOTIONAL**

16 TU	17 W	18 TH	19 F	20 SA	21 SU	22 M	23 TU	24 W	25 TH	26 F	27 SA	28 SU	29 M	30 TU	31 W	
C	D	E	F	G	H	J	K	L	M	N	O	P	Q	R	S	
DB	EC	FD	GE	HF	JG	KH	LJ	MK	NL	OM	PN	QO	RP	SQ	TR	
EA	FB	GC	HD	JE	KF	LG	MH	NJ	OK	PL	QM	RN	SO	TP	UQ	
F3	GA	HB	JC	KD	LE	MF	NG	OH	PJ	QK	RL	SM	TN	UO	VP	
G2	H3	JA	KB	LC	MD	NE	OF	PG	QH	RJ	SK	TL	UM	VN	WO	+
H1	J2	K3	LA	MB	NC	OD	PE	QF	RG	SH	TJ	UK	VL	WM	XN	
JZ	K1	L2	M3	NA	OB	PC	QD	RE	SF	TG	UH	VJ	WK	XL	YM	
KY	LZ	M1	N2	O3	PA	QB	RC	SD	TE	UF	VG	WH	XJ	YK	ZL	o
LX	MY	NZ	O1	P2	Q3	RA	SB	TC	UD	VE	WF	XG	YH	ZJ	1K	
MW	NX	OY	PZ	Q1	R2	S3	TA	UB	VC	WD	XE	YF	ZG	1H	2J	
NV	OW	PX	QY	RZ	S1	T2	U3	VA	WB	XC	YD	ZE	1F	2G	3H	–
OU	PV	QW	RX	SY	TZ	U1	V2	W3	XA	YB	ZC	1D	2E	3F	AG	
PT	QU	RV	SW	TX	UY	VZ	W1	X2	Y3	ZA	1B	2C	3D	AE	BF	
QS	RT	SU	TV	UW	VX	WY	XZ	Y1	Z2	13	2A	3B	AC	BD	CE	
R	S	T	U	V	W	X	Y	Z	1	2	3	A	B	C	D	

OCTOBER 1979 **INTELLECTUAL**

16 TU	17 W	18 TH	19 F	20 SA	21 SU	22 M	23 TU	24 W	25 TH	26 F	27 SA	28 SU	29 M	30 TU	31 W
J	K	L	M	N	O	P	Q	R	S	T	U	V	W	X	Y
KH	LJ	MK	NL	OM	PN	QO	RP	SQ	TR	US	VT	WU	XV	YW	ZX
LG	MH	NJ	OK	PL	QM	RN	SO	TP	UQ	VR	WS	XT	YU	ZV	1W
MF	NG	OH	PJ	QK	RL	SM	TN	UO	VP	WQ	XR	YS	ZT	1U	2V
NE	OF	PG	QH	RJ	SK	TL	UM	VN	WO	XP	YQ	ZR	1S	2T	3U
OD	PE	QF	RG	SH	TJ	UK	VL	WM	XN	YO	ZP	1Q	2R	3S	4T
PC	QD	RE	SF	TG	UH	VJ	WK	XL	YM	ZN	1O	2P	3Q	4R	5S
QB	RC	SD	TE	UF	VG	WH	XJ	YK	ZL	1M	2N	3O	4P	5Q	6R
RA	SB	TC	UD	VE	WF	XG	YH	ZJ	1K	2L	3M	4N	5O	6P	7Q
S8	TA	UB	VC	WD	XE	YF	ZG	1H	2J	3K	4L	5M	6N	7O	8P
T7	U8	VA	WB	XC	YD	ZE	1F	2G	3H	4J	5K	6L	7M	8N	AO
U6	V7	W8	XA	YB	ZC	1D	2E	3F	4G	5H	6J	7K	8L	AM	BN
V5	W6	X7	Y8	ZA	1B	2C	3D	4E	5F	6G	7H	8J	AK	BL	CM
W4	X5	Y6	Z7	18	2A	3B	4C	5D	6E	7F	8G	AH	BJ	CK	DL
X3	Y4	Z5	16	27	38	4A	5B	6C	7D	8E	AF	BG	CH	DJ	EK
Y2	Z3	14	25	36	47	58	6A	7B	8C	AD	BE	CF	DG	EH	FJ
Z1	12	23	34	45	56	67	78	8A	AB	BC	CD	DE	EF	FG	GH

NOVEMBER 1979 PHYSICAL

1 TH	2 F	3 SA	4 SU	5 M	6 TU	7 W	8 TH	9 F	10 SA	11 SU	12 M	13 TU	14 W	15 TH
CB	DC	ED	FE	GF	HG	JH	KJ	LK	ML	NM	ON	PO	QP	RQ
DA	EB	FC	GD	HE	JF	KG	LH	MJ	NK	OL	PM	QN	RO	SP
EX	FA	GB	HC	JD	KE	LF	MG	NH	OJ	PK	QL	RM	SN	TO
FW	GX	HA	JB	KC	LD	ME	NF	OG	PH	QJ	RK	SL	TM	UN
GV	HW	JX	KA	LB	MC	ND	OE	PF	QG	RH	SJ	TK	UL	VM
HU	JV	KW	LX	MA	NB	OC	PD	QE	RF	SG	TH	UJ	VK	WL
JT	KU	LV	MW	NX	OA	PB	QC	RD	SE	TF	UG	VH	WJ	XK
KS	LT	MU	NV	OW	PX	QA	RB	SC	TD	UE	VF	WG	XH	AJ
LR	MS	NT	OU	PV	QW	RX	SA	TB	UC	VD	WE	XF	AG	BH
MQ	NR	OS	PT	QU	RV	SW	TX	UA	VB	WC	XD	AE	BF	CG
NP	OQ	PR	QS	RT	SU	TV	UW	VX	WA	XB	AC	BD	CE	DF
O	P	Q	R	S	T	U	V	W	X	A	B	C	D	E

NOVEMBER 1979 EMOTIONAL

1 TH	2 F	3 SA	4 SU	5 M	6 TU	7 W	8 TH	9 F	10 SA	11 SU	12 M	13 TU	14 W	15 TH	
T	U	V	W	X	Y	Z	1	2	3	A	B	C	D	E	
US	VT	WU	XV	YW	ZX	1Y	2Z	31	A2	B3	CA	DB	EC	FD	
VR	WS	XT	YU	ZV	1W	2X	3Y	AZ	B1	C2	D3	EA	FB	GC	
WQ	XR	YS	ZT	1U	2V	3W	AX	BY	CZ	D1	E2	F3	GA	HB	
XP	YQ	ZR	1S	2T	3U	AV	BW	CX	DY	EZ	F1	G2	H3	JA	+
YO	ZP	1Q	2R	3S	AT	BU	CV	DW	EX	FY	GZ	H1	J2	K3	
ZN	1O	2P	3Q	AR	BS	CT	DU	EV	FW	GX	HY	JZ	K1	L2	
1M	2N	3O	AP	BQ	CR	DS	ET	FU	GV	HW	JX	KY	LZ	M1	o
2L	3M	AN	BO	CP	DQ	ER	FS	GT	HU	JV	KW	LX	MY	NZ	
3K	AL	BM	CN	DO	EP	FQ	GR	HS	JT	KU	LV	MW	NX	OY	
AJ	BK	CL	DM	EN	FO	GP	HQ	JR	KS	LT	MU	NV	OW	PX	−
BH	CJ	DK	EL	FM	GN	HO	JP	KQ	LR	MS	NT	OU	PV	QW	
CG	DH	EJ	FK	GL	HM	JN	KO	LP	MQ	NR	OS	PT	QU	RV	
DF	EG	FH	GJ	HK	JL	KM	LN	MO	NP	OQ	PR	QS	RT	SU	
E	F	G	H	J	K	L	M	N	O	P	Q	R	S	T	

NOVEMBER 1979 INTELLECTUAL

1 TH	2 F	3 SA	4 SU	5 M	6 TU	7 W	8 TH	9 F	10 SA	11 SU	12 M	13 TU	14 W	15 TH
Z	1	2	3	4	5	6	7	8	A	B	C	D	E	F
1Y	2Z	31	42	53	64	75	86	A7	B8	CA	DB	EC	FD	GE
2X	3Y	4Z	51	62	73	84	A5	B6	C7	D8	EA	FB	GC	HD
3W	4X	5Y	6Z	71	82	A3	B4	C5	D6	E7	F8	GA	HB	JC
4V	5W	6X	7Y	8Z	A1	B2	C3	D4	E5	F6	G7	H8	JA	KB
5U	6V	7W	8X	AY	BZ	C1	D2	E3	F4	G5	H6	J7	K8	LA
6T	7U	8V	AW	BX	CY	DZ	E1	F2	G3	H4	J5	K6	L7	M8
7S	8T	AU	BV	CW	DX	EY	FZ	G1	H2	J3	K4	L5	M6	N7
8R	AS	BT	CU	DV	EW	FX	GY	HZ	J1	K2	L3	M4	N5	O6
AQ	BR	CS	DT	EU	FV	GW	HX	JY	KZ	L1	M2	N3	O4	P5
BP	CQ	DR	ES	FT	GU	HV	JW	KX	LY	MZ	N1	O2	P3	Q4
CO	DP	EQ	FR	GS	HT	JU	KV	LW	MX	NY	OZ	P1	Q2	R3
DN	EO	FP	GQ	HR	JS	KT	LU	MV	NW	OX	PY	QZ	R1	S2
EM	FN	GO	HP	JQ	KR	LS	MT	NU	OV	PW	QX	RY	SZ	T1
FL	GM	HN	JO	KP	LQ	MR	NS	OT	PU	QV	RW	SX	TY	UZ
GK	HL	JM	KN	LO	MP	NQ	OR	PS	QT	RU	SV	TW	UX	VY
HJ	JK	KL	LM	MN	NO	OP	PQ	QR	RS	ST	TU	UV	VW	WX

NOVEMBER 1979 PHYSICAL

16 F	17 SA	18 SU	19 M	20 TU	21 W	22 TH	23 F	24 SA	25 SU	26 M	27 TU	28 W	29 TH	30 F
SR	TS	UT	VU	WV	XW	AX	BA	CB	DC	ED	FE	GF	HG	JH
TQ	UR	VS	WT	XU	AV	BW	CX	DA	EB	FC	GD	HE	JF	KG
UP	VQ	WR	XS	AT	BU	CV	DW	EX	FA	GB	HC	JD	KE	LF
VO	WP	XQ	AR	BS	CT	DU	EV	FW	GX	HA	JB	KC	LD	ME
WN	XO	AP	BQ	CR	DS	ET	FU	GV	HW	JX	KA	LB	MC	ND
XM	AN	BO	CP	DQ	ER	FS	GT	HU	JV	KW	LX	MA	NB	OC
AL	BM	CN	DO	EP	FQ	GR	HS	JT	KU	LV	MW	NX	OA	PB
BK	CL	DM	EN	FO	GP	HQ	JR	KS	LT	MU	NV	OW	PX	QA
CJ	DK	EL	FM	GN	HO	JP	KQ	LR	MS	NT	OU	PV	QW	RX
DH	EJ	FK	GL	HM	JN	KO	LP	MQ	NR	OS	PT	QU	RV	SW
EG	FH	GJ	HK	JL	KM	LN	MO	NP	OQ	PR	QS	RT	SU	TV
F	G	H	J	K	L	M	N	O	P	Q	R	S	T	U

NOVEMBER 1979 EMOTIONAL

16 F	17 SA	18 SU	19 M	20 TU	21 W	22 TH	23 F	24 SA	25 SU	26 M	27 TU	28 W	29 TH	30 F	
F	G	H	J	K	L	M	N	O	P	Q	R	S	T	U	
GE	HF	JG	KH	LJ	MK	NL	OM	PN	QO	RP	SQ	TR	US	VT	
HD	JE	KF	LG	MH	NJ	OK	PL	QM	RN	SO	TP	UQ	VR	WS	
JC	KD	LE	MF	NG	OH	PJ	QK	RL	SM	TN	UO	VP	WQ	XR	
KB	LC	MD	NE	OF	PG	QH	RJ	SK	TL	UM	VN	WO	XP	YQ	+
LA	MB	NC	OD	PE	QF	RG	SH	TJ	UK	VL	WM	XN	YO	ZP	
M3	NA	OB	PC	QD	RE	SF	TG	UH	VJ	WK	XL	YM	ZN	1O	
N2	O3	PA	QB	RC	SD	TE	UF	VG	WH	XJ	YK	ZL	1M	2N	o
O1	P2	Q3	RA	SB	TC	UD	VE	WF	XG	YH	ZJ	1K	2L	3M	
PZ	Q1	R2	S3	TA	UB	VC	WD	XE	YF	ZG	1H	2J	3K	AL	
QY	RZ	S1	T2	U3	VA	WB	XC	YD	ZE	1F	2G	3H	AJ	BK	–
RX	SY	TZ	U1	V2	W3	XA	YB	ZC	1D	2E	3F	AG	BH	CJ	
SW	TX	UY	VZ	W1	X2	Y3	ZA	1B	2C	3D	AE	BF	CG	DH	
TV	UW	VX	WY	XZ	Y1	Z2	13	2A	3B	AC	BD	CE	DF	EG	
U	V	W	X	Y	Z	1	2	3	A	B	C	D	E	F	

NOVEMBER 1979 INTELLECTUAL

16 F	17 SA	18 SU	19 M	20 TU	21 W	22 TH	23 F	24 SA	25 SU	26 M	27 TU	28 W	29 TH	30 F
G	H	J	K	L	M	N	O	P	Q	R	S	T	U	V
HF	JG	KH	LJ	MK	NL	OM	PN	QO	RP	SQ	TR	US	VT	WU
JE	KF	LG	MH	NJ	OK	PL	QM	RN	SO	TP	UQ	VR	WS	XT
KD	LE	MF	NG	OH	PJ	QK	RL	SM	TN	UO	VP	WQ	XR	YS
LC	MD	NE	OF	PG	QH	RJ	SK	TL	UM	VN	WO	XP	YQ	ZR
MB	NC	OD	PE	QF	RG	SH	TJ	UK	VL	WM	XN	YO	ZP	1Q
NA	OB	PC	QD	RE	SF	TG	UH	VJ	WK	XL	YM	ZN	1O	2P
O8	PA	QB	RC	SD	TE	UF	VG	WH	XJ	YK	ZL	1M	2N	3O
P7	Q8	RA	SB	TC	UD	VE	WF	XG	YH	ZJ	1K	2L	3M	4N
Q6	R7	S8	TA	UB	VC	WD	XE	YF	ZG	1H	2J	3K	4L	5M
R5	S6	T7	U8	VA	WB	XC	YD	ZE	1F	2G	3H	4J	5K	6L
S4	T5	U6	V7	W8	XA	YB	ZC	1D	2E	3F	4G	5H	6J	7K
T3	U4	V5	W6	X7	Y8	ZA	1B	2C	3D	4E	5F	6G	7H	8J
U2	V3	W4	X5	Y6	Z7	18	2A	3B	4C	5D	6E	7F	8G	AH
V1	W2	X3	Y4	Z5	16	27	38	4A	5B	6C	7D	8E	AF	BG
WZ	X1	Y2	Z3	14	25	36	47	58	6A	7B	8C	AD	BE	CF
XY	YZ	Z1	12	23	34	45	56	67	78	8A	AB	BC	CD	DE

DECEMBER 1979 PHYSICAL

1 SA	2 SU	3 M	4 TU	5 W	6 TH	7 F	8 SA	9 SU	10 M	11 TU	12 W	13 TH	14 F	15 SA
KJ	LK	ML	NM	ON	PO	QP	RQ	SR	TS	UT	VU	WV	XW	AX
LH	MJ	NK	OL	PM	QN	RO	SP	TQ	UR	VS	WT	XU	AV	BW
MG	NH	OJ	PK	QL	RM	SN	TO	UP	VQ	WR	XS	AT	BU	CV
NF	OG	PH	QJ	RK	SL	TM	UN	VO	WP	XQ	AR	BS	CT	DU
OE	PF	QG	RH	SJ	TK	UL	VM	WN	XO	AP	BQ	CR	DS	ET
PD	QE	RF	SG	TH	UJ	VK	WL	XM	AN	BO	CP	DQ	ER	FS
QC	RD	SE	TF	UG	VH	WJ	XK	AL	BM	CN	DO	EP	FQ	GR
RB	SC	TD	UE	VF	WG	XH	AJ	BK	CL	DM	EN	FO	GP	HQ
SA	TB	UC	VD	WE	XF	AG	BH	CJ	DK	EL	FM	GN	HO	JP
TX	UA	VB	WC	XD	AE	BF	CG	DH	EJ	FK	GL	HM	JN	KO
UW	VX	WA	XB	AC	BD	CE	DF	EG	FH	GJ	HK	JL	KM	LN
V	W	X	A	B	C	D	E	F	G	H	J	K	L	M

DECEMBER 1979 EMOTIONAL

1 SA	2 SU	3 M	4 TU	5 W	6 TH	7 F	8 SA	9 SU	10 M	11 TU	12 W	13 TH	14 F	15 SA	
V	W	X	Y	Z	1	2	3	A	B	C	D	E	F	G	
WU	XV	YW	ZX	1Y	2Z	31	A2	B3	CA	DB	EC	FD	GE	HF	
XT	YU	ZV	1W	2X	3Y	AZ	B1	C2	D3	EA	FB	GC	HD	JE	
YS	ZT	1U	2V	3W	AX	BY	CZ	D1	E2	F3	GA	HB	JC	KD	
ZR	1S	2T	3U	AV	BW	CX	DY	EZ	F1	G2	H3	JA	KB	LC	+
1Q	2R	3S	AT	BU	CV	DW	EX	FY	GZ	H1	J2	K3	LA	MB	
2P	3Q	AR	BS	CT	DU	EV	FW	GX	HY	JZ	K1	L2	M3	NA	
3O	AP	BQ	CR	DS	ET	FU	GV	HW	JX	KY	LZ	M1	N2	O3	o
AN	BO	CP	DQ	ER	FS	GT	HU	JV	KW	LX	MY	NZ	O1	P2	
BM	CN	DO	EP	FQ	GR	HS	JT	KU	LV	MW	NX	OY	PZ	Q1	
CL	DM	EN	FO	GP	HQ	JR	KS	LT	MU	NV	OW	PX	QY	RZ	−
DK	EL	FM	GN	HO	JP	KQ	LR	MS	NT	OU	PV	QW	RX	SY	
EJ	FK	GL	HM	JN	KO	LP	MQ	NR	OS	PT	QU	RV	SW	TX	
FH	GJ	HK	JL	KM	LN	MO	NP	OQ	PR	QS	RT	SU	TV	UW	
G	H	J	K	L	M	N	O	P	Q	R	S	T	U	V	

DECEMBER 1979 INTELLECTUAL

1 SA	2 SU	3 M	4 TU	5 W	6 TH	7 F	8 SA	9 SU	10 M	11 TU	12 W	13 TH	14 F	15 SA
W	X	Y	Z	1	2	3	4	5	6	7	8	A	B	C
XV	YW	ZX	1Y	2Z	31	42	53	64	75	86	A7	B8	CA	DB
YU	ZV	1W	2X	3Y	4Z	51	62	73	84	A5	B6	C7	D8	EA
ZT	1U	2V	3W	4X	5Y	6Z	71	82	A3	B4	C5	D6	E7	F8
1S	2T	3U	4V	5W	6X	7Y	8Z	A1	B2	C3	D4	E5	F6	G7
2R	3S	4T	5U	6V	7W	8X	AY	BZ	C1	D2	E3	F4	G5	H6
3Q	4R	5S	6T	7U	8V	AW	BX	CY	DZ	E1	F2	G3	H4	J5
4P	5Q	6R	7S	8T	AU	BV	CW	DX	EY	FZ	G1	H2	J3	K4
5O	6P	7Q	8R	AS	BT	CU	DV	EW	FX	GY	HZ	J1	K2	L3
6N	7O	8P	AQ	BR	CS	DT	EU	FV	GW	HX	JY	KZ	L1	M2
7M	8N	AO	BP	CQ	DR	ES	FT	GU	HV	JW	KX	LY	MZ	N1
8L	AM	BN	CO	DP	EQ	FR	GS	HT	JU	KV	LW	MX	NY	OZ
AK	BL	CM	DN	EO	FP	GQ	HR	JS	KT	LU	MV	NW	OX	PY
BJ	CK	DL	EM	FN	GO	HP	JQ	KR	LS	MT	NU	OV	PW	QX
CH	DJ	EK	FL	GM	HN	JO	KP	LQ	MR	NS	OT	PU	QV	RW
DG	EH	FJ	GK	HL	JM	KN	LO	MP	NQ	OR	PS	QT	RU	SV
EF	FG	GH	HJ	JK	KL	LM	MN	NO	OP	PQ	QR	RS	ST	TU

DECEMBER 1979 — PHYSICAL

16 SU	17 M	18 TU	19 W	20 TH	21 F	22 SA	23 SU	24 M	25 TU	26 W	27 TH	28 F	29 SA	30 SU	31 M
BA	CB	DC	ED	FE	GF	HG	JH	KJ	LK	ML	NM	ON	PO	QP	RQ
CX	DA	EB	FC	GD	HE	JF	KG	LH	MJ	NK	OL	PM	QN	RO	SP
DW	EX	FA	GB	HC	JD	KE	LF	MG	NH	OJ	PK	QL	RM	SN	TO
EV	FW	GX	HA	JB	KC	LD	ME	NF	OG	PH	QJ	RK	SL	TM	UN
FU	GV	HW	JX	KA	LB	MC	ND	OE	PF	QG	RH	SJ	TK	UL	VM
GT	HU	JV	KW	LX	MA	NB	OC	PD	QE	RF	SG	TH	UJ	VK	WL
HS	JT	KU	LV	MW	NX	OA	PB	QC	RD	SE	TF	UG	VH	WJ	XK
JR	KS	LT	MU	NV	OW	PX	QA	RB	SC	TD	UE	VF	WG	XH	AJ
KQ	LR	MS	NT	OU	PV	QW	RX	SA	TB	UC	VD	WE	XF	AG	BH
LP	MQ	NR	OS	PT	QU	RV	SW	TX	UA	VB	WC	XD	AE	BF	CG
MO	NP	OQ	PR	QS	RT	SU	TV	UW	VX	WA	XB	AC	BD	CE	DF
N	O	P	Q	R	S	T	U	V	W	X	A	B	C	D	E

DECEMBER 1979 — EMOTIONAL

16 SU	17 M	18 TU	19 W	20 TH	21 F	22 SA	23 SU	24 M	25 TU	26 W	27 TH	28 F	29 SA	30 SU	31 M	
H	J	K	L	M	N	O	P	Q	R	S	T	U	V	W	X	
JG	KH	LJ	MK	NL	OM	PN	QO	RP	SQ	TR	US	VT	WU	XV	YW	
KF	LG	MH	NJ	OK	PL	QM	RN	SO	TP	UQ	VR	WS	XT	YU	ZV	
LE	MF	NG	OH	PJ	QK	RL	SM	TN	UO	VP	WQ	XR	YS	ZT	1U	
MD	NE	OF	PG	QH	RJ	SK	TL	UM	VN	WO	XP	YQ	ZR	1S	2T	+
NC	OD	PE	QF	RG	SH	TJ	UK	VL	WM	XN	YO	ZP	1Q	2R	3S	
OB	PC	QD	RE	SF	TG	UH	VJ	WK	XL	YM	ZN	1O	2P	3Q	AR	
PA	QB	RC	SD	TE	UF	VG	WH	XJ	YK	ZL	1M	2N	3O	AP	BQ	o
Q3	RA	SB	TC	UD	VE	WF	XG	YH	ZJ	1K	2L	3M	AN	BO	CP	
R2	S3	TA	UB	VC	WD	XE	YF	ZG	1H	2J	3K	AL	BM	CN	DO	
S1	T2	U3	VA	WB	XC	YD	ZE	1F	2G	3H	AJ	BK	CL	DM	EN	−
TZ	U1	V2	W3	XA	YB	ZC	1D	2E	3F	AG	BH	CJ	DK	EL	FM	
UY	VZ	W1	X2	Y3	ZA	1B	2C	3D	AE	BF	CG	DH	EJ	FK	GL	
VX	WY	XZ	Y1	Z2	13	2A	3B	AC	BD	CE	DF	EG	FH	GJ	HK	
W	X	Y	Z	1	2	3	A	B	C	D	E	F	G	H	J	

DECEMBER 1979 — INTELLECTUAL

16 SU	17 M	18 TU	19 W	20 TH	21 F	22 SA	23 SU	24 M	25 TU	26 W	27 TH	28 F	29 SA	30 SU	31 M
D	E	F	G	H	J	K	L	M	N	O	P	Q	R	S	T
EC	FD	GE	HF	JG	KH	LJ	MK	NL	OM	PN	QO	RP	SQ	TR	US
FB	GC	HD	JE	KF	LG	MH	NJ	OK	PL	QM	RN	SO	TP	UQ	VR
GA	HB	JC	KD	LE	MF	NG	OH	PJ	QK	RL	SM	TN	UO	VP	WQ
H8	JA	KB	LC	MD	NE	OF	PG	QH	RJ	SK	TL	UM	VN	WO	XP
J7	K8	LA	MB	NC	OD	PE	QF	RG	SH	TJ	UK	VL	WM	XN	YO
K6	L7	M8	NA	OB	PC	QD	RE	SF	TG	UH	VJ	WK	XL	YM	ZN
L5	M6	N7	O8	PA	QB	RC	SD	TE	UF	VG	WH	XJ	YK	ZL	1M
M4	N5	O6	P7	Q8	RA	SB	TC	UD	VE	WF	XG	YH	ZJ	1K	2L
N3	O4	P5	Q6	R7	S8	TA	UB	VC	WD	XE	YF	ZG	1H	2J	3K
O2	P3	Q4	R5	S6	T7	U8	VA	WB	XC	YD	ZE	1F	2G	3H	4J
P1	Q2	R3	S4	T5	U6	V7	W8	XA	YB	ZC	1D	2E	3F	4G	5H
QZ	R1	S2	T3	U4	V5	W6	X7	Y8	ZA	1B	2C	3D	4E	5F	6G
RY	SZ	T1	U2	V3	W4	X5	Y6	Z7	18	2A	3B	4C	5D	6E	7F
SX	TY	UZ	V1	W2	X3	Y4	Z5	16	27	38	4A	5B	6C	7D	8E
TW	UX	VY	WZ	X1	Y2	Z3	14	25	36	47	58	6A	7B	8C	AD
UV	VW	WX	XY	YZ	Z1	12	23	34	45	56	67	78	8A	AB	BC

JANUARY 1980 PHYSICAL

1 TU	2 W	3 TH	4 F	5 SA	6 SU	7 M	8 TU	9 W	10 TH	11 F	12 SA	13 SU	14 M	15 TU
SR	TS	UT	VU	WV	XW	AX	BA	CB	DC	ED	FE	GF	HG	JH
TQ	UR	VS	WT	XU	AV	BW	CX	DA	EB	FC	GD	HE	JF	KG
UP	VQ	WR	XS	AT	BU	CV	DW	EX	FA	GB	HC	JD	KE	LF
VO	WP	XQ	AR	BS	CT	DU	EV	FW	GX	HA	JB	KC	LD	ME
WN	XO	AP	BQ	CR	DS	ET	FU	GV	HW	JX	KA	LB	MC	ND
XM	AN	BO	CP	DQ	ER	FS	GT	HU	JV	KW	LX	MA	NB	OC
AL	BM	CN	DO	EP	FQ	GR	HS	JT	KU	LV	MW	NX	OA	PB
BK	CL	DM	EN	FO	GP	HQ	JR	KS	LT	MU	NV	OW	PX	QA
CJ	DK	EL	FM	GN	HO	JP	KQ	LR	MS	NT	OU	PV	QW	RX
DH	EJ	FK	GL	HM	JN	KO	LP	MQ	NR	OS	PT	QU	RV	SW
EG	FH	GJ	HK	JL	KM	LN	MO	NP	OQ	PR	QS	RT	SU	TV
F	G	H	J	K	L	M	N	O	P	Q	R	S	T	U

JANUARY 1980 EMOTIONAL

1 TU	2 W	3 TH	4 F	5 SA	6 SU	7 M	8 TU	9 W	10 TH	11 F	12 SA	13 SU	14 M	15 TU	
Y	Z	1	2	3	A	B	C	D	E	F	G	H	J	K	
ZX	1Y	2Z	31	A2	B3	CA	DB	EC	FD	GE	HF	JG	KH	LJ	
1W	2X	3Y	AZ	B1	C2	D3	EA	FB	GC	HD	JE	KF	LG	MH	
2V	3W	AX	BY	CZ	D1	E2	F3	GA	HB	JC	KD	LE	MF	NG	
3U	AV	BW	CX	DY	EZ	F1	G2	H3	JA	KB	LC	MD	NE	OF	+
AT	BU	CV	DW	EX	FY	GZ	H1	J2	K3	LA	MB	NC	OD	PE	
BS	CT	DU	EV	FW	GX	HY	JZ	K1	L2	M3	NA	OB	PC	QD	
CR	DS	ET	FU	GV	HW	JX	KY	LZ	M1	N2	O3	PA	QB	RC	o
DQ	ER	FS	GT	HU	JV	KW	LX	MY	NZ	O1	P2	Q3	RA	SB	
EP	FQ	GR	HS	JT	KU	LV	MW	NX	OY	PZ	Q1	R2	S3	TA	
FO	GP	HQ	JR	KS	LT	MU	NV	OW	PX	QY	RZ	S1	T2	U3	–
GN	HO	JP	KQ	LR	MS	NT	OU	PV	QW	RX	SY	TZ	U1	V2	
HM	JN	KO	LP	MQ	NR	OS	PT	QU	RV	SW	TX	UY	VZ	W1	
JL	KM	LN	MO	NP	OQ	PR	QS	RT	SU	TV	UW	VX	WY	XZ	
K	L	M	N	O	P	Q	R	S	T	U	V	W	X	Y	

JANUARY 1980 INTELLECTUAL

1 TU	2 W	3 TH	4 F	5 SA	6 SU	7 M	8 TU	9 W	10 TH	11 F	12 SA	13 SU	14 M	15 TU
U	V	W	X	Y	Z	1	2	3	4	5	6	7	8	A
VT	WU	XV	YW	ZX	1Y	2Z	31	42	53	64	75	86	A7	B8
WS	XT	YU	ZV	1W	2X	3Y	4Z	51	62	73	84	A5	B6	C7
XR	YS	ZT	1U	2V	3W	4X	5Y	6Z	71	82	A3	B4	C5	D6
YQ	ZR	1S	2T	3U	4V	5W	6X	7Y	8Z	A1	B2	C3	D4	E5
ZP	1Q	2R	3S	4T	5U	6V	7W	8X	AY	BZ	C1	D2	E3	F4
1O	2P	3Q	4R	5S	6T	7U	8V	AW	BX	CY	DZ	E1	F2	G3
2N	3O	4P	5Q	6R	7S	8T	AU	BV	CW	DX	EY	FZ	G1	H2
3M	4N	5O	6P	7Q	8R	AS	BT	CU	DV	EW	FX	GY	HZ	J1
4L	5M	6N	7O	8P	AQ	BR	CS	DT	EU	FV	GW	HX	JY	KZ
5K	6L	7M	8N	AO	BP	CQ	DR	ES	FT	GU	HV	JW	KX	LY
6J	7K	8L	AM	BN	CO	DP	EQ	FR	GS	HT	JU	KV	LW	MX
7H	8J	AK	BL	CM	DN	EO	FP	GQ	HR	JS	KT	LU	MV	NW
8G	AH	BJ	CK	DL	EM	FN	GO	HP	JQ	KR	LS	MT	NU	OV
AF	BG	CH	DJ	EK	FL	GM	HN	JO	KP	LQ	MR	NS	OT	PU
BE	CF	DG	EH	FJ	GK	HL	JM	KN	LO	MP	NQ	OR	PS	QT
CD	DE	EF	FG	GH	HJ	JK	KL	LM	MN	NO	OP	PQ	QR	RS

JANUARY 1980 — PHYSICAL

16 W	17 TH	18 F	19 SA	20 SU	21 M	22 TU	23 W	24 TH	25 F	26 SA	27 SU	28 M	29 TU	30 W	31 TH
KJ	LK	ML	NM	ON	PO	QP	RQ	SR	TS	UT	VU	WV	XW	AX	BA
LH	MJ	NK	OL	PM	QN	RO	SP	TQ	UR	VS	WT	XU	AV	BW	CX
MG	NH	OJ	PK	QL	RM	SN	TO	UP	VQ	WR	XS	AT	BU	CV	DW
NF	OG	PH	QJ	RK	SL	TM	UN	VO	WP	XQ	AR	BS	CT	DU	EV
OE	PF	QG	RH	SJ	TK	UL	VM	WN	XO	AP	BQ	CR	DS	ET	FU
PD	QE	RF	SG	TH	UJ	VK	WL	XM	AN	BO	CP	DQ	ER	FS	GT
QC	RD	SE	TF	UG	VH	WJ	XK	AL	BM	CN	DO	EP	FQ	GR	HS
RB	SC	TD	UE	VF	WG	XH	AJ	BK	CL	DM	EN	FO	GP	HQ	JR
SA	TB	UC	VD	WE	XF	AG	BH	CJ	DK	EL	FM	GN	HO	JP	KQ
TX	UA	VB	WC	XD	AE	BF	CG	DH	EJ	FK	GL	HM	JN	KO	LP
UW	VX	WA	XB	AC	BD	CE	DF	EG	FH	GJ	HK	JL	KM	LN	MO
V	W	X	A	B	C	D	E	F	G	H	J	K	L	M	N

JANUARY 1980 — EMOTIONAL

16 W	17 TH	18 F	19 SA	20 SU	21 M	22 TU	23 W	24 TH	25 F	26 SA	27 SU	28 M	29 TU	30 W	31 TH	
L	M	N	O	P	Q	R	S	T	U	V	W	X	Y	Z	1	
MK	NL	OM	PN	QO	RP	SQ	TR	US	VT	WU	XV	YW	ZX	1Y	2Z	
NJ	OK	PL	QM	RN	SO	TP	UQ	VR	WS	XT	YU	ZV	1W	2X	3Y	
OH	PJ	QK	RL	SM	TN	UO	VP	WQ	XR	YS	ZT	1U	2V	3W	AX	
PG	QH	RJ	SK	TL	UM	VN	WO	XP	YQ	ZR	1S	2T	3U	AV	BW	+
QF	RG	SH	TJ	UK	VL	WM	XN	YO	ZP	1Q	2R	3S	AT	BU	CV	
RE	SF	TG	UH	VJ	WK	XL	YM	ZN	1O	2P	3Q	AR	BS	CT	DU	
SD	TE	UF	VG	WH	XJ	YK	ZL	1M	2N	3O	AP	BQ	CR	DS	ET	o
TC	UD	VE	WF	XG	YH	ZJ	1K	2L	3M	AN	BO	CP	DQ	ER	FS	
UB	VC	WD	XE	YF	ZG	1H	2J	3K	AL	BM	CN	DO	EP	FQ	GR	
VA	WB	XC	YD	ZE	1F	2G	3H	AJ	BK	CL	DM	EN	FO	GP	HQ	−
W3	XA	YB	ZC	1D	2E	3F	AG	BH	CJ	DK	EL	FM	GN	HO	JP	
X2	Y3	ZA	1B	2C	3D	AE	BF	CG	DH	EJ	FK	GL	HM	JN	KO	
Y1	Z2	13	2A	3B	AC	BD	CE	DF	EG	FH	GJ	HK	JL	KM	LN	
Z	1	2	3	A	B	C	D	E	F	G	H	J	K	L	M	

JANUARY 1980 — INTELLECTUAL

16 W	17 TH	18 F	19 SA	20 SU	21 M	22 TU	23 W	24 TH	25 F	26 SA	27 SU	28 M	29 TU	30 W	31 TH
B	C	D	E	F	G	H	J	K	L	M	N	O	P	Q	R
CA	DB	EC	FD	GE	HF	JG	KH	LJ	MK	NL	OM	PN	QO	RP	SQ
D8	EA	FB	GC	HD	JE	KF	LG	MH	NJ	OK	PL	QM	RN	SO	TP
E7	F8	GA	HB	JC	KD	LE	MF	NG	OH	PJ	QK	RL	SM	TN	UO
F6	G7	H8	JA	KB	LC	MD	NE	OF	PG	QH	RJ	SK	TL	UM	VN
G5	H6	J7	K8	LA	MB	NC	OD	PE	QF	RG	SH	TJ	UK	VL	WM
H4	J5	K6	L7	M8	NA	OB	PC	QD	RE	SF	TG	UH	VJ	WK	XL
J3	K4	L5	M6	N7	O8	PA	QB	RC	SD	TE	UF	VG	WH	XJ	YK
K2	L3	M4	N5	O6	P7	Q8	RA	SB	TC	UD	VE	WF	XG	YH	ZJ
L1	M2	N3	O4	P5	Q6	R7	S8	TA	UB	VC	WD	XE	YF	ZG	1H
MZ	N1	O2	P3	Q4	R5	S6	T7	U8	VA	WB	XC	YD	ZE	1F	2G
NY	OZ	P1	Q2	R3	S4	T5	U6	V7	W8	XA	YB	ZC	1D	2E	3F
OX	PY	QZ	R1	S2	T3	U4	V5	W6	X7	Y8	ZA	1B	2C	3D	4E
PW	QX	RY	SZ	T1	U2	V3	W4	X5	Y6	Z7	18	2A	3B	4C	5D
QV	RW	SX	TY	UZ	V1	W2	X3	Y4	Z5	16	27	38	4A	5B	6C
RU	SV	TW	UX	VY	WZ	X1	Y2	Z3	14	25	36	47	58	6A	7B
ST	TU	UV	VW	WX	XY	YZ	Z1	12	23	34	45	56	67	78	8A

FEBRUARY 1980 **PHYSICAL**

1 F	2 SA	3 SU	4 M	5 TU	6 W	7 TH	8 F	9 SA	10 SU	11 M	12 TU	13 W	14 TH	15 F
CB	DC	ED	FE	GF	HG	JH	KJ	LK	ML	NM	ON	PO	QP	RQ
DA	EB	FC	GD	HE	JF	KG	LH	MJ	NK	OL	PM	QN	RO	SP
EX	FA	GB	HC	JD	KE	LF	MG	NH	OJ	PK	QL	RM	SN	TO
FW	GX	HA	JB	KC	LD	ME	NF	OG	PH	QJ	RK	SL	TM	UN
GV	HW	JX	KA	LB	MC	ND	OE	PF	QG	RH	SJ	TK	UL	VM
HU	JV	KW	LX	MA	NB	OC	PD	QE	RF	SG	TH	UJ	VK	WL
JT	KU	LV	MW	NX	OA	PB	QC	RD	SE	TF	UG	VH	WJ	XK
KS	LT	MU	NV	OW	PX	QA	RB	SC	TD	UE	VF	WG	XH	AJ
LR	MS	NT	OU	PV	QW	RX	SA	TB	UC	VD	WE	XF	AG	BH
MQ	NR	OS	PT	QU	RV	SW	TX	UA	VB	WC	XD	AE	BF	CG
NP	OQ	PR	QS	RT	SU	TV	UW	VX	WA	XB	AC	BD	CE	DF
O	P	Q	R	S	T	U	V	W	X	A	B	C	D	E

FEBRUARY 1980 **EMOTIONAL**

1 F	2 SA	3 SU	4 M	5 TU	6 W	7 TH	8 F	9 SA	10 SU	11 M	12 TU	13 W	14 TH	15 F	
2	3	A	B	C	D	E	F	G	H	J	K	L	M	N	
31	A2	B3	CA	DB	EC	FD	GE	HF	JG	KH	LJ	MK	NL	OM	
AZ	B1	C2	D3	EA	FB	GC	HD	JE	KF	LG	MH	NJ	OK	PL	
BY	CZ	D1	E2	F3	GA	HB	JC	KD	LE	MF	NG	OH	PJ	QK	
CX	DY	EZ	F1	G2	H3	JA	KB	LC	MD	NE	OF	PG	QH	RJ	+
DW	EX	FY	GZ	H1	J2	K3	LA	MB	NC	OD	PE	QF	RG	SH	
EV	FW	GX	HY	JZ	K1	L2	M3	NA	OB	PC	QD	RE	SF	TG	
FU	GV	HW	JX	KY	LZ	M1	N2	O3	PA	QB	RC	SD	TE	UF	o
GT	HU	JV	KW	LX	MY	NZ	O1	P2	Q3	RA	SB	TC	UD	VE	
HS	JT	KU	LV	MW	NX	OY	PZ	Q1	R2	S3	TA	UB	VC	WD	
JR	KS	LT	MU	NV	OW	PX	QY	RZ	S1	T2	U3	VA	WB	XC	–
KQ	LR	MS	NT	OU	PV	QW	RX	SY	TZ	U1	V2	W3	XA	YB	
LP	MQ	NR	OS	PT	QU	RV	SW	TX	UY	VZ	W1	X2	Y3	ZA	
MO	NP	OQ	PR	QS	RT	SU	TV	UW	VX	WY	XZ	Y1	Z2	13	
N	O	P	Q	R	S	T	U	V	W	X	Y	Z	1	2	

FEBRUARY 1980 **INTELLECTUAL**

1 F	2 SA	3 SU	4 M	5 TU	6 W	7 TH	8 F	9 SA	10 SU	11 M	12 TU	13 W	14 TH	15 F
S	T	U	V	W	X	Y	Z	1	2	3	4	5	6	7
TR	US	VT	WU	XV	YW	ZX	1Y	2Z	31	42	53	64	75	86
UQ	VR	WS	XT	YU	ZV	1W	2X	3Y	4Z	51	62	73	84	A5
VP	WQ	XR	YS	ZT	1U	2V	3W	4X	5Y	6Z	71	82	A3	B4
WO	XP	YQ	ZR	1S	2T	3U	4V	5W	6X	7Y	8Z	A1	B2	C3
XN	YO	ZP	1Q	2R	3S	4T	5U	6V	7W	8X	AY	BZ	C1	D2
YM	ZN	1O	2P	3Q	4R	5S	6T	7U	8V	AW	BX	CY	DZ	E1
ZL	1M	2N	3O	4P	5Q	6R	7S	8T	AU	BV	CW	DX	EY	FZ
1K	2L	3M	4N	5O	6P	7Q	8R	AS	BT	CU	DV	EW	FX	GY
2J	3K	4L	5M	6N	7O	8P	AQ	BR	CS	DT	EU	FV	GW	HX
3H	4J	5K	6L	7M	8N	AO	BP	CQ	DR	ES	FT	GU	HV	JW
4G	5H	6J	7K	8L	AM	BN	CO	DP	EQ	FR	GS	HT	JU	KV
5F	6G	7H	8J	AK	BL	CM	DN	EO	FP	GQ	HR	JS	KT	LU
6E	7F	8G	AH	BJ	CK	DL	EM	FN	GO	HP	JQ	KR	LS	MT
7D	8E	AF	BG	CH	DJ	EK	FL	GM	HN	JO	KP	LQ	MR	NS
8C	AD	BE	CF	DG	EH	FJ	GK	HL	JM	KN	LO	MP	NQ	OR
AB	BC	CD	DE	EF	FG	GH	HJ	JK	KL	LM	MN	NO	OP	PQ

FEBRUARY 1980 **PHYSICAL**

16 SA	17 SU	18 M	19 TU	20 W	21 TH	22 F	23 SA	24 SU	25 M	26 TU	27 W	28 TH	29 F
SR	TS	UT	VU	WV	XW	AX	BA	CB	DC	ED	FE	GF	HG
TQ	UR	VS	WT	XU	AV	BW	CX	DA	EB	FC	GD	HE	JF
UP	VQ	WR	XS	AT	BU	CV	DW	EX	FA	GB	HC	JD	KE
VO	WP	XQ	AR	BS	CT	DU	EV	FW	GX	HA	JB	KC	LD
WN	XO	AP	BQ	CR	DS	ET	FU	GV	HW	JX	KA	LB	MC
XM	AN	BO	CP	DQ	ER	FS	GT	HU	JV	KW	LX	MA	NB
AL	BM	CN	DO	EP	FQ	GR	HS	JT	KU	LV	MW	NX	OA
BK	CL	DM	EN	FO	GP	HQ	JR	KS	LT	MU	NV	OW	PX
CJ	DK	EL	FM	GN	HO	JP	KQ	LR	MS	NT	OU	PV	QW
DH	EJ	FK	GL	HM	JN	KO	LP	MQ	NR	OS	PT	QU	RV
EG	FH	GJ	HK	JL	KM	LN	MO	NP	OQ	PR	QS	RT	SU
F	G	H	J	K	L	M	N	O	P	Q	R	S	T

FEBRUARY 1980 **EMOTIONAL**

16 SA	17 SU	18 M	19 TU	20 W	21 TH	22 F	23 SA	24 SU	25 M	26 TU	27 W	28 TH	29 F	
O	P	Q	R	S	T	U	V	W	X	Y	Z	1	2	
PN	QO	RP	SQ	TR	US	VT	WU	XV	YW	ZX	1Y	2Z	31	
QM	RN	SO	TP	UQ	VR	WS	XT	YU	ZV	1W	2X	3Y	AZ	
RL	SM	TN	UO	VP	WQ	XR	YS	ZT	1U	2V	3W	AX	BY	
SK	TL	UM	VN	WO	XP	YQ	ZR	1S	2T	3U	AV	BW	CX	+
TJ	UK	VL	WM	XN	YO	ZP	1Q	2R	3S	AT	BU	CV	DW	
UH	VJ	WK	XL	YM	ZN	1O	2P	3Q	AR	BS	CT	DU	EV	
VG	WH	XJ	YK	ZL	1M	2N	3O	AP	BQ	CR	DS	ET	FU	o
WF	XG	YH	ZJ	1K	2L	3M	AN	BO	CP	DQ	ER	FS	GT	
XE	YF	ZG	1H	2J	3K	AL	BM	CN	DO	EP	FQ	GR	HS	
YD	ZE	1F	2G	3H	AJ	BK	CL	DM	EN	FO	GP	HQ	JR	−
ZC	1D	2E	3F	AG	BH	CJ	DK	EL	FM	GN	HO	JP	KQ	
1B	2C	3D	AE	BF	CG	DH	EJ	FK	GL	HM	JN	KO	LP	
2A	3B	AC	BD	CE	DF	EG	FH	GJ	HK	JL	KM	LN	MO	
3	A	B	C	D	E	F	G	H	J	K	L	M	N	

FEBRUARY 1980 **INTELLECTUAL**

16 SA	17 SU	18 M	19 TU	20 W	21 TH	22 F	23 SA	24 SU	25 M	26 TU	27 W	28 TH	29 F
8	A	B	C	D	E	F	G	H	J	K	L	M	N
A7	B8	CA	DB	EC	FD	GE	HF	JG	KH	LJ	MK	NL	OM
B6	C7	D8	EA	FB	GC	HD	JE	KF	LG	MH	NJ	OK	PL
C5	D6	E7	F8	GA	HB	JC	KD	LE	MF	NG	OH	PJ	QK
D4	E5	F6	G7	H8	JA	KB	LC	MD	NE	OF	PG	QH	RJ
E3	F4	G5	H6	J7	K8	LA	MB	NC	OD	PE	QF	RG	SH
F2	G3	H4	J5	K6	L7	M8	NA	OB	PC	QD	RE	SF	TG
G1	H2	J3	K4	L5	M6	N7	O8	PA	QB	RC	SD	TE	UF
HZ	J1	K2	L3	M4	N5	O6	P7	Q8	RA	SB	TC	UD	VE
JY	KZ	L1	M2	N3	O4	P5	Q6	R7	S8	TA	UB	VC	WD
KX	LY	MZ	N1	O2	P3	Q4	R5	S6	T7	U8	VA	WB	XC
LW	MX	NY	OZ	P1	Q2	R3	S4	T5	U6	V7	W8	XA	YB
MV	NW	OX	PY	QZ	R1	S2	T3	U4	V5	W6	X7	Y8	ZA
NU	OV	PW	QX	RY	SZ	T1	U2	V3	W4	X5	Y6	Z7	18
OT	PU	QV	RW	SX	TY	UZ	V1	W2	X3	Y4	Z5	16	27
PS	QT	RU	SV	TW	UX	VY	WZ	X1	Y2	Z3	14	25	36
QR	RS	ST	TU	UV	VW	WX	XY	YZ	Z1	12	23	34	45

MARCH 1980 — PHYSICAL

1 SA	2 SU	3 M	4 TU	5 W	6 TH	7 F	8 SA	9 SU	10 M	11 TU	12 W	13 TH	14 F	15 SA
JH	KJ	LK	ML	NM	ON	PO	QP	RQ	SR	TS	UT	VU	WV	XW
KG	LH	MJ	NK	OL	PM	QN	RO	SP	TQ	UR	VS	WT	XU	AV
LF	MG	NH	OJ	PK	QL	RM	SN	TO	UP	VQ	WR	XS	AT	BU
ME	NF	OG	PH	QJ	RK	SL	TM	UN	VO	WP	XQ	AR	BS	CT
ND	OE	PF	QG	RH	SJ	TK	UL	VM	WN	XO	AP	BQ	CR	DS
OC	PD	QE	RF	SG	TH	UJ	VK	WL	XM	AN	BO	CP	DQ	ER
PB	QC	RD	SE	TF	UG	VH	WJ	XK	AL	BM	CN	DO	EP	FQ
QA	RB	SC	TD	UE	VF	WG	XH	AJ	BK	CL	DM	EN	FO	GP
RX	SA	TB	UC	VD	WE	XF	AG	BH	CJ	DK	EL	FM	GN	HO
SW	TX	UA	VB	WC	XD	AE	BF	CG	DH	EJ	FK	GL	HM	JN
TV	UW	VX	WA	XB	AC	BD	CE	DF	EG	FH	GJ	HK	JL	KM
U	V	W	X	A	B	C	D	E	F	G	H	J	K	L

MARCH 1980 — EMOTIONAL

1 SA	2 SU	3 M	4 TU	5 W	6 TH	7 F	8 SA	9 SU	10 M	11 TU	12 W	13 TH	14 F	15 SA	
3	A	B	C	D	E	F	G	H	J	K	L	M	N	O	
A2	B3	CA	DB	EC	FD	GE	HF	JG	KH	LJ	MK	NL	OM	PN	
B1	C2	D3	EA	FB	GC	HD	JE	KF	LG	MH	NJ	OK	PL	QM	
CZ	D1	E2	F3	GA	HB	JC	KD	LE	MF	NG	OH	PJ	QK	RL	
DY	EZ	F1	G2	H3	JA	KB	LC	MD	NE	OF	PG	QH	RJ	SK	+
EX	FY	GZ	H1	J2	K3	LA	MB	NC	OD	PE	QF	RG	SH	TJ	
FW	GX	HY	JZ	K1	L2	M3	NA	OB	PC	QD	RE	SF	TG	UH	
GV	HW	JX	KY	LZ	M1	N2	O3	PA	QB	RC	SD	TE	UF	VG	0
HU	JV	KW	LX	MY	NZ	O1	P2	Q3	RA	SB	TC	UD	VE	WF	
JT	KU	LV	MW	NX	OY	PZ	Q1	R2	S3	TA	UB	VC	WD	XE	
KS	LT	MU	NV	OW	PX	QY	RZ	S1	T2	U3	VA	WB	XC	YD	−
LR	MS	NT	OU	PV	QW	RX	SY	TZ	U1	V2	W3	XA	YB	ZC	
MQ	NR	OS	PT	QU	RV	SW	TX	UY	VZ	W1	X2	Y3	ZA	1B	
NP	OQ	PR	QS	RT	SU	TV	UW	VX	WY	XZ	Y1	Z2	13	2A	
O	P	Q	R	S	T	U	V	W	X	Y	Z	1	2	3	

MARCH 1980 — INTELLECTUAL

1 SA	2 SU	3 M	4 TU	5 W	6 TH	7 F	8 SA	9 SU	10 M	11 TU	12 W	13 TH	14 F	15 SA
O	P	Q	R	S	T	U	V	W	X	Y	Z	1	2	3
PN	QO	RP	SQ	TR	US	VT	WU	XV	YW	ZX	1Y	2Z	31	42
QM	RN	SO	TP	UQ	VR	WS	XT	YU	ZV	1W	2X	3Y	4Z	51
RL	SM	TN	UO	VP	WQ	XR	YS	ZT	1U	2V	3W	4X	5Y	6Z
SK	TL	UM	VN	WO	XP	YQ	ZR	1S	2T	3U	4V	5W	6X	7Y
TJ	UK	VL	WM	XN	YO	ZP	1Q	2R	3S	4T	5U	6V	7W	8X
UH	VJ	WK	XL	YM	ZN	10	2P	3Q	4R	5S	6T	7U	8V	AW
VG	WH	XJ	YK	ZL	1M	2N	3O	4P	5Q	6R	7S	8T	AU	BV
WF	XG	YH	ZJ	1K	2L	3M	4N	50	6P	7Q	8R	AS	BT	CU
XE	YF	ZG	1H	2J	3K	4L	5M	6N	7O	8P	AQ	BR	CS	DT
YD	ZE	1F	2G	3H	4J	5K	6L	7M	8N	AO	BP	CQ	DR	ES
ZC	1D	2E	3F	4G	5H	6J	7K	8L	AM	BN	CO	DP	EQ	FR
1B	2C	3D	4E	5F	6G	7H	8J	AK	BL	CM	DN	EO	FP	GQ
2A	3B	4C	5D	6E	7F	8G	AH	BJ	CK	DL	EM	FN	GO	HP
38	4A	5B	6C	7D	8E	AF	BG	CH	DJ	EK	FL	GM	HN	JO
47	58	6A	7B	8C	AD	BE	CF	DG	EH	FJ	GK	HL	JM	KN
56	67	78	8A	AB	BC	CD	DE	EF	FG	GH	HJ	JK	KL	LM

MARCH 1980 **PHYSICAL**

16 SU	17 M	18 TU	19 W	20 TH	21 F	22 SA	23 SU	24 M	25 TU	26 W	27 TH	28 F	29 SA	30 SU	31 M
AX	BA	CB	DC	ED	FE	GF	HG	JH	KJ	LK	ML	NM	ON	PO	QP
BW	CX	DA	EB	FC	GD	HE	JF	KG	LH	MJ	NK	OL	PM	QN	RO
CV	DW	EX	FA	GB	HC	JD	KE	LF	MG	NH	OJ	PK	QL	RM	SN
DU	EV	FW	GX	HA	JB	KC	LD	ME	NF	OG	PH	QJ	RK	SL	TM
ET	FU	GV	HW	JX	KA	LB	MC	ND	OE	PF	QG	RH	SJ	TK	UL
FS	GT	HU	JV	KW	LX	MA	NB	OC	PD	QE	RF	SG	TH	UJ	VK
GR	HS	JT	KU	LV	MW	NX	OA	PB	QC	RD	SE	TF	UG	VH	WJ
HQ	JR	KS	LT	MU	NV	OW	PX	QA	RB	SC	TD	UE	VF	WG	XH
JP	KQ	LR	MS	NT	OU	PV	QW	RX	SA	TB	UC	VD	WE	XF	AG
KO	LP	MQ	NR	OS	PT	QU	RV	SW	TX	UA	VB	WC	XD	AE	BF
LN	MO	NP	OQ	PR	QS	RT	SU	TV	UW	VX	WA	XB	AC	BD	CE
M	N	O	P	Q	R	S	T	U	V	W	X	A	B	C	D

MARCH 1980 **EMOTIONAL**

16 SU	17 M	18 TU	19 W	20 TH	21 F	22 SA	23 SU	24 M	25 TU	26 W	27 TH	28 F	29 SA	30 SU	31 M	
P	Q	R	S	T	U	V	W	X	Y	Z	1	2	3	A	B	
QO	RP	SQ	TR	US	VT	WU	XV	YW	ZX	1Y	2Z	31	A2	B3	CA	
RN	SO	TP	UQ	VR	WS	XT	YU	ZV	1W	2X	3Y	AZ	B1	C2	D3	
SM	TN	UO	VP	WQ	XR	YS	ZT	1U	2V	3W	AX	BY	CZ	D1	E2	
TL	UM	VN	WO	XP	YQ	ZR	1S	2T	3U	AV	BW	CX	DY	EZ	F1	+
UK	VL	WM	XN	YO	ZP	1Q	2R	3S	AT	BU	CV	DW	EX	FY	GZ	
VJ	WK	XL	YM	ZN	1O	2P	3Q	AR	BS	CT	DU	EV	FW	GX	HY	
WH	XJ	YK	ZL	1M	2N	3O	AP	BQ	CR	DS	ET	FU	GV	HW	JX	o
XG	YH	ZJ	1K	2L	3M	AN	BO	CP	DQ	ER	FS	GT	HU	JV	KW	
YF	ZG	1H	2J	3K	AL	BM	CN	DO	EP	FQ	GR	HS	JT	KU	LV	
ZE	1F	2G	3H	AJ	BK	CL	DM	EN	FO	GP	HQ	JR	KS	LT	MU	–
1D	2E	3F	AG	BH	CJ	DK	EL	FM	GN	HO	JP	KQ	LR	MS	NT	
2C	3D	AE	BF	CG	DH	EJ	FK	GL	HM	JN	KO	LP	MQ	NR	OS	
3B	AC	BD	CE	DF	EG	FH	GJ	HK	JL	KM	LN	MO	NP	OQ	PR	
A	B	C	D	E	F	G	H	J	K	L	M	N	O	P	Q	

MARCH 1980 **INTELLECTUAL**

16 SU	17 M	18 TU	19 W	20 TH	21 F	22 SA	23 SU	24 M	25 TU	26 W	27 TH	28 F	29 SA	30 SU	31 M
4	5	6	7	8	A	B	C	D	E	F	G	H	J	K	L
53	64	75	86	A7	B8	CA	DB	EC	FD	GE	HF	JG	KH	LJ	MK
62	73	84	A5	B6	C7	D8	EA	FB	GC	HD	JE	KF	LG	MH	NJ
71	82	A3	B4	C5	D6	E7	F8	GA	HB	JC	KD	LE	MF	NG	OH
8Z	A1	B2	C3	D4	E5	F6	G7	H8	JA	KB	LC	MD	NE	OF	PG
AY	BZ	C1	D2	E3	F4	G5	H6	J7	K8	LA	MB	NC	OD	PE	QF
BX	CY	DZ	E1	F2	G3	H4	J5	K6	L7	M8	NA	OB	PC	QD	RE
CW	DX	EY	FZ	G1	H2	J3	K4	L5	M6	N7	O8	PA	QB	RC	SD
DV	EW	FX	GY	HZ	J1	K2	L3	M4	N5	O6	P7	Q8	RA	SB	TC
EU	FV	GW	HX	JY	KZ	L1	M2	N3	O4	P5	Q6	R7	S8	TA	UB
FT	GU	HV	JW	KX	LY	MZ	N1	O2	P3	Q4	R5	S6	T7	U8	VA
GS	HT	JU	KV	LW	MX	NY	OZ	P1	Q2	R3	S4	T5	U6	V7	W8
HR	JS	KT	LU	MV	NW	OX	PY	QZ	R1	S2	T3	U4	V5	W6	X7
JQ	KR	LS	MT	NU	OV	PW	QX	RY	SZ	T1	U2	V3	W4	X5	Y6
KP	LQ	MR	NS	OT	PU	QV	RW	SX	TY	UZ	V1	W2	X3	Y4	Z5
LO	MP	NQ	OR	PS	QT	RU	SV	TW	UX	VY	WZ	X1	Y2	Z3	14
MN	NO	OP	PQ	QR	RS	ST	TU	UV	VW	WX	XY	YZ	Z1	12	23

APRIL 1980 **PHYSICAL**

1 TU	2 W	3 TH	4 F	5 SA	6 SU	7 M	8 TU	9 W	10 TH	11 F	12 SA	13 SU	14 M	15 TU
RQ	SR	TS	UT	VU	WV	XW	AX	BA	CB	DC	ED	FE	GF	HG
SP	TQ	UR	VS	WT	XU	AV	BW	CX	DA	EB	FC	GD	HE	JF
TO	UP	VQ	WR	XS	AT	BU	CV	DW	EX	FA	GB	HC	JD	KE
UN	VO	WP	XQ	AR	BS	CT	DU	EV	FW	GX	HA	JB	KC	LD
VM	WN	XO	AP	BQ	CR	DS	ET	FU	GV	HW	JX	KA	LB	MC
WL	XM	AN	BO	CP	DQ	ER	FS	GT	HU	JV	KW	LX	MA	NB
XK	AL	BM	CN	DO	EP	FQ	GR	HS	JT	KU	LV	MW	NX	OA
AJ	BK	CL	DM	EN	FO	GP	HQ	JR	KS	LT	MU	NV	OW	PX
BH	CJ	DK	EL	FM	GN	HO	JP	KQ	LR	MS	NT	OU	PV	QW
CG	DH	EJ	FK	GL	HM	JN	KO	LP	MQ	NR	OS	PT	QU	RV
DF	EG	FH	GJ	HK	JL	KM	LN	MO	NP	OQ	PR	QS	RT	SU
E	F	G	H	J	K	L	M	N	O	P	Q	R	S	T

APRIL 1980 **EMOTIONAL**

1 TU	2 W	3 TH	4 F	5 SA	6 SU	7 M	8 TU	9 W	10 TH	11 F	12 SA	13 SU	14 M	15 TU	
C	D	E	F	G	H	J	K	L	M	N	O	P	Q	R	
DB	EC	FD	GE	HF	JG	KH	LJ	MK	NL	OM	PN	QO	RP	SQ	
EA	FB	GC	HD	JE	KF	LG	MH	NJ	OK	PL	QM	RN	SO	TP	
F3	GA	HB	JC	KD	LE	MF	NG	OH	PJ	QK	RL	SM	TN	UO	
G2	H3	JA	KB	LC	MD	NE	OF	PG	QH	RJ	SK	TL	UM	VN	+
H1	J2	K3	LA	MB	NC	OD	PE	QF	RG	SH	TJ	UK	VL	WM	
JZ	K1	L2	M3	NA	OB	PC	QD	RE	SF	TG	UH	VJ	WK	XL	
KY	LZ	M1	N2	O3	PA	QB	RC	SD	TE	UF	VG	WH	XJ	YK	o
LX	MY	NZ	O1	P2	Q3	RA	SB	TC	UD	VE	WF	XG	YH	ZJ	
MW	NX	OY	PZ	Q1	R2	S3	TA	UB	VC	WD	XE	YF	ZG	1H	
NV	OW	PX	QY	RZ	S1	T2	U3	VA	WB	XC	YD	ZE	1F	2G	−
OU	PV	QW	RX	SY	TZ	U1	V2	W3	XA	YB	ZC	1D	2E	3F	
PT	QU	RV	SW	TX	UY	VZ	W1	X2	Y3	ZA	1B	2C	3D	AE	
QS	RT	SU	TV	UW	VX	WY	XZ	Y1	Z2	13	2A	3B	AC	BD	
R	S	T	U	V	W	X	Y	Z	1	2	3	A	B	C	

APRIL 1980 **INTELLECTUAL**

1 TU	2 W	3 TH	4 F	5 SA	6 SU	7 M	8 TU	9 W	10 TH	11 F	12 SA	13 SU	14 M	15 TU
M	N	O	P	Q	R	S	T	U	V	W	X	Y	Z	1
NL	OM	PN	QO	RP	SQ	TR	US	VT	WU	XV	YW	ZX	1Y	2Z
OK	PL	QM	RN	SO	TP	UQ	VR	WS	XT	YU	ZV	1W	2X	3Y
PJ	QK	RL	SM	TN	UO	VP	WQ	XR	YS	ZT	1U	2V	3W	4X
QH	RJ	SK	TL	UM	VN	WO	XP	YQ	ZR	1S	2T	3U	4V	5W
RG	SH	TJ	UK	VL	WM	XN	YO	ZP	1Q	2R	3S	4T	5U	6V
SF	TG	UH	VJ	WK	XL	YM	ZN	1O	2P	3Q	4R	5S	6T	7U
TE	UF	VG	WH	XJ	YK	ZL	1M	2N	3O	4P	5Q	6R	7S	8T
UD	VE	WF	XG	YH	ZJ	1K	2L	3M	4N	5O	6P	7Q	8R	AS
VC	WD	XE	YF	ZG	1H	2J	3K	4L	5M	6N	7O	8P	AQ	BR
WB	XC	YD	ZE	1F	2G	3H	4J	5K	6L	7M	8N	AO	BP	CQ
XA	YB	ZC	1D	2E	3F	4G	5H	6J	7K	8L	AM	BN	CO	DP
Y8	ZA	1B	2C	3D	4E	5F	6G	7H	8J	AK	BL	CM	DN	EO
Z7	18	2A	3B	4C	5D	6E	7F	8G	AH	BJ	CK	DL	EM	FN
16	27	38	4A	5B	6C	7D	8E	AF	BG	CH	DJ	EK	FL	GM
25	36	47	58	6A	7B	8C	AD	BE	CF	DG	EH	FJ	GK	HL
34	45	56	67	78	8A	AB	BC	CD	DE	EF	FG	GH	HJ	JK

APRIL 1980 — PHYSICAL

16 W	17 TH	18 F	19 SA	20 SU	21 M	22 TU	23 W	24 TH	25 F	26 SA	27 SU	28 M	29 TU	30 W
JH	KJ	LK	ML	NM	ON	PO	QP	RQ	SR	TS	UT	VU	WV	XW
KG	LH	MJ	NK	OL	PM	QN	RO	SP	TQ	UR	VS	WT	XU	AV
LF	MG	NH	OJ	PK	QL	RM	SN	TO	UP	VQ	WR	XS	AT	BU
ME	NF	OG	PH	QJ	RK	SL	TM	UN	VO	WP	XQ	AR	BS	CT
ND	OE	PF	QG	RH	SJ	TK	UL	VM	WN	XO	AP	BQ	CR	DS
OC	PD	QE	RF	SG	TH	UJ	VK	WL	XM	AN	BO	CP	DQ	ER
PB	QC	RD	SE	TF	UG	VH	WJ	XK	AL	BM	CN	DO	EP	FQ
QA	RB	SC	TD	UE	VF	WG	XH	AJ	BK	CL	DM	EN	FO	GP
RX	SA	TB	UC	VD	WE	XF	AG	BH	CJ	DK	EL	FM	GN	HO
SW	TX	UA	VB	WC	XD	AE	BF	CG	DH	EJ	FK	GL	HM	JN
TV	UW	VX	WA	XB	AC	BD	CE	DF	EG	FH	GJ	HK	JL	KM
U	V	W	X	A	B	C	D	E	F	G	H	J	K	L

APRIL 1980 — EMOTIONAL

16 W	17 TH	18 F	19 SA	20 SU	21 M	22 TU	23 W	24 TH	25 F	26 SA	27 SU	28 M	29 TU	30 W	
S	T	U	V	W	X	Y	Z	1	2	3	A	B	C	D	
TR	US	VT	WU	XV	YW	ZX	1Y	2Z	31	A2	B3	CA	DB	EC	
UQ	VR	WS	XT	YU	ZV	1W	2X	3Y	AZ	B1	C2	D3	EA	FB	
VP	WQ	XR	YS	ZT	1U	2V	3W	AX	BY	CZ	D1	E2	F3	GA	
WO	XP	YQ	ZR	1S	2T	3U	AV	BW	CX	DY	EZ	F1	G2	H3	+
XN	YO	ZP	1Q	2R	3S	AT	BU	CV	DW	EX	FY	GZ	H1	J2	
YM	ZN	1O	2P	3Q	AR	BS	CT	DU	EV	FW	GX	HY	JZ	K1	
ZL	1M	2N	3O	AP	BQ	CR	DS	ET	FU	GV	HW	JX	KY	LZ	o
1K	2L	3M	AN	BO	CP	DQ	ER	FS	GT	HU	JV	KW	LX	MY	
2J	3K	AL	BM	CN	DO	EP	FQ	GR	HS	JT	KU	LV	MW	NX	
3H	AJ	BK	CL	DM	EN	FO	GP	HQ	JR	KS	LT	MU	NV	OW	−
AG	BH	CJ	DK	EL	FM	GN	HO	JP	KQ	LR	MS	NT	OU	PV	
BF	CG	DH	EJ	FK	GL	HM	JN	KO	LP	MQ	NR	OS	PT	QU	
CE	DF	EG	FH	GJ	HK	JL	KM	LN	MO	NP	OQ	PR	QS	RT	
D	E	F	G	H	J	K	L	M	N	O	P	Q	R	S	

APRIL 1980 — INTELLECTUAL

16 W	17 TH	18 F	19 SA	20 SU	21 M	22 TU	23 W	24 TH	25 F	26 SA	27 SU	28 M	29 TU	30 W
2	3	4	5	6	7	8	A	B	C	D	E	F	G	H
31	42	53	64	75	86	A7	B8	CA	DB	EC	FD	GE	HF	JG
4Z	51	62	73	84	A5	B6	C7	D8	EA	FB	GC	HD	JE	KF
5Y	6Z	71	82	A3	B4	C5	D6	E7	F8	GA	HB	JC	KD	LE
6X	7Y	8Z	A1	B2	C3	D4	E5	F6	G7	H8	JA	KB	LC	MD
7W	8X	AY	BZ	C1	D2	E3	F4	G5	H6	J7	K8	LA	MB	NC
8V	AW	BX	CY	DZ	E1	F2	G3	H4	J5	K6	L7	M8	NA	OB
AU	BV	CW	DX	EY	FZ	G1	H2	J3	K4	L5	M6	N7	O8	PA
BT	CU	DV	EW	FX	GY	HZ	J1	K2	L3	M4	N5	O6	P7	Q8
CS	DT	EU	FV	GW	HX	JY	KZ	L1	M2	N3	O4	P5	Q6	R7
DR	ES	FT	GU	HV	JW	KX	LY	MZ	N1	O2	P3	Q4	R5	S6
EQ	FR	GS	HT	JU	KV	LW	MX	NY	OZ	P1	Q2	R3	S4	T5
FP	GQ	HR	JS	KT	LU	MV	NW	OX	PY	QZ	R1	S2	T3	U4
GO	HP	JQ	KR	LS	MT	NU	OV	PW	QX	RY	SZ	T1	U2	V3
HN	JO	KP	LQ	MR	NS	OT	PU	QV	RW	SX	TY	UZ	V1	W2
JM	KN	LO	MP	NQ	OR	PS	QT	RU	SV	TW	UX	VY	WZ	X1
KL	LM	MN	NO	OP	PQ	QR	RS	ST	TU	UV	VW	WX	XY	YZ

MAY 1980 **PHYSICAL**

1 TH	2 F	3 SA	4 SU	5 M	6 TU	7 W	8 TH	9 F	10 SA	11 SU	12 M	13 TU	14 W	15 TH
AX	BA	CB	DC	FD	FF	GF	HG	JH	KJ	LK	ML	NM	ON	PO
BW	CX	DA	EB	FC	GD	HE	JF	KG	LH	MJ	NK	OL	PM	QN
CV	DW	EX	FA	GB	HC	JD	KE	LF	MG	NH	OJ	PK	QL	RM
DU	EV	FW	GX	HA	JB	KC	LD	ME	NF	OG	PH	QJ	RK	SL
ET	FU	GV	HW	JX	KA	LB	MC	ND	OE	PF	QG	RH	SJ	TK
FS	GT	HU	JV	KW	LX	MA	NB	OC	PD	QE	RF	SG	TH	UJ
GR	HS	JT	KU	LV	MW	NX	OA	PB	QC	RD	SE	TF	UG	VH
HQ	JR	KS	LT	MU	NV	OW	PX	QA	RB	SC	TD	UE	VF	WG
JP	KQ	LR	MS	NT	OU	PV	QW	RX	SA	TB	UC	VD	WE	XF
KO	LP	MQ	NR	OS	PT	QU	RV	SW	TX	UA	VB	WC	XD	AE
LN	MO	NP	OQ	PR	QS	RT	SU	TV	UW	VX	WA	XB	AC	BD
M	N	O	P	Q	R	S	T	U	V	W	X	A	B	C

MAY 1980 **EMOTIONAL**

1 TH	2 F	3 SA	4 SU	5 M	6 TU	7 W	8 TH	9 F	10 SA	11 SU	12 M	13 TU	14 W	15 TH	
E	F	G	H	J	K	L	M	N	O	P	Q	R	S	T	
FD	GE	HF	JG	KH	LJ	MK	NL	OM	PN	QO	RP	SQ	TR	US	
GC	HD	JE	KF	LG	MH	NJ	OK	PL	QM	RN	SO	TP	UQ	VR	
HB	JC	KD	LE	MF	NG	OH	PJ	QK	RL	SM	TN	UO	VP	WQ	
JA	KB	LC	MD	NE	OF	PG	QH	RJ	SK	TL	UM	VN	WO	XP	+
K3	LA	MB	NC	OD	PE	QF	RG	SH	TJ	UK	VL	WM	XN	YO	
L2	M3	NA	OB	PC	QD	RE	SF	TG	UH	VJ	WK	XL	YM	ZN	
M1	N2	O3	PA	QB	RC	SD	TE	UF	VG	WH	XJ	YK	ZL	1M	o
NZ	O1	P2	Q3	RA	SB	TC	UD	VE	WF	XG	YH	ZJ	1K	2L	
OY	PZ	Q1	R2	S3	TA	UB	VC	WD	XE	YF	ZG	1H	2J	3K	
PX	QY	RZ	S1	T2	U3	VA	WB	XC	YD	ZE	1F	2G	3H	AJ	−
QW	RX	SY	TZ	U1	V2	W3	XA	YB	ZC	1D	2E	3F	AG	BH	
RV	SW	TX	UY	VZ	W1	X2	Y3	ZA	1B	2C	3D	AE	BF	CG	
SU	TV	UW	VX	WY	XZ	Y1	Z2	13	2A	3B	AC	BD	CE	DF	
T	U	V	W	X	Y	Z	1	2	3	A	B	C	D	E	

MAY 1980 **INTELLECTUAL**

1 TH	2 F	3 SA	4 SU	5 M	6 TU	7 W	8 TH	9 F	10 SA	11 SU	12 M	13 TU	14 W	15 TH
J	K	L	M	N	O	P	Q	R	S	T	U	V	W	X
KH	LJ	MK	NL	OM	PN	QO	RP	SQ	TR	US	VT	WU	XV	YW
LG	MH	NJ	OK	PL	QM	RN	SO	TP	UQ	VR	WS	XT	YU	ZV
MF	NG	OH	PJ	QK	RL	SM	TN	UO	VP	WQ	XR	YS	ZT	1U
NE	OF	PG	QH	RJ	SK	TL	UM	VN	WO	XP	YQ	ZR	1S	2T
OD	PE	QF	RG	SH	TJ	UK	VL	WM	XN	YO	ZP	1Q	2R	3S
PC	QD	RE	SF	TG	UH	VJ	WK	XL	YM	ZN	1O	2P	3Q	4R
QB	RC	SD	TE	UF	VG	WH	XJ	YK	ZL	1M	2N	3O	4P	5Q
RA	SB	TC	UD	VE	WF	XG	YH	ZJ	1K	2L	3M	4N	5O	6P
S8	TA	UB	VC	WD	XE	YF	ZG	1H	2J	3K	4L	5M	6N	7O
T7	U8	VA	WB	XC	YD	ZE	1F	2G	3H	4J	5K	6L	7M	8N
U6	V7	W8	XA	YB	ZC	1D	2E	3F	4G	5H	6J	7K	8L	AM
V5	W6	X7	Y8	ZA	1B	2C	3D	4E	5F	6G	7H	8J	AK	BL
W4	X5	Y6	Z7	18	2A	3B	4C	5D	6E	7F	8G	AH	BJ	CK
X3	Y4	Z5	16	27	38	4A	5B	6C	7D	8E	AF	BG	CH	DJ
Y2	Z3	14	25	36	47	58	6A	7B	8C	AD	BE	CF	DG	EH
Z1	12	23	34	45	56	67	78	8A	AB	BC	CD	DE	EF	FG

MAY 1980 — PHYSICAL

16 F	17 SA	18 SU	19 M	20 TU	21 W	22 TH	23 F	24 SA	25 SU	26 M	27 TU	28 W	29 TH	30 F	31 SA
QP	RQ	SR	TS	UT	VU	WV	XW	AX	BA	CB	DC	ED	FE	GF	HG
RO	SP	TQ	UR	VS	WT	XU	AV	BW	CX	DA	EB	FC	GD	HE	JF
SN	TO	UP	VQ	WR	XS	AT	BU	CV	DW	EX	FA	GB	HC	JD	KE
TM	UN	VO	WP	XQ	AR	BS	CT	DU	EV	FW	GX	HA	JB	KC	LD
UL	VM	WN	XO	AP	BQ	CR	DS	ET	FU	GV	HW	JX	KA	LB	MC
VK	WL	XM	AN	BO	CP	DQ	ER	FS	GT	HU	JV	KW	LX	MA	NB
WJ	XK	AL	BM	CN	DO	EP	FQ	GR	HS	JT	KU	LV	MW	NX	OA
XH	AJ	BK	CL	DM	EN	FO	GP	HQ	JR	KS	LT	MU	NV	OW	PX
AG	BH	CJ	DK	EL	FM	GN	HO	JP	KQ	LR	MS	NT	OU	PV	QW
BF	CG	DH	EJ	FK	GL	HM	JN	KO	LP	MQ	NR	OS	PT	QU	RV
CE	DF	EG	FH	GJ	HK	JL	KM	LN	MO	NP	OQ	PR	QS	RT	SU
D	E	F	G	H	J	K	L	M	N	O	P	Q	R	S	T

MAY 1980 — EMOTIONAL

16 F	17 SA	18 SU	19 M	20 TU	21 W	22 TH	23 F	24 SA	25 SU	26 M	27 TU	28 W	29 TH	30 F	31 SA	
U	V	W	X	Y	Z	1	2	3	A	B	C	D	E	F	G	
VT	WU	XV	YW	ZX	1Y	2Z	31	A2	B3	CA	DB	EC	FD	GE	HF	
WS	XT	YU	ZV	1W	2X	3Y	AZ	B1	C2	D3	EA	FB	GC	HD	JE	
XR	YS	ZT	1U	2V	3W	AX	BY	CZ	D1	E2	F3	GA	HB	JC	KD	
YQ	ZR	1S	2T	3U	AV	BW	CX	DY	EZ	F1	G2	H3	JA	KB	LC	+
ZP	1Q	2R	3S	AT	BU	CV	DW	EX	FY	GZ	H1	J2	K3	LA	MB	
1O	2P	3Q	AR	BS	CT	DU	EV	FW	GX	HY	JZ	K1	L2	M3	NA	
2N	3O	AP	BQ	CR	DS	ET	FU	GV	HW	JX	KY	LZ	M1	N2	O3	0
3M	AN	BO	CP	DQ	ER	FS	GT	HU	JV	KW	LX	MY	NZ	O1	P2	
AL	BM	CN	DO	EP	FQ	GR	HS	JT	KU	LV	MW	NX	OY	PZ	Q1	
BK	CL	DM	EN	FO	GP	HQ	JR	KS	LT	MU	NV	OW	PX	QY	RZ	−
CJ	DK	EL	FM	GN	HO	JP	KQ	LR	MS	NT	OU	PV	QW	RX	SY	
DH	EJ	FK	GL	HM	JN	KO	LP	MQ	NR	OS	PT	QU	RV	SW	TX	
EG	FH	GJ	HK	JL	KM	LN	MO	NP	OQ	PR	QS	RT	SU	TV	UW	
F	G	H	J	K	L	M	N	O	P	Q	R	S	T	U	V	

MAY 1980 — INTELLECTUAL

16 F	17 SA	18 SU	19 M	20 TU	21 W	22 TH	23 F	24 SA	25 SU	26 M	27 TU	28 W	29 TH	30 F	31 SA
Y	Z	1	2	3	4	5	6	7	8	A	B	C	D	E	F
ZX	1Y	2Z	31	42	53	64	75	86	A7	B8	CA	DB	EC	FD	GE
1W	2X	3Y	4Z	51	62	73	84	A5	B6	C7	D8	EA	FB	GC	HD
2V	3W	4X	5Y	6Z	71	82	A3	B4	C5	D6	E7	F8	GA	HB	JC
3U	4V	5W	6X	7Y	8Z	A1	B2	C3	D4	E5	F6	G7	H8	JA	KB
4T	5U	6V	7W	8X	AY	BZ	C1	D2	E3	F4	G5	H6	J7	K8	LA
5S	6T	7U	8V	AW	BX	CY	DZ	E1	F2	G3	H4	J5	K6	L7	M8
6R	7S	8T	AU	BV	CW	DX	EY	FZ	G1	H2	J3	K4	L5	M6	N7
7Q	8R	AS	BT	CU	DV	EW	FX	GY	HZ	J1	K2	L3	M4	N5	O6
8P	AQ	BR	CS	DT	EU	FV	GW	HX	JY	KZ	L1	M2	N3	O4	P5
AO	BP	CQ	DR	ES	FT	GU	HV	JW	KX	LY	MZ	N1	O2	P3	Q4
BN	CO	DP	EQ	FR	GS	HT	JU	KV	LW	MX	NY	OZ	P1	Q2	R3
CM	DN	EO	FP	GQ	HR	JS	KT	LU	MV	NW	OX	PY	QZ	R1	S2
DL	EM	FN	GO	HP	JQ	KR	LS	MT	NU	OV	PW	QX	RY	SZ	T1
EK	FL	GM	HN	JO	KP	LQ	MR	NS	OT	PU	QV	RW	SX	TY	UZ
FJ	GK	HL	JM	KN	LO	MP	NQ	OR	PS	QT	RU	SV	TW	UX	VY
GH	HJ	JK	KL	LM	MN	NO	OP	PQ	QR	RS	ST	TU	UV	VW	WX

JUNE 1980 — PHYSICAL

1 SU	2 M	3 TU	4 W	5 TH	6 F	7 SA	8 SU	9 M	10 TU	11 W	12 TH	13 F	14 SA	15 SU
JH	KJ	LK	ML	NM	ON	PO	QP	RQ	SR	TS	UT	VU	WV	XW
KG	LH	MJ	NK	OL	PM	QN	RO	SP	TQ	UR	VS	WT	XU	AV
LF	MG	NH	OJ	PK	QL	RM	SN	TO	UP	VQ	WR	XS	AT	BU
ME	NF	OG	PH	QJ	RK	SL	TM	UN	VO	WP	XQ	AR	BS	CT
ND	OE	PF	QG	RH	SJ	TK	UL	VM	WN	XO	AP	BQ	CR	DS
OC	PD	QE	RF	SG	TH	UJ	VK	WL	XM	AN	BO	CP	DQ	ER
PB	QC	RD	SE	TF	UG	VH	WJ	XK	AL	BM	CN	DO	EP	FQ
QA	RB	SC	TD	UE	VF	WG	XH	AJ	BK	CL	DM	EN	FO	GP
RX	SA	TB	UC	VD	WE	XF	AG	BH	CJ	DK	EL	FM	GN	HO
SW	TX	UA	VB	WC	XD	AE	BF	CG	DH	EJ	FK	GL	HM	JN
TV	UW	VX	WA	XB	AC	BD	CE	DF	EG	FH	GJ	HK	JL	KM
U	V	W	X	A	B	C	D	E	F	G	H	J	K	L

JUNE 1980 — EMOTIONAL

1 SU	2 M	3 TU	4 W	5 TH	6 F	7 SA	8 SU	9 M	10 TU	11 W	12 TH	13 F	14 SA	15 SU	
H	J	K	L	M	N	O	P	Q	R	S	T	U	V	W	
JG	KH	LJ	MK	NL	OM	PN	QO	RP	SQ	TR	US	VT	WU	XV	
KF	LG	MH	NJ	OK	PL	QM	RN	SO	TP	UQ	VR	WS	XT	YU	
LE	MF	NG	OH	PJ	QK	RL	SM	TN	UO	VP	WQ	XR	YS	ZT	
MD	NE	OF	PG	QH	RJ	SK	TL	UM	VN	WO	XP	YQ	ZR	1S	+
NC	OD	PE	QF	RG	SH	TJ	UK	VL	WM	XN	YO	ZP	1Q	2R	
OB	PC	QD	RE	SF	TG	UH	VJ	WK	XL	YM	ZN	1O	2P	3Q	
PA	QB	RC	SD	TE	UF	VG	WH	XJ	YK	ZL	1M	2N	3O	AP	o
Q3	RA	SB	TC	UD	VE	WF	XG	YH	ZJ	1K	2L	3M	AN	BO	
R2	S3	TA	UB	VC	WD	XE	YF	ZG	1H	2J	3K	AL	BM	CN	
S1	T2	U3	VA	WB	XC	YD	ZE	1F	2G	3H	AJ	BK	CL	DM	
TZ	U1	V2	W3	XA	YB	ZC	1D	2E	3F	AG	BH	CJ	DK	EL	−
UY	VZ	W1	X2	Y3	ZA	1B	2C	3D	AE	BF	CG	DH	EJ	FK	
VX	WY	XZ	Y1	Z2	13	2A	3B	AC	BD	CE	DF	EG	FH	GJ	
W	X	Y	Z	1	2	3	A	B	C	D	E	F	G	H	

JUNE 1980 — INTELLECTUAL

1 SU	2 M	3 TU	4 W	5 TH	6 F	7 SA	8 SU	9 M	10 TU	11 W	12 TH	13 F	14 SA	15 SU
G	H	J	K	L	M	N	O	P	Q	R	S	T	U	V
HF	JG	KH	LJ	MK	NL	OM	PN	QO	RP	SQ	TR	US	VT	WU
JE	KF	LG	MH	NJ	OK	PL	QM	RN	SO	TP	UQ	VR	WS	XT
KD	LE	MF	NG	OH	PJ	QK	RL	SM	TN	UO	VP	WQ	XR	YS
LC	MD	NE	OF	PG	QH	RJ	SK	TL	UM	VN	WO	XP	YQ	ZR
MB	NC	OD	PE	QF	RG	SH	TJ	UK	VL	WM	XN	YO	ZP	1Q
NA	OB	PC	QD	RE	SF	TG	UH	VJ	WK	XL	YM	ZN	1O	2P
O8	PA	QB	RC	SD	TE	UF	VG	WH	XJ	YK	ZL	1M	2N	3O
P7	Q8	RA	SB	TC	UD	VE	WF	XG	YH	ZJ	1K	2L	3M	4N
Q6	R7	S8	TA	UB	VC	WD	XE	YF	ZG	1H	2J	3K	4L	5M
R5	S6	T7	U8	VA	WB	XC	YD	ZE	1F	2G	3H	4J	5K	6L
S4	T5	U6	V7	W8	XA	YB	ZC	1D	2E	3F	4G	5H	6J	7K
T3	U4	V5	W6	X7	Y8	ZA	1B	2C	3D	4E	5F	6G	7H	8J
U2	V3	W4	X5	Y6	Z7	18	2A	3B	4C	5D	6E	7F	8G	AH
V1	W2	X3	Y4	Z5	16	27	38	4A	5B	6C	7D	8E	AF	BG
WZ	X1	Y2	Z3	14	25	36	47	58	6A	7B	8C	AD	BE	CF
XY	YZ	Z1	12	23	34	45	56	67	78	8A	AB	BC	CD	DE

JUNE 1980 — PHYSICAL

16 M	17 TU	18 W	19 TH	20 F	21 SA	22 SU	23 M	24 TU	25 W	26 TH	27 F	28 SA	29 SU	30 M
AX	BA	CB	DC	ED	FE	GF	HG	JH	KJ	LK	ML	NM	ON	PO
BW	CX	DA	EB	FC	GD	HE	JF	KG	LH	MJ	NK	OL	PM	QN
CV	DW	EX	FA	GB	HC	JD	KE	LF	MG	NH	OJ	PK	QL	RM
DU	EV	FW	GX	HA	JB	KC	LD	ME	NF	OG	PH	QJ	RK	SL
ET	FU	GV	HW	JX	KA	LB	MC	ND	OE	PF	QG	RH	SJ	TK
FS	GT	HU	JV	KW	LX	MA	NB	OC	PD	QE	RF	SG	TH	UJ
GR	HS	JT	KU	LV	MW	NX	OA	PB	QC	RD	SE	TF	UG	VH
HQ	JR	KS	LT	MU	NV	OW	PX	QA	RB	SC	TD	UE	VF	WG
JP	KQ	LR	MS	NT	OU	PV	QW	RX	SA	TB	UC	VD	WE	XF
KO	LP	MQ	NR	OS	PT	QU	RV	SW	TX	UA	VB	WC	XD	AE
LN	MO	NP	OQ	PR	QS	RT	SU	TV	UW	VX	WA	XB	AC	BD
M	N	O	P	Q	R	S	T	U	V	W	X	A	B	C

JUNE 1980 — EMOTIONAL

16 M	17 TU	18 W	19 TH	20 F	21 SA	22 SU	23 M	24 TU	25 W	26 TH	27 F	28 SA	29 SU	30 M	
X	Y	Z	1	2	3	A	B	C	D	E	F	G	H	J	
YW	ZX	1Y	2Z	31	A2	B3	CA	DB	EC	FD	GE	HF	JG	KH	
ZV	1W	2X	3Y	AZ	B1	C2	D3	EA	FB	GC	HD	JE	KF	LG	
1U	2V	3W	AX	BY	CZ	D1	E2	F3	GA	HB	JC	KD	LE	MF	
2T	3U	AV	BW	CX	DY	EZ	F1	G2	H3	JA	KB	LC	MD	NE	+
3S	AT	BU	CV	DW	EX	FY	GZ	H1	J2	K3	LA	MB	NC	OD	
AR	BS	CT	DU	EV	FW	GX	HY	JZ	K1	L2	M3	NA	OB	PC	
BQ	CR	DS	ET	FU	GV	HW	JX	KY	LZ	M1	N2	O3	PA	QB	o
CP	DQ	ER	FS	GT	HU	JV	KW	LX	MY	NZ	O1	P2	Q3	RA	
DO	EP	FQ	GR	HS	JT	KU	LV	MW	NX	OY	PZ	Q1	R2	S3	
EN	FO	GP	HQ	JR	KS	LT	MU	NV	OW	PX	QY	RZ	S1	T2	−
FM	GN	HO	JP	KQ	LR	MS	NT	OU	PV	QW	RX	SY	TZ	U1	
GL	HM	JN	KO	LP	MQ	NR	OS	PT	QU	RV	SW	TX	UY	VZ	
HK	JL	KM	LN	MO	NP	OQ	PR	QS	RT	SU	TV	UW	VX	WY	
J	K	L	M	N	O	P	Q	R	S	T	U	V	W	X	

JUNE 1980 — INTELLECTUAL

16 M	17 TU	18 W	19 TH	20 F	21 SA	22 SU	23 M	24 TU	25 W	26 TH	27 F	28 SA	29 SU	30 M
W	X	Y	Z	1	2	3	4	5	6	7	8	A	B	C
XV	YW	ZX	1Y	2Z	31	42	53	64	75	86	A7	B8	CA	DB
YU	ZV	1W	2X	3Y	4Z	51	62	73	84	A5	B6	C7	D8	EA
ZT	1U	2V	3W	4X	5Y	6Z	71	82	A3	B4	C5	D6	E7	F8
1S	2T	3U	4V	5W	6X	7Y	8Z	A1	B2	C3	D4	E5	F6	G7
2R	3S	4T	5U	6V	7W	8X	AY	BZ	C1	D2	E3	F4	G5	H6
3Q	4R	5S	6T	7U	8V	AW	BX	CY	DZ	E1	F2	G3	H4	J5
4P	5Q	6R	7S	8T	AU	BV	CW	DX	EY	FZ	G1	H2	J3	K4
5O	6P	7Q	8R	AS	BT	CU	DV	EW	FX	GY	HZ	J1	K2	L3
6N	7O	8P	AQ	BR	CS	DT	EU	FV	GW	HX	JY	KZ	L1	M2
7M	8N	AO	BP	CQ	DR	ES	FT	GU	HV	JW	KX	LY	MZ	N1
8L	AM	BN	CO	DP	EQ	FR	GS	HT	JU	KV	LW	MX	NY	OZ
AK	BL	CM	DN	EO	FP	GQ	HR	JS	KT	LU	MV	NW	OX	PY
BJ	CK	DL	EM	FN	GO	HP	JQ	KR	LS	MT	NU	OV	PW	QX
CH	DJ	EK	FL	GM	HN	JO	KP	LQ	MR	NS	OT	PU	QV	RW
DG	EH	FJ	GK	HL	JM	KN	LO	MP	NQ	OR	PS	QT	RU	SV
EF	FG	GH	HJ	JK	KL	LM	MN	NO	OP	PQ	QR	RS	ST	TU

JULY 1980 **PHYSICAL**

1 TU	2 W	3 TH	4 F	5 SA	6 SU	7 M	8 TU	9 W	10 TH	11 F	12 SA	13 SU	14 M	15 TU
QP	RQ	SR	TS	UT	VU	WV	XW	AX	BA	CB	DC	ED	FE	GF
RO	SP	TQ	UR	VS	WT	XU	AV	BW	CX	DA	EB	FC	GD	HE
SN	TO	UP	VQ	WR	XS	AT	BU	CV	DW	EX	FA	GB	HC	JD
TM	UN	VO	WP	XQ	AR	BS	CT	DU	EV	FW	GX	HA	JB	KC
UL	VM	WN	XO	AP	BQ	CR	DS	ET	FU	GV	HW	JX	KA	LB
VK	WL	XM	AN	BO	CP	DQ	ER	FS	GT	HU	JV	KW	LX	MA
WJ	XK	AL	BM	CN	DO	EP	FQ	GR	HS	JT	KU	LV	MW	NX
XH	AJ	BK	CL	DM	EN	FO	GP	HQ	JR	KS	LT	MU	NV	OW
AG	BH	CJ	DK	EL	FM	GN	HO	JP	KQ	LR	MS	NT	OU	PV
BF	CG	DH	EJ	FK	GL	HM	JN	KO	LP	MQ	NR	OS	PT	QU
CE	DF	EG	FH	GJ	HK	JL	KM	LN	MO	NP	OQ	PR	QS	RT
D	E	F	G	H	J	K	L	M	N	O	P	Q	R	S

JULY 1980 **EMOTIONAL**

1 TU	2 W	3 TH	4 F	5 SA	6 SU	7 M	8 TU	9 W	10 TH	11 F	12 SA	13 SU	14 M	15 TU	
K	L	M	N	O	P	Q	R	S	T	U	V	W	X	Y	
LJ	MK	NL	OM	PN	QO	RP	SQ	TR	US	VT	WU	XV	YW	ZX	
MH	NJ	OK	PL	QM	RN	SO	TP	UQ	VR	WS	XT	YU	ZV	1W	
NG	OH	PJ	QK	RL	SM	TN	UO	VP	WQ	XR	YS	ZT	1U	2V	
OF	PG	QH	RJ	SK	TL	UM	VN	WO	XP	YQ	ZR	1S	2T	3U	+
PE	QF	RG	SH	TJ	UK	VL	WM	XN	YO	ZP	1Q	2R	3S	AT	
QD	RE	SF	TG	UH	VJ	WK	XL	YM	ZN	1O	2P	3Q	AR	BS	
RC	SD	TE	UF	VG	WH	XJ	YK	ZL	1M	2N	3O	AP	BQ	CR	o
SB	TC	UD	VE	WF	XG	YH	ZJ	1K	2L	3M	AN	BO	CP	DQ	
TA	UB	VC	WD	XE	YF	ZG	1H	2J	3K	AL	BM	CN	DO	EP	
U3	VA	WB	XC	YD	ZE	1F	2G	3H	AJ	BK	CL	DM	EN	FO	–
V2	W3	XA	YB	ZC	1D	2E	3F	AG	BH	CJ	DK	EL	FM	GN	
W1	X2	Y3	ZA	1B	2C	3D	AE	BF	CG	DH	EJ	FK	GL	HM	
XZ	Y1	Z2	13	2A	3B	AC	BD	CE	DF	EG	FH	GJ	HK	JL	
Y	Z	1	2	3	A	B	C	D	E	F	G	H	J	K	

JULY 1980 **INTELLECTUAL**

1 TU	2 W	3 TH	4 F	5 SA	6 SU	7 M	8 TU	9 W	10 TH	11 F	12 SA	13 SU	14 M	15 TU
D	E	F	G	H	J	K	L	M	N	O	P	Q	R	S
EC	FD	GE	HF	JG	KH	LJ	MK	NL	OM	PN	QO	RP	SQ	TR
FB	GC	HD	JE	KF	LG	MH	NJ	OK	PL	QM	RN	SO	TP	UQ
GA	HB	JC	KD	LE	MF	NG	OH	PJ	QK	RL	SM	TN	UO	VP
H8	JA	KB	LC	MD	NE	OF	PG	QH	RJ	SK	TL	UM	VN	WO
J7	K8	LA	MB	NC	OD	PE	QF	RG	SH	TJ	UK	VL	WM	XN
K6	L7	M8	NA	OB	PC	QD	RE	SF	TG	UH	VJ	WK	XL	YM
L5	M6	N7	O8	PA	QB	RC	SD	TE	UF	VG	WH	XJ	YK	ZL
M4	N5	O6	P7	Q8	RA	SB	TC	UD	VE	WF	XG	YH	ZJ	1K'
N3	O4	P5	Q6	R7	S8	TA	UB	VC	WD	XE	YF	ZG	1H	2J
O2	P3	Q4	R5	S6	T7	U8	VA	WB	XC	YD	ZE	1F	2G	3H
P1	Q2	R3	S4	T5	U6	V7	W8	XA	YB	ZC	1D	2E	3F	4G
QZ	R1	S2	T3	U4	V5	W6	X7	Y8	ZA	1B	2C	3D	4E	5F
RY	SZ	T1	U2	V3	W4	X5	Y6	Z7	18	2A	3B	4C	5D	6E
SX	TY	UZ	V1	W2	X3	Y4	Z5	16	27	38	4A	5B	6C	7D
TW	UX	VY	WZ	X1	Y2	Z3	14	25	36	47	58	6A	7B	8C
UV	VW	WX	XY	YZ	Z1	12	23	34	45	56	67	78	8A	AB

JULY 1980 **PHYSICAL**

16 W	17 TH	18 F	19 SA	20 SU	21 M	22 TU	23 W	24 TH	25 F	26 SA	27 SU	28 M	29 TU	30 W	31 TH
HG	JH	KJ	LK	ML	NM	ON	PO	QP	RQ	SR	TS	UT	VU	WV	XW
JF	KG	LH	MJ	NK	OL	PM	QN	RO	SP	TQ	UR	VS	WT	XU	AV
KE	LF	MG	NH	OJ	PK	QL	RM	SN	TO	UP	VQ	WR	XS	AT	BU
LD	ME	NF	OG	PH	QJ	RK	SL	TM	UN	VO	WP	XQ	AR	BS	CT
MC	ND	OE	PF	QG	RH	SJ	TK	UL	VM	WN	XO	AP	BQ	CR	DS
NB	OC	PD	QE	RF	SG	TH	UJ	VK	WL	XM	AN	BO	CP	DQ	ER
OA	PB	QC	RD	SE	TF	UG	VH	WJ	XK	AL	BM	CN	DO	EP	FQ
PX	QA	RB	SC	TD	UE	VF	WG	XH	AJ	BK	CL	DM	EN	FO	GP
QW	RX	SA	TB	UC	VD	WE	XF	AG	BH	CJ	DK	EL	FM	GN	HO
RV	SW	TX	UA	VB	WC	XD	AE	BF	CG	DH	EJ	FK	GL	HM	JN
SU	TV	UW	VX	WA	XB	AC	BD	CE	DF	EG	FH	GJ	HK	JL	KM
T	U	V	W	X	A	B	C	D	E	F	G	H	J	K	L

JULY 1980 **EMOTIONAL**

16 W	17 TH	18 F	19 SA	20 SU	21 M	22 TU	23 W	24 TH	25 F	26 SA	27 SU	28 M	29 TU	30 W	31 TH	
Z	1	2	3	A	B	C	D	E	F	G	H	J	K	L	M	
1Y	2Z	31	A2	B3	CA	DB	EC	FD	GE	HF	JG	KH	LJ	MK	NL	
2X	3Y	AZ	B1	C2	D3	EA	FB	GC	HD	JE	KF	LG	MH	NJ	OK	
3W	AX	BY	CZ	D1	E2	F3	GA	HB	JC	KD	LE	MF	NG	OH	PJ	
AV	BW	CX	DY	EZ	F1	G2	H3	JA	KB	LC	MD	NE	OF	PG	QH	+
BU	CV	DW	EX	FY	GZ	H1	J2	K3	LA	MB	NC	OD	PE	QF	RG	
CT	DU	EV	FW	GX	HY	JZ	K1	L2	M3	NA	OB	PC	QD	RE	SF	
DS	ET	FU	GV	HW	JX	KY	LZ	M1	N2	O3	PA	QB	RC	SD	TE	0
ER	FS	GT	HU	JV	KW	LX	MY	NZ	O1	P2	Q3	RA	SB	TC	UD	
FQ	GR	HS	JT	KU	LV	MW	NX	OY	PZ	Q1	R2	S3	TA	UB	VC	
GP	HQ	JR	KS	LT	MU	NV	OW	PX	QY	RZ	S1	T2	U3	VA	WB	–
HO	JP	KQ	LR	MS	NT	OU	PV	QW	RX	SY	TZ	U1	V2	W3	XA	
JN	KO	LP	MQ	NR	OS	PT	QU	RV	SW	TX	UY	VZ	W1	X2	Y3	
KM	LN	MO	NP	OQ	PR	QS	RT	SU	TV	UW	VX	WY	XZ	Y1	Z2	
L	M	N	O	P	Q	R	S	T	U	V	W	X	Y	Z	1	

JULY 1980 **INTELLECTUAL**

16 W	17 TH	18 F	19 SA	20 SU	21 M	22 TU	23 W	24 TH	25 F	26 SA	27 SU	28 M	29 TU	30 W	31 TH
T	U	V	W	X	Y	Z	1	2	3	4	5	6	7	8	A
US	VT	WU	XV	YW	ZX	1Y	2Z	31	42	53	64	75	86	A7	B8
VR	WS	XT	YU	ZV	1W	2X	3Y	4Z	51	62	73	84	A5	B6	C7
WQ	XR	YS	ZT	1U	2V	3W	4X	5Y	6Z	71	82	A3	B4	C5	D6
XP	YQ	ZR	1S	2T	3U	4V	5W	6X	7Y	8Z	A1	B2	C3	D4	E5
YO	ZP	1Q	2R	3S	4T	5U	6V	7W	8X	AY	BZ	C1	D2	E3	F4
ZN	1O	2P	3Q	4R	5S	6T	7U	8V	AW	BX	CY	DZ	E1	F2	G3
1M	2N	3O	4P	5Q	6R	7S	8T	AU	BV	CW	DX	EY	FZ	G1	H2
2L	3M	4N	5O	6P	7Q	8R	AS	BT	CU	DV	EW	FX	GY	HZ	J1
3K	4L	5M	6N	7O	8P	AQ	BR	CS	DT	EU	FV	GW	HX	JY	KZ
4J	5K	6L	7M	8N	AO	BP	CQ	DR	ES	FT	GU	HV	JW	KX	LY
5H	6J	7K	8L	AM	BN	CO	DP	EQ	FR	GS	HT	JU	KV	LW	MX
6G	7H	8J	AK	BL	CM	DN	EO	FP	GQ	HR	JS	KT	LU	MV	NW
7F	8G	AH	BJ	CK	DL	EM	FN	GO	HP	JQ	KR	LS	MT	NU	OV
8E	AF	BG	CH	DJ	EK	FL	GM	HN	JO	KP	LQ	MR	NS	OT	PU
AD	BE	CF	DG	EH	FJ	GK	HL	JM	KN	LO	MP	NQ	OR	PS	QT
BC	CD	DE	EF	FG	GH	HJ	JK	KL	LM	MN	NO	OP	PQ	QR	RS

AUGUST 1980 **PHYSICAL**

1 F	2 SA	3 SU	4 M	5 TU	6 W	7 TH	8 F	9 SA	10 SU	11 M	12 TU	13 W	14 TH	15 F
AX	BA	CB	DC	ED	FE	GF	HG	JH	KJ	LK	ML	NM	ON	PO
BW	CX	DA	EB	FC	GD	HE	JF	KG	LH	MJ	NK	OL	PM	QN
CV	DW	EX	FA	GB	HC	JD	KE	LF	MG	NH	OJ	PK	QL	RM
DU	EV	FW	GX	HA	JB	KC	LD	ME	NF	OG	PH	QJ	RK	SL
ET	FU	GV	HW	JX	KA	LB	MC	ND	OE	PF	QG	RH	SJ	TK
FS	GT	HU	JV	KW	LX	MA	NB	OC	PD	QE	RF	SG	TH	UJ
GR	HS	JT	KU	LV	MW	NX	OA	PB	QC	RD	SE	TF	UG	VH
HQ	JR	KS	LT	MU	NV	OW	PX	QA	RB	SC	TD	UE	VF	WG
JP	KQ	LR	MS	NT	OU	PV	QW	RX	SA	TB	UC	VD	WE	XF
KO	LP	MQ	NR	OS	PT	QU	RV	SW	TX	UA	VB	WC	XD	AE
LN	MO	NP	OQ	PR	QS	RT	SU	TV	UW	VX	WA	XB	AC	BD
M	N	O	P	Q	R	S	T	U	V	W	X	A	B	C

AUGUST 1980 **EMOTIONAL**

1 F	2 SA	3 SU	4 M	5 TU	6 W	7 TH	8 F	9 SA	10 SU	11 M	12 TU	13 W	14 TH	15 F	
N	O	P	Q	R	S	T	U	V	W	X	Y	Z	1	2	
OM	PN	QO	RP	SQ	TR	US	VT	WU	XV	YW	ZX	1Y	2Z	31	
PL	QM	RN	SO	TP	UQ	VR	WS	XT	YU	ZV	1W	2X	3Y	AZ	
QK	RL	SM	TN	UO	VP	WQ	XR	YS	ZT	1U	2V	3W	AX	BY	
RJ	SK	TL	UM	VN	WO	XP	YQ	ZR	1S	2T	3U	AV	BW	CX	+
SH	TJ	UK	VL	WM	XN	YO	ZP	1Q	2R	3S	AT	BU	CV	DW	
TG	UH	VJ	WK	XL	YM	ZN	1O	2P	3Q	AR	BS	CT	DU	EV	
UF	VG	WH	XJ	YK	ZL	1M	2N	3O	AP	BQ	CR	DS	ET	FU	o
VE	WF	XG	YH	ZJ	1K	2L	3M	AN	BO	CP	DQ	ER	FS	GT	
WD	XE	YF	ZG	1H	2J	3K	AL	BM	CN	DO	EP	FQ	GR	HS	
XC	YD	ZE	1F	2G	3H	AJ	BK	CL	DM	EN	FO	GP	HQ	JR	
YB	ZC	1D	2E	3F	AG	BH	CJ	DK	EL	FM	GN	HO	JP	KQ	−
ZA	1B	2C	3D	AE	BF	CG	DH	EJ	FK	GL	HM	JN	KO	LP	
13	2A	3B	AC	BD	CE	DF	EG	FH	GJ	HK	JL	KM	LN	MO	
2	3	A	B	C	D	E	F	G	H	J	K	L	M	N	

AUGUST 1980 **INTELLECTUAL**

1 F	2 SA	3 SU	4 M	5 TU	6 W	7 TH	8 F	9 SA	10 SU	11 M	12 TU	13 W	14 TH	15 F
B	C	D	E	F	G	H	J	K	L	M	N	O	P	Q
CA	DB	EC	FD	GE	HF	JG	KH	LJ	MK	NL	OM	PN	QO	RP
D8	EA	FB	GC	HD	JE	KF	LG	MH	NJ	OK	PL	QM	RN	SO
E7	F8	GA	HB	JC	KD	LE	MF	NG	OH	PJ	QK	RL	SM	TN
F6	G7	H8	JA	KB	LC	MD	NE	OF	PG	QH	RJ	SK	TL	UM
G5	H6	J7	K8	LA	MB	NC	OD	PE	QF	RG	SH	TJ	UK	VL
H4	J5	K6	L7	M8	NA	OB	PC	QD	RE	SF	TG	UH	VJ	WK
J3	K4	L5	M6	N7	O8	PA	QB	RC	SD	TE	UF	VG	WH	XJ
K2	L3	M4	N5	O6	P7	Q8	RA	SB	TC	UD	VE	WF	XG	YH
L1	M2	N3	O4	P5	Q6	R7	S8	TA	UB	VC	WD	XE	YF	ZG
MZ	N1	O2	P3	Q4	R5	S6	T7	U8	VA	WB	XC	YD	ZE	1F
NY	OZ	P1	Q2	R3	S4	T5	U6	V7	W8	XA	YB	ZC	1D	2E
OX	PY	QZ	R1	S2	T3	U4	V5	W6	X7	Y8	ZA	1B	2C	3D
PW	QX	RY	SZ	T1	U2	V3	W4	X5	Y6	Z7	18	2A	3B	4C
QV	RW	SX	TY	UZ	V1	W2	X3	Y4	Z5	16	27	38	4A	5B
RU	SV	TW	UX	VY	WZ	X1	Y2	Z3	14	25	36	47	58	6A
ST	TU	UV	VW	WX	XY	YZ	Z1	12	23	34	45	56	67	78

AUGUST 1980 — PHYSICAL

16 SA	17 SU	18 M	19 TU	20 W	21 TH	22 F	23 SA	24 SU	25 M	26 TU	27 W	28 TH	29 F	30 SA	31 SU
QP	RQ	SR	TS	UT	VU	WV	XW	AX	BA	CB	DC	ED	FE	GF	HG
RO	SP	TQ	UR	VS	WT	XU	AV	BW	CX	DA	EB	FC	GD	HE	JF
SN	TO	UP	VQ	WR	XS	AT	BU	CV	DW	EX	FA	GB	HC	JD	KE
TM	UN	VO	WP	XQ	AR	BS	CT	DU	EV	FW	GX	HA	JB	KC	LD
UL	VM	WN	XO	AP	BQ	CR	DS	ET	FU	GV	HW	JX	KA	LB	MC
VK	WL	XM	AN	BO	CP	DQ	ER	FS	GT	HU	JV	KW	LX	MA	NB
WJ	XK	AL	BM	CN	DO	EP	FQ	GR	HS	JT	KU	LV	MW	NX	OA
XH	AJ	BK	CL	DM	EN	FO	GP	HQ	JR	KS	LT	MU	NV	OW	PX
AG	BH	CJ	DK	EL	FM	GN	HO	JP	KQ	LR	MS	NT	OU	PV	QW
BF	CG	DH	EJ	FK	GL	HM	JN	KO	LP	MQ	NR	OS	PT	QU	RV
CE	DF	EG	FH	GJ	HK	JL	KM	LN	MO	NP	OQ	PR	QS	RT	SU
D	E	F	G	H	J	K	L	M	N	O	P	Q	R	S	T

AUGUST 1980 — EMOTIONAL

16 SA	17 SU	18 M	19 TU	20 W	21 TH	22 F	23 SA	24 SU	25 M	26 TU	27 W	28 TH	29 F	30 SA	31 SU	
3	A	B	C	D	E	F	G	H	J	K	L	M	N	O	P	
A2	B3	CA	DB	EC	FD	GE	HF	JG	KH	LJ	MK	NL	OM	PN	QO	
B1	C2	D3	EA	FB	GC	HD	JE	KF	LG	MH	NJ	OK	PL	QM	RN	
CZ	D1	E2	F3	GA	HB	JC	KD	LE	MF	NG	OH	PJ	QK	RL	SM	
DY	EZ	F1	G2	H3	JA	KB	LC	MD	NE	OF	PG	QH	RJ	SK	TL	+
EX	FY	GZ	H1	J2	K3	LA	MB	NC	OD	PE	QF	RG	SH	TJ	UK	
FW	GX	HY	JZ	K1	L2	M3	NA	OB	PC	QD	RE	SF	TG	UH	VJ	
GV	HW	JX	KY	LZ	M1	N2	O3	PA	QB	RC	SD	TE	UF	VG	WH	0
HU	JV	KW	LX	MY	NZ	O1	P2	Q3	RA	SB	TC	UD	VE	WF	XG	
JT	KU	LV	MW	NX	OY	PZ	Q1	R2	S3	TA	UB	VC	WD	XE	YF	
KS	LT	MU	NV	OW	PX	QY	RZ	S1	T2	U3	VA	WB	XC	YD	ZE	−
LR	MS	NT	OU	PV	QW	RX	SY	TZ	U1	V2	W3	XA	YB	ZC	1D	
MQ	NR	OS	PT	QU	RV	SW	TX	UY	VZ	W1	X2	Y3	ZA	1B	2C	
NP	OQ	PR	QS	RT	SU	TV	UW	VX	WY	XZ	Y1	Z2	13	2A	3B	
O	P	Q	R	S	T	U	V	W	X	Y	Z	1	2	3	A	

AUGUST 1980 — INTELLECTUAL

16 SA	17 SU	18 M	19 TU	20 W	21 TH	22 F	23 SA	24 SU	25 M	26 TU	27 W	28 TH	29 F	30 SA	31 SU
R	S	T	U	V	W	X	Y	Z	1	2	3	4	5	6	7
SQ	TR	US	VT	WU	XV	YW	ZX	1Y	2Z	31	42	53	64	75	86
TP	UQ	VR	WS	XT	YU	ZV	1W	2X	3Y	4Z	51	62	73	84	A5
UO	VP	WQ	XR	YS	ZT	1U	2V	3W	4X	5Y	6Z	71	82	A3	B4
VN	WO	XP	YQ	ZR	1S	2T	3U	4V	5W	6X	7Y	8Z	A1	B2	C3
WM	XN	YO	ZP	1Q	2R	3S	4T	5U	6V	7W	8X	AY	BZ	C1	D2
XL	YM	ZN	1O	2P	3Q	4R	5S	6T	7U	8V	AW	BX	CY	DZ	E1
YK	ZL	1M	2N	3O	4P	5Q	6R	7S	8T	AU	BV	CW	DX	EY	FZ
ZJ	1K	2L	3M	4N	5O	6P	7Q	8R	AS	BT	CU	DV	EW	FX	GY
1H	2J	3K	4L	5M	6N	7O	8P	AQ	BR	CS	DT	EU	FV	GW	HX
2G	3H	4J	5K	6L	7M	8N	AO	BP	CQ	DR	ES	FT	GU	HV	JW
3F	4G	5H	6J	7K	8L	AM	BN	CO	DP	EQ	FR	GS	HT	JU	KV
4E	5F	6G	7H	8J	AK	BL	CM	DN	EO	FP	GQ	HR	JS	KT	LU
5D	6E	7F	8G	AH	BJ	CK	DL	EM	FN	GO	HP	JQ	KR	LS	MT
6C	7D	8E	AF	BG	CH	DJ	EK	FL	GM	HN	JO	KP	LQ	MR	NS
7B	8C	AD	BE	CF	DG	EH	FJ	GK	HL	JM	KN	LO	MP	NQ	OR
8A	AB	BC	CD	DE	EF	FG	GH	HJ	JK	KL	LM	MN	NO	OP	PQ

SEPTEMBER 1980 **PHYSICAL**

1 M	2 TU	3 W	4 TH	5 F	6 SA	7 SU	8 M	9 TU	10 W	11 TH	12 F	13 SA	14 SU	15 M
JH	KJ	LK	ML	NM	ON	PO	QP	RQ	SR	TS	UT	VU	WV	XW
KG	LH	MJ	NK	OL	PM	QN	RO	SP	TQ	UR	VS	WT	XU	AV
LF	MG	NH	OJ	PK	QL	RM	SN	TO	UP	VQ	WR	XS	AT	BU
ME	NF	OG	PH	QJ	RK	SL	TM	UN	VO	WP	XQ	AR	BS	CT
ND	OE	PF	QG	RH	SJ	TK	UL	VM	WN	XO	AP	BQ	CR	DS
OC	PD	QE	RF	SG	TH	UJ	VK	WL	XM	AN	BO	CP	DQ	ER
PB	QC	RD	SE	TF	UG	VH	WJ	XK	AL	BM	CN	DO	EP	FQ
QA	RB	SC	TD	UE	VF	WG	XH	AJ	BK	CL	DM	EN	FO	GP
RX	SA	TB	UC	VD	WE	XF	AG	BH	CJ	DK	EL	FM	GN	HO
SW	TX	UA	VB	WC	XD	AE	BF	CG	DH	EJ	FK	GL	HM	JN
TV	UW	VX	WA	XB	AC	BD	CE	DF	EG	FH	GJ	HK	JL	KM
U	V	W	X	A	B	C	D	E	F	G	H	J	K	L

SEPTEMBER 1980 **EMOTIONAL**

1 M	2 TU	3 W	4 TH	5 F	6 SA	7 SU	8 M	9 TU	10 W	11 TH	12 F	13 SA	14 SU	15 M	
Q	R	S	T	U	V	W	X	Y	Z	1	2	3	A	B	
RP	SQ	TR	US	VT	WU	XV	YW	ZX	1Y	2Z	31	A2	B3	CA	
SO	TP	UQ	VR	WS	XT	YU	ZV	1W	2X	3Y	AZ	B1	C2	D3	
TN	UO	VP	WQ	XR	YS	ZT	1U	2V	3W	AX	BY	CZ	D1	E2	
UM	VN	WO	XP	YQ	ZR	1S	2T	3U	AV	BW	CX	DY	EZ	F1	+
VL	WM	XN	YO	ZP	1Q	2R	3S	AT	BU	CV	DW	EX	FY	GZ	
WK	XL	YM	ZN	1O	2P	3Q	AR	BS	CT	DU	EV	FW	GX	HY	
XJ	YK	ZL	1M	2N	3O	AP	BQ	CR	DS	ET	FU	GV	HW	JX	o
YH	ZJ	1K	2L	3M	AN	BO	CP	DQ	ER	FS	GT	HU	JV	KW	
ZG	1H	2J	3K	AL	BM	CN	DO	EP	FQ	GR	HS	JT	KU	LV	
1F	2G	3H	AJ	BK	CL	DM	EN	FO	GP	HQ	JR	KS	LT	MU	−
2E	3F	AG	BH	CJ	DK	EL	FM	GN	HO	JP	KQ	LR	MS	NT	
3D	AE	BF	CG	DH	EJ	FK	GL	HM	JN	KO	LP	MQ	NR	OS	
AC	BD	CE	DF	EG	FH	GJ	HK	JL	KM	LN	MO	NP	OQ	PR	
B	C	D	E	F	G	H	J	K	L	M	N	O	P	Q	

SEPTEMBER 1980 **INTELLECTUAL**

1 M	2 TU	3 W	4 TH	5 F	6 SA	7 SU	8 M	9 TU	10 W	11 TH	12 F	13 SA	14 SU	15 M
8	A	B	C	D	E	F	G	H	J	K	L	M	N	O
A7	B8	CA	DB	EC	FD	GE	HF	JG	KH	LJ	MK	NL	OM	PN
B6	C7	D8	EA	FB	GC	HD	JE	KF	LG	MH	NJ	OK	PL	QM
C5	D6	E7	F8	GA	HB	JC	KD	LE	MF	NG	OH	PJ	QK	RL
D4	E5	F6	G7	H8	JA	KB	LC	MD	NE	OF	PG	QH	RJ	SK
E3	F4	G5	H6	J7	K8	LA	MB	NC	OD	PE	QF	RG	SH	TJ
F2	G3	H4	J5	K6	L7	M8	NA	OB	PC	QD	RE	SF	TG	UH
G1	H2	J3	K4	L5	M6	N7	O8	PA	QB	RC	SD	TE	UF	VG
HZ	J1	K2	L3	M4	N5	O6	P7	Q8	RA	SB	TC	UD	VE	WF
JY	KZ	L1	M2	N3	O4	P5	Q6	R7	S8	TA	UB	VC	WD	XE
KX	LY	MZ	N1	O2	P3	Q4	R5	S6	T7	U8	VA	WB	XC	YD
LW	MX	NY	OZ	P1	Q2	R3	S4	T5	U6	V7	W8	XA	YB	ZC
MV	NW	OX	PY	QZ	R1	S2	T3	U4	V5	W6	X7	Y8	ZA	1B
NU	OV	PW	QX	RY	SZ	T1	U2	V3	W4	X5	Y6	Z7	18	2A
OT	PU	QV	RW	SX	TY	UZ	V1	W2	X3	Y4	Z5	16	27	38
PS	QT	RU	SV	TW	UX	VY	WZ	X1	Y2	Z3	14	25	36	47
QR	RS	ST	TU	UV	VW	WX	XY	YZ	Z1	12	23	34	45	56

SEPTEMBER 1980 — PHYSICAL

16 TU	17 W	18 TH	19 F	20 SA	21 SU	22 M	23 TU	24 W	25 TH	26 F	27 SA	28 SU	29 M	30 TU
AX	BA	CB	DC	ED	FE	GF	HG	JH	KJ	LK	ML	NM	ON	PO
BW	CX	DA	EB	FC	GD	HE	JF	KG	LH	MJ	NK	OL	PM	QN
CV	DW	EX	FA	GB	HC	JD	KE	LF	MG	NH	OJ	PK	QL	RM
DU	EV	FW	GX	HA	JB	KC	LD	ME	NF	OG	PH	QJ	RK	SL
ET	FU	GV	HW	JX	KA	LB	MC	ND	OE	PF	QG	RH	SJ	TK
FS	GT	HU	JV	KW	LX	MA	NB	OC	PD	QE	RF	SG	TH	UJ
GR	HS	JT	KU	LV	MW	NX	OA	PB	QC	RD	SE	TF	UG	VH
HQ	JR	KS	LT	MU	NV	OW	PX	QA	RB	SC	TD	UE	VF	WG
JP	KQ	LR	MS	NT	OU	PV	QW	RX	SA	TB	UC	VD	WE	XF
KO	LP	MQ	NR	OS	PT	QU	RV	SW	TX	UA	VB	WC	XD	AE
LN	MO	NP	OQ	PR	QS	RT	SU	TV	UW	VX	WA	XB	AC	BD
M	N	O	P	Q	R	S	T	U	V	W	X	A	B	C

SEPTEMBER 1980 — EMOTIONAL

16 TU	17 W	18 TH	19 F	20 SA	21 SU	22 M	23 TU	24 W	25 TH	26 F	27 SA	28 SU	29 M	30 TU	
C	D	E	F	G	H	J	K	L	M	N	O	P	Q	R	
DB	EC	FD	GE	HF	JG	KH	LJ	MK	NL	OM	PN	QO	RP	SQ	
EA	FB	GC	HD	JE	KF	LG	MH	NJ	OK	PL	QM	RN	SO	TP	
F3	GA	HB	JC	KD	LE	MF	NG	OH	PJ	QK	RL	SM	TN	UO	
G2	H3	JA	KB	LC	MD	NE	OF	PG	QH	RJ	SK	TL	UM	VN	+
H1	J2	K3	LA	MB	NC	OD	PE	QF	RG	SH	TJ	UK	VL	WM	
JZ	K1	L2	M3	NA	OB	PC	QD	RE	SF	TG	UH	VJ	WK	XL	
KY	LZ	M1	N2	O3	PA	QB	RC	SD	TE	UF	VG	WH	XJ	YK	o
LX	MY	NZ	O1	P2	Q3	RA	SB	TC	UD	VE	WF	XG	YH	ZJ	
MW	NX	OY	PZ	Q1	R2	S3	TA	UB	VC	WD	XE	YF	ZG	1H	
NV	OW	PX	QY	RZ	S1	T2	U3	VA	WB	XC	YD	ZE	1F	2G	−
OU	PV	QW	RX	SY	TZ	U1	V2	W3	XA	YB	ZC	1D	2E	3F	
PT	QU	RV	SW	TX	UY	VZ	W1	X2	Y3	ZA	1B	2C	3D	AE	
QS	RT	SU	TV	UW	VX	WY	XZ	Y1	Z2	13	2A	3B	AC	BD	
R	S	T	U	V	W	X	Y	Z	1	2	3	A	B	C	

SEPTEMBER 1980 — INTELLECTUAL

16 TU	17 W	18 TH	19 F	20 SA	21 SU	22 M	23 TU	24 W	25 TH	26 F	27 SA	28 SU	29 M	30 TU
P	Q	R	S	T	U	V	W	X	Y	Z	1	2	3	4
QO	RP	SQ	TR	US	VT	WU	XV	YW	ZX	1Y	2Z	31	42	53
RN	SO	TP	UQ	VR	WS	XT	YU	ZV	1W	2X	3Y	4Z	51	62
SM	TN	UO	VP	WQ	XR	YS	ZT	1U	2V	3W	4X	5Y	6Z	71
TL	UM	VN	WO	XP	YQ	ZR	1S	2T	3U	4V	5W	6X	7Y	8Z
UK	VL	WM	XN	YO	ZP	1Q	2R	3S	4T	5U	6V	7W	8X	AY
VJ	WK	XL	YM	ZN	1O	2P	3Q	4R	5S	6T	7U	8V	AW	BX
WH	XJ	YK	ZL	1M	2N	3O	4P	5Q	6R	7S	8T	AU	BV	CW
XG	YH	ZJ	1K	2L	3M	4N	5O	6P	7Q	8R	AS	BT	CU	DV
YF	ZG	1H	2J	3K	4L	5M	6N	7O	8P	AQ	BR	CS	DT	EU
ZE	1F	2G	3H	4J	5K	6L	7M	8N	AO	BP	CQ	DR	ES	FT
1D	2E	3F	4G	5H	6J	7K	8L	AM	BN	CO	DP	EQ	FR	GS
2C	3D	4E	5F	6G	7H	8J	AK	BL	CM	DN	EO	FP	GQ	HR
3B	4C	5D	6E	7F	8G	AH	BJ	CK	DL	EM	FN	GO	HP	JQ
4A	5B	6C	7D	8E	AF	BG	CH	DJ	EK	FL	GM	HN	JO	KP
58	6A	7B	8C	AD	BE	CF	DG	EH	FJ	GK	HL	JM	KN	LO
67	78	8A	AB	BC	CD	DE	EF	FG	GH	HJ	JK	KL	LM	MN

OCTOBER 1980 — PHYSICAL

1 W	2 TH	3 F	4 SA	5 SU	6 M	7 TU	8 W	9 TH	10 F	11 SA	12 SU	13 M	14 TU	15 W
QP	RQ	SR	TS	UT	VU	WV	XW	AX	BA	CB	DC	ED	FE	GF
RO	SP	TQ	UR	VS	WT	XU	AV	BW	CX	DA	EB	FC	GD	HE
SN	TO	UP	VQ	WR	XS	AT	BU	CV	DW	EX	FA	GB	HC	JD
TM	UN	VO	WP	XQ	AR	BS	CT	DU	EV	FW	GX	HA	JB	KC
UL	VM	WN	XO	AP	BQ	CR	DS	ET	FU	GV	HW	JX	KA	LB
VK	WL	XM	AN	BO	CP	DQ	ER	FS	GT	HU	JV	KW	LX	MA
WJ	XK	AL	BM	CN	DO	EP	FQ	GR	HS	JT	KU	LV	MW	NX
XH	AJ	BK	CL	DM	EN	FO	GP	HQ	JR	KS	LT	MU	NV	OW
AG	BH	CJ	DK	EL	FM	GN	HO	JP	KQ	LR	MS	NT	OU	PV
BF	CG	DH	EJ	FK	GL	HM	JN	KO	LP	MQ	NR	OS	PT	QU
CE	DF	EG	FH	GJ	HK	JL	KM	LN	MO	NP	OQ	PR	QS	RT
D	E	F	G	H	J	K	L	M	N	O	P	Q	R	S

OCTOBER 1980 — EMOTIONAL

1 W	2 TH	3 F	4 SA	5 SU	6 M	7 TU	8 W	9 TH	10 F	11 SA	12 SU	13 M	14 TU	15 W	
S	T	U	V	W	X	Y	Z	1	2	3	A	B	C	D	
TR	US	VT	WU	XV	YW	ZX	1Y	2Z	31	A2	B3	CA	DB	EC	
UQ	VR	WS	XT	YU	ZV	1W	2X	3Y	AZ	B1	C2	D3	EA	FB	
VP	WQ	XR	YS	ZT	1U	2V	3W	AX	BY	CZ	D1	E2	F3	GA	
WO	XP	YQ	ZR	1S	2T	3U	AV	BW	CX	DY	EZ	F1	G2	H3	+
XN	YO	ZP	1Q	2R	3S	AT	BU	CV	DW	EX	FY	GZ	H1	J2	
YM	ZN	10	2P	3Q	AR	BS	CT	DU	EV	FW	GX	HY	JZ	K1	
ZL	1M	2N	30	AP	BQ	CR	DS	ET	FU	GV	HW	JX	KY	LZ	o
1K	2L	3M	AN	BO	CP	DQ	ER	FS	GT	HU	JV	KW	LX	MY	
2J	3K	AL	BM	CN	DO	EP	FQ	GR	HS	JT	KU	LV	MW	NX	
3H	AJ	BK	CL	DM	EN	FO	GP	HQ	JR	KS	LT	MU	NV	OW	–
AG	BH	CJ	DK	EL	FM	GN	HO	JP	KQ	LR	MS	NT	OU	PV	
BF	CG	DH	EJ	FK	GL	HM	JN	KO	LP	MQ	NR	OS	PT	QU	
CE	DF	EG	FH	GJ	HK	JL	KM	LN	MO	NP	OQ	PR	QS	RT	
D	E	F	G	H	J	K	L	M	N	O	P	Q	R	S	

OCTOBER 1980 — INTELLECTUAL

1 W	2 TH	3 F	4 SA	5 SU	6 M	7 TU	8 W	9 TH	10 F	11 SA	12 SU	13 M	14 TU	15 W
5	6	7	8	A	B	C	D	E	F	G	H	J	K	L
64	75	86	A7	B8	CA	DB	EC	FD	GE	HF	JG	KH	LJ	MK
73	84	A5	B6	C7	D8	EA	FB	GC	HD	JE	KF	LG	MH	NJ
82	A3	B4	C5	D6	E7	F8	GA	HB	JC	KD	LE	MF	NG	OH
A1	B2	C3	D4	E5	F6	G7	H8	JA	KB	LC	MD	NE	OF	PG
BZ	C1	D2	E3	F4	G5	H6	J7	K8	LA	MB	NC	OD	PE	QF
CY	DZ	E1	F2	G3	H4	J5	K6	L7	M8	NA	OB	PC	QD	RE
DX	EY	FZ	G1	H2	J3	K4	L5	M6	N7	O8	PA	QB	RC	SD
EW	FX	GY	HZ	J1	K2	L3	M4	N5	O6	P7	Q8	RA	SB	TC
FV	GW	HX	JY	KZ	L1	M2	N3	O4	P5	Q6	R7	S8	TA	UB
GU	HV	JW	KX	LY	MZ	N1	O2	P3	Q4	R5	S6	T7	U8	VA
HT	JU	KV	LW	MX	NY	OZ	P1	Q2	R3	S4	T5	U6	V7	W8
JS	KT	LU	MV	NW	OX	PY	QZ	R1	S2	T3	U4	V5	W6	X7
KR	LS	MT	NU	OV	PW	QX	RY	SZ	T1	U2	V3	W4	X5	Y6
LQ	MR	NS	OT	PU	QV	RW	SX	TY	UZ	V1	W2	X3	Y4	Z5
MP	NQ	OR	PS	QT	RU	SV	TW	UX	VY	WZ	X1	Y2	Z3	14
NO	OP	PQ	QR	RS	ST	TU	UV	VW	WX	XY	YZ	Z1	12	23

OCTOBER 1980 **PHYSICAL**

16 TH	17 F	18 SA	19 SU	20 M	21 TU	22 W	23 TH	24 F	25 SA	26 SU	27 M	28 TU	29 W	30 TH	31 F
HG	JH	KJ	LK	ML	NM	ON	PO	QP	RQ	SR	TS	UT	VU	WV	XW
JF	KG	LH	MJ	NK	OL	PM	QN	RO	SP	TQ	UR	VS	WT	XU	AV
KE	LF	MG	NH	OJ	PK	QL	RM	SN	TO	UP	VQ	WR	XS	AT	BU
LD	ME	NF	OG	PH	QJ	RK	SL	TM	UN	VO	WP	XQ	AR	BS	CT
MC	ND	OE	PF	QG	RH	SJ	TK	UL	VM	WN	XO	AP	BQ	CR	DS
NB	OC	PD	QE	RF	SG	TH	UJ	VK	WL	XM	AN	BO	CP	DQ	ER
OA	PB	QC	RD	SE	TF	UG	VH	WJ	XK	AL	BM	CN	DO	EP	FQ
PX	QA	RB	SC	TD	UE	VF	WG	XH	AJ	BK	CL	DM	EN	FO	GP
QW	RX	SA	TB	UC	VD	WE	XF	AG	BH	CJ	DK	EL	FM	GN	HO
RV	SW	TX	UA	VB	WC	XD	AE	BF	CG	DH	EJ	FK	GL	HM	JN
SU	TV	UW	VX	WA	XB	AC	BD	CE	DF	EG	FH	GJ	HK	JL	KM
T	U	V	W	X	A	B	C	D	E	F	G	H	J	K	L

OCTOBER 1980 **EMOTIONAL**

16 TH	17 F	18 SA	19 SU	20 M	21 TU	22 W	23 TH	24 F	25 SA	26 SU	27 M	28 TU	29 W	30 TH	31 F	
E	F	G	H	J	K	L	M	N	O	P	Q	R	S	T	U	
FD	GE	HF	JG	KH	LJ	MK	NL	OM	PN	QO	RP	SQ	TR	US	VT	
GC	HD	JE	KF	LG	MH	NJ	OK	PL	QM	RN	SO	TP	UQ	VR	WS	
HB	JC	KD	LE	MF	NG	OH	PJ	QK	RL	SM	TN	UO	VP	WQ	XR	
JA	KB	LC	MD	NE	OF	PG	QH	RJ	SK	TL	UM	VN	WO	XP	YQ	+
K3	LA	MB	NC	OD	PE	QF	RG	SH	TJ	UK	VL	WM	XN	YO	ZP	
L2	M3	NA	OB	PC	QD	RE	SF	TG	UH	VJ	WK	XL	YM	ZN	1O	
M1	N2	O3	PA	QB	RC	SD	TE	UF	VG	WH	XJ	YK	ZL	1M	2N	o
NZ	O1	P2	Q3	RA	SB	TC	UD	VE	WF	XG	YH	ZJ	1K	2L	3M	
OY	PZ	Q1	R2	S3	TA	UB	VC	WD	XE	YF	ZG	1H	2J	3K	AL	
PX	QY	RZ	S1	T2	U3	VA	WB	XC	YD	ZE	1F	2G	3H	AJ	BK	−
QW	RX	SY	TZ	U1	V2	W3	XA	YB	ZC	1D	2E	3F	AG	BH	CJ	
RV	SW	TX	UY	VZ	W1	X2	Y3	ZA	1B	2C	3D	AE	BF	CG	DH	
SU	TV	UW	VX	WY	XZ	Y1	Z2	13	2A	3B	AC	BD	CE	DF	EG	
T	U	V	W	X	Y	Z	1	2	3	A	B.	C	D	E	F	

OCTOBER 1980 **INTELLECTUAL**

16 TH	17 F	18 SA	19 SU	20 M	21 TU	22 W	23 TH	24 F	25 SA	26 SU	27 M	28 TU	29 W	30 TH	31 F
M	N	O	P	Q	R	S	T	U	V	W	X	Y	Z	1	2
NL	OM	PN	QO	RP	SQ	TR	US	VT	WU	XV	YW	ZX	1Y	2Z	31
OK	PL	QM	RN	SO	TP	UQ	VR	WS	XT	YU	ZV	1W	2X	3Y	4Z
PJ	QK	RL	SM	TN	UO	VP	WQ	XR	YS	ZT	1U	2V	3W	4X	5Y
QH	RJ	SK	TL	UM	VN	WO	XP	YQ	ZR	1S	2T	3U	4V	5W	6X
RG	SH	TJ	UK	VL	WM	XN	YO	ZP	1Q	2R	3S	4T	5U	6V	7W
SF	TG	UH	VJ	WK	XL	YM	ZN	1O	2P	3Q	4R	5S	6T	7U	8V
TE	UF	VG	WH	XJ	YK	ZL	1M	2N	3O	4P	5Q	6R	7S	8T	AU
UD	VE	WF	XG	YH	ZJ	1K	2L	3M	4N	5O	6P	7Q	8R	AS	BT
VC	WD	XE	YF	ZG	1H	2J	3K	4L	5M	6N	7O	8P	AQ	BR	CS
WB	XC	YD	ZE	1F	2G	3H	4J	5K	6L	7M	8N	AO	BP	CQ	DR
XA	YB	ZC	1D	2E	3F	4G	5H	6J	7K	8L	AM	BN	CO	DP	EQ
Y8	ZA	1B	2C	3D	4E	5F	6G	7H	8J	AK	BL	CM	DN	EO	FP
Z7	18	2A	3B	4C	5D	6E	7F	8G	AH	BJ	CK	DL	EM	FN	GO
16	27	38	4A	5B	6C	7D	8E	AF	BG	CH	DJ	EK	FL	GM	HN
25	36	47	58	6A	7B	8C	AD	BE	CF	DG	EH	FJ	GK	HL	JM
34	45	56	67	78	8A	AB	BC	CD	DE	EF	FG	GH	HJ	JK	KL

NOVEMBER 1980 — PHYSICAL

1 SA	2 SU	3 M	4 TU	5 W	6 TH	7 F	8 SA	9 SU	10 M	11 TU	12 W	13 TH	14 F	15 SA
AX	BA	CB	DC	ED	FE	GF	HG	JH	KJ	LK	ML	NM	ON	PO
BW	CX	DA	EB	FC	GD	HE	JF	KG	LH	MJ	NK	OL	PM	QN
CV	DW	EX	FA	GB	HC	JD	KE	LF	MG	NH	OJ	PK	QL	RM
DU	EV	FW	GX	HA	JB	KC	LD	ME	NF	OG	PH	QJ	RK	SL
ET	FU	GV	HW	JX	KA	LB	MC	ND	OE	PF	QG	RH	SJ	TK
FS	GT	HU	JV	KW	LX	MA	NB	OC	PD	QE	RF	SG	TH	UJ
GR	HS	JT	KU	LV	MW	NX	OA	PB	QC	RD	SE	TF	UG	VH
HQ	JR	KS	LT	MU	NV	OW	PX	QA	RB	SC	TD	UE	VF	WG
JP	KQ	LR	MS	NT	OU	PV	QW	RX	SA	TB	UC	VD	WE	XF
KO	LP	MQ	NR	OS	PT	QU	RV	SW	TX	UA	VB	WC	XD	AE
LN	MO	NP	OQ	PR	QS	RT	SU	TV	UW	VX	WA	XB	AC	BD
M	N	O	P	Q	R	S	T	U	V	W	X	A	B	C

NOVEMBER 1980 — EMOTIONAL

1 SA	2 SU	3 M	4 TU	5 W	6 TH	7 F	8 SA	9 SU	10 M	11 TU	12 W	13 TH	14 F	15 SA	
V	W	X	Y	Z	1	2	3	A	B	C	D	E	F	G	
WU	XV	YW	ZX	1Y	2Z	31	A2	B3	CA	DB	EC	FD	GE	HF	
XT	YU	ZV	1W	2X	3Y	AZ	B1	C2	D3	EA	FB	GC	HD	JE	
YS	ZT	1U	2V	3W	AX	BY	CZ	D1	E2	F3	GA	HB	JC	KD	
ZR	1S	2T	3U	AV	BW	CX	DY	EZ	F1	G2	H3	JA	KB	LC	+
1Q	2R	3S	AT	BU	CV	DW	EX	FY	GZ	H1	J2	K3	LA	MB	
2P	3Q	AR	BS	CT	DU	EV	FW	GX	HY	JZ	K1	L2	M3	NA	
3O	AP	BQ	CR	DS	ET	FU	GV	HW	JX	KY	LZ	M1	N2	O3	o
AN	BO	CP	DQ	ER	FS	GT	HU	JV	KW	LX	MY	NZ	O1	P2	
BM	CN	DO	EP	FQ	GR	HS	JT	KU	LV	MW	NX	OY	PZ	Q1	
CL	DM	EN	FO	GP	HQ	JR	KS	LT	MU	NV	OW	PX	QY	RZ	−
DK	EL	FM	GN	HO	JP	KQ	LR	MS	NT	OU	PV	QW	RX	SY	
EJ	FK	GL	HM	JN	KO	LP	MQ	NR	OS	PT	QU	RV	SW	TX	
FH	GJ	HK	JL	KM	LN	MO	NP	OQ	PR	QS	RT	SU	TV	UW	
G	H	J	K	L	M	N	O	P	Q	R	S	T	U	V	

NOVEMBER 1980 — INTELLECTUAL

1 SA	2 SU	3 M	4 TU	5 W	6 TH	7 F	8 SA	9 SU	10 M	11 TU	12 W	13 TH	14 F	15 SA
3	4	5	6	7	8	A	B	C	D	E	F	G	H	J
42	53	64	75	86	A7	B8	CA	DB	EC	FD	GE	HF	JG	KH
51	62	73	84	A5	B6	C7	D8	EA	FB	GC	HD	JE	KF	LG
6Z	71	82	A3	B4	C5	D6	E7	F8	GA	HB	JC	KD	LE	MF
7Y	8Z	A1	B2	C3	D4	E5	F6	G7	H8	JA	KB	LC	MD	NE
8X	AY	BZ	C1	D2	E3	F4	G5	H6	J7	K8	LA	MB	NC	OD
AW	BX	CY	DZ	E1	F2	G3	H4	J5	K6	L7	M8	NA	OB	PC
BV	CW	DX	EY	FZ	G1	H2	J3	K4	L5	M6	N7	O8	PA	QB
CU	DV	EW	FX	GY	HZ	J1	K2	L3	M4	N5	O6	P7	Q8	RA
DT	EU	FV	GW	HX	JY	KZ	L1	M2	N3	O4	P5	Q6	R7	S8
ES	FT	GU	HV	JW	KX	LY	MZ	N1	O2	P3	Q4	R5	S6	T7
FR	GS	HT	JU	KV	LW	MX	NY	OZ	P1	Q2	R3	S4	T5	U6
GQ	HR	JS	KT	LU	MV	NW	OX	PY	QZ	R1	S2	T3	U4	V5
HP	JQ	KR	LS	MT	NU	OV	PW	QX	RY	SZ	T1	U2	V3	W4
JO	KP	LQ	MR	NS	OT	PU	QV	RW	SX	TY	UZ	V1	W2	X3
KN	LO	MP	NQ	OR	PS	QT	RU	SV	TW	UX	VY	WZ	X1	Y2
LM	MN	NO	OP	PQ	QR	RS	ST	TU	UV	VW	WX	XY	YZ	Z1

NOVEMBER 1980 **PHYSICAL**

16 SU	17 M	18 TU	19 W	20 TH	21 F	22 SA	23 SU	24 M	25 TU	26 W	27 TH	28 F	29 SA	30 SU
QP	RQ	SR	TS	UT	VU	WV	XW	AX	BA	CB	DC	ED	FE	GF
RO	SP	TQ	UR	VS	WT	XU	AV	BW	CX	DA	EB	FC	GD	HE
SN	TO	UP	VQ	WR	XS	AT	BU	CV	DW	EX	FA	GB	HC	JD
TM	UN	VO	WP	XQ	AR	BS	CT	DU	EV	FW	GX	HA	JB	KC
UL	VM	WN	XO	AP	BQ	CR	DS	ET	FU	GV	HW	JX	KA	LB
VK	WL	XM	AN	BO	CP	DQ	ER	FS	GT	HU	JV	KW	LX	MA
WJ	XK	AL	BM	CN	DO	EP	FQ	GR	HS	JT	KU	LV	MW	NX
XH	AJ	BK	CL	DM	EN	FO	GP	HQ	JR	KS	LT	MU	NV	OW
AG	BH	CJ	DK	EL	FM	GN	HO	JP	KQ	LR	MS	NT	OU	PV
BF	CG	DH	EJ	FK	GL	HM	JN	KO	LP	MQ	NR	OS	PT	QU
CE	DF	EG	FH	GJ	HK	JL	KM	LN	MO	NP	OQ	PR	QS	RT
D	E	F	G	H	J	K	L	M	N	O	P	Q	R	S

NOVEMBER 1980 **EMOTIONAL**

16 SU	17 M	18 TU	19 W	20 TH	21 F	22 SA	23 SU	24 M	25 TU	26 W	27 TH	28 F	29 SA	30 SU	
H	J	K	L	M	N	O	P	Q	R	S	T	U	V	W	
JG	KH	LJ	MK	NL	OM	PN	QO	RP	SQ	TR	US	VT	WU	XV	
KF	LG	MH	NJ	OK	PL	QM	RN	SO	TP	UQ	VR	WS	XT	YU	
LE	MF	NG	OH	PJ	QK	RL	SM	TN	UO	VP	WQ	XR	YS	ZT	
MD	NE	OF	PG	QH	RJ	SK	TL	UM	VN	WO	XP	YQ	ZR	1S	+
NC	OD	PE	QF	RG	SH	TJ	UK	VL	WM	XN	YO	ZP	1Q	2R	
OB	PC	QD	RE	SF	TG	UH	VJ	WK	XL	YM	ZN	1O	2P	3Q	
PA	QB	RC	SD	TE	UF	VG	WH	XJ	YK	ZL	1M	2N	3O	AP	0
Q3	RA	SB	TC	UD	VE	WF	XG	YH	ZJ	1K	2L	3M	AN	BO	
R2	S3	TA	UB	VC	WD	XE	YF	ZG	1H	2J	3K	AL	BM	CN	
S1	T2	U3	VA	WB	XC	YD	ZE	1F	2G	3H	AJ	BK	CL	DM	−
TZ	U1	V2	W3	XA	YB	ZC	1D	2E	3F	AG	BH	CJ	DK	EL	
UY	VZ	W1	X2	Y3	ZA	1B	2C	3D	AE	BF	CG	DH	EJ	FK	
VX	WY	XZ	Y1	Z2	13	2A	3B	AC	BD	CE	DF	EG	FH	GJ	
W	X	Y	Z	1	2	3	A	B	C	D	E	F	.G	H	

NOVEMBER 1980 **INTELLECTUAL**

16 SU	17 M	18 TU	19 W	20 TH	21 F	22 SA	23 SU	24 M	25 TU	26 W	27 TH	28 F	29 SA	30 SU
K	L	M	N	O	P	Q	R	S	T	U	V	W	X	Y
LJ	MK	NL	OM	PN	QO	RP	SQ	TR	US	VT	WU	XV	YW	ZX
MH	NJ	OK	PL	QM	RN	SO	TP	UQ	VR	WS	XT	YU	ZV	1W
NG	OH	PJ	QK	RL	SM	TN	UO	VP	WQ	XR	YS	ZT	1U	2V
OF	PG	QH	RJ	SK	TL	UM	VN	WO	XP	YQ	ZR	1S	2T	3U
PE	QF	RG	SH	TJ	UK	VL	WM	XN	YO	ZP	1Q	2R	3S	4T
QD	RE	SF	TG	UH	VJ	WK	XL	YM	ZN	1O	2P	3Q	4R	5S
RC	SD	TE	UF	VG	WH	XJ	YK	ZL	1M	2N	3O	4P	5Q	6R
SB	TC	UD	VE	WF	XG	YH	ZJ	1K	2L	3M	4N	5O	6P	7Q
TA	UB	VC	WD	XE	YF	ZG	1H	2J	3K	4L	5M	6N	7O	8P
U8	VA	WB	XC	YD	ZE	1F	2G	3H	4J	5K	6L	7M	8N	AO
V7	W8	XA	YB	ZC	1D	2E	3F	4G	5H	6J	7K	8L	AM	BN
W6	X7	Y8	ZA	1B	2C	3D	4E	5F	6G	7H	8J	AK	BL	CM
X5	Y6	Z7	18	2A	3B	4C	5D	6E	7F	8G	AH	BJ	CK	DL
Y4	Z5	16	27	38	4A	5B	6C	7D	8E	AF	BG	CH	DJ	EK
Z3	14	25	36	47	58	6A	7B	8C	AD	BE	CF	DG	EH	FJ
12	·23	34	45	56	67	78	8A	AB	BC	CD	DE	EF	FG	GH

DECEMBER 1980 **PHYSICAL**

1 M	2 TU	3 W	4 TH	5 F	6 SA	7 SU	8 M	9 TU	10 W	11 TH	12 F	13 SA	14 SU	15 M
HG	JH	KJ	LK	ML	NM	ON	PO	QP	RQ	SR	TS	UT	VU	WV
JF	KG	LH	MJ	NK	OL	PM	QN	RO	SP	TQ	UR	VS	WT	XU
KE	LF	MG	NH	OJ	PK	QL	RM	SN	TO	UP	VQ	WR	XS	AT
LD	ME	NF	OG	PH	QJ	RK	SL	TM	UN	VO	WP	XQ	AR	BS
MC	ND	OE	PF	QG	RH	SJ	TK	UL	VM	WN	XO	AP	BQ	CR
NB	OC	PD	QE	RF	SG	TH	UJ	VK	WL	XM	AN	BO	CP	DQ
OA	PB	QC	RD	SE	TF	UG	VH	WJ	XK	AL	BM	CN	DO	EP
PX	QA	RB	SC	TD	UE	VF	WG	XH	AJ	BK	CL	DM	EN	FO
QW	RX	SA	TB	UC	VD	WE	XF	AG	BH	CJ	DK	EL	FM	GN
RV	SW	TX	UA	VB	WC	XD	AE	BF	CG	DH	EJ	FK	GL	HM
SU	TV	UW	VX	WA	XB	AC	BD	CE	DF	EG	FH	GJ	HK	JL
T	U	V	W	X	A	B	C	D	E	F	G	H	J	K

DECEMBER 1980 **EMOTIONAL**

1 M	2 TU	3 W	4 TH	5 F	6 SA	7 SU	8 M	9 TU	10 W	11 TH	12 F	13 SA	14 SU	15 M	
X	Y	Z	1	2	3	A	B	C	D	E	F	G	H	J	
YW	ZX	1Y	2Z	31	A2	B3	CA	DB	EC	FD	GE	HF	JG	KH	
ZV	1W	2X	3Y	AZ	B1	C2	D3	EA	FB	GC	HD	JE	KF	LG	
1U	2V	3W	AX	BY	CZ	D1	E2	F3	GA	HB	JC	KD	LE	MF	
2T	3U	AV	BW	CX	DY	EZ	F1	G2	H3	JA	KB	LC	MD	NE	+
3S	AT	BU	CV	DW	EX	FY	GZ	H1	J2	K3	LA	MB	NC	OD	
AR	BS	CT	DU	EV	FW	GX	HY	JZ	K1	L2	M3	NA	OB	PC	
BQ	CR	DS	ET	FU	GV	HW	JX	KY	LZ	M1	N2	O3	PA	QB	o
CP	DQ	ER	FS	GT	HU	JV	KW	LX	MY	NZ	O1	P2	Q3	RA	
DO	EP	FQ	GR	HS	JT	KU	LV	MW	NX	OY	PZ	Q1	R2	S3	
EN	FO	GP	HQ	JR	KS	LT	MU	NV	OW	PX	QY	RZ	S1	T2	−
FM	GN	HO	JP	KQ	LR	MS	NT	OU	PV	QW	RX	SY	TZ	U1	
GL	HM	JN	KO	LP	MQ	NR	OS	PT	QU	RV	SW	TX	UY	VZ	
HK	JL	KM	LN	MO	NP	OQ	PR	QS	RT	SU	TV	UW	VX	WY	
J	K	L	M	N	O	P	Q	R	S	T	U	V	W	X	

DECEMBER 1980 **INTELLECTUAL**

1 M	2 TU	3 W	4 TH	5 F	6 SA	7 SU	8 M	9 TU	10 W	11 TH	12 F	13 SA	14 SU	15 M
Z	1	2	3	4	5	6	7	8	A	B	C	D	E	F
1Y	2Z	31	42	53	64	75	86	A7	B8	CA	DB	EC	FD	GE
2X	3Y	4Z	51	62	73	84	A5	B6	C7	D8	EA	FB	GC	HD
3W	4X	5Y	6Z	71	82	A3	B4	C5	D6	E7	F8	GA	HB	JC
4V	5W	6X	7Y	8Z	A1	B2	C3	D4	E5	F6	G7	H8	JA	KB
5U	6V	7W	8X	AY	BZ	C1	D2	E3	F4	G5	H6	J7	K8	LA
6T	7U	8V	AW	BX	CY	DZ	E1	F2	G3	H4	J5	K6	L7	M8
7S	8T	AU	BV	CW	DX	EY	FZ	G1	H2	J3	K4	L5	M6	N7
8R	AS	BT	CU	DV	EW	FX	GY	HZ	J1	K2	L3	M4	N5	O6
AQ	BR	CS	DT	EU	FV	GW	HX	JY	KZ	L1	M2	N3	O4	P5
BP	CQ	DR	ES	FT	GU	HV	JW	KX	LY	MZ	N1	O2	P3	Q4
CO	DP	EQ	FR	GS	HT	JU	KV	LW	MX	NY	OZ	P1	Q2	R3
DN	EO	FP	GQ	HR	JS	KT	LU	MV	NW	OX	PY	QZ	R1	S2
EM	FN	GO	HP	JQ	KR	LS	MT	NU	OV	PW	QX	RY	SZ	T1
FL	GM	HN	JO	KP	LQ	MR	NS	OT	PU	QV	RW	SX	TY	UZ
GK	HL	JM	KN	LO	MP	NQ	OR	PS	QT	RU	SV	TW	UX	VY
HJ	JK	KL	LM	MN	NO	OP	PQ	QR	RS	ST	TU	UV	VW	WX

DECEMBER 1980 PHYSICAL

16 TU	17 W	18 TH	19 F	20 SA	21 SU	22 M	23 TU	24 W	25 TH	26 F	27 SA	28 SU	29 M	30 TU	31 W
XW	AX	BA	CB	DC	ED	FE	GF	HG	JH	KJ	LK	ML	NM	ON	PO
AV	BW	CX	DA	EB	FC	GD	HE	JF	KG	LH	MJ	NK	OL	PM	QN
BU	CV	DW	EX	FA	GB	HC	JD	KE	LF	MG	NH	OJ	PK	QL	RM
CT	DU	EV	FW	GX	HA	JB	KC	LD	ME	NF	OG	PH	QJ	RK	SL
DS	ET	FU	GV	HW	JX	KA	LB	MC	ND	OE	PF	QG	RH	SJ	TK
ER	FS	GT	HU	JV	KW	LX	MA	NB	OC	PD	QE	RF	SG	TH	UJ
FQ	GR	HS	JT	KU	LV	MW	NX	OA	PB	QC	RD	SE	TF	UG	VH
GP	HQ	JR	KS	LT	MU	NV	OW	PX	QA	RB	SC	TD	UE	VF	WG
HO	JP	KQ	LR	MS	NT	OU	PV	QW	RX	SA	TB	UC	VD	WE	XF
JN	KO	LP	MQ	NR	OS	PT	QU	RV	SW	TX	UA	VB	WC	XD	AE
KM	LN	MO	NP	OQ	PR	QS	RT	SU	TV	UW	VX	WA	XB	AC	BD
L	M	N	O	P	Q	R	S	T	U	V	W	X	A	B	C

DECEMBER 1980 EMOTIONAL

16 TU	17 W	18 TH	19 F	20 SA	21 SU	22 M	23 TU	24 W	25 TH	26 F	27 SA	28 SU	29 M	30 TU	31 W	
K	L	M	N	O	P	Q	R	S	T	U	V	W	X	Y	Z	
LJ	MK	NL	OM	PN	QO	RP	SQ	TR	US	VT	WU	XV	YW	ZX	1Y	
MH	NJ	OK	PL	QM	RN	SO	TP	UQ	VR	WS	XT	YU	ZV	1W	2X	
NG	OH	PJ	QK	RL	SM	TN	UO	VP	WQ	XR	YS	ZT	1U	2V	3W	
OF	PG	QH	RJ	SK	TL	UM	VN	WO	XP	YQ	ZR	1S	2T	3U	AV	+
PE	QF	RG	SH	TJ	UK	VL	WM	XN	YO	ZP	1Q	2R	3S	AT	BU	
QD	RE	SF	TG	UH	VJ	WK	XL	YM	ZN	1O	2P	3Q	AR	BS	CT	
RC	SD	TE	UF	VG	WH	XJ	YK	ZL	1M	2N	3O	AP	BQ	CR	DS	o
SB	TC	UD	VE	WF	XG	YH	ZJ	1K	2L	3M	AN	BO	CP	DQ	ER	
TA	UB	VC	WD	XE	YF	ZG	1H	2J	3K	AL	BM	CN	DO	EP	FQ	
U3	VA	WB	XC	YD	ZE	1F	2G	3H	AJ	BK	CL	DM	EN	FO	GP	−
V2	W3	XA	YB	ZC	1D	2E	3F	AG	BH	CJ	DK	EL	FM	GN	HO	
W1	X2	Y3	ZA	1B	2C	3D	AE	BF	CG	DH	EJ	FK	GL	HM	JN	
XZ	Y1	Z2	13	2A	3B	AC	BD	CE	DF	EG	FH	GJ	HK	JL	KM	
Y	Z	1	2	3	A	B	C	D	E	F	G	H	J	K	L	

DECEMBER 1980 INTELLECTUAL

16 TU	17 W	18 TH	19 F	20 SA	21 SU	22 M	23 TU	24 W	25 TH	26 F	27 SA	28 SU	29 M	30 TU	31 W
G	H	J	K	L	M	N	O	P	Q	R	S	T	U	V	W
HF	JG	KH	LJ	MK	NL	OM	PN	QO	RP	SQ	TR	US	VT	WU	XV
JE	KF	LG	MH	NJ	OK	PL	QM	RN	SO	TP	UQ	VR	WS	XT	YU
KD	LE	MF	NG	OH	PJ	QK	RL	SM	TN	UO	VP	WQ	XR	YS	ZT
LC	MD	NE	OF	PG	QH	RJ	SK	TL	UM	VN	WO	XP	YQ	ZR	1S
MB	NC	OD	PE	QF	RG	SH	TJ	UK	VL	WM	XN	YO	ZP	1Q	2R
NA	OB	PC	QD	RE	SF	TG	UH	VJ	WK	XL	YM	ZN	1O	2P	3Q
O8	PA	QB	RC	SD	TE	UF	VG	WH	XJ	YK	ZL	1M	2N	3O	4P
P7	Q8	RA	SB	TC	UD	VE	WF	XG	YH	ZJ	1K	2L	3M	4N	5O
Q6	R7	S8	TA	UB	VC	WD	XE	YF	ZG	1H	2J	3K	4L	5M	6N
R5	S6	T7	U8	VA	WB	XC	YD	ZE	1F	2G	3H	4J	5K	6L	7M
S4	T5	U6	V7	W8	XA	YB	ZC	1D	2E	3F	4G	5H	6J	7K	8L
T3	U4	V5	W6	X7	Y8	ZA	1B	2C	3D	4E	5F	6G	7H	8J	AK
U2	V3	W4	X5	Y6	Z7	18	2A	3B	4C	5D	6E	7F	8G	AH	BJ
V1	W2	X3	Y4	Z5	16	27	38	4A	5B	6C	7D	8E	AF	BG	CH
WZ	X1	Y2	Z3	14	25	36	47	58	6A	7B	8C	AD	BE	CF	DG
XY	YZ	Z1	12	23	34	45	56	67	78	8A	AB	BC	CD	DE	EF

JANUARY 1981 PHYSICAL

1 TH	2 F	3 SA	4 SU	5 M	6 TU	7 W	8 TH	9 F	10 SA	11 SU	12 M	13 TU	14 W	15 TH	
QP	RQ	SR	TS	UT	VU	WV	XW	AX	BA	CB	DC	ED	FE	GF	
RO	SP	TQ	UR	VS	WT	XU	AV	BW	CX	DA	EB	FC	GD	HE	
SN	TO	UP	VQ	WR	XS	AT	BU	CV	DW	EX	FA	GB	HC	JD	+
TM	UN	VO	WP	XQ	AR	BS	CT	DU	EV	FW	GX	HA	JB	KC	
UL	VM	WN	XO	AP	BQ	CR	DS	ET	FU	GV	HW	JX	KA	LB	
VK	WL	XM	AN	BO	CP	DQ	ER	FS	GT	HU	JV	KW	LX	MA	0
WJ	XK	AL	BM	CN	DO	EP	FQ	GR	HS	JT	KU	LV	MW	NX	
XH	AJ	BK	CL	DM	EN	FO	GP	HQ	JR	KS	LT	MU	NV	OW	
AG	BH	CJ	DK	EL	FM	GN	HO	JP	KQ	LR	MS	NT	OU	PV	−
BF	CG	DH	EJ	FK	GL	HM	JN	KO	LP	MQ	NR	OS	PT	QU	
CE	DF	EG	FH	GJ	HK	JL	KM	LN	MO	NP	OQ	PR	QS	RT	
D	E	F	G	H	J	K	L	M	N	O	P	Q	R	S	

JANUARY 1981 EMOTIONAL

1 TH	2 F	3 SA	4 SU	5 M	6 TU	7 W	8 TH	9 F	10 SA	11 SU	12 M	13 TU	14 W	15 TH	
1	2	3	A	B	C	D	E	F	G	H	J	K	L	M	
2Z	31	A2	B3	CA	DB	EC	FD	GE	HF	JG	KH	LJ	MK	NL	
3Y	AZ	B1	C2	D3	EA	FB	GC	HD	JE	KF	LG	MH	NJ	OK	
AX	BY	CZ	D1	E2	F3	GA	HB	JC	KD	LE	MF	NG	OH	PJ	
BW	CX	DY	EZ	F1	G2	H3	JA	KB	LC	MD	NE	OF	PG	QH	+
CV	DW	EX	FY	GZ	H1	J2	K3	LA	MB	NC	OD	PE	QF	RG	
DU	EV	FW	GX	HY	JZ	K1	L2	M3	NA	OB	PC	QD	RE	SF	
ET	FU	GV	HW	JX	KY	LZ	M1	N2	O3	PA	QB	RC	SD	TE	0
FS	GT	HU	JV	KW	LX	MY	NZ	O1	P2	Q3	RA	SB	TC	UD	
GR	HS	JT	KU	LV	MW	NX	OY	PZ	Q1	R2	S3	TA	UB	VC	
HQ	JR	KS	LT	MU	NV	OW	PX	QY	RZ	S1	T2	U3	VA	WB	−
JP	KQ	LR	MS	NT	OU	PV	QW	RX	SY	TZ	U1	V2	W3	XA	
KO	LP	MQ	NR	OS	PT	QU	RV	SW	TX	UY	VZ	W1	X2	Y3	
LN	MO	NP	OQ	PR	QS	RT	SU	TV	UW	VX	WY	XZ	Y1	Z2	
M	N	O	P	Q	R	S	T	U	V	W	X	Y	Z	1	

JANUARY 1981 INTELLECTUAL

1 TH	2 F	3 SA	4 SU	5 M	6 TU	7 W	8 TH	9 F	10 SA	11 SU	12 M	13 TU	14 W	15 TH	
X	Y	Z	1	2	3	4	5	6	7	8	A	B	C	D	
YW	ZX	1Y	2Z	31	42	53	64	75	86	A7	B8	CA	DB	EC	
ZV	1W	2X	3Y	4Z	51	62	73	84	A5	B6	C7	D8	EA	FB	
1U	2V	3W	4X	5Y	6Z	71	82	A3	B4	C5	D6	E7	F8	GA	
2T	3U	4V	5W	6X	7Y	8Z	A1	B2	C3	D4	E5	F6	G7	H8	
3S	4T	5U	6V	7W	8X	AY	BZ	C1	D2	E3	F4	G5	H6	J7	+
4R	5S	6T	7U	8V	AW	BX	CY	DZ	E1	F2	G3	H4	J5	K6	
5Q	6R	7S	8T	AU	BV	CW	DX	EY	FZ	G1	H2	J3	K4	L5	
6P	7Q	8R	AS	BT	CU	DV	EW	FX	GY	HZ	J1	K2	L3	M4	0
7O	8P	AQ	BR	CS	DT	EU	FV	GW	HX	JY	KZ	L1	M2	N3	
8N	AO	BP	CQ	DR	ES	FT	GU	HV	JW	KX	LY	MZ	N1	O2	
AM	BN	CO	DP	EQ	FR	GS	HT	JU	KV	LW	MX	NY	OZ	P1	−
BL	CM	DN	EO	FP	GQ	HR	JS	KT	LU	MV	NW	OX	PY	QZ	
CK	DL	EM	FN	GO	HP	JQ	KR	LS	MT	NU	OV	PW	QX	RY	
DJ	EK	FL	GM	HN	JO	KP	LQ	MR	NS	OT	PU	QV	RW	SX	
EH	FJ	GK	HL	JM	KN	LO	MP	NQ	OR	PS	QT	RU	SV	TW	
FG	GH	HJ	JK	KL	LM	MN	NO	OP	PQ	QR	RS	ST	TU	UV	

JANUARY 1981 **PHYSICAL**

16 F	17 SA	18 SU	19 M	20 TU	21 W	22 TH	23 F	24 SA	25 SU	26 M	27 TU	28 W	29 TH	30 F	31 SA	
HG	JH	KJ	LK	ML	NM	ON	PO	QP	RQ	SR	TS	UT	VU	WV	XW	
JF	KG	LH	MJ	NK	OL	PM	QN	RO	SP	TQ	UR	VS	WT	XU	AV	
KE	LF	MG	NH	OJ	PK	QL	RM	SN	TO	UP	VQ	WR	XS	AT	BU	+
LD	ME	NF	OG	PH	QJ	RK	SL	TM	UN	VO	WP	XQ	AR	BS	CT	
MC	ND	OE	PF	QG	RH	SJ	TK	UL	VM	WN	XO	AP	BQ	CR	DS	
NB	OC	PD	QE	RF	SG	TH	UJ	VK	WL	XM	AN	BO	CP	DQ	ER	0
OA	PB	QC	RD	SE	TF	UG	VH	WJ	XK	AL	BM	CN	DO	EP	FQ	
PX	QA	RB	SC	TD	UE	VF	WG	XH	AJ	BK	CL	DM	EN	FO	GP	
QW	RX	SA	TB	UC	VD	WE	XF	AG	BH	CJ	DK	EL	FM	GN	HO	−
RV	SW	TX	UA	VB	WC	XD	AE	BF	CG	DH	EJ	FK	GL	HM	JN	
SU	TV	UW	VX	WA	XB	AC	BD	CE	DF	EG	FH	GJ	HK	JL	KM	
T	U	V	W	X	A	B	C	D	E	F	G	H	J	K	L	

JANUARY 1981 **EMOTIONAL**

16 F	17 SA	18 SU	19 M	20 TU	21 W	22 TH	23 F	24 SA	25 SU	26 M	27 TU	28 W	29 TH	30 F	31 SA	
N	O	P	Q	R	S	T	U	V	W	X	Y	Z	1	2	3	
OM	PN	QO	RP	SQ	TR	US	VT	WU	XV	YW	ZX	1Y	2Z	31	A2	
PL	QM	RN	SO	TP	UQ	VR	WS	XT	YU	ZV	1W	2X	3Y	AZ	B1	
QK	RL	SM	TN	UO	VP	WQ	XR	YS	ZT	1U	2V	3W	AX	BY	CZ	
RJ	SK	TL	UM	VN	WO	XP	YQ	ZR	1S	2T	3U	AV	BW	CX	DY	+
SH	TJ	UK	VL	WM	XN	YO	ZP	1Q	2R	3S	AT	BU	CV	DW	EX	
TG	UH	VJ	WK	XL	YM	ZN	1O	2P	3Q	AR	BS	CT	DU	EV	FW	
UF	VG	WH	XJ	YK	ZL	1M	2N	3O	AP	BQ	CR	DS	ET	FU	GV	0
VE	WF	XG	YH	ZJ	1K	2L	3M	AN	BO	CP	DQ	ER	FS	GT	HU	
WD	XE	YF	ZG	1H	2J	3K	AL	BM	CN	DO	EP	FQ	GR	HS	JT	
XC	YD	ZE	1F	2G	3H	AJ	BK	CL	DM	EN	FO	GP	HQ	JR	KS	−
YB	ZC	1D	2E	3F	AG	BH	CJ	DK	EL	FM	GN	HO	JP	KQ	LR	
ZA	1B	2C	3D	AE	BF	CG	DH	EJ	FK	GL	HM	JN	KO	LP	MQ	
13	2A	3B	AC	BD	CE	DF	EG	FH	GJ	HK	JL	KM	LN	MO	NP	
2	3	A	B	C	D	E	F	G	H	J	K	L	M	N	O	

JANUARY 1981 **INTELLECTUAL**

16 F	17 SA	18 SU	19 M	20 TU	21 W	22 TH	23 F	24 SA	25 SU	26 M	27 TU	28 W	29 TH	30 F	31 SA	
E	F	G	H	J	K	L	M	N	O	P	Q	R	S	T	U	
FD	GE	HF	JG	KH	LJ	MK	NL	OM	PN	QO	RP	SQ	TR	US	VT	
GC	HD	JE	KF	LG	MH	NJ	OK	PL	QM	RN	SO	TP	UQ	VR	WS	
HB	JC	KD	LE	MF	NG	OH	PJ	QK	RL	SM	TN	UO	VP	WQ	XR	
JA	KB	LC	MD	NE	OF	PG	QH	RJ	SK	TL	UM	VN	WO	XP	YQ	
K8	LA	MB	NC	OD	PE	QF	RG	SH	TJ	UK	VL	WM	XN	YO	ZP	+
L7	M8	NA	OB	PC	QD	RE	SF	TG	UH	VJ	WK	XL	YM	ZN	1O	
M6	N7	O8	PA	QB	RC	SD	TE	UF	VG	WH	XJ	YK	ZL	1M	2N	
N5	O6	P7	Q8	RA	SB	TC	UD	VE	WF	XG	YH	ZJ	1K	2L	3M	0
O4	P5	Q6	R7	S8	TA	UB	VC	WD	XE	YF	ZG	1H	2J	3K	4L	
P3	Q4	R5	S6	T7	U8	VA	WB	XC	YD	ZE	1F	2G	3H	4J	5K	−
Q2	R3	S4	T5	U6	V7	W8	XA	YB	ZC	1D	2E	3F	4G	5H	6J	
R1	S2	T3	U4	V5	W6	X7	Y8	ZA	1B	2C	3D	4E	5F	6G	7H	
SZ	T1	U2	V3	W4	X5	Y6	Z7	18	2A	3B	4C	5D	6E	7F	8G	
TY	UZ	V1	W2	X3	Y4	Z5	16	27	38	4A	5B	6C	7D	8E	AF	
UX	VY	WZ	X1	Y2	Z3	14	25	36	47	58	6A	7B	8C	AD	BE	
VW	WX	XY	YZ	Z1	12	23	34	45	56	67	78	8A	AB	BC	CD	

FEBRUARY 1981 PHYSICAL

1	2	3	4	5	6	7	8	9	10	11	12	13	14	15	
SU	M	TU	W	TH	F	SA	SU	M	TU	W	TH	F	SA	SU	
AX	BA	CB	DC	ED	FE	GF	HG	JH	KJ	LK	ML	NM	ON	PO	
BW	CX	DA	EB	FC	GD	HE	JF	KG	LH	MJ	NK	OL	PM	QN	
CV	DW	EX	FA	GB	HC	JD	KE	LF	MG	NH	OJ	PK	QL	RM	+
DU	EV	FW	GX	HA	JB	KC	LD	ME	NF	OG	PH	QJ	RK	SL	
ET	FU	GV	HW	JX	KA	LB	MC	ND	OE	PF	QG	RH	SJ	TK	
FS	GT	HU	JV	KW	LX	MA	NB	OC	PD	QE	RF	SG	TH	UJ	0
GR	HS	JT	KU	LV	MW	NX	OA	PB	QC	RD	SE	TF	UG	VH	
HQ	JR	KS	LT	MU	NV	OW	PX	QA	RB	SC	TD	UE	VF	WG	
JP	KQ	LR	MS	NT	OU	PV	QW	RX	SA	TB	UC	VD	WE	XF	−
KO	LP	MQ	NR	OS	PT	QU	RV	SW	TX	UA	VB	WC	XD	AE	
LN	MO	NP	OQ	PR	QS	RT	SU	TV	UW	VX	WA	XB	AC	BD	
M	N	O	P	Q	R	S	T	U	V	W	X	A	B	C	

FEBRUARY 1981 EMOTIONAL

1	2	3	4	5	6	7	8	9	10	11	12	13	14	15	
SU	M	TU	W	TH	F	SA	SU	M	TU	W	TH	F	SA	SU	
A	B	C	D	E	F	G	H	J	K	L	M	N	O	P	
B3	CA	DB	EC	FD	GE	HF	JG	KH	LJ	MK	NL	OM	PN	QO	
C2	D3	EA	FB	GC	HD	JE	KF	LG	MH	NJ	OK	PL	QM	RN	
D1	E2	F3	GA	HB	JC	KD	LE	MF	NG	OH	PJ	QK	RL	SM	
EZ	F1	G2	H3	JA	KB	LC	MD	NE	OF	PG	QH	RJ	SK	TL	+
FY	GZ	H1	J2	K3	LA	MB	NC	OD	PE	QF	RG	SH	TJ	UK	
GX	HY	JZ	K1	L2	M3	NA	OB	PC	QD	RE	SF	TG	UH	VJ	
HW	JX	KY	LZ	M1	N2	O3	PA	QB	RC	SD	TE	UF	VG	WH	0
JV	KW	LX	MY	NZ	O1	P2	Q3	RA	SB	TC	UD	VE	WF	XG	
KU	LV	MW	NX	OY	PZ	Q1	R2	S3	TA	UB	VC	WD	XE	YF	
LT	MU	NV	OW	PX	QY	RZ	S1	T2	U3	VA	WB	XC	YD	ZE	−
MS	NT	OU	PV	QW	RX	SY	TZ	U1	V2	W3	XA	YB	ZC	1D	
NR	OS	PT	QU	RV	SW	TX	UY	VZ	W1	X2	Y3	ZA	1B	2C	
OQ	PR	QS	RT	SU	TV	UW	VX	WY	XZ	Y1	Z2	13	2A	3B	
P	Q	R	S	T	U	V	W	X	Y	Z	1	2	3	A	

FEBRUARY 1981 INTELLECTUAL

1	2	3	4	5	6	7	8	9	10	11	12	13	14	15	
SU	M	TU	W	TH	F	SA	SU	M	TU	W	TH	F	SA	SU	
V	W	X	Y	Z	1	2	3	4	5	6	7	8	A	B	
WU	XV	YW	ZX	1Y	2Z	31	42	53	64	75	86	A7	B8	CA	
XT	YU	ZV	1W	2X	3Y	4Z	51	62	73	84	A5	B6	C7	D8	
YS	ZT	1U	2V	3W	4X	5Y	6Z	71	82	A3	B4	C5	D6	E7	
ZR	1S	2T	3U	4V	5W	6X	7Y	8Z	A1	B2	C3	D4	E5	F6	
1Q	2R	3S	4T	5U	6V	7W	8X	AY	BZ	C1	D2	E3	F4	G5	+
2P	3Q	4R	5S	6T	7U	8V	AW	BX	CY	DZ	E1	F2	G3	H4	
3O	4P	5Q	6R	7S	8T	AU	BV	CW	DX	EY	FZ	G1	H2	J3	
4N	5O	6P	7Q	8R	AS	BT	CU	DV	EW	FX	GY	HZ	J1	K2	0
5M	6N	7O	8P	AQ	BR	CS	DT	EU	FV	GW	HX	JY	KZ	L1	
6L	7M	8N	AO	BP	CQ	DR	ES	FT	GU	HV	JW	KX	LY	MZ	
7K	8L	AM	BN	CO	DP	EQ	FR	GS	HT	JU	KV	LW	MX	NY	−
8J	AK	BL	CM	DN	EO	FP	GQ	HR	JS	KT	LU	MV	NW	OX	
AH	BJ	CK	DL	EM	FN	GO	HP	JQ	KR	LS	MT	NU	OV	PW	
BG	CH	DJ	EK	FL	GM	HN	JO	KP	LQ	MR	NS	OT	PU	QV	
CF	DG	EH	FJ	GK	HL	JM	KN	LO	MP	NQ	OR	PS	QT	RU	
DE	EF	FG	GH	HJ	JK	KL	LM	MN	NO	OP	PQ	QR	RS	ST	

FEBRUARY 1981 **PHYSICAL**

16 M	17 TU	18 W	19 TH	20 F	21 SA	22 SU	23 M	24 TU	25 W	26 TH	27 F	28 SA	
QP	RQ	SR	TS	UT	VU	WV	XW	AX	BA	CB	DC	ED	
RO	SP	TQ	UR	VS	WT	XU	AV	BW	CX	DA	EB	FC	
SN	TO	UP	VQ	WR	XS	AT	BU	CV	DW	EX	FA	GB	+
TM	UN	VO	WP	XQ	AR	BS	CT	DU	EV	FW	GX	HA	
UL	VM	WN	XO	AP	BQ	CR	DS	ET	FU	GV	HW	JX	
VK	WL	XM	AN	BO	CP	DQ	ER	FS	GT	HU	JV	KW	0
WJ	XK	AL	BM	CN	DO	EP	FQ	GR	HS	JT	KU	LV	
XH	AJ	BK	CL	DM	EN	FO	GP	HQ	JR	KS	LT	MU	
AG	BH	CJ	DK	EL	FM	GN	HO	JP	KQ	LR	MS	NT	−
BF	CG	DH	EJ	FK	GL	HM	JN	KO	LP	MQ	NR	OS	
CE	DF	EG	FH	GJ	HK	JL	KM	LN	MO	NP	OQ	PR	
D	E	F	G	H	J	K	L	M	N	O	P	Q	

FEBRUARY 1981 **EMOTIONAL**

16 M	17 TU	18 W	19 TH	20 F	21 SA	22 SU	23 M	24 TU	25 W	26 TH	27 F	28 SA	
Q	R	S	T	U	V	W	X	Y	Z	1	2	3	
RP	SQ	TR	US	VT	WU	XV	YW	ZX	1Y	2Z	31	A2	
SO	TP	UQ	VR	WS	XT	YU	ZV	1W	2X	3Y	AZ	B1	
TN	UO	VP	WQ	XR	YS	ZT	1U	2V	3W	AX	BY	CZ	
UM	VN	WO	XP	YQ	ZR	1S	2T	3U	AV	BW	CX	DY	+
VL	WM	XN	YO	ZP	1Q	2R	3S	AT	BU	CV	DW	EX	
WK	XL	YM	ZN	1O	2P	3Q	AR	BS	CT	DU	EV	FW	
XJ	YK	ZL	1M	2N	3O	AP	BQ	CR	DS	ET	FU	GV	0
YH	ZJ	1K	2L	3M	AN	BO	CP	DQ	ER	FS	GT	HU	
ZG	1H	2J	3K	AL	BM	CN	DO	EP	FQ	GR	HS	JT	
1F	2G	3H	AJ	BK	CL	DM	EN	FO	GP	HQ	JR	KS	−
2E	3F	AG	BH	CJ	DK	EL	FM	GN	HO	JP	KQ	LR	
3D	AE	BF	CG	DH	EJ	FK	GL	HM	JN	KO	LP	MQ	
AC	BD	CE	DF	EG	FH	GJ	HK	JL	KM	LN	MO	NP	
B	C	D	E	F	G	H	J	K	L	M	N	O	

FEBRUARY 1981 **INTELLECTUAL**

16 M	17 TU	18 W	19 TH	20 F	21 SA	22 SU	23 M	24 TU	25 W	26 TH	27 F	28 SA	
C	D	E	F	G	H	J	K	L	M	N	O	P	
DB	EC	FD	GE	HF	JG	KH	LJ	MK	NL	OM	PN	QO	
EA	FB	GC	HD	JE	KF	LG	MH	NJ	OK	PL	QM	RN	
F8	GA	HB	JC	KD	LE	MF	NG	OH	PJ	QK	RL	SM	
G7	H8	JA	KB	LC	MD	NE	OF	PG	QH	RJ	SK	TL	
H6	J7	K8	LA	MB	NC	OD	PE	QF	RG	SH	TJ	UK	+
J5	K6	L7	M8	NA	OB	PC	QD	RE	SF	TG	UH	VJ	
K4	L5	M6	N7	O8	PA	QB	RC	SD	TE	UF	VG	WH	
L3	M4	N5	O6	P7	Q8	RA	SB	TC	UD	VE	WF	XG	0
M2	N3	O4	P5	Q6	R7	S8	TA	UB	VC	WD	XE	YF	
N1	O2	P3	Q4	R5	S6	T7	U8	VA	WB	XC	YD	ZE	−
OZ	P1	Q2	R3	S4	T5	U6	V7	W8	XA	YB	ZC	1D	
PY	QZ	R1	S2	T3	U4	V5	W6	X7	Y8	ZA	1B	2C	
QX	RY	SZ	T1	U2	V3	W4	X5	Y6	Z7	18	2A	3B	
RW	SX	TY	UZ	V1	W2	X3	Y4	Z5	16	27	38	4A	
SV	TW	UX	VY	WZ	X1	Y2	Z3	14	25	36	47	58	
TU	UV	VW	WX	XY	YZ	Z1	12	23	34	45	56	67	

MARCH 1981 **PHYSICAL**

1 SU	2 M	3 TU	4 W	5 TH	6 F	7 SA	8 SU	9 M	10 TU	11 W	12 TH	13 F	14 SA	15 SU	
FE	GF	HG	JH	KJ	LK	ML	NM	ON	PO	QP	RQ	SR	TS	UT	
GD	HE	JF	KQ	LH	MJ	NK	OL	PM	QN	RO	SP	TQ	UR	VS	
HC	JD	KE	LF	MG	NH	OJ	PK	QL	RM	SN	TO	UP	VQ	WR	+
JB	KC	LD	ME	NF	OG	PH	QJ	RK	SL	TM	UN	VO	WP	XQ	
KA	LB	MC	ND	OE	PF	QG	RH	SJ	TK	UL	VM	WN	XO	AP	
LX	MA	NB	OC	PD	QE	RF	SG	TH	UJ	VK	WL	XM	AN	BO	0
MW	NX	OA	PB	QC	RD	SE	TF	UG	VH	WJ	XK	AL	BM	CN	
NV	OW	PX	QA	RB	SC	TD	UE	VF	WG	XH	AJ	BK	CL	DM	
OU	PV	QW	RX	SA	TB	UC	VD	WE	XF	AG	BH	CJ	DK	EL	−
PT	QU	RV	SW	TX	UA	VB	WC	XD	AE	BF	CG	DH	EJ	FK	
QS	RT	SU	TV	UW	VX	WA	XB	AC	BD	CE	DF	EG	FH	GJ	
R	S	T	U	V	W	X	A	B	C	D	E	F	G	H	

MARCH 1981 **EMOTIONAL**

1 SU	2 M	3 TU	4 W	5 TH	6 F	7 SA	8 SU	9 M	10 TU	11 W	12 TH	13 F	14 SA	15 SU	
A	B	C	D	E	F	G	H	J	K	L	M	N	O	P	
B3	CA	DB	EC	FD	GE	HF	JG	KH	LJ	MK	NL	OM	PN	QO	
C2	D3	EA	FB	GC	HD	JE	KF	LG	MH	NJ	OK	PL	QM	RN	
D1	E2	F3	GA	HB	JC	KD	LE	MF	NG	OH	PJ	QK	RL	SM	
EZ	F1	G2	H3	JA	KB	LC	MD	NE	OF	PG	QH	RJ	SK	TL	+
FY	GZ	H1	J2	K3	LA	MB	NC	OD	PE	QF	RG	SH	TJ	UK	
GX	HY	JZ	K1	L2	M3	NA	OB	PC	QD	RE	SF	TG	UH	VJ	
HW	JX	KY	LZ	M1	N2	O3	PA	QB	RC	SD	TE	UF	VG	WH	0
JV	KW	LX	MY	NZ	O1	P2	Q3	RA	SB	TC	UD	VE	WF	XG	
KU	LV	MW	NX	OY	PZ	Q1	R2	S3	TA	UB	VC	WD	XE	YF	
LT	MU	NV	OW	PX	QY	RZ	S1	T2	U3	VA	WB	XC	YD	ZE	−
MS	NT	OU	PV	QW	RX	SY	TZ	U1	V2	W3	XA	YB	ZC	1D	
NR	OS	PT	QU	RV	SW	TX	UY	VZ	W1	X2	Y3	ZA	1B	2C	
OQ	PR	QS	RT	SU	TV	UW	VX	WY	XZ	Y1	Z2	13	2A	3B	
P	Q	R	S	T	U	V	W	X	Y	Z	1	2	3	A	

MARCH 1981 **INTELLECTUAL**

1 SU	2 M	3 TU	4 W	5 TH	6 F	7 SA	8 SU	9 M	10 TU	11 W	12 TH	13 F	14 SA	15 SU	
Q	R	S	T	U	V	W	X	Y	Z	1	2	3	4	5	
RP	SQ	TR	US	VT	WU	XV	YW	ZX	1Y	2Z	31	42	53	64	
SO	TP	UQ	VR	WS	XT	YU	ZV	1W	2X	3Y	4Z	51	62	73	
TN	UO	VP	WQ	XR	YS	ZT	1U	2V	3W	4X	5Y	6Z	71	82	
UM	VN	WO	XP	YQ	ZR	1S	2T	3U	4V	5W	6X	7Y	8Z	A1	
VL	WM	XN	YO	ZP	1Q	2R	3S	4T	5U	6V	7W	8X	AY	BZ	+
WK	XL	YM	ZN	1O	2P	3Q	4R	5S	6T	7U	8V	AW	BX	CY	
XJ	YK	ZL	1M	2N	3O	4P	5Q	6R	7S	8T	AU	BV	CW	DX	
YH	ZJ	1K	2L	3M	4N	5O	6P	7Q	8R	AS	BT	CU	DV	EW	0
ZG	1H	2J	3K	4L	5M	6N	7O	8P	AQ	BR	CS	DT	EU	FV	
1F	2G	3H	4J	5K	6L	7M	8N	AO	BP	CQ	DR	ES	FT	GU	
2E	3F	4G	5H	6J	7K	8L	AM	BN	CO	DP	EQ	FR	GS	HT	−
3D	4E	5F	6G	7H	8J	AK	BL	CM	DN	EO	FP	GQ	HR	JS	
4C	5D	6E	7F	8G	AH	BJ	CK	DL	EM	FN	GO	HP	JQ	KR	
5B	6C	7D	8E	AF	BG	CH	DJ	EK	FL	GM	HN	JO	KP	LQ	
6A	7B	8C	AD	BE	CF	DG	EH	FJ	GK	HL	JM	KN	LO	MP	
78	8A	AB	BC	CD	DE	EF	FG	GH	HJ	JK	KL	LM	MN	NO	

MARCH 1981 **PHYSICAL**

16 M	17 TU	18 W	19 TH	20 F	21 SA	22 SU	23 M	24 TU	25 W	26 TH	27 F	28 SK	29 SU	30 M	31 TU	
VU	WV	XW	AX	BA	CB	DC	ED	FE	GF	HG	JH	KJ	LK	ML	NM	
WT	XU	AV	BW	CX	DA	EB	FC	GD	HE	JF	KG	LH	MJ	NK	OL	
XS	AT	BU	CV	DW	EX	FA	GB	HC	JD	KE	LF	MG	NH	OJ	PK	+
AR	BS	CT	DU	EV	FW	GX	HA	JB	KC	LD	ME	NF	OG	PH	QJ	
BQ	CR	DS	ET	FU	GV	HW	JX	KA	LB	MC	ND	OE	PF	QG	RH	
CP	DQ	ER	FS	GT	HU	JV	KW	LX	MA	NB	OC	PD	QE	RF	SG	0
DO	EP	FQ	GR	HS	JT	KU	LV	MW	NX	OA	PB	QC	RD	SE	TF	
EN	FO	GP	HQ	JR	KS	LT	MU	NV	OW	PX	QA	RB	SC	TD	UE	
FM	GN	HO	JP	KQ	LR	MS	NT	OU	PV	QW	RX	SA	TB	UC	VD	−
GL	HM	JN	KO	LP	MQ	NR	OS	PT	QU	RV	SW	TX	UA	VB	WC	
HK	JL	KM	LN	MO	NP	OQ	PR	QS	RT	SU	TV	UW	VX	WA	XB	
J	K	L	M	N	O	P	Q	R	S	T	U	V	W	X	A	

MARCH 1981 **EMOTIONAL**

16 M	17 TU	18 W	19 TH	20 F	21 SA	22 SU	23 M	24 TU	25 W	26 TH	27 F	28 SA	29 SU	30 M	31 TU	
Q	R	S	T	U	V	W	X	Y	Z	1	2	3	A	B	C	
RP	SQ	TR	US	VT	WU	XV	YW	ZX	1Y	2Z	31	A2	B3	CA	DB	
SO	TP	UQ	VR	WS	XT	YU	ZV	1W	2X	3Y	AZ	B1	C2	D3	EA	
TN	UO	VP	WQ	XR	YS	ZT	1U	2V	3W	AX	BY	CZ	D1	E2	F3	
UM	VN	WO	XP	YQ	ZR	1S	2T	3U	AV	BW	CX	DY	EZ	F1	G2	+
VL	WM	XN	YO	ZP	1Q	2R	3S	AT	BU	CV	DW	EX	FY	GZ	H1	
WK	XL	YM	ZN	1O	2P	3Q	AR	BS	CT	DU	EV	FW	GX	HY	JZ	
XJ	YK	ZL	1M	2N	3O	AP	BQ	CR	DS	ET	FU	GV	HW	JX	KY	0
YH	ZJ	1K	2L	3M	AN	BO	CP	DQ	ER	FS	GT	HU	JV	KW	LX	
ZG	1H	2J	3K	AL	BM	CN	DO	EP	FQ	GR	HS	JT	KU	LV	MW	
1F	2G	3H	AJ	BK	CL	DM	EN	FO	GP	HQ	JR	KS	LT	MU	NV	−
2E	3F	AG	BH	CJ	DK	EL	FM	GN	HO	JP	KQ	LR	MS	NT	OU	
3D	AE	BF	CG	DH	EJ	FK	GL	HM	JN	KO	LP	MQ	NR	OS	PT	
AC	BD	CE	DF	EG	FH	GJ	HK	JL	KM	LN	MO	NP	OQ	PR	QS	
B	C	D	E	F	G	H	J	K	L	M	N	O	P	Q	R	

MARCH 1981 **INTELLECTUAL**

16 M	17 TU	18 W	19 TH	20 F	21 SA	22 SU	23 M	24 TU	25 W	26 TH	27 F	28 SA	29 SU	30 M	31 TU	
6	7	8	A	B	C	D	E	F	G	H	J	K	L	M	N	
75	86	A7	B8	CA	DB	EC	FD	GE	HF	JG	KH	LJ	MK	NL	OM	
84	A5	B6	C7	D8	EA	FB	GC	HD	JE	KF	LG	MH	NJ	OK	PL	
A3	B4	C5	D6	E7	F8	GA	HB	JC	KD	LE	MF	NG	OH	PJ	QK	
B2	C3	D4	E5	F6	G7	H8	JA	KB	LC	MD	NE	OF	PG	QH	RJ	
C1	D2	E3	F4	G5	H6	J7	K8	LA	MB	NC	OD	PE	QF	RG	SH	+
DZ	E1	F2	G3	H4	J5	K6	L7	M8	NA	OB	PC	QD	RE	SF	TG	
EY	FZ	G1	H2	J3	K4	L5	M6	N7	O8	PA	QB	RC	SD	TE	UF	
FX	GY	HZ	J1	K2	L3	M4	N5	O6	P7	Q8	RA	SB	TC	UD	VE	0
GW	HX	JY	KZ	L1	M2	N3	O4	P5	Q6	R7	S8	TA	UB	VC	WD	
HV	JW	KX	LY	MZ	N1	O2	P3	Q4	R5	S6	T7	U8	VA	WB	XC	
JU	KV	LW	MX	NY	OZ	P1	Q2	R3	S4	T5	U6	V7	W8	XA	YB	−
KT	LU	MV	NW	OX	PY	QZ	R1	S2	T3	U4	V5	W6	X7	Y8	ZA	
LS	MT	NU	OV	PW	QX	RY	SZ	T1	U2	V3	W4	X5	Y6	Z7	18	
MR	NS	OT	PU	QV	RW	SX	TY	UZ	V1	W2	X3	Y4	Z5	16	27	
NQ	OR	PS	QT	RU	SV	TW	UX	VY	WZ	X1	Y2	Z3	14	25	36	
OP	PQ	QR	RS	ST	TU	UV	VW	WX	XY	YZ	Z1	12	23	34	45	

APRIL 1981 **PHYSICAL**

1 W	2 TH	3 F	4 SA	5 SU	6 M	7 TU	8 W	9 TH	10 F	11 SA	12 SU	13 M	14 TU	15 W	
ON	PO	QP	RQ	SR	TS	UT	VU	WV	XW	AX	BA	CB	DC	ED	
PM	QN	RO	SP	TQ	UR	VS	WT	XU	AV	BW	CX	DA	EB	FC	
QL	RM	SN	TO	UP	VQ	WR	XS	AT	BU	CV	DW	EX	FA	GB	+
RK	SL	TM	UN	VO	WP	XQ	AR	BS	CT	DU	EV	FW	GX	HA	
SJ	TK	UL	VM	WN	XO	AP	BQ	CR	DS	ET	FU	GV	HW	JX	
TH	UJ	VK	WL	XM	AN	BO	CP	DQ	ER	FS	GT	HU	JV	KW	0
UG	VH	WJ	XK	AL	BM	CN	DO	EP	FQ	GR	HS	JT	KU	LV	
VF	WG	XH	AJ	BK	CL	DM	EN	FO	GP	HQ	JR	KS	LT	MU	
WE	XF	AG	BH	CJ	DK	EL	FM	GN	HO	JP	KQ	LR	MS	NT	−
XD	AE	BF	CG	DH	EJ	FK	GL	HM	JN	KO	LP	MQ	NR	OS	
AC	BD	CE	DF	EG	FH	GJ	HK	JL	KM	LN	MO	NP	OQ	PR	
B	C	D	E	F	G	H	J	K	L	M	N	O	P	Q	

APRIL 1981 **EMOTIONAL**

1 W	2 TH	3 F	4 SA	5 SU	6 M	7 TU	8 W	9 TH	10 F	11 SA	12 SU	13 M	14 TU	15 W	
D	E	F	G	H	J	K	L	M	N	O	P	Q	R	S	
EC	FD	GE	HF	JG	KH	LJ	MK	NL	OM	PN	QO	RP	SQ	TR	
FB	GC	HD	JE	KF	LG	MH	NJ	OK	PL	QM	RN	SO	TP	UQ	
GA	HB	JC	KD	LE	MF	NG	OH	PJ	QK	RL	SM	TN	UO	VP	
H3	JA	KB	LC	MD	NE	OF	PG	QH	RJ	SK	TL	UM	VN	WO	+
J2	K3	LA	MB	NC	OD	PE	QF	RG	SH	TJ	UK	VL	WM	XN	
K1	L2	M3	NA	OB	PC	QD	RE	SF	TG	UH	VJ	WK	XL	YM	
LZ	M1	N2	O3	PA	QB	RC	SD	TE	UF	VG	WH	XJ	YK	ZL	0
MY	NZ	O1	P2	Q3	RA	SB	TC	UD	VE	WF	XG	YH	ZJ	1K	
NX	OY	PZ	Q1	R2	S3	TA	UB	VC	WD	XE	YF	ZG	1H	2J	
OW	PX	QY	RZ	S1	T2	U3	VA	WB	XC	YD	ZE	1F	2G	3H	−
PV	QW	RX	SY	TZ	U1	V2	W3	XA	YB	ZC	1D	2E	3F	AG	
QU	RV	SW	TX	UY	VZ	W1	X2	Y3	ZA	1B	2C	3D	AE	BF	
RT	SU	TV	UW	VX	WY	XZ	Y1	Z2	13	2A	3B	AC	BD	CE	
S	T	U	V	W	X	Y	Z	1	2	3	A	B	C	D	

APRIL 1981 **INTELLECTUAL**

1 W	2 TH	3 F	4 SA	5 SU	6 M	7 TU	8 W	9 TH	10 F	11 SA	12 SU	13 M	14 TU	15 W	
O	P	Q	R	S	T	U	V	W	X	Y	Z	1	2	3	
PN	QO	RP	SQ	TR	US	VT	WU	XV	YW	ZX	1Y	2Z	31	42	
QM	RN	SO	TP	UQ	VR	WS	XT	YU	ZV	1W	2X	3Y	4Z	51	
RL	SM	TN	UO	VP	WQ	XR	YS	ZT	1U	2V	3W	4X	5Y	6Z	
SK	TL	UM	VN	WO	XP	YQ	ZR	1S	2T	3U	4V	5W	6X	7Y	
TJ	UK	VL	WM	XN	YO	ZP	1Q	2R	3S	4T	5U	6V	7W	8X	+
UH	VJ	WK	XL	YM	ZN	1O	2P	3Q	4R	5S	6T	7U	8V	AW	
VG	WH	XJ	YK	ZL	1M	2N	3O	4P	5Q	6R	7S	8T	AU	BV	
WF	XG	YH	ZJ	1K	2L	3M	4N	5O	6P	7Q	8R	AS	BT	CU	0
XE	YF	ZG	1H	2J	3K	4L	5M	6N	7O	8P	AQ	BR	CS	DT	
YD	ZE	1F	2G	3H	4J	5K	6L	7M	8N	AO	BP	CQ	DR	ES	
ZC	1D	2E	3F	4G	5H	6J	7K	8L	AM	BN	CO	DP	EQ	FR	−
1B	2C	3D	4E	5F	6G	7H	8J	AK	BL	CM	DN	EO	FP	GQ	
2A	3B	4C	5D	6E	7F	8G	AH	BJ	CK	DL	EM	FN	GO	HP	
38	4A	5B	6C	7D	8E	AF	BG	CH	DJ	EK	FL	GM	HN	JO	
47	58	6A	7B	8C	AD	BE	CF	DG	EH	FJ	GK	HL	JM	KN	
56	67	78	8A	AB	BC	CD	DE	EF	FG	GH	HJ	JK	KL	LM	

APRIL 1981 — PHYSICAL

16	17	18	19	20	21	22	23	24	25	26	27	28	29	30	
TH	F	SA	SU	M	TU	W	TH	F	SA	SU	M	TU	W	TH	
FE	GF	HG	JH	KJ	LK	ML	NM	ON	PO	QP	RQ	SR	TS	UT	
GD	HE	JF	KG	LH	MJ	NK	OL	PM	QN	RO	SP	TQ	UR	VS	
HC	JD	KE	LF	MG	NH	OJ	PK	QL	RM	SN	TO	UP	VQ	WR	+
JB	KC	LD	ME	NF	OG	PH	QJ	RK	SL	TM	UN	VO	WP	XQ	
KA	LB	MC	ND	OE	PF	QG	RH	SJ	TK	UL	VM	WN	XO	AP	
LX	MA	NB	OC	PD	QE	RF	SG	TH	UJ	VK	WL	XM	AN	BO	0
MW	NX	OA	PB	QC	RD	SE	TF	UG	VH	WJ	XK	AL	BM	CN	
NV	OW	PX	QA	RB	SC	TD	UE	VF	WG	XH	AJ	BK	CL	DM	
OU	PV	QW	RX	SA	TB	UC	VD	WE	XF	AG	BH	CJ	DK	EL	−
PT	QU	RV	SW	TX	UA	VB	WC	XD	AE	BF	CG	DH	EJ	FK	
QS	RT	SU	TV	UW	VX	WA	XB	AC	BD	CE	DF	EG	FH	GJ	
R	S	T	U	V	W	X	A	B	Ç	D	E	F	G	H	

APRIL 1981 — EMOTIONAL

16	17	18	19	20	21	22	23	24	25	26	27	28	29	30	
TH	F	SA	SU	M	TU	W	TH	F	SA	SU	M	TU	W	TH	
T	U	V	W	X	Y	Z	1	2	3	A	B	C	D	E	
US	VT	WU	XV	YW	ZX	1Y	2Z	31	A2	B3	CA	DB	EC	FD	
VR	WS	XT	YU	ZV	1W	2X	3Y	AZ	B1	C2	D3	EA	FB	GC	
WQ	XR	YS	ZT	1U	2V	3W	AX	BY	CZ	D1	E2	F3	GA	HB	
XP	YQ	ZR	1S	2T	3U	AV	BW	CX	DY	EZ	F1	G2	H3	JA	+
YO	ZP	1Q	2R	3S	AT	BU	CV	DW	EX	FY	GZ	H1	J2	K3	
ZN	1O	2P	3Q	AR	BS	CT	DU	EV	FW	GX	HY	JZ	K1	L2	
1M	2N	3O	AP	BQ	CR	DS	ET	FU	GV	HW	JX	KY	LZ	M1	0
2L	3M	AN	BO	CP	DQ	ER	FS	GT	HU	JV	KW	LX	MY	NZ	
3K	AL	BM	CN	DO	EP	FQ	GR	HS	JT	KU	LV	MW	NX	OY	
AJ	BK	CL	DM	EN	FO	GP	HQ	JR	KS	LT	MU	NV	OW	PX	−
BH	CJ	DK	EL	FM	GN	HO	JP	KQ	LR	MS	NT	OU	PV	QW	
CG	DH	EJ	FK	GL	HM	JN	KO	LP	MQ	NR	OS	PT	QU	RV	
DF	EG	FH	GJ	HK	JL	KM	LN	MO	NP	OQ	PR	QS	RT	SU	
E	F	G	H	J	K	L	M	N	O	P	Q	R	S	T	

APRIL 1981 — INTELLECTUAL

16	17	18	19	20	21	22	23	24	25	26	27	28	29	30	
TH	F	SA	SU	M	TU	W	TH	F	SA	SU	M	TU	W	TH	
4	5	6	7	8	A	B	C	D	E	F	G	H	J	K	
53	64	75	86	A7	B8	CA	DB	EC	FD	GE	HF	JG	KH	LJ	
62	73	84	A5	B6	C7	D8	EA	FB	GC	HD	JE	KF	LG	MH	
71	82	A3	B4	C5	D6	E7	F8	GA	HB	JC	KD	LE	MF	NG	
8Z	A1	B2	C3	D4	E5	F6	G7	H8	JA	KB	LC	MD	NE	OF	
AY	BZ	C1	D2	E3	F4	G5	H6	J7	K8	LA	MB	NC	OD	PE	+
BX	CY	DZ	E1	F2	G3	H4	J5	K6	L7	M8	NA	OB	PC	QD	
CW	DX	EY	FZ	G1	H2	J3	K4	L5	M6	N7	O8	PA	QB	RC	
DV	EW	FX	GY	HZ	J1	K2	L3	M4	N5	O6	P7	Q8	RA	SB	0
EU	FV	GW	HX	JY	KZ	L1	M2	N3	O4	P5	Q6	R7	S8	TA	
FT	GU	HV	JW	KX	LY	MZ	N1	O2	P3	Q4	R5	S6	T7	U8	
GS	HT	JU	KV	LW	MX	NY	OZ	P1	Q2	R3	S4	T5	U6	V7	−
HR	JS	KT	LU	MV	NW	OX	PY	QZ	R1	S2	T3	U4	V5	W6	
JQ	KR	LS	MT	NU	OV	PW	QX	RY	SZ	T1	U2	V3	W4	X5	
KP	LQ	MR	NS	OT	PU	QV	RW	SX	TY	UZ	V1	W2	X3	Y4	
LO	MP	NQ	OR	PS	QT	RU	SV	TW	UX	VY	WZ	X1	Y2	Z3	
MN	NO	OP	PQ	QR	RS	ST	TU	UV	VW	WX	XY	YZ	Z1	12	

MAY 1981 **PHYSICAL**

1 F	2 SA	3 SU	4 M	5 TU	6 W	7 TH	8 F	9 SA	10 SU	11 M	12 TU	13 W	14 TH	15 F	
VU	WV	XW	AX	BA	CB	DC	ED	FE	GF	HG	JH	KJ	LK	ML	
WT	XU	AV	BW	CX	DA	EB	FC	GD	HE	JF	KG	LH	MJ	NK	
XS	AT	BU	CV	DW	EX	FA	GB	HC	JD	KE	LF	MG	NH	OJ	+
AR	BS	CT	DU	EV	FW	GX	HA	JB	KC	LD	ME	NF	OG	PH	
BQ	CR	DS	ET	FU	GV	HW	JX	KA	LB	MC	ND	OE	PF	QG	
CP	DQ	ER	FS	GT	HU	JV	KW	LX	MA	NB	OC	PD	QE	RF	0
DO	EP	FQ	GR	HS	JT	KU	LV	MW	NX	OA	PB	QC	RD	SE	
EN	FO	GP	HQ	JR	KS	LT	MU	NV	OW	PX	QA	RB	SC	TD	
FM	GN	HO	JP	KQ	LR	MS	NT	OU	PV	QW	RX	SA	TB	UC	−
GL	HM	JN	KO	LP	MQ	NR	OS	PT	QU	RV	SW	TX	UA	VB	
HK	JL	KM	LN	MO	NP	OQ	PR	QS	RT	SU	TV	UW	VX	WA	
J	K	L	M	N	O	P	Q	R	S	T	U	V	W	X	

MAY 1981 **EMOTIONAL**

1 F	2 SA	3 SU	4 M	5 TU	6 W	7 TH	8 F	9 SA	10 SU	11 M	12 TU	13 W	14 TH	15 F	
F	G	H	J	K	L	M	N	O	P	Q	R	S	T	U	
GE	HF	JG	KH	LJ	MK	NL	OM	PN	QO	RP	SQ	TR	US	VT	
HD	JE	KF	LG	MH	NJ	OK	PL	QM	RN	SO	TP	UQ	VR	WS	
JC	KD	LE	MF	NG	OH	PJ	QK	RL	SM	TN	UO	VP	WQ	XR	
KB	LC	MD	NE	OF	PG	QH	RJ	SK	TL	UM	VN	WO	XP	YQ	+
LA	MB	NC	OD	PE	QF	RG	SH	TJ	UK	VL	WM	XN	YO	ZP	
M3	NA	OB	PC	QD	RE	SF	TG	UH	VJ	WK	XL	YM	ZN	1O	
N2	O3	PA	QB	RC	SD	TE	UF	VG	WH	XJ	YK	ZL	1M	2N	0
O1	P2	Q3	RA	SB	TC	UD	VE	WF	XG	YH	ZJ	1K	2L	3M	
PZ	Q1	R2	S3	TA	UB	VC	WD	XE	YF	ZG	1H	2J	3K	AL	
QY	RZ	S1	T2	U3	VA	WB	XC	YD	ZE	1F	2G	3H	AJ	BK	−
RX	SY	TZ	U1	V2	W3	XA	YB	ZC	1D	2E	3F	AG	BH	CJ	
SW	TX	UY	VZ	W1	X2	Y3	ZA	1B	2C	3D	AE	BF	CG	DH	
TV	UW	VX	WY	XZ	Y1	Z2	13	2A	3B	AC	BD	CE	DF	EG	
U	V	W	X	Y	Z	1	2	3	A	B	C	D	E	F	

MAY 1981 **INTELLECTUAL**

1 F	2 SA	3 SU	4 M	5 TU	6 W	7 TH	8 F	9 SA	10 SU	11 M	12 TU	13 W	14 TH	15 F	
L	M	N	O	P	Q	R	S	T	U	V	W	X	Y	Z	
MK	NL	OM	PN	QO	RP	SQ	TR	US	VT	WU	XV	YW	ZX	1Y	
NJ	OK	PL	QM	RN	SO	TP	UQ	VR	WS	XT	YU	ZV	1W	2X	
OH	PJ	QK	RL	SM	TN	UO	VP	WQ	XR	YS	ZT	1U	2V	3W	
PG	QH	RJ	SK	TL	UM	VN	WO	XP	YQ	ZR	1S	2T	3U	4V	
QF	RG	SH	TJ	UK	VL	WM	XN	YO	ZP	1Q	2R	3S	4T	5U	+
RE	SF	TG	UH	VJ	WK	XL	YM	ZN	1O	2P	3Q	4R	5S	6T	
SD	TE	UF	VG	WH	XJ	YK	ZL	1M	2N	3O	4P	5Q	6R	7S	
TC	UD	VE	WF	XG	YH	ZJ	1K	2L	3M	4N	5O	6P	7Q	8R	0
UB	VC	WD	XE	YF	ZG	1H	2J	3K	4L	5M	6N	7O	8P	AQ	
VA	WB	XC	YD	ZE	1F	2G	3H	4J	5K	6L	7M	8N	AO	BP	
W8	XA	YB	ZC	1D	2E	3F	4G	5H	6J	7K	8L	AM	BN	CO	−
X7	Y8	ZA	1B	2C	3D	4E	5F	6G	7H	8J	AK	BL	CM	DN	
Y6	Z7	18	2A	3B	4C	5D	6E	7F	8G	AH	BJ	CK	DL	EM	
Z5	16	27	38	4A	5B	6C	7D	8E	AF	BG	CH	DJ	EK	FL	
14	25	36	47	58	6A	7B	8C	AD	BE	CF	DG	EH	FJ	GK	
23	34	45	56	67	78	8A	AB	BC	CD	DE	EF	FG	GH	HJ	

MAY 1981 **PHYSICAL**

16 SA	17 SU	18 M	19 TU	20 W	21 TH	22 F	23 SA	24 SU	25 M	26 TU	27 W	28 TH	29 F	30 SA	31 SU	
NM	ON	PO	QP	RQ	SR	TS	UT	VU	WV	XW	AX	BA	CB	DC	ED	
OL	PM	QN	RO	SP	TQ	UR	VS	WT	XU	AV	BW	CX	DA	EB	FC	
PK	QL	RM	SN	TO	UP	VQ	WR	XS	AT	BU	CV	DW	EX	FA	GB	+
QJ	RK	SL	TM	UN	VO	WP	XQ	AR	BS	CT	DU	EV	FW	GX	HA	
RH	SJ	TK	UL	VM	WN	XO	AP	BQ	CR	DS	ET	FU	GV	HW	JX	
SG	TH	UJ	VK	WL	XM	AN	BO	CP	DQ	ER	FS	GT	HU	JV	KW	0
TF	UG	VH	WJ	XK	AL	BM	CN	DO	EP	FQ	GR	HS	JT	KU	LV	
UE	VF	WG	XH	AJ	BK	CL	DM	EN	FO	GP	HQ	JR	KS	LT	MU	
VD	WE	XF	AG	BH	CJ	DK	EL	FM	GN	HO	JP	KQ	LR	MS	NT	−
WC	XD	AE	BF	CG	DH	EJ	FK	GL	HM	JN	KO	LP	MQ	NR	OS	
XB	AC	BD	CE	DF	EG	FH	GJ	HK	JL	KM	LN	MO	NP	OQ	PR	
A	B	C	D	E	F	G	H	J	K	L	M	N	O	P	Q	

MAY 1981 **EMOTIONAL**

16 SA	17 SU	18 M	19 TU	20 W	21 TH	22 F	23 SA	24 SU	25 M	26 TU	27 W	28 TH	29 F	30 SA	31 SU	
V	W	X	Y	Z	1	2	3	A	B	C	D	E	F	G	H	
WU	XV	YW	ZX	1Y	2Z	31	A2	B3	CA	DB	EC	FD	GE	HF	JG	
XT	YU	ZV	1W	2X	3Y	AZ	B1	C2	D3	EA	FB	GC	HD	JE	KF	
YS	ZT	1U	2V	3W	AX	BY	CZ	D1	E2	F3	GA	HB	JC	KD	LE	
ZR	1S	2T	3U	AV	BW	CX	DY	EZ	F1	G2	H3	JA	KB	LC	MD	+
1Q	2R	3S	AT	BU	CV	DW	EX	FY	GZ	H1	J2	K3	LA	MB	NC	
2P	3Q	AR	BS	CT	DU	EV	FW	GX	HY	JZ	K1	L2	M3	NA	OB	
3O	AP	BQ	CR	DS	ET	FU	GV	HW	JX	KY	LZ	M1	N2	O3	PA	0
AN	BO	CP	DQ	ER	FS	GT	HU	JV	KW	LX	MY	NZ	O1	P2	Q3	
BM	CN	DO	EP	FQ	GR	HS	JT	KU	LV	MW	NX	OY	PZ	Q1	R2	
CL	DM	EN	FO	GP	HQ	JR	KS	LT	MU	NV	OW	PX	QY	RZ	S1	−
DK	EL	FM	GN	HO	JP	KQ	LR	MS	NT	OU	PV	QW	RX	SY	TZ	
EJ	FK	GL	HM	JN	KO	LP	MQ	NR	OS	PT	QU	RV	SW	TX	UY	
FH	GJ	HK	JL	KM	LN	MO	NP	OQ	PR	QS	RT	SU	TV	UW	VX	
G	H	J	K	L	M	N	O	P	Q	R	S	T	U	V	W	

MAY 1981 **INTELLECTUAL**

16 SA	17 SU	18 M	19 TU	20 W	21 TH	22 F	23 SA	24 SU	25 M	26 TU	27 W	28 TH	29 F	30 SA	31 SU	
1	2	3	4	5	6	7	8	A	B	C	D	E	F	G	H	
2Z	31	42	53	64	75	86	A7	B8	CA	DB	EC	FD	GE	HF	JG	
3Y	4Z	51	62	73	84	A5	B6	C7	D8	EA	FB	GC	HD	JE	KF	
4X	5Y	6Z	71	82	A3	B4	C5	D6	E7	F8	GA	HB	JC	KD	LE	
5W	6X	7Y	8Z	A1	B2	C3	D4	E5	F6	G7	H8	JA	KB	LC	MD	
6V	7W	8X	AY	BZ	C1	D2	E3	F4	G5	H6	J7	K8	LA	MB	NC	+
7U	8V	AW	BX	CY	DZ	E1	F2	G3	H4	J5	K6	L7	M8	NA	OB	
8T	AU	BV	CW	DX	EY	FZ	G1	H2	J3	K4	L5	M6	N7	O8	PA	
AS	BT	CU	DV	EW	FX	GY	HZ	J1	K2	L3	M4	N5	O6	P7	Q8	0
BR	CS	DT	EU	FV	GW	HX	JY	KZ	L1	M2	N3	O4	P5	Q6	R7	
CQ	DR	ES	FT	GU	HV	JW	KX	LY	MZ	N1	O2	P3	Q4	R5	S6	
DP	EQ	FR	GS	HT	JU	KV	LW	MX	NY	OZ	P1	Q2	R3	S4	T5	−
EO	FP	GQ	HR	JS	KT	LU	MV	NW	OX	PY	QZ	R1	S2	T3	U4	
FN	GO	HP	JQ	KR	LS	MT	NU	OV	PW	QX	RY	SZ	T1	U2	V3	
GM	HN	JO	KP	LQ	MR	NS	OT	PU	QV	RW	SX	TY	UZ	V1	W2	
HL	JM	KN	LO	MP	NQ	OR	PS	QT	RU	SV	TW	UX	VY	WZ	X1	
JK	KL	LM	MN	NO	OP	PQ	QR	RS	ST	TU	UV	VW	WX	XY	YZ	

JUNE 1981 — PHYSICAL

1 M	2 TU	3 W	4 TH	5 F	6 SA	7 SU	8 M	9 TU	10 W	11 TH	12 F	13 SA	14 SU	15 M	
FE	GF	HG	JH	KJ	LK	ML	NM	ON	PO	QP	RQ	SR	TS	UT	
GD	HE	JF	KG	LH	MJ	NK	OL	PM	QN	RO	SP	TQ	UR	VS	
HC	JD	KE	LF	MG	NH	OJ	PK	QL	RM	SN	TO	UP	VQ	WR	+
JB	KC	LD	ME	NF	OG	PH	QJ	RK	SL	TM	UN	VO	WP	XQ	
KA	LB	MC	ND	OE	PF	QG	RH	SJ	TK	UL	VM	WN	XO	AP	
LX	MA	NB	OC	PD	QE	RF	SG	TH	UJ	VK	WL	XM	AN	BO	0
MW	NX	OA	PB	QC	RD	SE	TF	UG	VH	WJ	XK	AL	BM	CN	
NV	OW	PX	QA	RB	SC	TD	UE	VF	WG	XH	AJ	BK	CL	DM	
OU	PV	QW	RX	SA	TB	UC	VD	WE	XF	AG	BH	CJ	DK	EL	−
PT	QU	RV	SW	TX	UA	VB	WC	XD	AE	BF	CG	DH	EJ	FK	
QS	RT	SU	TV	UW	VX	WA	XB	AC	BD	CE	DF	EG	FH	GJ	
R	S	T	U	V	W	X	A	B	C	D	E	F	G	H	

JUNE 1981 — EMOTIONAL

1 M	2 TU	3 W	4 TH	5 F	6 SA	7 SU	8 M	9 TU	10 W	11 TH	12 F	13 SA	14 SU	15 M	
J	K	L	M	N	O	P	Q	R	S	T	U	V	W	X	
KH	LJ	MK	NL	OM	PN	QO	RP	SQ	TR	US	VT	WU	XV	YW	
LG	MH	NJ	OK	PL	QM	RN	SO	TP	UQ	VR	WS	XT	YU	ZV	
MF	NG	OH	PJ	QK	RL	SM	TN	UO	VP	WQ	XR	YS	ZT	1U	
NE	OF	PG	QH	RJ	SK	TL	UM	VN	WO	XP	YQ	ZR	1S	2T	+
OD	PE	QF	RG	SH	TJ	UK	VL	WM	XN	YO	ZP	1Q	2R	3S	
PC	QD	RE	SF	TG	UH	VJ	WK	XL	YM	ZN	1O	2P	3Q	AR	
QB	RC	SD	TE	UF	VG	WH	XJ	YK	ZL	1M	2N	3O	AP	BQ	0
RA	SB	TC	UD	VE	WF	XG	YH	ZJ	1K	2L	3M	AN	BO	CP	
S3	TA	UB	VC	WD	XE	YF	ZG	1H	2J	3K	AL	BM	CN	DO	
T2	U3	VA	WB	XC	YD	ZE	1F	2G	3H	AJ	BK	CL	DM	EN	−
U1	V2	W3	XA	YB	ZC	1D	2E	3F	AG	BH	CJ	DK	EL	FM	
VZ	W1	X2	Y3	ZA	1B	2C	3D	AE	BF	CG	DH	EJ	FK	GL	
WY	XZ	Y1	Z2	13	2A	3B	AC	BD	CE	DF	EG	FH	GJ	HK	
X	Y	Z	1	2	3	A	B	C	D	E	F	G	H	J	

JUNE 1981 — INTELLECTUAL

1 M	2 TU	3 W	4 TH	5 F	6 SA	7 SU	8 M	9 TU	10 W	11 TH	12 F	13 SA	14 SU	15 M	
J	K	L	M	N	O	P	Q	R	S	T	U	V	W	X	
KH	LJ	MK	NL	OM	PN	QO	RP	SQ	TR	US	VT	WU	XV	YW	
LG	MH	NJ	OK	PL	QM	RN	SO	TP	UQ	VR	WS	XT	YU	ZV	
MF	NG	OH	PJ	QK	RL	SM	TN	UO	VP	WQ	XR	YS	ZT	1U	
NE	OF	PG	QH	RJ	SK	TL	UM	VN	WO	XP	YQ	ZR	1S	2T	
OD	PE	QF	RG	SH	TJ	UK	VL	WM	XN	YO	ZP	1Q	2R	3S	+
PC	QD	RE	SF	TG	UH	VJ	WK	XL	YM	ZN	1O	2P	3Q	4R	
QB	RC	SD	TE	UF	VG	WH	XJ	YK	ZL	1M	2N	3O	4P	5Q	
RA	SB	TC	UD	VE	WF	XG	YH	ZJ	1K	2L	3M	4N	5O	6P	0
S8	TA	UB	VC	WD	XE	YF	ZG	1H	2J	3K	4L	5M	6N	7O	
T7	U8	VA	WB	XC	YD	ZE	1F	2G	3H	4J	5K	6L	7M	8N	
U6	V7	W8	XA	YB	ZC	1D	2E	3F	4G	5H	6J	7K	8L	AM	−
V5	W6	X7	Y8	ZA	1B	2C	3D	4E	5F	6G	7H	8J	AK	BL	
W4	X5	Y6	Z7	18	2A	3B	4C	5D	6E	7F	8G	AH	BJ	CK	
X3	Y4	Z5	16	27	38	4A	5B	6C	7D	8E	AF	BG	CH	DJ	
Y2	Z3	14	25	36	47	58	6A	7B	8C	AD	BE	CF	DG	EH	
Z1	12	23	34	45	56	67	78	8A	AB	BC	CD	DE	EF	FG	

JUNE 1981 — PHYSICAL

16 TU	17 W	18 TH	19 F	20 SA	21 SU	22 M	23 TU	24 W	25 TH	26 F	27 SA	28 SU	29 M	30 TU	
VU	WV	XW	AX	BA	CB	DC	ED	FE	GF	HG	JH	KJ	LK	ML	
WT	XU	AV	BW	CX	DA	EB	FC	GD	HE	JF	KG	LH	MJ	NK	
XS	AT	BU	CV	DW	EX	FA	GB	HC	JD	KE	LF	MG	NH	OJ	+
AR	BS	CT	DU	EV	FW	GX	HA	JB	KC	LD	ME	NF	OG	PH	
BQ	CR	DS	ET	FU	GV	HW	JX	KA	LB	MC	ND	OE	PF	QG	
CP	DQ	ER	FS	GT	HU	JV	KW	LX	MA	NB	OC	PD	QE	RF	0
DO	EP	FQ	GR	HS	JT	KU	LV	MW	NX	OA	PB	QC	RD	SE	
EN	FO	GP	HQ	JR	KS	LT	MU	NV	OW	PX	QA	RB	SC	TD	
FM	GN	HO	JP	KQ	LR	MS	NT	OU	PV	QW	RX	SA	TB	UC	−
GL	HM	JN	KO	LP	MQ	NR	OS	PT	QU	RV	SW	TX	UA	VB	
HK	JL	KM	LN	MO	NP	OQ	PR	QS	RT	SU	TV	UW	VX	WA	
J	K	L	M	N	O	P	Q	R	S	T	U	V	W	X	

JUNE 1981 — EMOTIONAL

16 TU	17 W	18 TH	19 F	20 SA	21 SU	22 M	23 TU	24 W	25 TH	26 F	27 SA	28 SU	29 M	30 TU	
Y	Z	1	2	3	A	B	C	D	E	F	G	H	J	K	
ZX	1Y	2Z	31	A2	B3	CA	DB	EC	FD	GE	HF	JG	KH	LJ	
1W	2X	3Y	AZ	B1	C2	D3	EA	FB	GC	HD	JE	KF	LG	MH	
2V	3W	AX	BY	CZ	D1	E2	F3	GA	HB	JC	KD	LE	MF	NG	
3U	AV	BW	CX	DY	EZ	F1	G2	H3	JA	KB	LC	MD	NE	OF	+
AT	BU	CV	DW	EX	FY	GZ	H1	J2	K3	LA	MB	NC	OD	PE	
BS	CT	DU	EV	FW	GX	HY	JZ	K1	L2	M3	NA	OB	PC	QD	
CR	DS	ET	FU	GV	HW	JX	KY	LZ	M1	N2	O3	PA	QB	RC	0
DQ	ER	FS	GT	HU	JV	KW	LX	MY	NZ	O1	P2	Q3	RA	SB	
EP	FQ	GR	HS	JT	KU	LV	MW	NX	OY	PZ	Q1	R2	S3	TA	
FO	GP	HQ	JR	KS	LT	MU	NV	OW	PX	QY	RZ	S1	T2	U3	−
GN	HO	JP	KQ	LR	MS	NT	OU	PV	QW	RX	SY	TZ	U1	V2	
HM	JN	KO	LP	MQ	NR	OS	PT	QU	RV	SW	TX	UY	VZ	W1	
JL	KM	LN	MO	NP	OQ	PR	QS	RT	SU	TV	UW	VX	WY	XZ	
K	L	M	N	O	P	Q	R	S	T	U	V	W	X	Y	

JUNE 1981 — INTELLECTUAL

16 TU	17 W	18 TH	19 F	20 SA	21 SU	22 M	23 TU	24 W	25 TH	26 F	27 SA	28 SU	29 M	30 TU	
Y	Z	1	2	3	4	5	6	7	8	A	B	C	D	E	
ZX	1Y	2Z	31	42	53	64	75	86	A7	B8	CA	DB	EC	FD	
1W	2X	3Y	4Z	51	62	73	84	A5	B6	C7	D8	EA	FB	GC	
2V	3W	4X	5Y	6Z	71	82	A3	B4	C5	D6	E7	F8	GA	HB	
3U	4V	5W	6X	7Y	8Z	A1	B2	C3	D4	E5	F6	G7	H8	JA	
4T	5U	6V	7W	8X	AY	BZ	C1	D2	E3	F4	G5	H6	J7	K8	+
5S	6T	7U	8V	AW	BX	CY	DZ	E1	F2	G3	H4	J5	K6	L7	
6R	7S	8T	AU	BV	CW	DX	EY	FZ	G1	H2	J3	K4	L5	M6	
7Q	8R	AS	BT	CU	DV	EW	FX	GY	HZ	J1	K2	L3	M4	N5	0
8P	AQ	BR	CS	DT	EU	FV	GW	HX	JY	KZ	L1	M2	N3	O4	
AO	BP	CQ	DR	ES	FT	GU	HV	JW	KX	LY	MZ	N1	O2	P3	
BN	CO	DP	EQ	FR	GS	HT	JU	KV	LW	MX	NY	OZ	P1	Q2	−
CM	DN	EO	FP	GQ	HR	JS	KT	LU	MV	NW	OX	PY	QZ	R1	
DL	EM	FN	GO	HP	JQ	KR	LS	MT	NU	OV	PW	QX	RY	SZ	
EK	FL	GM	HN	JO	KP	LQ	MR	NS	OT	PU	QV	RW	SX	TY	
FJ	GK	HL	JM	KN	LO	MP	NQ	OR	PS	QT	RU	SV	TW	UX	
GH	HJ	JK	KL	LM	MN	NO	OP	PQ	QR	RS	ST	TU	UV	VW	

JULY 1981 **PHYSICAL**

1 W	2 TH	3 F	4 SA	5 SU	6 M	7 TU	8 W	9 TH	10 F	11 SA	12 SU	13 M	14 TU	15 W	
NM	ON	PO	QP	RQ	SR	TS	UT	VU	WV	XW	AX	BA	CB	DC	
OL	PM	QN	RO	SP	TQ	UR	VS	WT	XU	AV	BW	CX	DA	EB	
PK	QL	RM	SN	TO	UP	VQ	WR	XS	AT	BU	CV	DW	EX	FA	+
QJ	RK	SL	TM	UN	VO	WP	XQ	AR	BS	CT	DU	EV	FW	GX	
RH	SJ	TK	UL	VM	WN	XO	AP	BQ	CR	DS	ET	FU	GV	HW	
SG	TH	UJ	VK	WL	XM	AN	BO	CP	DQ	ER	FS	GT	HU	JV	0
TF	UG	VH	WJ	XK	AL	BM	CN	DO	EP	FQ	GR	HS	JT	KU	
UE	VF	WG	XH	AJ	BK	CL	DM	EN	FO	GP	HQ	JR	KS	LT	
VD	WE	XF	AG	BH	CJ	DK	EL	FM	GN	HO	JP	KQ	LR	MS	−
WC	XD	AE	BF	CG	DH	EJ	FK	GL	HM	JN	KO	LP	MQ	NR	
XB	AC	BD	CE	DF	EG	FH	GJ	HK	JL	KM	LN	MO	NP	OQ	
A	B	C	D	E	F	G	H	J	K	L	M	N	O	P	

JULY 1981 **EMOTIONAL**

1 W	2 TH	3 F	4 SA	5 SU	6 M	7 TU	8 W	9 TH	10 F	11 SA	12 SU	13 M	14 TU	15 W	
L	M	N	O	P	Q	R	S	T	U	V	W	X	Y	Z	
MK	NL	OM	PN	QO	RP	SQ	TR	US	VT	WU	XV	YW	ZX	1Y	
NJ	OK	PL	QM	RN	SO	TP	UQ	VR	WS	XT	YU	ZV	1W	2X	
OH	PJ	QK	RL	SM	TN	UO	VP	WQ	XR	YS	ZT	1U	2V	3W	
PG	QH	RJ	SK	TL	UM	VN	WO	XP	YQ	ZR	1S	2T	3U	AV	+
QF	RG	SH	TJ	UK	VL	WM	XN	YO	ZP	1Q	2R	3S	AT	BU	
RE	SF	TG	UH	VJ	WK	XL	YM	ZN	1O	2P	3Q	AR	BS	CT	
SD	TE	UF	VG	WH	XJ	YK	ZL	1M	2N	3O	AP	BQ	CR	DS	0
TC	UD	VE	WF	XG	YH	ZJ	1K	2L	3M	AN	BO	CP	DQ	ER	
UB	VC	WD	XE	YF	ZG	1H	2J	3K	AL	BM	CN	DO	EP	FQ	
VA	WB	XC	YD	ZE	1F	2G	3H	AJ	BK	CL	DM	EN	FO	GP	−
W3	XA	YB	ZC	1D	2E	3F	AG	BH	CJ	DK	EL	FM	GN	HO	
X2	Y3	ZA	1B	2C	3D	AE	BF	CG	DH	EJ	FK	GL	HM	JN	
Y1	Z2	13	2A	3B	AC	BD	CE	DF	EG	FH	GJ	HK	JL	KM	
Z	1	2	3	A	B	C	D	E	F	G	H	J	K	L	

JULY 1981 **INTELLECTUAL**

1 W	2 TH	3 F	4 SA	5 SU	6 M	7 TU	8 W	9 TH	10 F	11 SA	12 SU	13 M	14 TU	15 W	
F	G	H	J	K	L	M	N	O	P	Q	R	S	T	U	
GE	HF	JG	KH	LJ	MK	NL	OM	PN	QO	RP	SQ	TR	US	VT	
HD	JE	KF	LG	MH	NJ	OK	PL	QM	RN	SO	TP	UQ	VR	WS	
JC	KD	LE	MF	NG	OH	PJ	QK	RL	SM	TN	UO	VP	WQ	XR	
KB	LC	MD	NE	OF	PG	QH	RJ	SK	TL	UM	VN	WO	XP	YQ	
LA	MB	NC	OD	PE	QF	RG	SH	TJ	UK	VL	WM	XN	YO	ZP	+
M8	NA	OB	PC	QD	RE	SF	TG	UH	VJ	WK	XL	YM	ZN	1O	
N7	O8	PA	QB	RC	SD	TE	UF	VG	WH	XJ	YK	ZL	1M	2N	
O6	P7	Q8	RA	SB	TC	UD	VE	WF	XG	YH	ZJ	1K	2L	3M	0
P5	Q6	R7	S8	TA	UB	VC	WD	XE	YF	ZG	1H	2J	3K	4L	
Q4	R5	S6	T7	U8	VA	WB	XC	YD	ZE	1F	2G	3H	4J	5K	
R3	S4	T5	U6	V7	W8	XA	YB	ZC	1D	2E	3F	4G	5H	6J	−
S2	T3	U4	V5	W6	X7	Y8	ZA	1B	2C	3D	4E	5F	6G	7H	
T1	U2	V3	W4	X5	Y6	Z7	18	2A	3B	4C	5D	6E	7F	8G	
UZ	V1	W2	X3	Y4	Z5	16	27	38	4A	5B	6C	7D	8E	AF	
VY	WZ	X1	Y2	Z3	14	25	36	47	58	6A	7B	8C	AD	BE	
WX	XY	YZ	Z1	12	23	34	45	56	67	78	8A	AB	BC	CD	

JULY 1981 — PHYSICAL

16 TH	17 F	18 SA	19 SU	20 M	21 TU	22 W	23 TH	24 F	25 SA	26 SU	27 M	28 TU	29 W	30 TH	31 F	
ED	FE	GF	HG	JH	KJ	LK	ML	NM	ON	PO	QP	RQ	SR	TS	UT	
FC	GD	HE	JF	KG	LH	MJ	NK	OL	PM	QN	RO	SP	TQ	UR	VS	
GB	HC	JD	KE	LF	MG	NH	OJ	PK	QL	RM	SN	TO	UP	VQ	WR	+
HA	JB	KC	LD	ME	NF	OG	PH	QJ	RK	SL	TM	UN	VO	WP	XQ	
JX	KA	LB	MC	ND	OE	PF	QG	RH	SJ	TK	UL	VM	WN	XO	AP	
KW	LX	MA	NB	OC	PD	QE	RF	SG	TH	UJ	VK	WL	XM	AN	BO	0
LV	MW	NX	OA	PB	QC	RD	SE	TF	UG	VH	WJ	XK	AL	BM	CN	
MU	NV	OW	PX	QA	RB	SC	TD	UE	VF	WG	XH	AJ	BK	CL	DM	
NT	OU	PV	QW	RX	SA	TB	UC	VD	WE	XF	AG	BH	CJ	DK	EL	−
OS	PT	QU	RV	SW	TX	UA	VB	WC	XD	AE	BF	CG	DH	EJ	FK	
PR	QS	RT	SU	TV	UW	VX	WA	XB	AC	BD	CE	DF	EG	FH	GJ	
Q	R	S	T	U	V	W	X	A	B	C	D	E	F	G	H	

JULY 1981 — EMOTIONAL

16 TH	17 F	18 SA	19 SU	20 M	21 TU	22 W	23 TH	24 F	25 SA	26 SU	27 M	28 TU	29 W	30 TH	31 F	
1	2	3	A	B	C	D	E	F	G	H	J	K	L	M	N	
2Z	31	A2	B3	CA	DB	EC	FD	GE	HF	JG	KH	LJ	MK	NL	OM	
3Y	AZ	B1	C2	D3	EA	FB	GC	HD	JE	KF	LG	MH	NJ	OK	PL	
AX	BY	CZ	D1	E2	F3	GA	HB	JC	KD	LE	MF	NG	OH	PJ	QK	
BW	CX	DY	EZ	F1	G2	H3	JA	KB	LC	MD	NE	OF	PG	QH	RJ	+
CV	DW	EX	FY	GZ	H1	J2	K3	LA	MB	NC	OD	PE	QF	RG	SH	
DU	EV	FW	GX	HY	JZ	K1	L2	M3	NA	OB	PC	QD	RE	SF	TG	
ET	FU	GV	HW	JX	KY	LZ	M1	N2	O3	PA	QB	RC	SD	TE	UF	0
FS	GT	HU	JV	KW	LX	MY	NZ	O1	P2	Q3	RA	SB	TC	UD	VE	
GR	HS	JT	KU	LV	MW	NX	OY	PZ	Q1	R2	S3	TA	UB	VC	WD	
HQ	JR	KS	LT	MU	NV	OW	PX	QY	RZ	S1	T2	U3	VA	WB	XC	−
JP	KQ	LR	MS	NT	OU	PV	QW	RX	SY	TZ	U1	V2	W3	XA	YB	
KO	LP	MQ	NR	OS	PT	QU	RV	SW	TX	UY	VZ	W1	X2	Y3	ZA	
LN	MO	NP	OQ	PR	QS	RT	SU	TV	UW	VX	WY	XZ	Y1	Z2	13	
M	N	O	P	Q	R	S	T	U	V	W	X	Y	Z	1	2	

JULY 1981 — INTELLECTUAL

16 TH	17 F	18 SA	19 SU	20 M	21 TU	22 W	23 TH	24 F	25 SA	26 SU	27 M	28 TU	29 W	30 TH	31 F	
V	W	X	Y	Z	1	2	3	4	5	6	7	8	A	B	C	
WU	XV	YW	ZX	1Y	2Z	31	42	53	64	75	86	A7	B8	CA	DB	
XT	YU	ZV	1W	2X	3Y	4Z	51	62	73	84	A5	B6	C7	D8	EA	
YS	ZT	1U	2V	3W	4X	5Y	6Z	71	82	A3	B4	C5	D6	E7	F8	
ZR	1S	2T	3U	4V	5W	6X	7Y	8Z	A1	B2	C3	D4	E5	F6	G7	
1Q	2R	3S	4T	5U	6V	7W	8X	AY	BZ	C1	D2	E3	F4	G5	H6	+
2P	3Q	4R	5S	6T	7U	8V	AW	BX	CY	DZ	E1	F2	G3	H4	J5	
3O	4P	5Q	6R	7S	8T	AU	BV	CW	DX	EY	FZ	G1	H2	J3	K4	
4N	5O	6P	7Q	8R	AS	BT	CU	DV	EW	FX	GY	HZ	J1	K2	L3	0
5M	6N	7O	8P	AQ	BR	CS	DT	EU	FV	GW	HX	JY	KZ	L1	M2	
6L	7M	8N	AO	BP	CQ	DR	ES	FT	GU	HV	JW	KX	LY	MZ	N1	
7K	8L	AM	BN	CO	DP	EQ	FR	GS	HT	JU	KV	LW	MX	NY	OZ	−
8J	AK	BL	CM	DN	EO	FP	GQ	HR	JS	KT	LU	MV	NW	OX	PY	
AH	BJ	CK	DL	EM	FN	GO	HP	JQ	KR	LS	MT	NU	OV	PW	QX	
BG	CH	DJ	EK	FL	GM	HN	JO	KP	LQ	MR	NS	OT	PU	QV	RW	
CF	DG	EH	FJ	GK	HL	JM	KN	LO	MP	NQ	OR	PS	QT	RU	SV	
DE	EF	FG	GH	HJ	JK	KL	LM	MN	NO	OP	PQ	QR	RS	ST	TU	

AUGUST 1981 **PHYSICAL**

1 SA	2 SU	3 M	4 TU	5 W	6 TH	7 F	8 SA	9 SU	10 M	11 TU	12 W	13 TH	14 F	15 SA	
VU	WV	XW	AX	BA	CB	DC	ED	FE	GF	HG	JH	KJ	LK	ML	
WT	XU	AV	BW	CX	DA	EB	FC	GD	HE	JF	KG	LH	MJ	NK	
XS	AT	BU	CV	DW	EX	FA	GB	HC	JD	KE	LF	MG	NH	OJ	+
AR	BS	CT	DU	EV	FW	GX	HA	JB	KC	LD	ME	NF	OG	PH	
BQ	CR	DS	ET	FU	GV	HW	JX	KA	LB	MC	ND	OE	PF	QG	
CP	DQ	ER	FS	GT	HU	JV	KW	LX	MA	NB	OC	PD	QE	RF	0
DO	EP	FQ	GR	HS	JT	KU	LV	MW	NX	OA	PB	QC	RD	SE	
EN	FO	GP	HQ	JR	KS	LT	MU	NV	OW	PX	QA	RB	SC	TD	
FM	GN	HO	JP	KQ	LR	MS	NT	OU	PV	QW	RX	SA	TB	UC	−
GL	HM	JN	KO	LP	MQ	NR	OS	PT	QU	RV	SW	TX	UA	VB	
HK	JL	KM	LN	MO	NP	OQ	PR	QS	RT	SU	TV	UW	VX	WA	
J	K	L	M	N	O	P	Q	R	S	T	U	V	W	X	

AUGUST 1981 **EMOTIONAL**

1 SA	2 SU	3 M	4 TU	5 W	6 TH	7 F	8 SA	9 SU	10 M	11 TU	12 W	13 TH	14 F	15 SA	
O	P	Q	R	S	T	U	V	W	X	Y	Z	1	2	3	
PN	QO	RP	SQ	TR	US	VT	WU	XV	YW	ZX	1Y	2Z	31	A2	
QM	RN	SO	TP	UQ	VR	WS	XT	YU	ZV	1W	2X	3Y	AZ	B1	
RL	SM	TN	UO	VP	WQ	XR	YS	ZT	1U	2V	3W	AX	BY	CZ	
SK	TL	UM	VN	WO	XP	YQ	ZR	1S	2T	3U	AV	BW	CX	DY	+
TJ	UK	VL	WM	XN	YO	ZP	1Q	2R	3S	AT	BU	CV	DW	EX	
UH	VJ	WK	XL	YM	ZN	1O	2P	3Q	AR	BS	CT	DU	EV	FW	
VG	WH	XJ	YK	ZL	1M	2N	3O	AP	BQ	CR	DS	ET	FU	GV	0
WF	XG	YH	ZJ	1K	2L	3M	AN	BO	CP	DQ	ER	FS	GT	HU	
XE	YF	ZG	1H	2J	3K	AL	BM	CN	DO	EP	FQ	GR	HS	JT	
YD	ZE	1F	2G	3H	AJ	BK	CL	DM	EN	FO	GP	HQ	JR	KS	−
ZC	1D	2E	3F	AG	BH	CJ	DK	EL	FM	GN	HO	JP	KQ	LR	
1B	2C	3D	AE	BF	CG	DH	EJ	FK	GL	HM	JN	KO	LP	MQ	
2A	3B	AC	BD	CE	DF	EG	FH	GJ	HK	JL	KM	LN	MO	NP	
3	A	B	C	D	E	F	G	H	J	K	L	M	N	O	

AUGUST 1981 **INTELLECTUAL**

1 SA	2 SU	3 M	4 TU	5 W	6 TH	7 F	8 SA	9 SU	10 M	11 TU	12 W	13 TH	14 F	15 SA	
D	E	F	G	H	J	K	L	M	N	O	P	Q	R	S	
EC	FD	GE	HF	JG	KH	LJ	MK	NL	OM	PN	QO	RP	SQ	TR	
FB	GC	HD	JE	KF	LG	MH	NJ	OK	PL	QM	RN	SO	TP	UQ	
GA	HB	JC	KD	LE	MF	NG	OH	PJ	QK	RL	SM	TN	UO	VP	
H8	JA	KB	LC	MD	NE	OF	PG	QH	RJ	SK	TL	UM	VN	WO	
J7	K8	LA	MB	NC	OD	PE	QF	RG	SH	TJ	UK	VL	WM	XN	+
K6	L7	M8	NA	OB	PC	QD	RE	SF	TG	UH	VJ	WK	XL	YM	
L5	M6	N7	O8	PA	QB	RC	SD	TE	UF	VG	WH	XJ	YK	ZL	
M4	N5	O6	P7	Q8	RA	SB	TC	UD	VE	WF	XG	YH	ZJ	1K	0
N3	O4	P5	Q6	R7	S8	TA	UB	VC	WD	XE	YF	ZG	1H	2J	
O2	P3	Q4	R5	S6	T7	U8	VA	WB	XC	YD	ZE	1F	2G	3H	
P1	Q2	R3	S4	T5	U6	V7	W8	XA	YB	ZC	1D	2E	3F	4G	−
QZ	R1	S2	T3	U4	V5	W6	X7	Y8	ZA	1B	2C	3D	4E	5F	
RY	SZ	T1	U2	V3	W4	X5	Y6	Z7	18	2A	3B	4C	5D	6E	
SX	TY	UZ	V1	W2	X3	Y4	Z5	16	27	38	4A	5B	6C	7D	
TW	UX	VY	WZ	X1	Y2	Z3	14	25	36	47	58	6A	7B	8C	
UV	VW	WX	XY	YZ	Z1	12	23	34	45	56	67	78	8A	AB	

AUGUST 1981 **PHYSICAL**

16 SU	17 M	18 TU	19 W	20 TH	21 F	22 SA	23 SU	24 M	25 TU	26 W	27 TH	28 F	29 SA	30 SU	31 M	
NM	ON	PO	QP	RQ	SR	TS	UT	VU	WV	XW	AX	BA	CB	DC	ED	
OL	PM	QN	RO	SP	TQ	UR	VS	WT	XU	AV	BW	CX	DA	EB	FC	
PK	QL	RM	SN	TO	UP	VQ	WR	XS	AT	BU	CV	DW	EX	FA	GB	+
QJ	RK	SL	TM	UN	VO	WP	XQ	AR	BS	CT	DU	EV	FW	GX	HA	
RH	SJ	TK	UL	VM	WN	XO	AP	BQ	CR	DS	ET	FU	GV	HW	JX	
SG	TH	UJ	VK	WL	XM	AN	BO	CP	DQ	ER	FS	GT	HU	JV	KW	0
TF	UG	VH	WJ	XK	AL	BM	CN	DO	EP	FQ	GR	HS	JT	KU	LV	
UE	VF	WG	XH	AJ	BK	CL	DM	EN	FO	GP	HQ	JR	KS	LT	MU	
VD	WE	XF	AG	BH	CJ	DK	EL	FM	GN	HO	JP	KQ	LR	MS	NT	−
WC	XD	AE	BF	CG	DH	EJ	FK	GL	HM	JN	KO	LP	MQ	NR	OS	
XB	AC	BD	CE	DF	EG	FH	GJ	HK	JL	KM	LN	MO	NP	OQ	PR	
A	B	C	D	E	F	G	H	J	K	L	M	N	O	P	Q	

AUGUST 1981 **EMOTIONAL**

16 SU	17 M	18 TU	19 W	20 TH	21 F	22 SA	23 SU	24 M	25 TU	26 W	27 TH	28 F	29 SA	30 SU	31 M	
A	B	C	D	E	F	G	H	J	K	L	M	N	O	P	Q	
B3	CA	DB	EC	FD	GE	HF	JG	KH	LJ	MK	NL	OM	PN	QO	RP	
C2	D3	EA	FB	GC	HD	JE	KF	LG	MH	NJ	OK	PL	QM	RN	SO	
D1	E2	F3	GA	HB	JC	KD	LE	MF	NG	OH	PJ	QK	RL	SM	TN	
EZ	F1	G2	H3	JA	KB	LC	MD	NE	OF	PG	QH	RJ	SK	TL	UM	+
FY	GZ	H1	J2	K3	LA	MB	NC	OD	PE	QF	RG	SH	TJ	UK	VL	
GX	HY	JZ	K1	L2	M3	NA	OB	PC	QD	RE	SF	TG	UH	VJ	WK	
HW	JX	KY	LZ	M1	N2	O3	PA	QB	RC	SD	TE	UF	VG	WH	XJ	0
JV	KW	LX	MY	NZ	O1	P2	Q3	RA	SB	TC	UD	VE	WF	XG	YH	
KU	LV	MW	NX	OY	PZ	Q1	R2	S3	TA	UB	VC	WD	XE	YF	ZG	
LT	MU	NV	OW	PX	QY	RZ	S1	T2	U3	VA	WB	XC	YD	ZE	1F	−
MS	NT	OU	PV	QW	RX	SY	TZ	U1	V2	W3	XA	YB	ZC	1D	2E	
NR	OS	PT	QU	RV	SW	TX	UY	VZ	W1	X2	Y3	ZA	1B	2C	3D	
OQ	PR	QS	RT	SU	TV	UW	VX	WY	XZ	Y1	Z2	13	2A	3B	AC	
P	Q	R	S	T	U	V	W	X	Y	Z	1	2	3	A	B	

AUGUST 1981 **INTELLECTUAL**

16 SU	17 M	18 TU	19 W	20 TH	21 F	22 SA	23 SU	24 M	25 TU	26 W	27 TH	28 F	29 SA	30 SU	31 M	
T	U	V	W	X	Y	Z	1	2	3	4	5	6	7	8	A	
US	VT	WU	XV	YW	ZX	1Y	2Z	31	42	53	64	75	86	A7	B8	
VR	WS	XT	YU	ZV	1W	2X	3Y	4Z	51	62	73	84	A5	B6	C7	
WQ	XR	YS	ZT	1U	2V	3W	4X	5Y	6Z	71	82	A3	B4	C5	D6	
XP	YQ	ZR	1S	2T	3U	4V	5W	6X	7Y	8Z	A1	B2	C3	D4	E5	
YO	ZP	1Q	2R	3S	4T	5U	6V	7W	8X	AY	BZ	C1	D2	E3	F4	+
ZN	1O	2P	3Q	4R	5S	6T	7U	8V	AW	BX	CY	DZ	E1	F2	G3	
1M	2N	3O	4P	5Q	6R	7S	8T	AU	BV	CW	DX	EY	FZ	G1	H2	
2L	3M	4N	5O	6P	7Q	8R	AS	BT	CU	DV	EW	FX	GY	HZ	J1	0
3K	4L	5M	6N	7O	8P	AQ	BR	CS	DT	EU	FV	GW	HX	JY	KZ	
4J	5K	6L	7M	8N	AO	BP	CQ	DR	ES	FT	GU	HV	JW	KX	LY	
5H	6J	7K	8L	AM	BN	CO	DP	EQ	FR	GS	HT	JU	KV	LW	MX	−
6G	7H	8J	AK	BL	CM	DN	EO	FP	GQ	HR	JS	KT	LU	MV	NW	
7F	8G	AH	BJ	CK	DL	EM	FN	GO	HP	JQ	KR	LS	MT	NU	OV	
8E	AF	BG	CH	DJ	EK	FL	GM	HN	JO	KP	LQ	MR	NS	OT	PU	
AD	BE	CF	DG	EH	FJ	GK	HL	JM	KN	LO	MP	NQ	OR	PS	QT	
BC	CD	DE	EF	FG	GH	HJ	JK	KL	LM	MN	NO	OP	PQ	QR	RS	

SEPTEMBER 1981 PHYSICAL

1 TU	2 W	3 TH	4 F	5 SA	6 SU	7 M	8 TU	9 W	10 TH	11 F	12 SA	13 SU	14 M	15 TU	
FE	GF	HG	JH	KJ	LK	ML	NM	ON	PO	QP	RQ	SR	TS	UT	
GD	HE	JF	KG	LH	MJ	NK	OL	PM	QN	RO	SP	TQ	UR	VS	
HC	JD	KE	LF	MG	NH	OJ	PK	QL	RM	SN	TO	UP	VQ	WR	+
JB	KC	LD	ME	NF	OG	PH	QJ	RK	SL	TM	UN	VO	WP	XQ	
KA	LB	MC	ND	OE	PF	QG	RH	SJ	TK	UL	VM	WN	XO	AP	
LX	MA	NB	OC	PD	QE	RF	SG	TH	UJ	VK	WL	XM	AN	BO	0
MW	NX	OA	PB	QC	RD	SE	TF	UG	VH	WJ	XK	AL	BM	CN	
NV	OW	PX	QA	RB	SC	TD	UE	VF	WG	XH	AJ	BK	CL	DM	
OU	PV	QW	RX	SA	TB	UC	VD	WE	XF	AG	BH	CJ	DK	EL	−
PT	QU	RV	SW	TX	UA	VB	WC	XD	AE	BF	CG	DH	EJ	FK	
QS	RT	SU	TV	UW	VX	WA	XB	AC	BD	CE	DF	EG	FH	GJ	
R	S	T	U	V	W	X	A	B	C	D	E	F	G	H	

SEPTEMBER 1981 EMOTIONAL

1 TU	2 W	3 TH	4 F	5 SA	6 SU	7 M	8 TU	9 W	10 TH	11 F	12 SA	13 SU	14 M	15 TU	
R	S.	T	U	V	W	X	Y	Z	1	2	3	A	B	C	
SQ	TR	US	VT	WU	XV	YW	ZX	1Y	2Z	31	A2	B3	CA	DB	
TP	UQ	VR	WS	XT	YU	ZV	1W	2X	3Y	AZ	B1	C2	D3	EA	
UO	VP	WQ	XR	YS	ZT	1U	2V	3W	AX	BY	CZ	D1	E2	F3	
VN	WO	XP	YQ	ZR	1S	2T	3U	AV	BW	CX	DY	EZ	F1	G2	+
WM	XN	YO	ZP	1Q	2R	3S	AT	BU	CV	DW	EX	FY	GZ	H1	
XL	YM	ZN	1O	2P	3Q	AR	BS	CT	DU	EV	FW	GX	HY	JZ	
YK	ZL	1M	2N	3O	AP	BQ	CR	DS	ET	FU	GV	HW	JX	KY	0
ZJ	1K	2L	3M	AN	BO	CP	DQ	ER	FS	GT	HU	JV	KW	LX	
1H	2J	3K	AL	BM	CN	DO	EP	FQ	GR	HS	JT	KU	LV	MW	
2G	3H	AJ	BK	CL	DM	EN	FO	GP	HQ	JR	KS	LT	MU	NV	−
3F	AG	BH	CJ	DK	EL	FM	GN	HO	JP	KQ	LR	MS	NT	OU	
AE	BF	CG	DH	EJ	FK	GL	HM	JN	KO	LP	MQ	NR	OS	PT	
BD	CE	DF	EG	FH	GJ	HK	JL	KM	LN	MO	NP	OQ	PR	QS	
C	D	E	F	G	H	J	K	L	M	N	O	P	Q	R	

SEPTEMBER 1981 INTELLECTUAL

1 TU	2 W	3 TH	4 F	5 SA	6 SU	7 M	8 TU	9 W	10 TH	11 F	12 SA	13 SU	14 M	15 TU	
B	C	D	E	F	G	H	J	K	L	M	N	O	P	Q	
CA	DB	EC	FD	GE	HF	JG	KH	LJ	MK	NL	OM	PN	QO	RP	
D8	EA	FB	GC	HD	JE	KF	LG	MH	NJ	OK	PL	QM	RN	SO	
E7	F8	GA	HB	JC	KD	LE	MF	NG	OH	PJ	QK	RL	SM	TN	
F6	G7	H8	JA	KB	LC	MD	NE	OF	PG	QH	RJ	SK	TL	UM	
G5	H6	J7	K8	LA	MB	NC	OD	PE	QF	RG	SH	TJ	UK	VL	+
H4	J5	K6	L7	M8	NA	OB	PC	QD	RE	SF	TG	UH	VJ	WK	
J3	K4	L5	M6	N7	O8	PA	QB	RC	SD	TE	UF	VG	WH	XJ	
K2	L3	M4	N5	O6	P7	Q8	RA	SB	TC	UD	VE	WF	XG	YH	0
L1	M2	N3	O4	P5	Q6	R7	S8	TA	UB	VC	WD	XE	YF	ZG	
MZ	N1	O2	P3	Q4	R5	S6	T7	U8	VA	WB	XC	YD	ZE	1F	
NY	OZ	P1	Q2	R3	S4	T5	U6	V7	W8	XA	YB	ZC	1D	2E	−
OX	PY	QZ	R1	S2	T3	U4	V5	W6	X7	Y8	ZA	1B	2C	3D	
PW	QX	RY	SZ	T1	U2	V3	W4	X5	Y6	Z7	18	2A	3B	4C	
QV	RW	SX	TY	UZ	V1	W2	X3	Y4	Z5	16	27	38	4A	5B	
RU	SV	TW	UX	VY	WZ	X1	Y2	Z3	14	25	36	47	58	6A	
ST	TU	UV	VW	WX	XY	YZ	Z1	12	23	34	45	56	67	78	

SEPTEMBER 1981 **PHYSICAL**

16 W	17 TH	18 F	19 SA	20 SU	21 M	22 TU	23 W	24 TH	25 F	26 SA	27 SU	28 M	29 TU	30 W	
VU	WV	XW	AX	BA	CB	DC	ED	FE	GF	HG	JH	KJ	LK	ML	
WT	XU	AV	BW	CX	DA	EB	FC	GD	HE	JF	KG	LH	MJ	NK	
XS	AT	BU	CV	DW	EX	FA	GB	HC	JD	KE	LF	MG	NH	OJ	+
AR	BS	CT	DU	EV	FW	GX	HA	JB	KC	LD	ME	NF	OG	PH	
BQ	CR	DS	ET	FU	GV	HW	JX	KA	LB	MC	ND	OE	PF	QG	
CP	DQ	ER	FS	GT	HU	JV	KW	LX	MA	NB	OC	PD	QE	RF	0
DO	EP	FQ	GR	HS	JT	KU	LV	MW	NX	OA	PB	QC	RD	SE	
EN	FO	GP	HQ	JR	KS	LT	MU	NV	OW	PX	QA	RB	SC	TD	
FM	GN	HO	JP	KQ	LR	MS	NT	OU	PV	QW	RX	SA	TB	UC	−
GL	HM	JN	KO	LP	MQ	NR	OS	PT	QU	RV	SW	TX	UA	VB	
HK	JL	KM	LN	MO	NP	OQ	PR	QS	RT	SU	TV	UW	VX	WA	
J	K	L	M	N	O	P	Q	R	S	T	U	V	W	X	

SEPTEMBER 1981 **EMOTIONAL**

16 W	17 TH	18 F	19 SA	20 SU	21 M	22 TU	23 W	24 TH	25 F	26 SA	27 SU	28 M	29 TU	30 W	
D	E	F	G	H	J	K	L	M	N	O	P	Q	R	S	
EC	FD	GE	HF	JG	KH	LJ	MK	NL	OM	PN	QO	RP	SQ	TR	
FB	GC	HD	JE	KF	LG	MH	NJ	OK	PL	QM	RN	SO	TP	UQ	
GA	HB	JC	KD	LE	MF	NG	OH	PJ	QK	RL	SM	TN	UO	YP	
H3	JA	KB	LC	MD	NE	OF	PG	QH	RJ	SK	TL	UM	VN	WO	+
J2	K3	LA	MB	NC	OD	PE	QF	RG	SH	TJ	UK	VL	WM	XN	
K1	L2	M3	NA	OB	PC	QD	RE	SF	TG	UH	VJ	WK	XL	YM	
LZ	M1	N2	O3	PA	QB	RC	SD	TE	UF	VG	WH	XJ	YK	ZL	0
MY	NZ	O1	P2	Q3	RA	SB	TC	UD	VE	WF	XG	YH	ZJ	1K	
NX	OY	PZ	Q1	R2	S3	TA	UB	VC	WD	XE	YF	ZG	1H	2J	
OW	PX	QY	RZ	S1	T2	U3	VA	WB	XC	YD	ZE	1F	2G	3H	−
PV	QW	RX	SY	TZ	U1	V2	W3	XA	YB	ZC	1D	2E	3F	AG	
QU	RV	SW	TX	UY	VZ	W1	X2	Y3	ZA	1B	2C	3D	AE	BF	
RT	SU	TV	UW	VX	WY	XZ	Y1	Z2	13	2A	3B	AC	BD	CE	
S	T	U	V	W	X	Y	Z	1	2	3	A	B	C	D	

SEPTEMBER 1981 **INTELLECTUAL**

16 W	17 TH	18 F	19 SA	20 SU	21 M	22 TU	23 W	24 TH	25 F	26 SA	27 SU	28 M	29 TU	30 W	
R	S	T	U	V	W	X	Y	Z	1	2	3	4	5	6	
SQ	TR	US	VT	WU	XV	YW	ZX	1Y	2Z	31	42	53	64	75	
TP	UQ	VR	WS	XT	YU	ZV	1W	2X	3Y	4Z	51	62	73	84	
UO	VP	WQ	XR	YS	ZT	1U	2V	3W	4X	5Y	6Z	71	82	A3	
VN	WO	XP	YQ	ZR	1S	2T	3U	4V	5W	6X	7Y	8Z	A1	B2	
WM	XN	YO	ZP	1Q	2R	3S	4T	5U	6V	7W	8X	AY	BZ	C1	+
XL	YM	ZN	1O	2P	3Q	4R	5S	6T	7U	8V	AW	BX	CY	DZ	
YK	ZL	1M	2N	3O	4P	5Q	6R	7S	8T	AU	BV	CW	DX	EY	
ZJ	1K	2L	3M	4N	5O	6P	7Q	8R	AS	BT	CU	DV	EW	FX	0
1H	2J	3K	4L	5M	6N	7O	8P	AQ	BR	CS	DT	EU	FV	GW	
2G	3H	4J	5K	6L	7M	8N	AO	BP	CQ	DR	ES	FT	GU	HV	
3F	4G	5H	6J	7K	8L	AM	BN	CO	DP	EQ	FR	GS	HT	JU	−
4E	5F	6G	7H	8J	AK	BL	CM	DN	EO	FP	GQ	HR	JS	KT	
5D	6E	7F	8G	AH	BJ	CK	DL	EM	FN	GO	HP	JQ	KR	LS	
6C	7D	8E	AF	BG	CH	DJ	EK	FL	GM	HN	JO	KP	LQ	MR	
7B	8C	AD	BE	CF	DG	EH	FJ	GK	HL	JM	KN	LO	MP	NQ	
8A	AB	BC	CD	DE	EF	FG	GH	HJ	JK	KL	LM	MN	NO	OP	

OCTOBER 1981 PHYSICAL

1 TH	2 F	3 SA	4 SU	5 M	6 TU	7 W	8 TH	9 F	10 SA	11 SU	12 M	13 TU	14 W	15 TH	
NM	ON	PO	QP	RQ	SR	TS	UT	VU	WV	XW	AX	BA	CB	DC	
OL	PM	QN	RO	SP	TQ	UR	VS	WT	XU	AV	BW	CX	DA	EB	
PK	QL	RM	SN	TO	UP	VQ	WR	XS	AT	BU	CV	DW	EX	FA	+
QJ	RK	SL	TM	UN	VO	WP	XQ	AR	BS	CT	DU	EV	FW	GX	
RH	SJ	TK	UL	VM	WN	XO	AP	BQ	CR	DS	ET	FU	GV	HW	
SG	TH	UJ	VK	WL	XM	AN	BO	CP	DQ	ER	FS	GT	HU	JV	0
TF	UG	VH	WJ	XK	AL	BM	CN	DO	EP	FQ	GR	HS	JT	KU	
UE	VF	WG	XH	AJ	BK	CL	DM	EN	FO	GP	HQ	JR	KS	LT	
VD	WE	XF	AG	BH	CJ	DK	EL	FM	GN	HO	JP	KQ	LR	MS	−
WC	XD	AE	BF	CG	DH	EJ	FK	GL	HM	JN	KO	LP	MQ	NR	
XB	AC	BD	CE	DF	EG	FH	GJ	HK	JL	KM	LN	MO	NP	OQ	
A	B	C	D	E	F	G	H	J	K	L	M	N	O	P	

OCTOBER 1981 EMOTIONAL

1 TH	2 F	3 SA	4 SU	5 M	6 TU	7 W	8 TH	9 F	10 SA	11 SU	12 M	13 TU	14 W	15 TH	
T	U	V	W	X	Y	Z	1	2	3	A	B	C	D	E	
US	VT	WU	XV	YW	ZX	1Y	2Z	31	A2	B3	CA	DB	EC	FD	
VR	WS	XT	YU	ZV	1W	2X	3Y	AZ	B1	C2	D3	EA	FB	GC	
WQ	XR	YS	ZT	1U	2V	3W	AX	BY	CZ	D1	E2	F3	GA	HB	
XP	YQ	ZR	1S	2T	3U	AV	BW	CX	DY	EZ	F1	G2	H3	JA	+
YO	ZP	1Q	2R	3S	AT	BU	CV	DW	EX	FY	GZ	H1	J2	K3	
ZN	1O	2P	3Q	AR	BS	CT	DU	EV	FW	GX	HY	JZ	K1	L2	
1M	2N	3O	AP	BQ	CR	DS	ET	FU	GV	HW	JX	KY	LZ	M1	0
2L	3M	AN	BO	CP	DQ	ER	FS	GT	HU	JV	KW	LX	MY	NZ	
3K	AL	BM	CN	DO	EP	FQ	GR	HS	JT	KU	LV	MW	NX	OY	
AJ	BK	CL	DM	EN	FO	GP	HQ	JR	KS	LT	MU	NV	OW	PX	−
BH	CJ	DK	EL	FM	GN	HO	JP	KQ	LR	MS	NT	OU	PV	QW	
CG	DH	EJ	FK	GL	HM	JN	KO	LP	MQ	NR	OS	PT	QU	RV	
DF	EG	FH	GJ	HK	JL	KM	LN	MO	NP	OQ	PR	QS	RT	SU	
E	F	G	H	J	K	L	M	N	O	P	Q	R	S	T	

OCTOBER 1981 INTELLECTUAL

1 TH	2 F	3 SA	4 SU	5 M	6 TU	7 W	8 TH	9 F	10 SA	11 SU	12 M	13 TU	14 W	15 TH	
7	8	A	B	C	D	E	F	G	H	J	K	L	M	N	
86	A7	B8	CA	DB	EC	FD	GE	HF	JG	KH	LJ	MK	NL	OM	
A5	B6	C7	D8	EA	FB	GC	HD	JE	KF	LG	MH	NJ	OK	PL	
B4	C5	D6	E7	F8	GA	HB	JC	KD	LE	MF	NG	OH	PJ	QK	
C3	D4	E5	F6	G7	H8	JA	KB	LC	MD	NE	OF	PG	QH	RJ	
D2	E3	F4	G5	H6	J7	K8	LA	MB	NC	OD	PE	QF	RG	SH	+
E1	F2	G3	H4	J5	K6	L7	M8	NA	OB	PC	QD	RE	SF	TG	
FZ	G1	H2	J3	K4	L5	M6	N7	O8	PA	QB	RC	SD	TE	UF	
GY	HZ	J1	K2	L3	M4	N5	O6	P7	Q8	RA	SB	TC	UD	VE	0
HX	JY	KZ	L1	M2	N3	O4	P5	Q6	R7	S8	TA	UB	VC	WD	
JW	KX	LY	MZ	N1	O2	P3	Q4	R5	S6	T7	U8	VA	WB	XC	
KV	LW	MX	NY	OZ	P1	Q2	R3	S4	T5	U6	V7	W8	XA	YB	−
LU	MV	NW	OX	PY	QZ	R1	S2	T3	U4	V5	W6	X7	Y8	ZA	
MT	NU	OV	PW	QX	RY	SZ	T1	U2	V3	W4	X5	Y6	Z7	18	
NS	OT	PU	QV	RW	SX	TY	UZ	V1	W2	X3	Y4	Z5	16	27	
OR	PS	QT	RU	SV	TW	UX	VY	WZ	X1	Y2	Z3	14	25	36	
PQ	QR	RS	ST	TU	UV	VW	WX	XY	YZ	Z1	12	23	34	45	

OCTOBER 1981 — PHYSICAL

16 F	17 SA	18 SU	19 M	20 TU	21 W	22 TH	23 F	24 SA	25 SU	26 M	27 TU	28 W	29 TH	30 F	31 SA	
ED	FE	GF	HG	JH	KJ	LK	ML	NM	ON	PO	QP	RQ	SR	TS	UT	
FC	GD	HE	JF	KG	LH	MJ	NK	OL	PM	QN	RO	SP	TQ	UR	VS	
GB	HC	JD	KE	LF	MG	NH	OJ	PK	QL	RM	SN	TO	UP	VQ	WR	+
HA	JB	KC	LD	ME	NF	OG	PH	QJ	RK	SL	TM	UN	VO	WP	XQ	
JX	KA	LB	MC	ND	OE	PF	QG	RH	SJ	TK	UL	VM	WN	XO	AP	
KW	LX	MA	NB	OC	PD	QE	RF	SG	TH	UJ	VK	WL	XM	AN	BO	0
LV	MW	NX	OA	PB	QC	RD	SE	TF	UG	VH	WJ	XK	AL	BM	CN	
MU	NV	OW	PX	QA	RB	SC	TD	UE	VF	WG	XH	AJ	BK	CL	DM	
NT	OU	PV	QW	RX	SA	TB	UC	VD	WE	XF	AG	BH	CJ	DK	EL	−
OS	PT	QU	RV	SW	TX	UA	VB	WC	XD	AE	BF	CG	DH	EJ	FK	
PR	QS	RT	SU	TV	UW	VX	WA	XB	AC	BD	CE	DF	EG	FH	GJ	
Q	R	S	T	U	V	W	X	A	B	C	D	E	F	G	H	

OCTOBER 1981 — EMOTIONAL

16 F	17 SA	18 SU	19 M	20 TU	21 W	22 TH	23 F	24 SA	25 SU	26 M	27 TU	28 W	29 TH	30 F	31 SA	
F	G	H	J	K	L	M	N	O	P	Q	R	S	T	U	V	
GE	HF	JG	KH	LJ	MK	NL	OM	PN	QO	RP	SQ	TR	US	VT	WU	
HD	JE	KF	LG	MH	NJ	OK	PL	QM	RN	SO	TP	UQ	VR	WS	XT	
JC	KD	LE	MF	NG	OH	PJ	QK	RL	SM	TN	UO	VP	WQ	XR	YS	
KB	LC	MD	NE	OF	PG	QH	RJ	SK	TL	UM	VN	WO	XP	YQ	ZR	+
LA	MB	NC	OD	PE	QF	RG	SH	TJ	UK	VL	WM	XN	YO	ZP	1Q	
M3	NA	OB	PC	QD	RE	SF	TG	UH	VJ	WK	XL	YM	ZN	1O	2P	
N2	O3	PA	QB	RC	SD	TE	UF	VG	WH	XJ	YK	ZL	1M	2N	3O	0
O1	P2	Q3	RA	SB	TC	UD	VE	WF	XG	YH	ZJ	1K	2L	3M	AN	
PZ	Q1	R2	S3	TA	UB	VC	WD	XE	YF	ZG	1H	2J	3K	AL	BM	
QY	RZ	S1	T2	U3	VA	WB	XC	YD	ZE	1F	2G	3H	AJ	BK	CL	−
RX	SY	TZ	U1	V2	W3	XA	YB	ZC	1D	2E	3F	AG	BH	CJ	DK	
SW	TX	UY	VZ	W1	X2	Y3	ZA	1B	2C	3D	AE	BF	CG	DH	EJ	
TV	UW	VX	WY	XZ	Y1	Z2	13	2A	3B	AC	BD	CE	DF	EG	FH	
U	V	W	X	Y	Z	1	2	3	A	B	C	D	E	F	G	

OCTOBER 1981 — INTELLECTUAL

16 F	17 SA	18 SU	19 M	20 TU	21 W	22 TH	23 F	24 SA	25 SU	26 M	27 TU	28 W	29 TH	30 F	31 SA	
O	P	Q	R	S	T	U	V	W	X	Y	Z	1	2	3	4	
PN	QO	RP	SQ	TR	US	VT	WU	XV	YW	ZX	1Y	2Z	31	42	53	
QM	RN	SO	TP	UQ	VR	WS	XT	YU	ZV	1W	2X	3Y	4Z	51	62	
RL	SM	TN	UO	VP	WQ	XR	YS	ZT	1U	2V	3W	4X	5Y	6Z	71	
SK	TL	UM	VN	WO	XP	YQ	ZR	1S	2T	3U	4V	5W	6X	7Y	8Z	
TJ	UK	VL	WM	XN	YO	ZP	1Q	2R	3S	4T	5U	6V	7W	8X	AY	+
UH	VJ	WK	XL	YM	ZN	1O	2P	3Q	4R	5S	6T	7U	8V	AW	BX	
VG	WH	XJ	YK	ZL	1M	2N	3O	4P	5Q	6R	7S	8T	AU	BV	CW	
WF	XG	YH	ZJ	1K	2L	3M	4N	5O	6P	7Q	8R	AS	BT	CU	DV	0
XE	YF	ZG	1H	2J	3K	4L	5M	6N	7O	8P	AQ	BR	CS	DT	EU	
YD	ZE	1F	2G	3H	4J	5K	6L	7M	8N	AO	BP	CQ	DR	ES	FT	
ZC	1D	2E	3F	4G	5H	6J	7K	8L	AM	BN	CO	DP	EQ	FR	GS	−
1B	2C	3D	4E	5F	6G	7H	8J	AK	BL	CM	DN	EO	FP	GQ	HR	
2A	3B	4C	5D	6E	7F	8G	AH	BJ	CK	DL	EM	FN	GO	HP	JQ	
38	4A	5B	6C	7D	8E	AF	BG	CH	DJ	EK	FL	GM	HN	JO	KP	
47	58	6A	7B	8C	AD	BE	CF	DG	EH	FJ	GK	HL	JM	KN	LO	
56	67	78	8A	AB	BC	CD	DE	EF	FG	GH	HJ	JK	KL	LM	MN	

NOVEMBER 1981 PHYSICAL

1	2	3	4	5	6	7	8	9	10	11	12	13	14	15	
SU	M	TU	W	TH	F	SA	SU	M	TU	W	TH	F	SA	SU	
VU	WV	XW	AX	BA	CB	DC	ED	FE	GF	HG	JH	KJ	LK	ML	
WT	XU	AV	BW	CX	DA	EB	FC	GD	HE	JF	KG	LH	MJ	NK	
XS	AT	BU	CV	DW	EX	FA	GB	HC	JD	KE	LF	MG	NH	OJ	+
AR	BS	CT	DU	EV	FW	GX	HA	JB	KC	LD	ME	NF	OG	PH	
BQ	CR	DS	ET	FU	GV	HW	JX	KA	LB	MC	ND	OE	PF	QG	
CP	DQ	ER	FS	GT	HU	JV	KW	LX	MA	NB	OC	PD	QE	RF	0
DO	EP	FQ	GR	HS	JT	KU	LV	MW	NX	OA	PB	QC	RD	SE	
EN	FO	GP	HQ	JR	KS	LT	MU	NV	OW	PX	QA	RB	SC	TD	
FM	GN	HO	JP	KQ	LR	MS	NT	OU	PV	QW	RX	SA	TB	UC	−
GL	HM	JN	KO	LP	MQ	NR	OS	PT	QU	RV	SW	TX	UA	VB	
HK	JL	KM	LN	MO	NP	OQ	PR	QS	RT	SU	TV	UW	VX	WA	
J	K	L	M	N	O	P	Q	R	S	T	U	V	W	X	

NOVEMBER 1981 EMOTIONAL

1	2	3	4	5	6	7	8	9	10	11	12	13	14	15	
SU	M	TU	W	TH	F	SA	SU	M	TU	W	TH	F	SA	SU	
W	X	Y	Z	1	2	3	A	B	C	D	E	F	G	H	
XV	YW	ZX	1Y	2Z	31	A2	B3	CA	DB	EC	FD	GE	HF	JG	
YU	ZV	1W	2X	3Y	AZ	B1	C2	D3	EA	FB	GC	HD	JE	KF	
ZT	1U	2V	3W	AX	BY	CZ	D1	E2	F3	GA	HB	JC	KD	LE	
1S	2T	3U	AV	BW	CX	DY	EZ	F1	G2	H3	JA	KB	LC	MD	+
2R	3S	AT	BU	CV	DW	EX	FY	GZ	H1	J2	K3	LA	MB	NC	
3Q	AR	BS	CT	DU	EV	FW	GX	HY	JZ	K1	L2	M3	NA	OB	
AP	BQ	CR	DS	ET	FU	GV	HW	JX	KY	LZ	M1	N2	O3	PA	0
BO	CP	DQ	ER	FS	GT	HU	JV	KW	LX	MY	NZ	O1	P2	Q3	
CN	DO	EP	FQ	GR	HS	JT	KU	LV	MW	NX	OY	PZ	Q1	R2	
DM	EN	FO	GP	HQ	JR	KS	LT	MU	NV	OW	PX	QY	RZ	S1	−
EL	FM	GN	HO	JP	KQ	LR	MS	NT	OU	PV	QW	RX	SY	TZ	
FK	GL	HM	JN	KO	LP	MQ	NR	OS	PT	QU	RV	SW	TX	UY	
GJ	HK	JL	KM	LN	MO	NP	OQ	PR	QS	RT	SU	TV	UW	VX	
H	J	K	L	M	N	O	P	Q	R	S	T	U	V	W	

NOVEMBER 1981 INTELLECTUAL

1	2	3	4	5	6	7	8	9	10	11	12	13	14	15	
SU	M	TU	W	TH	F	SA	SU	M	TU	W	TH	F	SA	SU	
5	6	7	8	A	B	C	D	E	F	G	H	J	K	L	
64	75	86	A7	B8	CA	DB	EC	FD	GE	HF	JG	KH	LJ	MK	
73	84	A5	B6	C7	D8	EA	FB	GC	HD	JE	KF	LG	MH	NJ	
82	A3	B4	C5	D6	E7	F8	GA	HB	JC	KD	LE	MF	NG	OH	
A1	B2	C3	D4	E5	F6	G7	H8	JA	KB	LC	MD	NE	OF	PG	
BZ	C1	D2	E3	F4	G5	H6	J7	K8	LA	MB	NC	OD	PE	QF	+
CY	DZ	E1	F2	G3	H4	J5	K6	L7	M8	NA	OB	PC	QD	RE	
DX	EY	FZ	G1	H2	J3	K4	L5	M6	N7	O8	PA	QB	RC	SD	
EW	FX	GY	HZ	J1	K2	L3	M4	N5	O6	P7	Q8	RA	SB	TC	0
FV	GW	HX	JY	KZ	L1	M2	N3	O4	P5	Q6	R7	S8	TA	UB	
GU	HV	JW	KX	LY	MZ	N1	O2	P3	Q4	R5	S6	T7	U8	VA	
HT	JU	KV	LW	MX	NY	OZ	P1	Q2	R3	S4	T5	U6	V7	W8	−
JS	KT	LU	MV	NW	OX	PY	QZ	R1	S2	T3	U4	V5	W6	X7	
KR	LS	MT	NU	OV	PW	QX	RY	SZ	T1	U2	V3	W4	X5	Y6	
LQ	MR	NS	OT	PU	QV	RW	SX	TY	UZ	V1	W2	X3	Y4	Z5	
MP	NQ	OR	PS	QT	RU	SV	TW	UX	VY	WZ	X1	Y2	Z3	14	
NO	OP	PQ	QR	RS	ST	TU	UV	VW	WX	XY	YZ	Z1	12	23	

NOVEMBER 1981 **PHYSICAL**

16 M	17 TU	18 W	19 TH	20 F	21 SA	22 SU	23 M	24 TU	25 W	26 TH	27 F	28 SA	29 SU	30 M	
NM	ON	PO	QP	RQ	SR	TS	UT	VU	WV	XW	AX	BA	CB	DC	
OL	PM	QN	RO	SP	TQ	UR	VS	WT	XU	AV	BW	CX	DA	EB	
PK	QL	RM	SN	TO	UP	VQ	WR	XS	AT	BU	CV	DW	EX	FA	+
QJ	RK	SL	TM	UN	VO	WP	XQ	AR	BS	CT	DU	EV	FW	GX	
RH	SJ	TK	UL	VM	WN	XO	AP	BQ	CR	DS	ET	FU	GV	HW	
SG	TH	UJ	VK	WL	XM	AN	BO	CP	DQ	ER	FS	GT	HU	JV	0
TF	UG	VH	WJ	XK	AL	BM	CN	DO	EP	FQ	GR	HS	JT	KU	
UE	VF	WG	XH	AJ	BK	CL	DM	EN	FO	GP	HQ	JR	KS	LT	
VD	WE	XF	AG	BH	CJ	DK	EL	FM	GN	HO	JP	KQ	LR	MS	–
WC	XD	AE	BF	CG	DH	EJ	FK	GL	HM	JN	KO	LP	MQ	NR	
XB	AC	BD	CE	DF	EG	FH	GJ	HK	JL	KM	LN	MO	NP	OQ	
A	B	C	D	E	F	G	H	J	K	L	M	N	O	P	

NOVEMBER 1981 **EMOTIONAL**

16 M	17 TU	18 W	19 TH	20 F	21 SA	22 SU	23 M	24 TU	25 W	26 TH	27 F	28 SA	29 SU	30 M	
J	K	L	M	N	O	P	Q	R	S	T	U	V	W	X	
KH	LJ	MK	NL	OM	PN	QO	RP	SQ	TR	US	VT	WU	XV	YW	
LG	MH	NJ	OK	PL	QM	RN	SO	TP	UQ	VR	WS	XT	YU	ZV	
MF	NG	OH	PJ	QK	RL	SM	TN	UO	VP	WQ	XR	YS	ZT	1U	
NE	OF	PG	QH	RJ	SK	TL	UM	VN	WO	XP	YQ	ZR	1S	2T	+
OD	PE	QF	RG	SH	TJ	UK	VL	WM	XN	YO	ZP	1Q	2R	3S	
PC	QD	RE	SF	TG	UH	VJ	WK	XL	YM	ZN	1O	2P	3Q	AR	
QB	RC	SD	TE	UF	VG	WH	XJ	YK	ZL	1M	2N	3O	AP	BQ	0
RA	SB	TC	UD	VE	WF	XG	YH	ZJ	1K	2L	3M	AN	BO	CP	
S3	TA	UB	VC	WD	XE	YF	ZG	1H	2J	3K	AL	BM	CN	DO	
T2	U3	VA	WB	XC	YD	ZE	1F	2G	3H	AJ	BK	CL	DM	EN	–
U1	V2	W3	XA	YB	ZC	1D	2E	3F	AG	BH	CJ	DK	EL	FM	
VZ	W1	X2	Y3	ZA	1B	2C	3D	AE	BF	CG	DH	EJ	FK	GL	
WY	XZ	Y1	Z2	13	2A	3B	AC	BD	CE	DF	EG	FH	GJ	HK	
X	Y	Z	1	2	3	A	B	C	D	E	F	G	H	J	

NOVEMBER 1981 **INTELLECTUAL**

16 M	17 TU	18 W	19 TH	20 F	21 SA	22 SU	23 M	24 TU	25 W	26 TH	27 F	28 SA	29 SU	30 M	
M	N	O	P	Q	R	S	T	U	V	W	X	Y	Z	1	
NL	OM	PN	QO	RP	SQ	TR	US	VT	WU	XV	YW	ZX	1Y	2Z	
OK	PL	QM	RN	SO	TP	UQ	VR	WS	XT	YU	ZV	1W	2X	3Y	
PJ	QK	RL	SM	TN	UO	VP	WQ	XR	YS	ZT	1U	2V	3W	4X	
QH	RJ	SK	TL	UM	VN	WO	XP	YQ	ZR	1S	2T	3U	4V	5W	
RG	SH	TJ	UK	VL	WM	XN	YO	ZP	1Q	2R	3S	4T	5U	6V	+
SF	TG	UH	VJ	WK	XL	YM	ZN	1O	2P	3Q	4R	5S	6T	7U	
TE	UF	VG	WH	XJ	YK	ZL	1M	2N	3O	4P	5Q	6R	7S	8T	
UD	VE	WF	XG	YH	ZJ	1K	2L	3M	4N	5O	6P	7Q	8R	AS	0
VC	WD	XE	YF	ZG	1H	2J	3K	4L	5M	6N	7O	8P	AQ	BR	
WB	XC	YD	ZE	1F	2G	3H	4J	5K	6L	7M	8N	AO	BP	CQ	
XA	YB	ZC	1D	2E	3F	4G	5H	6J	7K	8L	AM	BN	CO	DP	–
Y8	ZA	1B	2C	3D	4E	5F	6G	7H	8J	AK	BL	CM	DN	EO	
Z7	18	2A	3B	4C	5D	6E	7F	8G	AH	BJ	CK	DL	EM	FN	
16	27	38	4A	5B	6C	7D	8E	AF	BG	CH	DJ	EK	FL	GM	
25	36	47	58	6A	7B	8C	AD	BE	CF	DG	EH	FJ	GK	HL	
34	45	56	67	78	8A	AB	BC	CD	DE	EF	FG	GH	HJ	JK	

DECEMBER 1981 **PHYSICAL**

1 TU	2 W	3 TH	4 F	5 SA	6 SU	7 M	8 TU	9 W	10 TH	11 F	12 SA	13 SU	14 M	15 TU	
ED	FE	GF	HG	JH	KJ	LK	ML	NM	ON	PO	QP	RQ	SR	TS	
FC	GD	HE	JF	KG	LH	MJ	NK	OL	PM	QN	RO	SP	TQ	UR	
GB	HC	JD	KE	LF	MG	NH	OJ	PK	QL	RM	SN	TO	UP	VQ	+
HA	JB	KC	LD	ME	NF	OG	PH	QJ	RK	SL	TM	UN	VO	WP	
JX	KA	LB	MC	ND	OE	PF	QG	RH	SJ	TK	UL	VM	WN	XO	
KW	LX	MA	NB	OC	PD	QE	RF	SG	TH	UJ	VK	WL	XM	AN	0
LV	MW	NX	OA	PB	QC	RD	SE	TF	UG	VH	WJ	XK	AL	BM	
MU	NV	OW	PX	QA	RB	SC	TD	UE	VF	WG	XH	AJ	BK	CL	
NT	OU	PV	QW	RX	SA	TB	UC	VD	WE	XF	AG	BH	CJ	DK	−
OS	PT	QU	RV	SW	TX	UA	VB	WC	XD	AE	BF	CG	DH	EJ	
PR	QS	RT	SU	TV	UW	VX	WA	XB	AC	BD	CE	DF	EG	FH	
Q	R	S	T	U	V	W	X	A	B	C	D	E	F	G	

DECEMBER 1981 **EMOTIONAL**

1 TU	2 W	3 TH	4 F	5 SA	6 SU	7 M	8 TU	9 W	10 TH	11 F	12 SA	13 SU	14 M	15 TU	
Y	Z	1	2	3	A	B	C	D	E	F	G	H	J	K	
ZX	1Y	2Z	31	A2	B3	CA	DB	EC	FD	GE	HF	JG	KH	LJ	
1W	2X	3Y	AZ	B1	C2	D3	EA	FB	GC	HD	JE	KF	LG	MH	
2V	3W	AX	BY	CZ	D1	E2	F3	GA	HB	JC	KD	LE	MF	NG	
3U	AV	BW	CX	DY	EZ	F1	G2	H3	JA	KB	LC	MD	NE	OF	+
AT	BU	CV	DW	EX	FY	GZ	H1	J2	K3	LA	MB	NC	OD	PE	
BS	CT	DU	EV	FW	GX	HY	JZ	K1	L2	M3	NA	OB	PC	QD	
CR	DS	ET	FU	GV	HW	JX	KY	LZ	M1	N2	O3	PA	QB	RC	0
DQ	ER	FS	GT	HU	JV	KW	LX	MY	NZ	O1	P2	Q3	RA	SB	
EP	FQ	GR	HS	JT	KU	LV	MW	NX	OY	PZ	Q1	R2	S3	TA	
FO	GP	HQ	JR	KS	LT	MU	NV	OW	PX	QY	RZ	S1	T2	U3	−
GN	HO	JP	KQ	LR	MS	NT	OU	PV	QW	RX	SY	TZ	U1	V2	
HM	JN	KO	LP	MQ	NR	OS	PT	QU	RV	SW	TX	UY	VZ	W1	
JL	KM	LN	MO	NP	OQ	PR	QS	RT	SU	TV	UW	VX	WY	XZ	
K	L	M	N	O	P	Q	R	S	T	U	V	W	X	Y	

DECEMBER 1981 **INTELLECTUAL**

1 TU	2 W	3 TH	4 F	5 SA	6 SU	7 M	8 TU	9 W	10 TH	11 F	12 SA	13 SU	14 M	15 TU	
2	3	4	5	6	7	8	A	B	C	D	E	F	G	H	
31	42	53	64	75	86	A7	B8	CA	DB	EC	FD	GE	HF	JG	
4Z	51	62	73	84	A5	B6	C7	D8	EA	FB	GC	HD	JE	KF	
5Y	6Z	71	82	A3	B4	C5	D6	E7	F8	GA	HB	JC	KD	LE	
6X	7Y	8Z	A1	B2	C3	D4	E5	F6	G7	H8	JA	KB	LC	MD	
7W	8X	AY	BZ	C1	D2	E3	F4	G5	H6	J7	K8	LA	MB	NC	+
8V	AW	BX	CY	DZ	E1	F2	G3	H4	J5	K6	L7	M8	NA	OB	
AU	BV	CW	DX	EY	FZ	G1	H2	J3	K4	L5	M6	N7	O8	PA	
BT	CU	DV	EW	FX	GY	HZ	J1	K2	L3	M4	N5	O6	P7	Q8	0
CS	DT	EU	FV	GW	HX	JY	KZ	L1	M2	N3	O4	P5	Q6	R7	
DR	ES	FT	GU	HV	JW	KX	LY	MZ	N1	O2	P3	Q4	R5	S6	
EQ	FR	GS	HT	JU	KV	LW	MX	NY	OZ	P1	Q2	R3	S4	T5	−
FP	GQ	HR	JS	KT	LU	MV	NW	OX	PY	QZ	R1	S2	T3	U4	
GO	HP	JQ	KR	LS	MT	NU	OV	PW	QX	RY	SZ	T1	U2	V3	
HN	JO	KP	LQ	MR	NS	OT	PU	QV	RW	SX	TY	UZ	V1	W2	
JM	KN	LO	MP	NQ	OR	PS	QT	RU	SV	TW	UX	VY	WZ	X1	
KL	LM	MN	NO	OP	PQ	QR	RS	ST	TU	UV	VW	WX	XY	YZ	

DECEMBER 1981 **PHYSICAL**

16 W	17 TH	18 F	19 SA	20 SU	21 M	22 TU	23 W	24 TH	25 F	26 SA	27 SU	28 M	29 TU	30 W	31 TH	
UT	VU	WV	XW	AX	BA	CB	DC	ED	FE	GF	HG	JH	KJ	LK	ML	
VS	WT	XU	AV	BW	CX	DA	EB	FC	GD	HE	JF	KG	LH	MJ	NK	
WR	XS	AT	BU	CV	DW	EX	FA	GB	HC	JD	KE	LF	MG	NH	OJ	+
XQ	AR	BS	CT	DU	EV	FW	GX	HA	JB	KC	LD	ME	NF	OG	PH	
AP	BQ	CR	DS	ET	FU	GV	HW	JX	KA	LB	MC	ND	OE	PF	QG	
BO	CP	DQ	ER	FS	GT	HU	JV	KW	LX	MA	NB	OC	PD	QE	RF	0
CN	DO	EP	FQ	GR	HS	JT	KU	LV	MW	NX	OA	PB	QC	RD	SE	
DM	EN	FO	GP	HQ	JR	KS	LT	MU	NV	OW	PX	QA	RB	SC	TD	
EL	FM	GN	HO	JP	KQ	LR	MS	NT	OU	PV	QW	RX	SA	TB	UC	−
FK	GL	HM	JN	KO	LP	MQ	NR	OS	PT	QU	RV	SW	TX	UA	VB	
GJ	HK	JL	KM	LN	MO	NP	OQ	PR	QS	RT	SU	TV	UW	VX	WA	
H	J	K	L	M	N	O	P	Q	R	S	T	U	V	W	X	

DECEMBER 1981 **EMOTIONAL**

16 W	17 TH	18 F	19 SA	20 SU	21 M	22 TU	23 W	24 TH	25 F	26 SA	27 SU	28 M	29 TU	30 W	31 TH	
L	M	N	O	P	Q	R	S	T	U	V	W	X	Y	Z	1	
MK	NL	OM	PN	QO	RP	SQ	TR	US	VT	WU	XV	YW	ZX	1Y	2Z	
NJ	OK	PL	QM	RN	SO	TP	UQ	VR	WS	XT	YU	ZV	1W	2X	3Y	
OH	PJ	QK	RL	SM	TN	UO	VP	WQ	XR	YS	ZT	1U	2V	3W	AX	
PG	QH	RJ	SK	TL	UM	VN	WO	XP	YQ	ZR	1S	2T	3U	AV	BW	+
QF	RG	SH	TJ	UK	VL	WM	XN	YO	ZP	1Q	2R	3S	AT	BU	CV	
RE	SF	TG	UH	VJ	WK	XL	YM	ZN	1O	2P	3Q	AR	BS	CT	DU	
SD	TE	UF	VG	WH	XJ	YK	ZL	1M	2N	3O	AP	BQ	CR	DS	ET	0
TC	UD	VE	WF	XG	YH	ZJ	1K	2L	3M	AN	BO	CP	DQ	ER	FS	
UB	VC	WD	XE	YF	ZG	1H	2J	3K	AL	BM	CN	DO	EP	FQ	GR	
VA	WD	XC	YD	ZE	1F	2G	3H	AJ	BK	CL	DM	EN	FO	GP	HQ	−
W3	XA	YB	ZC	1D	2E	3F	AG	BH	CJ	DK	EL	FM	GN	HO	JP	
X2	Y3	ZA	1B	2C	3D	AE	BF	CG	DH	EJ	FK	GL	HM	JN	KO	
Y1	Z2	13	2A	3B	AC	BD	CE	DF	EG	FH	GJ	HK	JL	KM	LN	
Z	1	2	3	A	B	C	D	E	F	G	H	J	K	L	M	

DECEMBER 1981 **INTELLECTUAL**

16 W	17 TH	18 F	19 SA	20 SU	21 M	22 TU	23 W	24 TH	25 F	26 SA	27 SU	28 M	29 TU	30 W	31 TH	
J	K	L	M	N	O	P	Q	R	S	T	U	V	W	X	Y	
KH	LJ	MK	NL	OM	PN	QO	RP	SQ	TR	US	VT	WU	XV	YW	ZX	
LG	MH	NJ	OK	PL	QM	RN	SO	TP	UQ	VR	WS	XT	YU	ZV	1W	
MF	NG	OH	PJ	QK	RL	SM	TN	UO	VP	WQ	XR	YS	ZT	1U	2V	
NE	OF	PG	QH	RJ	SK	TL	UM	VN	WO	XP	YQ	ZR	1S	2T	3U	
OD	PE	QF	RG	SH	TJ	UK	VL	WM	XN	YO	ZP	1Q	2R	3S	4T	+
PC	QD	RE	SF	TG	UH	VJ	WK	XL	YM	ZN	1O	2P	3Q	4R	5S	
QB	RC	SD	TE	UF	VG	WH	XJ	YK	ZL	1M	2N	3O	4P	5Q	6R	
RA	SB	TC	UD	VE	WF	XG	YH	ZJ	1K	2L	3M	4N	5O	6P	7Q	0
S8	TA	UB	VC	WD	XE	YF	ZG	1H	2J	3K	4L	5M	6N	7O	8P	
T7	U8	VA	WB	XC	YD	ZE	1F	2G	3H	4J	5K	6L	7M	8N	AO	
U6	V7	W8	XA	YB	ZC	1D	2E	3F	4G	5H	6J	7K	8L	AM	BN	−
V5	W6	X7	Y8	ZA	1B	2C	3D	4E	5F	6G	7H	8J	AK	BL	CM	
W4	X5	Y6	Z7	18	2A	3B	4C	5D	6E	7F	8G	AH	BJ	CK	DL	
X3	Y4	Z5	16	27	38	4A	5B	6C	7D	8E	AF	BG	CH	DJ	EK	
Y2	Z3	14	25	36	47	58	6A	7B	8C	AD	BE	CF	DG	EH	FJ	
Z1	12	23	34	45	56	67	78	8A	AB	BC	CD	DE	EF	FG	GH	

PEI

THN = M.

E4K = F

JC6 = L

QCT J